4th International Congress on Sustainability Science and Engineering (ICOSSE '15)

Balatonfured, Hungary
26-29 May 2015

ISBN: 978-1-5108-1774-6

Printed from e-media with permission by:

Curran Associates, Inc.
57 Morehouse Lane
Red Hook, NY 12571

Printed by Curran Associates, Inc. (2016)

AIChE
120 Wall Street, FL 23
New York, NY 10005-4020

Phone: (800) 242-4363
Fax: (203) 775-5177

www.aiche.org

Additional copies of this publication are available from:

Curran Associates, Inc.
57 Morehouse Lane
Red Hook, NY 12571 USA
Phone: 845-758-0400
Fax: 845-758-2634
Email: curran@proceedings.com
Web: www.proceedings.com

TABLE OF CONTENTS

Innovations in Water Reuse to Enable a Sustainable Circular Economy

- **Conference:** International Congress on Sustainability Science Engineering ICOSSE
- **Year:** 2015
- **Proceeding:** ICOSSE 2015
- **Group:** Abstract Submissions
- **Session:**

Plenary session: Water

- **Time:**

Wednesday, May 27, 2015 - 8:20am-8:50am

Authors:

Young, T., The Dow Chemical Company

With the world population estimated to grow to more than 9 billion in 2050, the stress on limited fresh water resources is rapidly becoming unsustainable. Beyond the profound impact on our water supply by our world's burgeoning population, other issues, such as rapid urbanization, industrialization, pollution, and climate change are putting pressure on available fresh water resources worldwide. And while water is an important ingredient for the sustainability of towns and cities, it is also used in refining processes for food, fuels, chemicals, and other products. Over the next five years, the technologies associated with salt removal and turning low-quality wastewater and raw water sources into high quality process water will become an essential part of sustainable economic growth. As a leading water purification technology provider, Dow technologies today are providing 15 million gallons of clean water globally each minute. To address future challenges, further breakthrough innovations will be critical.

Supporting Decision-Making for the Nexus

- **Conference:** International Congress on Sustainability Science Engineering ICOSSE
- **Year:** 2015
- **Proceeding:** ICOSSE 2015
- **Group:** Abstract Submissions
- **Session:**

Plenary session: Framing the Nexus

- **Time:**

Wednesday, May 27, 2015 - 8:50am-9:20am

Authors:

Irabien, A., Cantabria University

A Sustainable Water-Food-Energy Nexus must connect the well-being demand of a community with its water and energy needs. A qualitative approach based on a holistic view should be able to identify the "bottlenecks" in the system in order to get a technical view and scope the analysis for the optimization and decision-making procedures to be applied.

We propose a methodology based on two steps: (a) Qualitative description of the system and (b) Quantitative assessment of the system behaviour. This enables a hierarchical methodology to support decision-making in different scenarios.

The Process Analysis Method (PAM) allows the identification of: a Guiding Process, Core Processes and Enabling Processes in order to develop the appropriate framework (1)

REFERENCES

Darton, R. (2015). Setting a policy for Sustainability: the importance of measurement in, "Assessing and Measuring Environmental Impact and Sustainability". Chapter 14 Ed. Jiri Klemes. Butterworth-Heinemann

Incorporating Sustainability into Engineering Education in the United States

David Allen
University of Texas at Austin

The University of Texas at Austin
Center for Energy and Environmental Resources

Benchmarking Sustainable Engineering Research & Education

How are University engineering programs around the U.S. implementing sustainability into curricula?

What are opportunities for practicing engineers?

What are opportunities for high school students?

CSE

The University of Texas at Austin
Center for Energy and Environmental Resources

Benchmarking Sustainable Engineering Research & Education

How are University engineering programs around the U.S. Implementing sustainability into curricula?

C S E

Murphy, C., D.T. Allen, B. Allenby, J. Crittenden, C. Davidson, C. Hendrickson and S. Matthews "*Environmental Science and Technology*, 43, 5558-5564 (2009).

www.csengin.org

The University of Texas at Austin
Center for Energy and Environmental Resources

Benchmarking process

- Contact more than 1500 department chairs and program heads
- Based on information from department heads, and other contacts, identify more than 350 sustainable engineering faculty champions
- Sustainability champions described 155 courses and hundreds of research projects; more than 80% of the top 100 engineering programs (USN&WR) reported some activity

The University of Texas at Austin
Center for Energy and Environmental Resources

Benchmarking Sustainable Engineering Research & Education

How are University engineering programs around the U.S. Implementing sustainability into curricula?

- Courses
- Programs

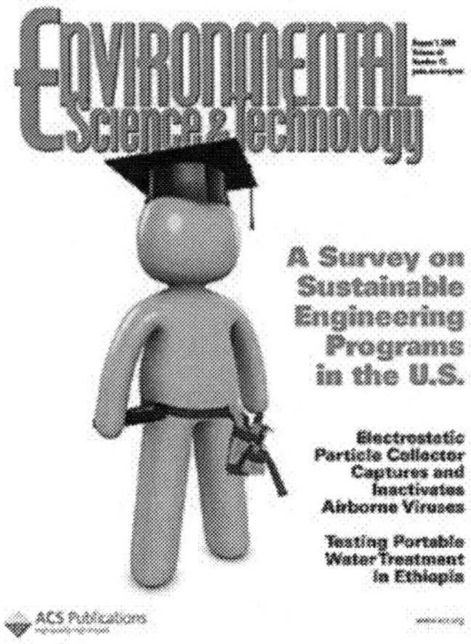

CSE

Murphy, C., D.T. Allen, B. Allenby, J. Crittenden, C. Davidson, C. Hendrickson and S. Matthews "*Environmental Science and Technology*, 43, 5558-5564 (2009).

www.csengin.org

The University of Texas at Austin
Center for Energy and Environmental Resources

Categories of courses and role in the curriculum

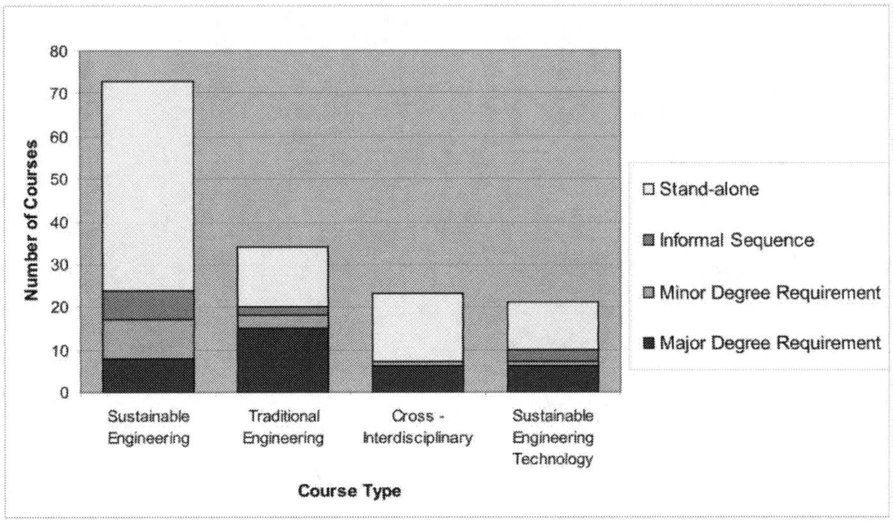

The University of Texas at Austin
Center for Energy and Environmental Resources

Who takes the courses?

- Sustainable engineering: mixed upper division and graduate
- Traditional engineering: upper division
- Cross/Inter-disciplinary: often graduate
- Sustainable engineering technology: mixed upper division and graduate

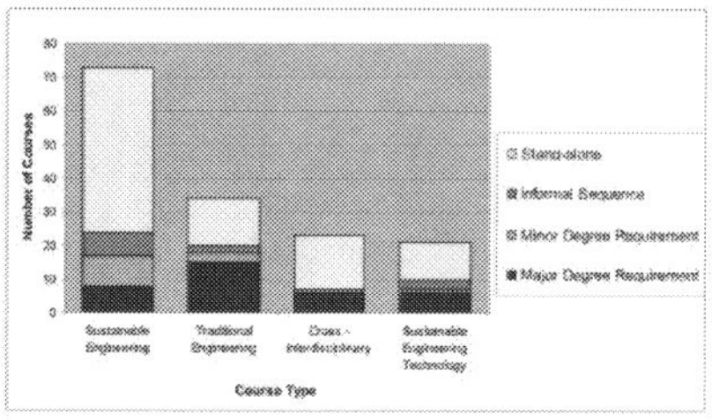

The University of Texas at Austin
Center for Energy and Environmental Resources

Overall synthesis

- A student in an engineering course covering sustainability is most likely to be an upper division undergraduate or graduate student taking a stand alone course that is not part of a formal sequence

The University of Texas at Austin
Center for Energy and Environmental Resources

What's in these courses?

A structure for categorizing content

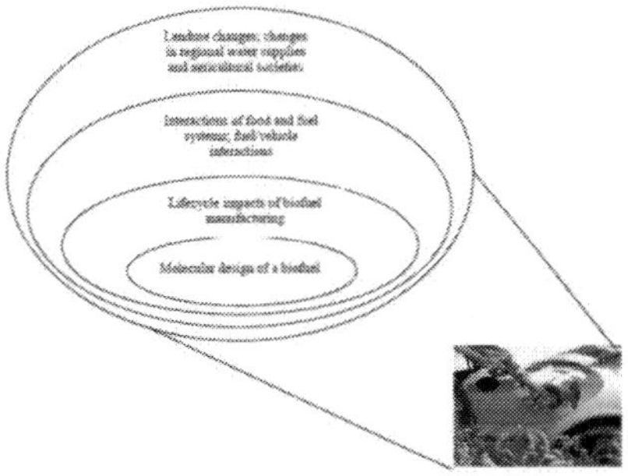

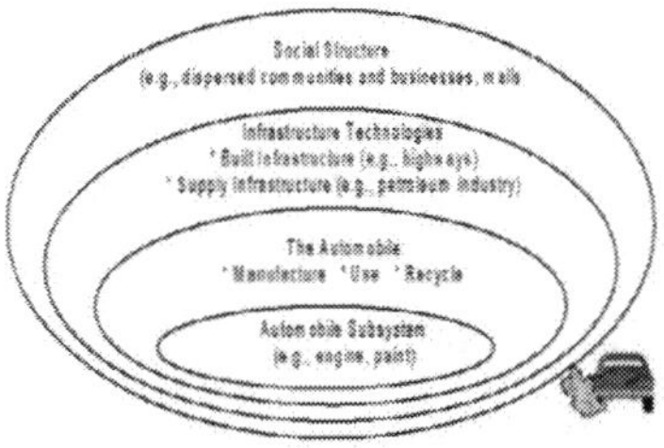

The University of Texas at Austin
Center for Energy and Environmental Resources

What system sizes are covered in courses?

Content level low

Content level high

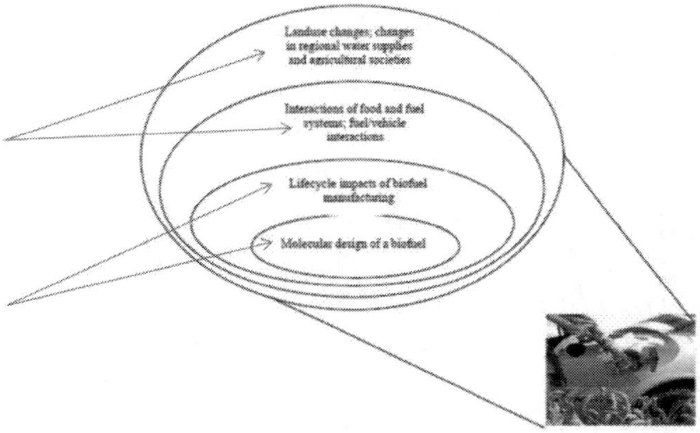

The University of Texas at Austin
Center for Energy and Environmental Resources

Benchmarking Sustainable Engineering Research & Education

How are University engineering programs around the U.S. Implementing sustainability into curricula?

- Courses
- <u>Programs</u>

C S E

Murphy, C., D.T. Allen, B. Allenby, J. Crittenden, C. Davidson, C. Hendrickson and S. Matthews "*Environmental Science and Technology*, 43, 5558-5564 (2009).

www.csengin.org

The University of Texas at Austin
Center for Energy and Environmental Resources

Exemplary program level initiatives

- **Engineers teaching about sustainability to a general undergraduate university population**: University of Texas "Signature Course" (Sustaining a Planet)
- **Sustainability concentration within a college of engineering**: Virginia Tech's concentration in sustainable (green) engineering
- **Graduate level programs across colleges**: Yale and Michigan offer joint MS programs (Engineering and Forestry and Environmental Studies; Engineering and the School of Natural Resources and the Environment); Carnegie Mellon offers Engineering and Public Policy programs within the College of Engineering
- **Course modules:** The Center for Sustainable Engineering and other groups have developed or are planning to develop course modules

The University of Texas at Austin
Center for Energy and Environmental Resources

Conclusions regarding university curricula

- Including sustainable engineering concepts into engineering programs is becoming widely accepted (80% of top 100 engineering programs are doing something)
- Wide range of content
- Some common elements:
  - Multiple scales
  - Systems perspective
  - Multi-objective product and process design
  - Life cycle methods
  - Green materials
  - Recycling and reuse

The University of Texas at Austin
Center for Energy and Environmental Resources

Content of university based courses can be codified into a "Body of knowledge"

- Three elements to the chemical engineering framework:
  - Framing the challenge
  - Assessment and design
    - Molecular
    - Process
    - System
  - Systems tools and perspectives

Sustainability in Chemical Engineering Education: Identifying a Core Body of Knowledge

David T. Allen
Dept. of Chemical Engineering, University of Texas, Austin, TX

David R. Shonnard
Dept. of Chemical Engineering, Michigan Technological University, Houghton, MI

Benchmarking Sustainable Engineering Research & Education

How are University engineering programs around the U.S. implementing sustainability into curricula?

What are opportunities for practicing engineers?

What are opportunities for high school students?

CSE

The University of Texas at Austin
Center for Energy and Environmental Resources

AIChE Credential for Sustainability Professionals

New certificate program offered by the American Institute of Chemical Engineers' Institute for Sustainability

www.aiche.org/ifs/resources/sustainability-credentials

The University of Texas at Austin
Center for Energy and Environmental Resources

Differences in emphasis in bodies of knowledge for university students and practicing professionals

Academic

- Three elements to the chemical engineering framework:
  - Framing the challenge
  - Assessment and design
    - Molecular
    - Process
    - System
  - Systems tools and perspectives

Practicing professional

AIChE Credential for Sustainability Professionals

- Strategic Commitment/Ethics
- Innovation
- Environmental Performance
- Safety Performance
- Product Stewardship
- Social Responsibility
- Value Chain Management

Differences in delivery for university students and practicing professionals

Academic

- Stand alone courses
- Materials embedded in required courses

Practicing professional

(AIChE Credential for Sustainability Professionals)

- Three university or AIChE Academy courses
- Portfolio/Project overseen by a professional mentor

Benchmarking Sustainable Engineering Research & Education

How are University engineering programs around the U.S. implementing sustainability into curricula?

What are opportunities for practicing engineers?

<u>What are opportunities for high school students?</u>

CSE

The University of Texas at Austin
Center for Energy and Environmental Resources

High School Engineering Education in the United States

- More high school engineering students than college engineering students in the United States
- Multiple models for curricula – most are at least year-long full courses, however, some opportunities for course modules in science courses
- Many year long courses are project based
- Still an emerging area

The University of Texas at Austin
Center for Energy and Environmental Resources

Case study of a 9 week project in a year long high school engineering course

- Design a water management system for a river system with multiple reservoirs with spatially distributed rainfall and seasonal water uses
- Emphasis on building dynamic flow models
- Emphasis on conserving a limited natural resource
- Both physical and mathematical representation of system

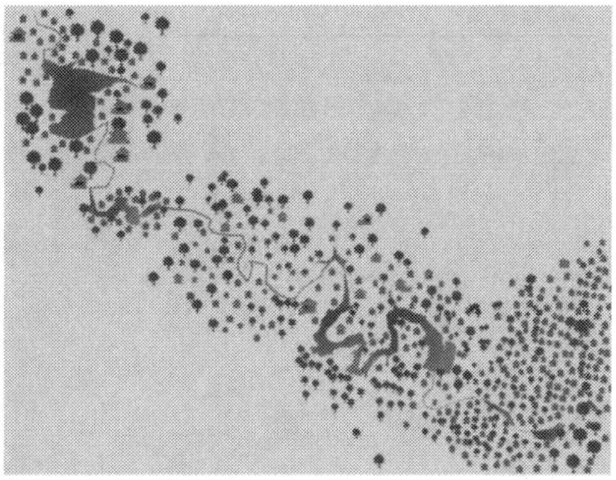

Many successful strategies

- Modules for existing courses
- Stand alone courses
- Minor/concentration program
- Degree programs
- Professional certification

In chemical engineering, efforts began with modules (in the 1990s) and have evolved to include the entire range of activities listed above

The University of Texas at Austin
Center for Energy and Environmental Resources

MANCHESTER
1824

The University of Manchester

Sustainable manufacturing: Integrating environmental, economic and social aspects of sustainability

Adisa Azapagic

The University of Manchester, UK

adisa.azapagic@manchester.ac.uk

Sustainable Industrial Systems

What is manufacturing?

- Mechanical, physical, or chemical transformation of materials, substances, or components into new products

(US Bureau of Labor Statistics)

Sustainable Industrial Systems

What is included in manufacturing?

- Food
- Beverage and Tobacco
- Textile
- Apparel
- Leather and Allied Products
- Wood Products
- Paper
- Printing and Related
- Petroleum and Coal Products
- Chemicals
- Plastics and Rubber Products
- Non-metallic Mineral Products
- Primary Metals
- Fabricated Metal Products
- Machinery
- Electrical and Electronic Equipment
- Transportation Equipment
- Furniture
- Miscellaneous

Sustainable Industrial Systems

What is sustainable manufacturing

- Manufacturing that meets the needs of the present without compromising the ability of future generations to meet their own needs
- By providing products and services that are
  - environmentally benign
  - economically viable and
  - socially beneficial

Sustainable Industrial Systems

Sustainability challenges for manufacturing

- Population growth
- Rapid urbanisation
- Food supply
- Water access
- Energy demand
- Climate change
- Resource scarcity
- Profligate consumption

Sustainable Industrial Systems

Systems approach and life cycle thinking: Integrating sustainability aspects

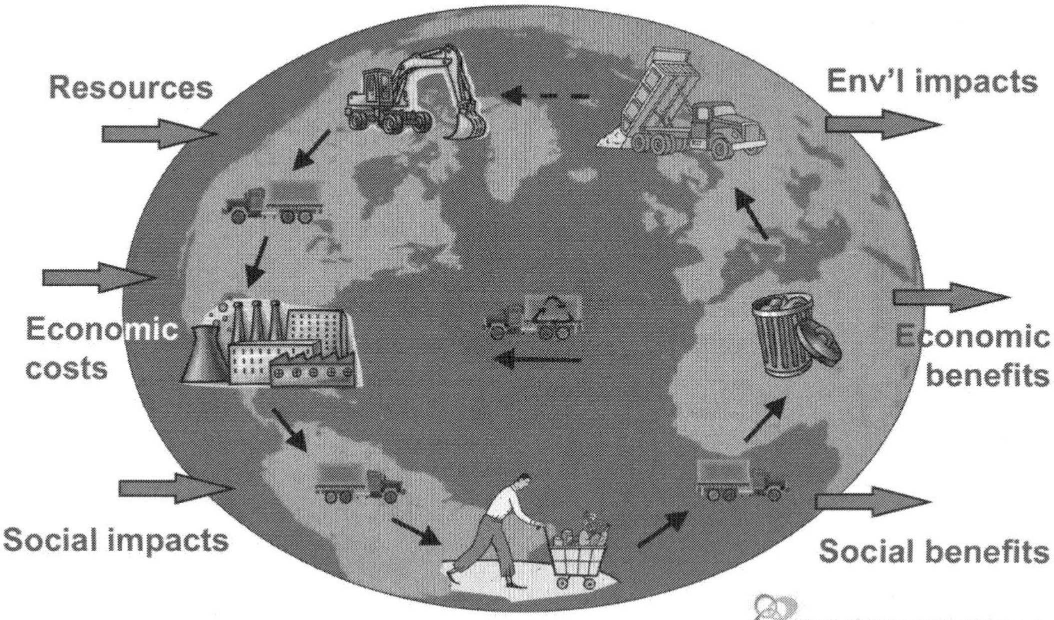

Sustainable Industrial Systems

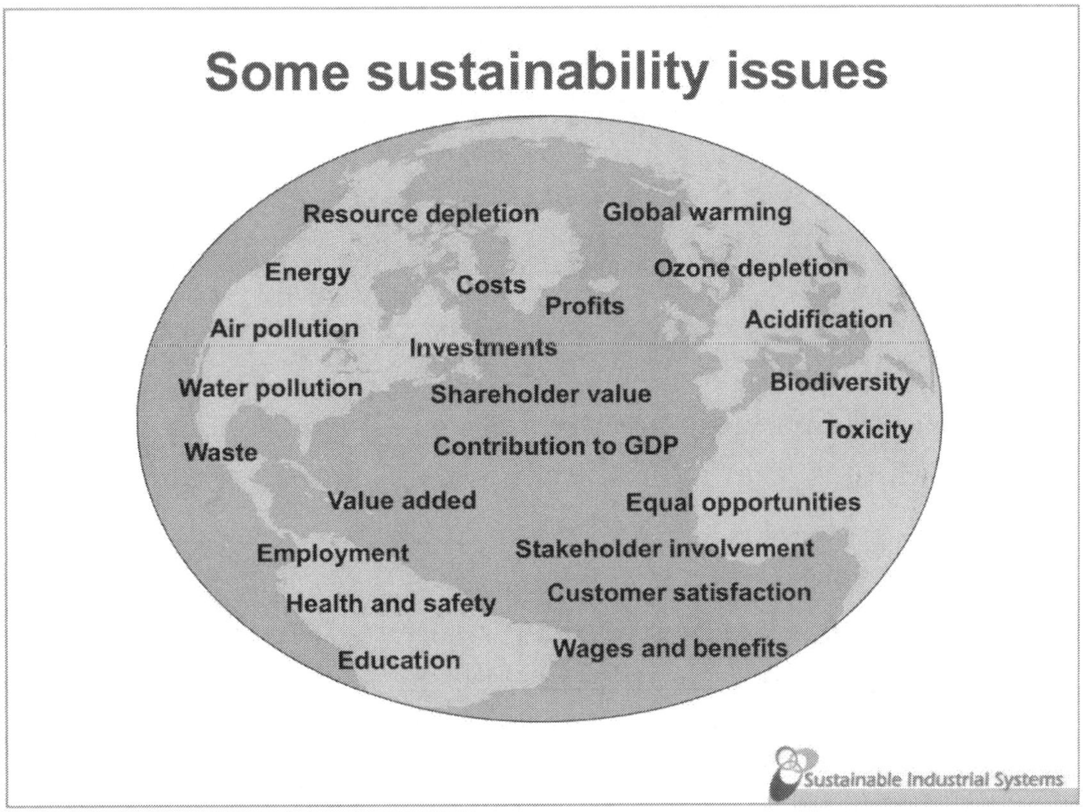
Some sustainability issues
Resource depletion
Global warming
Energy
Costs
Ozone depletion
Profits
Air pollution
Acidification
Investments
Water pollution
Shareholder value
Biodiversity
Toxicity
Waste
Contribution to GDP
Value added
Equal opportunities
Employment
Stakeholder involvement
Health and safety
Customer satisfaction
Education
Wages and benefits
Sustainable Industrial Systems

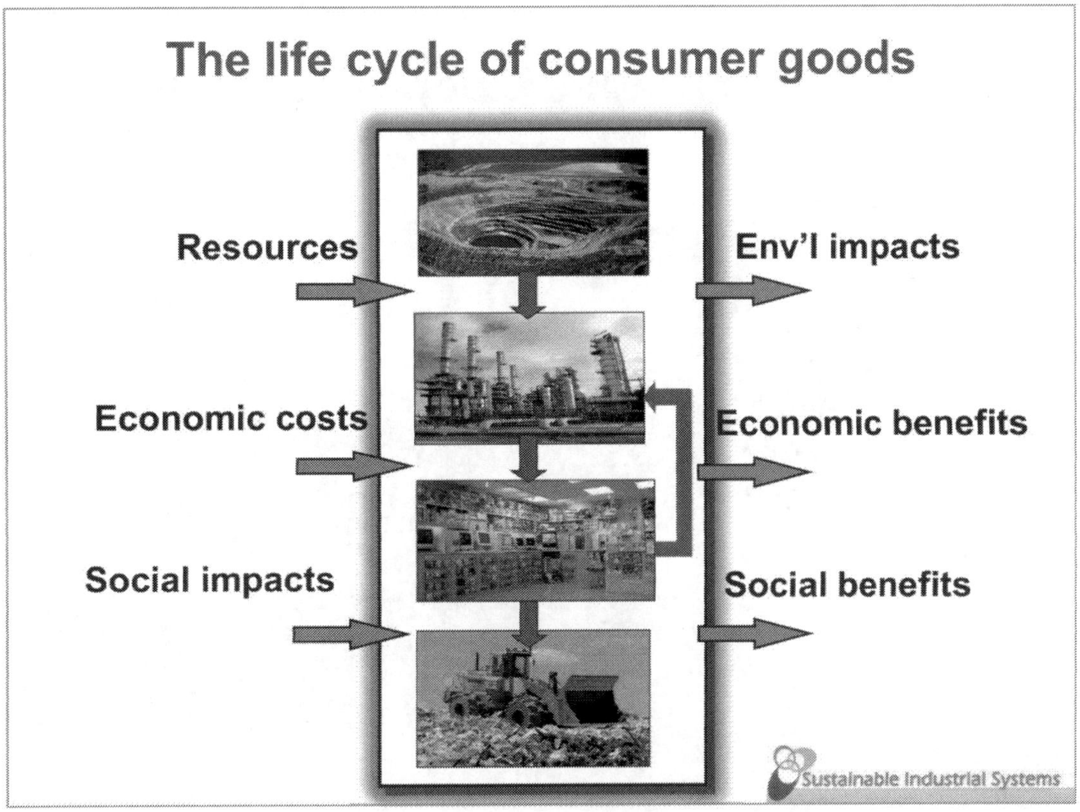
The life cycle of consumer goods
Resources
Env'l impacts
Economic costs
Economic benefits
Social impacts
Social benefits
Sustainable Industrial Systems

Key sustainability issues

- Profligate and inequitable use of resources
  - US 
  - UK 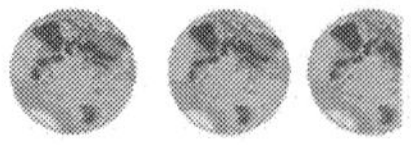
- Lack of/low reuse and recycling
  - 90% of products become waste within 6 months
- Increasing scarcity
  - The number of chemical companies affected by resource scarcity may triple by 2020 (IChemE)

Sustainable Industrial Systems

Our favourite products deplete annually...

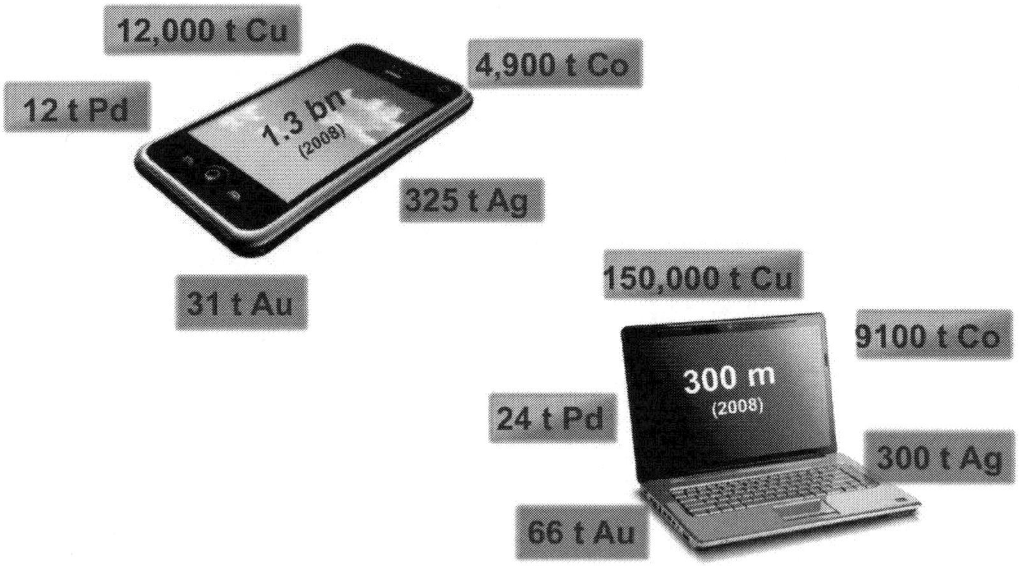

Adapted from Hunt et al. (2013)

Sustainable Industrial Systems

Sustainable resource management

- Catalysis: many of the critical elements used as catalysts: e.g. PGMs, Co, Ce, Ge, Sb, In
  - Use of non-critical metals
  - Use of non-metallic catalysts, e.g. enzymes
- Urban mining
  - WEEE (50 million t/yr)
  - MSW (1.6 bn t/yr)
  - Road dust (PGMs)
  - Landfills

- Delivering service rather than product

Sustainable Industrial Systems

Delivering service rather than product

'We don't sell a tin of paint – we sell an effect on a wall'
Dulux

'1,500 petabytes shipped in 2009'
Xyratex

'Power by the hour'
Rolls Royce

Courtesy of Mike Pitts, TSB/CIKTN

Sustainable Industrial Systems

Systems approach: Biofuels

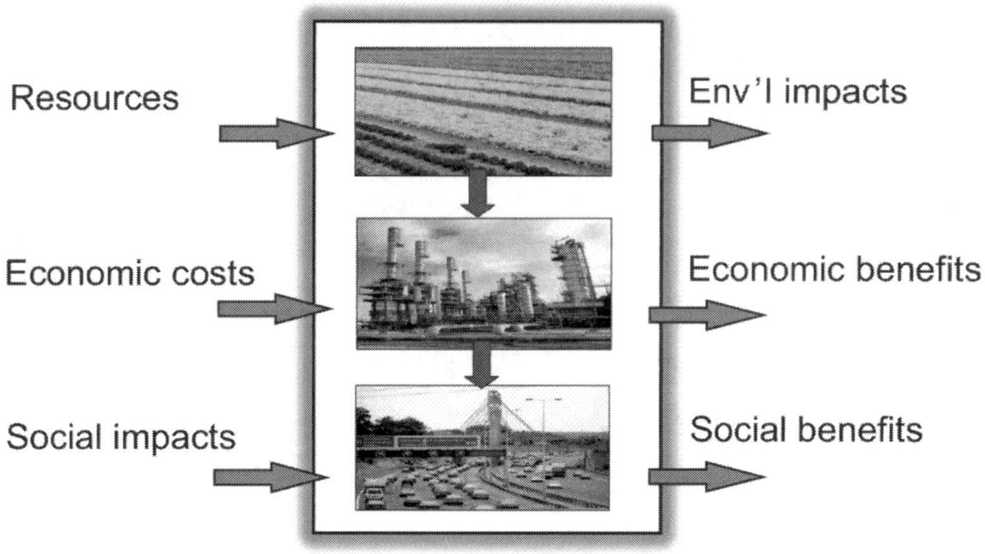

Sustainable Industrial Systems

Key sustainability issues

- Feedstocks
- GHG emissions
- Land-use change
- Biodiversity
- Competition with food

Sustainable Industrial Systems

Carbon footprint of ethanol

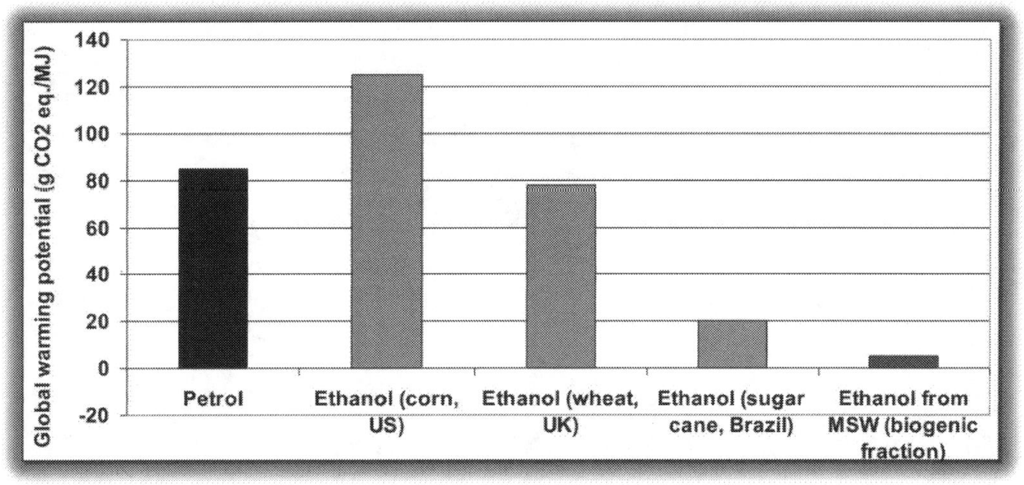

DfT. Carbon and sustainability reporting within the renewable transport fuel obligation requirements and guidance. London: Department for Transport; 2008.

Stichnothe & Azapagic, Resources, Conservation and Recycling 53 (2009) 624–630.

Sustainable Industrial Systems

Comparison of other environmental impacts

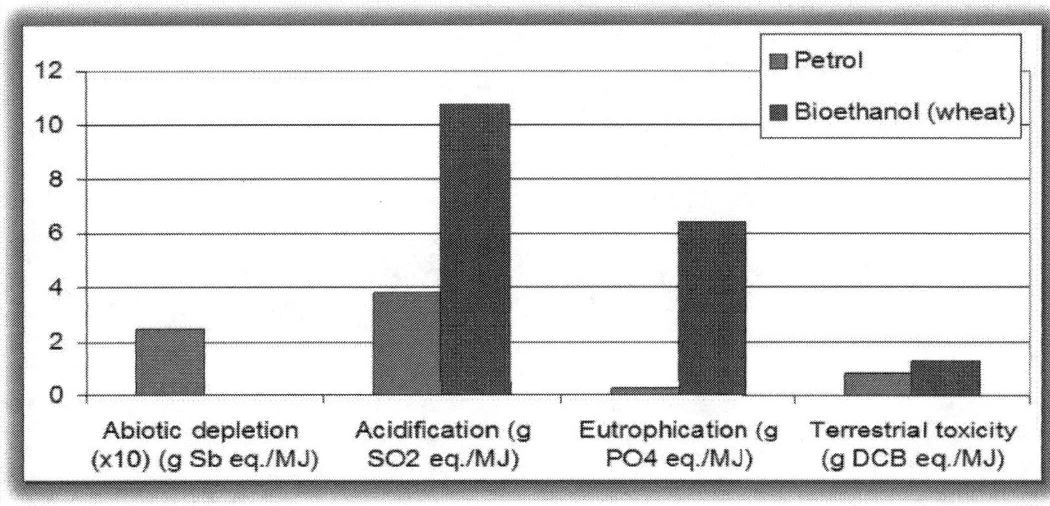

Ecoinvent V2.2

Sustainable Industrial Systems

2nd gen biofuels from integrated biorefineries?

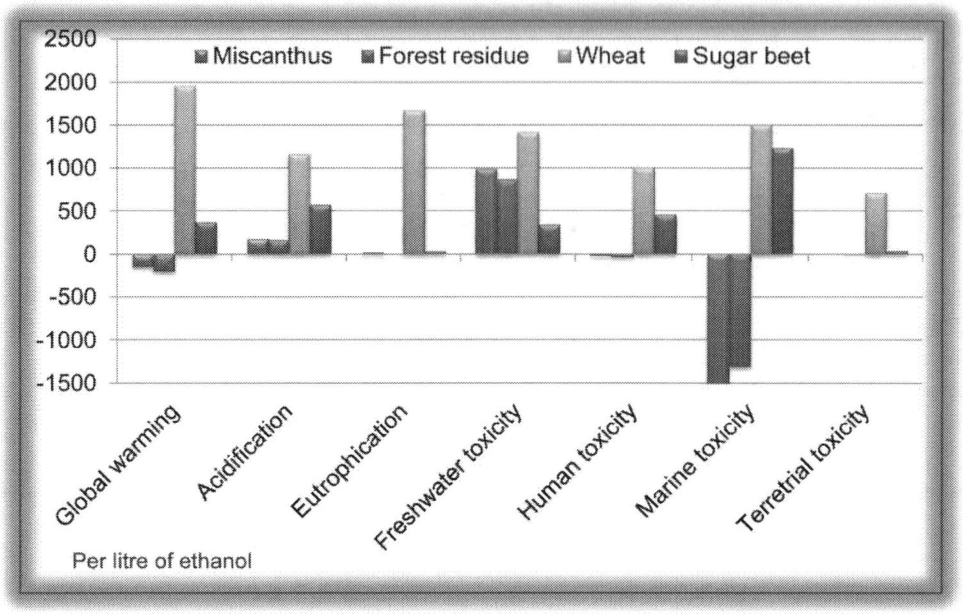

Temitope et al. (2014). *Biotechnology J.*, 9 753-765.

Sustainable Industrial Systems

Land use change and GHG emissions

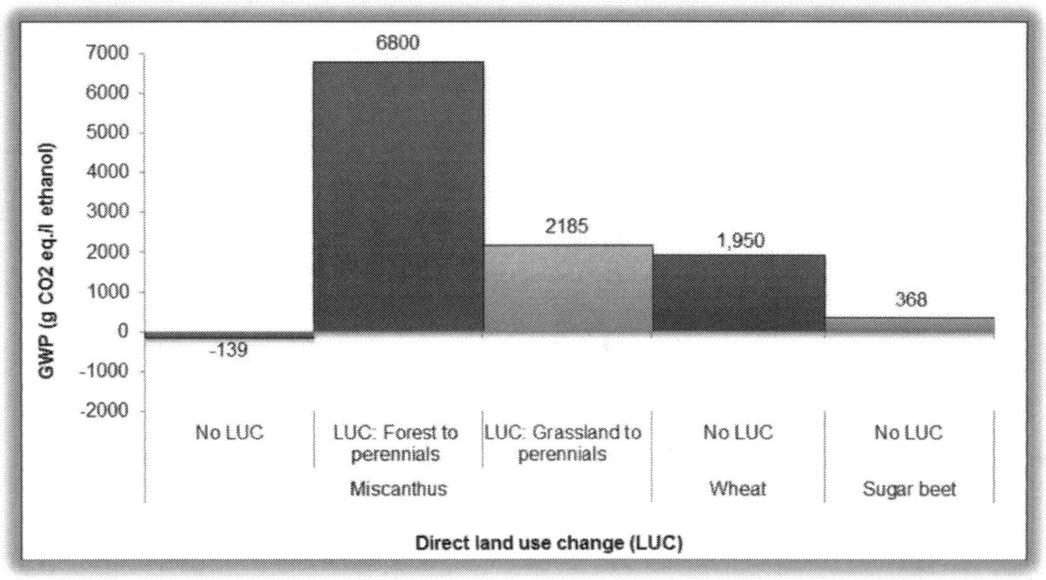

Temitope et al. (2014). *Biotechnology J.*, 9 753-765.

Sustainable Industrial Systems

Is it all worth it?

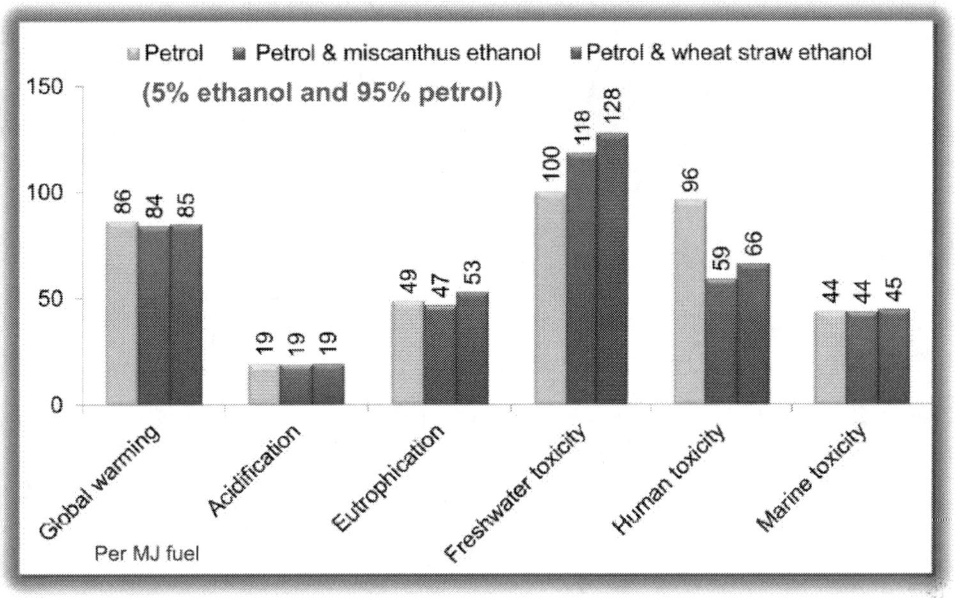

Temitope et al. (2014). *Biotechnology J.*, 9 753-765.

Sustainable Industrial Systems

Sustainable manufacturing

- We face unprecedented challenges
- Much of engineering today is needed to overcome the problems engineering created in the past
- We should see this as a great opportunity - not to repeat the mistakes from the past
- Addressing the challenges requires systems approach and life cycle thinking
- Understanding economic, environmental and social trade offs is essential

Sustainable Industrial Systems

Sustainable manufacturing

- Working with non-technical disciplines
- Understanding consumer behaviour
- Engaging with the public
- Providing robust evidence for policy makers
- Educating the next generation of engineer with sustainability in mind

Sustainable Industrial Systems

ELSEVIER

NEW FOR 2015
SUSTAINABLE PRODUCTION AND CONSUMPTION

A new journal at the intersection of engineering, natural and social sciences

Editor-in-Chief
Adisa Azapagic
School of Chemical Engineering & Analytical Science, University of Manchester

Sustainable Production and Consumption is a new journal aimed at engineers, natural and social scientists. Our multidisciplinary nature is reflected by our international Editorial Board whose members come from a wide spectrum of disciplines including engineering, sociology, economics, business, policy and environmental sciences.

If your research is concerned with life cycle thinking and any topics related to sustainable production and consumption then we would love to hear from you. Visit the journal homepage for the full Aims & Scope and topics and to submit your papers.

The Whole World Needs to Be Like Güssing

- **Conference:** International Congress on Sustainability Science Engineering ICOSSE
- **Year:** 2015
- **Proceeding:** ICOSSE 2015
- **Group:** Abstract Submissions
- **Session:**

Plenary session: Energy

- **Time:**

Thursday, May 28, 2015 - 9:00am-9:30am

Authors:

Koch, R.

This talk will provide an example of a sustainable renewable energy supply of a region by using local raw materials. The so-called "Güssing Model" is the strategy of de-centralised, local energy production with all available renewable resources in a region. This model was developed based on the necessity to stop depopulation, to increase economic growth and to be independent from fossil fuels and is the quintessence of what has been happening in Güssing since the early 1990s with significant involvement of the EEE. Since every region has certain resources in different measures, the model can serve as an example for many communities.

Ionic Liquids: Sustainable and designed solution for Process Industry and Green Environment

by
M I Abdul Mutalib, Tauqeer A, Shah N. S., Al Kaisy, G.M.J, Uoahid Ben Ghanem and Lethesh K.C.
Center for Ionic Liquids Research (CORIL), Universiti Teknologi PETRONAS,
32610 Sri Iskandar, Perak, MALAYSIA

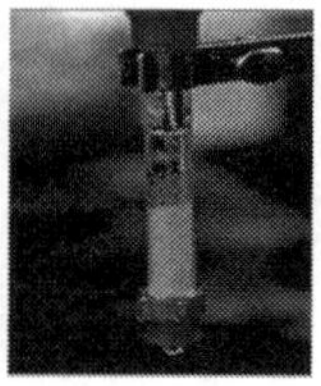

4th International Congress on Sustainability Science & Engineering, May 26-29th 2015, Balatonfüred, Hungary

Chemical Engineering
Inspiring Potential·Generating Futures

Overview of Ionic Liquids...

*"Ionic liquids (ILs) are the generic term for a class of materials, **consisting entirely of ions and being liquid below 100°C**, and if they are liquid at room temperature, we call them room temperature ionic liquids (RTILs)" –K. R. Seddon-*

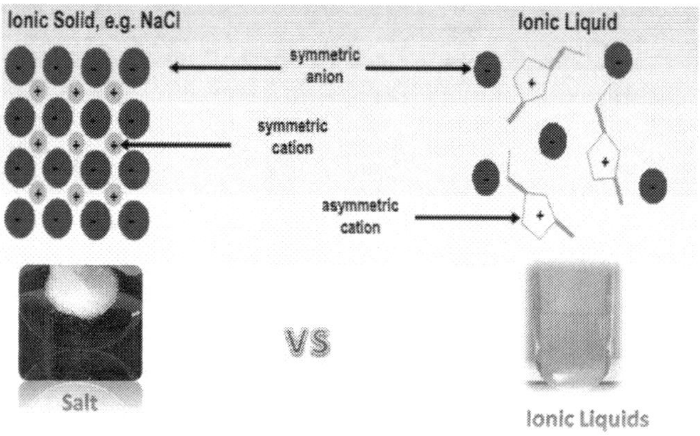

They are not an aqueous solution of salts... 2

Chemical Engineering
Inspiring Potential·Generating Futures

Often Ionic Liquids are also referred to as Designer Solvents due to the flexibility of choosing the functional group pairing(s) to form the ionic liquids

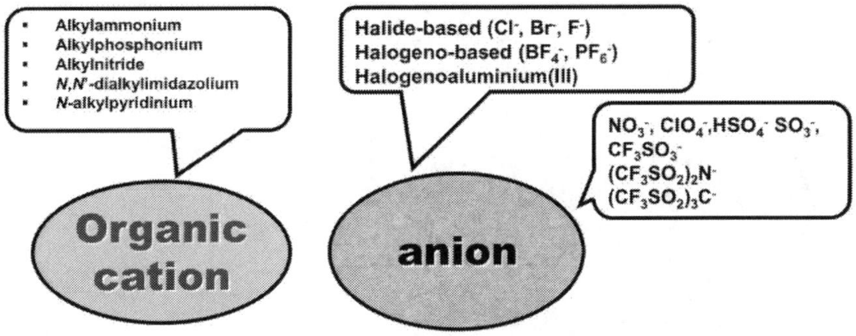

Because of the wide range of selection for possible cation and anion combinations, Ionic Liquids can be designed to suit desired applications and the overall properties are determined by the selected combination...

Ionic Liquid Vs Organic Solvents

Property	Organic Solvents	Ionic Liquid
Number of solvents	>1000	>1,000,000
Applicability	Single function	Multifunction
Catalytic ability	Rare	Common and tuneable
Chirality	Rare	Common and tuneable
Vapour pressure	Obeys the Clausius–Clapeyron equation	Negligible vapour pressure under normal conditions
Flammability	Usually flammable	Usually nonflammable
Polarity	Conventional polarity concepts apply	Polarity concept questionable
Tuneability	Limited range of solvents available	Virtually unlimited range means "designer solvents"
Cost	Normally cheap	Typically between 2 and 100 times the cost of organic solvents
Recyclability	Green imperative	Economic imperative
Viscosity, cP	0.2–100	22–40,000
Density, g/cm^3	0.6–1.7	0.8–3.3
Refractive index	1.3–1.6	1.5–2.2

More & Better Options

Greener Attributes

Concerns in process applications

Plechkova, N. V., & Seddon, K. R. (2008). Applications of ionic liquids in the chemical industry. *Chemical Society Reviews*, *37*(1), 123-150.

4

There is an increasing pressure for industry to adopt greener solutions whilst maintaining the economic competitiveness through ...

- developing more environmentally benign solvents or use solvent-less systems
- replacing toxic and environmentally hazardous and/or volatile solvents
- Developing multi-function solvents that can catalyse reactions whilst aiding separation simultaneously to allow for process intensification.

Blake Simmons, Ionic Liquid Pretreatment, EERE Webinar, June 24, 2013.

Company	Process	IL is ...	Scale
BASF	Acid Scavenging	Auxiliary	commercial
	Extractive Distillation	Extractant	pilot
	Chlorination	Solvent	pilot
IFP	Olefin Dimerization	Solvent	commercial
Degussa	Hydrosilylation	Solvent	pilot
	Compatibilizer	Performance Additive	commercial
Arkema	Fluorination	Solvent	pilot
Chevron Philips	Olefin Oligomerization	Catalyst	pilot
Eastman	Rearrangement	Catalyst	commercial
Eli Lilly	Cleavage of Ether	Catalyst / Reagent	pilot
Air Products	Storage of Gases	Liquid Support	commercial
Iolitec / Wandres	Cleaning Fluid	Performance Additive	commercial

Already many examples of industrial applications...

Ionic Liquids has the potential to offer solutions...

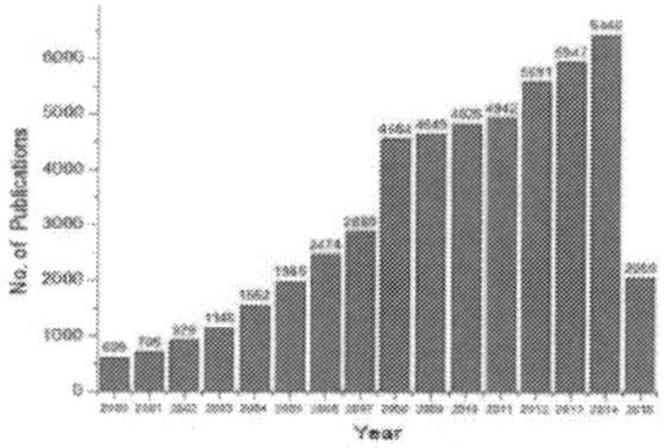

Also increasing no. of publications in open literature ...

5

There are already significant progress made in the applications of ionic liquids in other areas ...

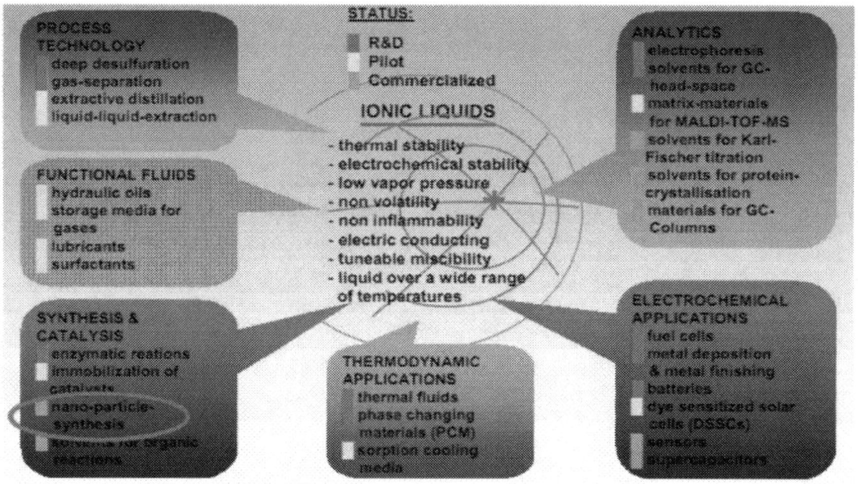

Xin Huang, Ionic Liquids, Green Chemistry presentation.

6

At the Center of Research in Ionic Liquids at UTP, there were many research towards the application of Ionic Liquids in oil and gas were conducted....

Selected Projects for today's sharing ...

- Naphthenic acid removal from oil
- Desulfurization and removal of aromatics
- Mercury Extraction from Natural Gas
- Toxicity of Ionic Liquids in aquatic environment

7

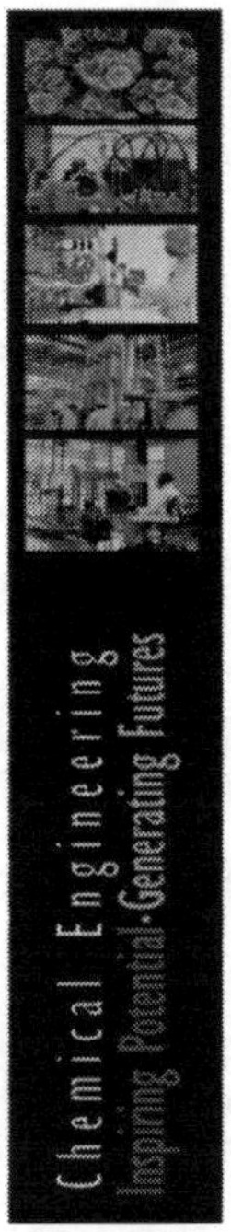

NAPHTHENIC ACID REMOVAL

Crude oil is a complex mixture containing organic and inorganic compounds i.e., hydrocarbons, species of sulfur & chloride, nitrogen compounds, trace metals, naphthenic acids etc.

Heavy Crude Oil

- Higher acid number (TAN)
- More corrosive for process

Increasing supply in the market
Cheaper !

Petroleum products
Bitumen
etc.

Higher Naphthenic Acids as by products

Presently can be sold at reasonably high price depending on purity.

Current acid recovery processes posed significant limitation especially from environmental perspective

Limitations of the current naphthenic acid recovery processes ....

Process Description	Short Comings
Ion Exchange	• Caused by electrostatic forces thus large ion with high valency in the eluted compounds takes longer retention time to desorb from the adsorbent column. • Adsorption of organic material with the anion beads for anion exchanger resin reduces the exchange capacity and lead to a reduction in treated crude quality.
Caustic Treatment	• Generates significant amounts of waste • Emulsions that are problematic to treat.
Fiber Film Contactor	• Only used for jet fuel, kerosene, diesel • It is a very expensive technology
Esterification	• Reaction kinetics need to be determined because of difference between molecular structure of acids
Decarboxylation	• Noticeable reduction is observed only at temperatures above 250°C which lead to corrosion problems • Catalyst preparation

Application of Ionic Liquids as greener solvent ....

9

The mechanism of the ionic liquids extraction for deacidification ...

Formation of Naphthenate salts of the Ionic Liquids from the neutralization reaction.

Direct extraction by mixing the ionic liquids with the crude oil

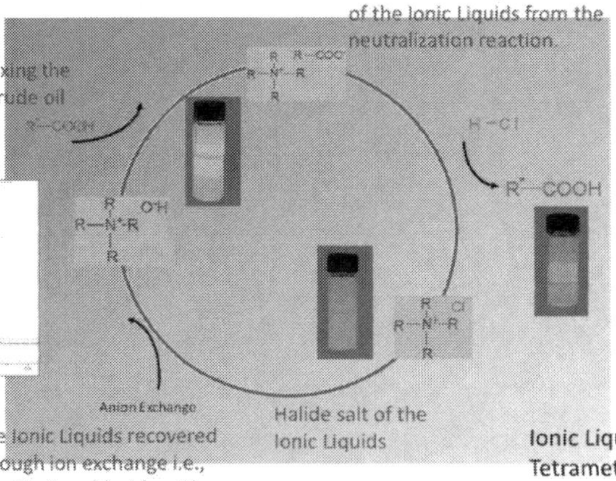

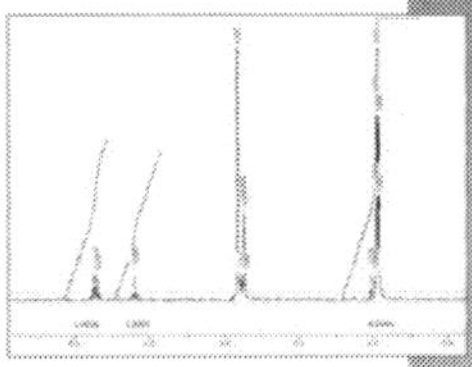

NMR Spectra of ILs

Reformation of Naphthenic Acid by reacting the naphthenate salts with cheap HCl.

- The Ionic Liquids recovered through ion exchange i.e., substituting chloride with hydroxide.

Ionic Liquids used.
Tetramethyl Ammonium Hydroxide
Choline Hydroxide

Some important results from the work ...

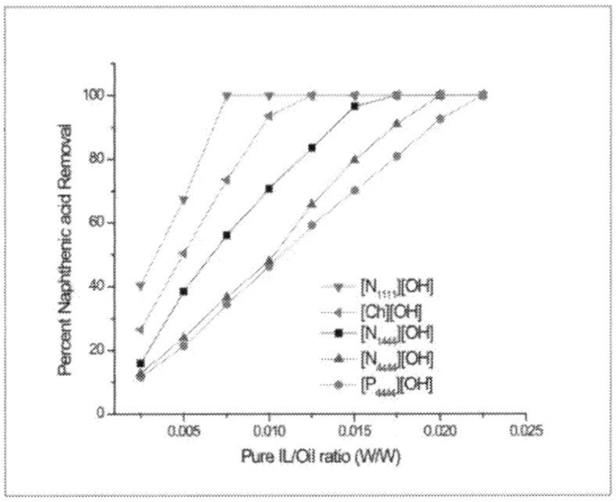

Extraction of Naphthenic Acid from Highly Acidic Oil Using Hydroxide-Based Ionic Liquids,Syed Nasir Shah, M. Ibrahim. A. Mutalib, Rashidah Binti Mohd Pilus, Kallidanthiyil Chellappan Lethesh, *Energy Fuels* 2015 29 (1), 106-111

- Low Ionic Liquids to crude oil ratio
- Able to recover naphthenic acid at high recovery factor
- Post extraction and other required separation processes were easy – phase separation.
- Purity of naphthenic acid recovered are relatively higher than from conventional process.

FTIR spectra of NA regen.

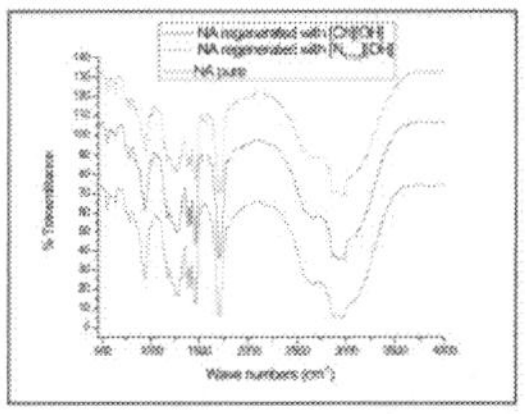

Application of Ionic Liquids in solution led to lower viscosity for feasible process application. Ease or recovery and recycle has made the ionic liquids competitive from cost perspective.

11

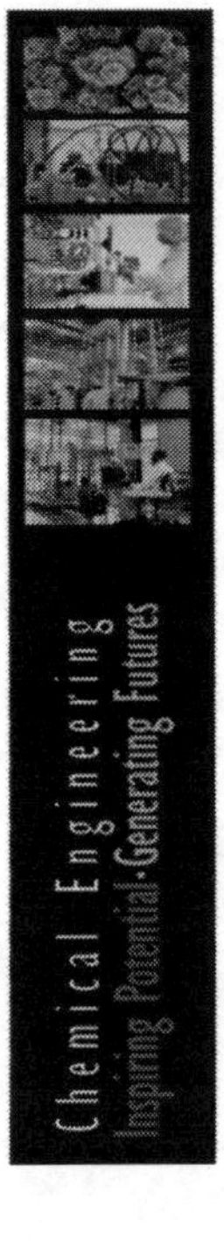

Proposed Pilot Plant Concept

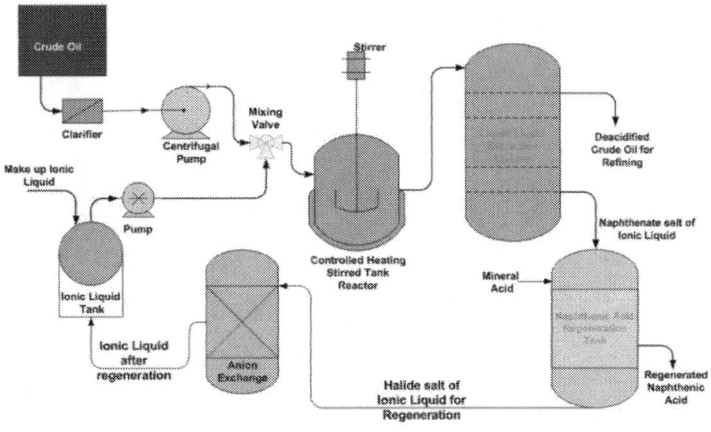

12

Chemical Engineering
Inspiring Potential·Generating Futures

UNIVERSITI TEKNOLOGI PETRONAS

DESULFURISATION AND REMOVAL OF AROMATICS FROM OIL

The requirements for higher class base lubricating oil has increased due to tighter environmental legislation coupled with the need for higher performance engine. This has led to the requirements of removing aromatics and sulfur content to a much lower level which currently are obtained through hydrocracking/hydrotreating .

Alternative through solvent extraction using Ionic Liquids would be able to reduce the severity extent of the process conditions (compared to hydrotreating/hydrocracking) whilst reducing the waste generation (compared to conventional solvent extraction).

Ionic Liquids Screening was conducted in order to determine the appropriate ones. The approach consists of ;

- Screening method using COSMO-RS software to select appropriate Ionic Liquids.
- The combinations of 7 cations and 23 anions were screened for the removal of naphthalene (aromatics representative) and dibenzothiophene (sulfur species representative) from C40 (Polyalphaolefin base oil).
- Activity coefficient values were determined from the software which then enabled the calculation for selectivity and capacity of the Ionic Liquids.
- Sigma Profile screening was also conducted to reinforce the screening method.

Chemical Engineering
Inspiring Potential·Generating Futures

UNIVERSITI TEKNOLOGI PETRONAS

Activity coefficient at Infinite Dilution for the various Ionic Liquids (Cation-Anion pairings)

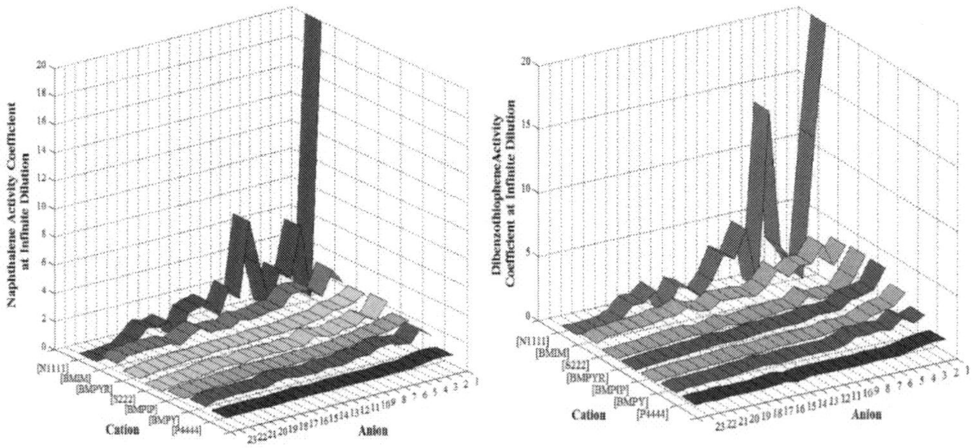

14

From the activity coefficient at infinite dilution, the selectivity and capacity were determined ...

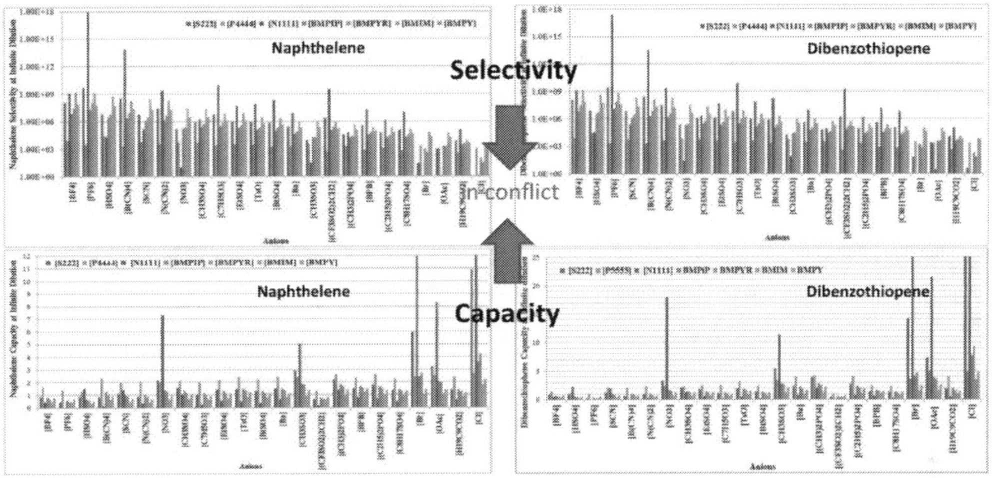

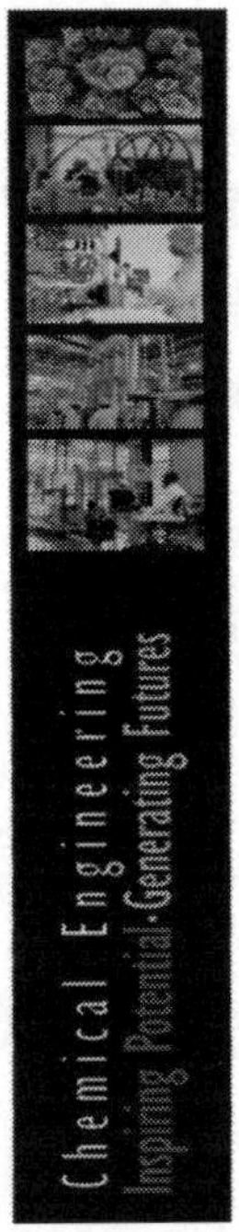

Selection of the potential Ionic Liquids (Pairing) is made through determination of removal of efficiency for sulfur and aromatics species...

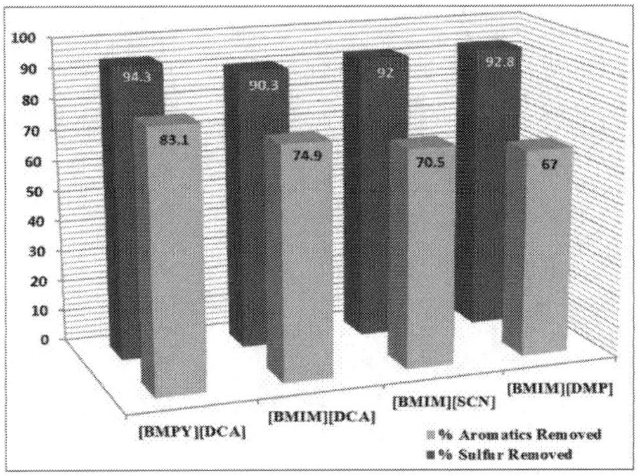

Sigma Profile Screening

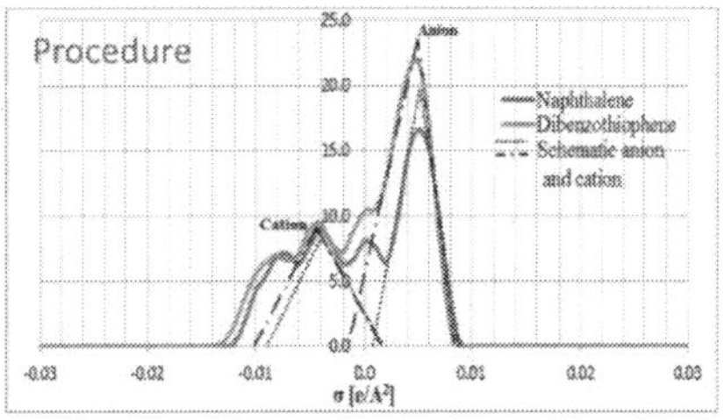

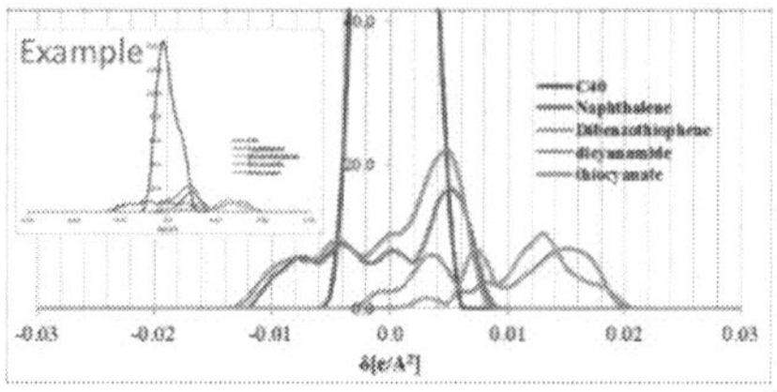

σ-Profiles of small linear anions with explicit charge distribution with (naphthalene, DBT and C40)

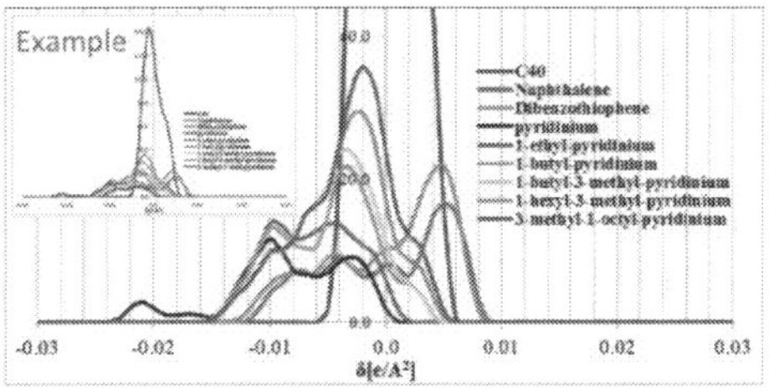

σ-Profiles of Pyridinium cations with naphthalene, DBT and C40

Selection of Ionic Liquids
Pairing of Cation and Anion

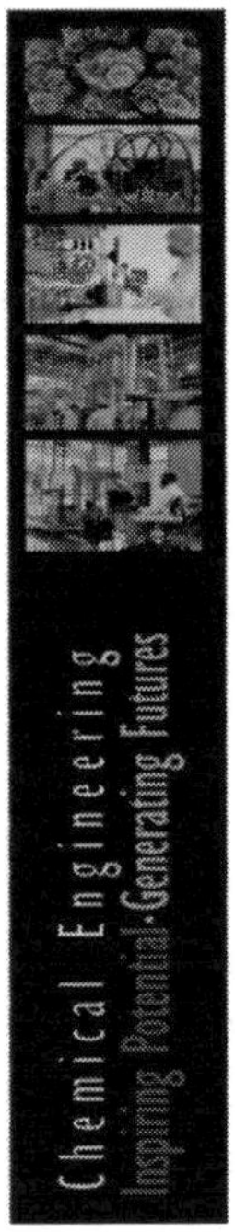

Experimental work was then conducted focusing on the selected Ionic Liquids (cation-anion pairing)

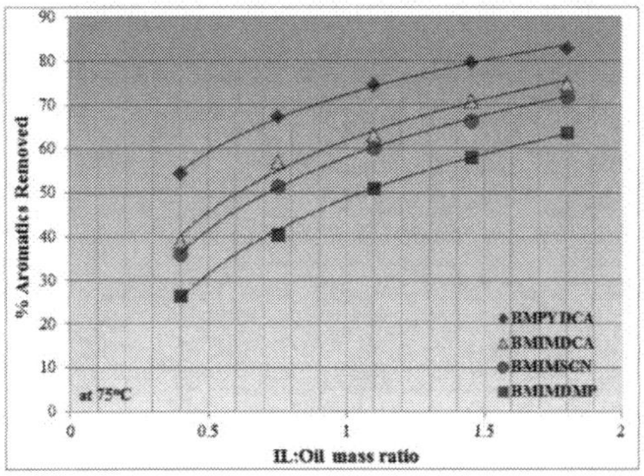

Effect of IL:Oil mass ratio at 75 C for Aromatics removal

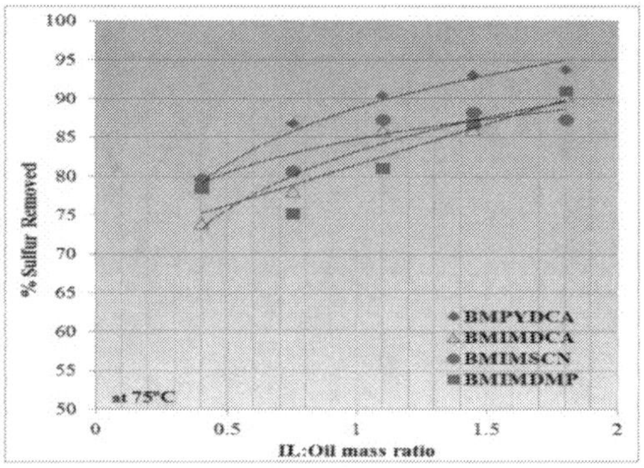

Effect of IL:Oil mass ratio at 75 C for Sulfur removal

- Reasonably high simultaneous removal of aromatics and sulfur was obtained
- Similar experimental work conducted on temperature reveals its much weaker effect on the removal efficiency

MERCURY REMOVAL FROM NATURAL GAS

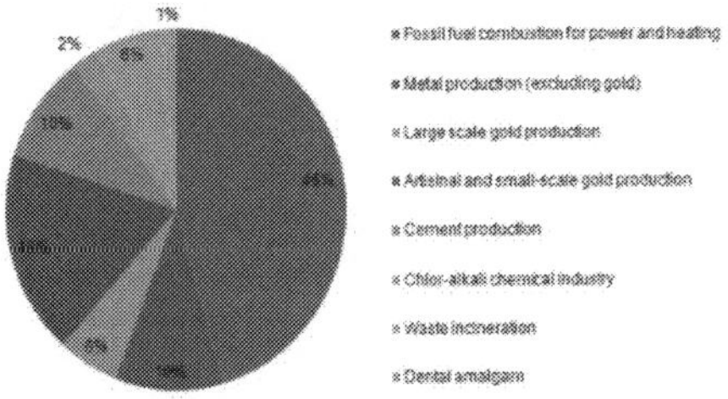

Source: AMAP/UNEP, 2008. Technical Background Report to the Global Atmospheric Mercury Assessment. Arctic Monitoring and Assessment Programme/UNEP Chemicals Branch. Page 38.

Anthropogenic Mercury Resources

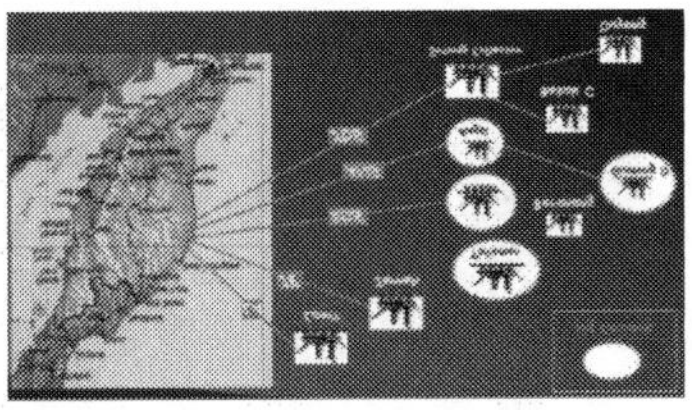

Region	Mercury Concentration (μg/m3)
Eastern Europe	1-2x103
Far East	0.02-193
Germany (Northern)	15-450
Malaysia	1-200
Germany (Southern)	<0.1-0.3
Indonesia (Sumatra)	200-300
Middle East	1-9
North America	0.005-40
South America	69-119

Mercury in Natural Gas by Region

Lang, D., Gardner, M., & Holmes, J. (2012). Mercury arising from oil and gas production in the United Kingdom and UK continental shelf. *A report by University of Oxford, Department of Earth Sciences, United Kingdom.*
Bigham, M.D., " Field Detection and Implications of Mercury in Natural Gas", SPE Prod. Eng., May 1990, 120-124.
Wilhelm, S.M., and McArthur, A., "Removal and Treatment of Mercury Contamination at Gas Processing Plant", Soc. of Pet. Eng., (SPE No. 29721) Sharizalazam, Mercury, PETRONAS Research Sdn. Bhd

There is a need to solve the increasing issue on Hg ...

Trend on mercury concentration from offshore hydrocarbon is increasing

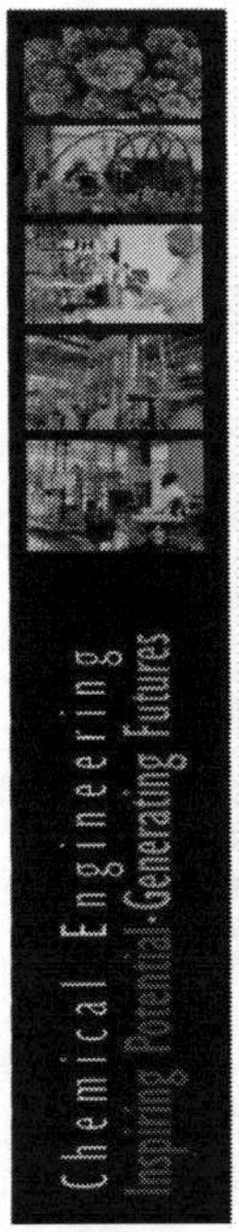

The threatening Effect of Mercury in Oil and Gas Industry

Equipment degradation
- Liquid metal embrittlement
- Stress cracking
- Aluminum corrosion & oxidation

Catalyst Poisoning
- Deactivation of catalyst
- Hg contaminated spent catalyst & absorbents

Contaminated Process & Product
- Decon requirements for maintenance
- HSE & pollution issues
- Higher cost for <Hg crude
- Lower value >Hg off spec product

Explosion Hazards
- Formation of Hg-N (mercury nitride) & Hg-N-0 (mercury nitrite) in ammonia plants

Mahmoud Abu El Ela et al., (2006) Egyptian gas plant employs absorbents for Hg removal, *Oil and Gas Journal*, 104(46), 52-58.
Abbas, T et al., (2014) Developments in Mercury Removal from Natural Gas-A Short Review. Applied Mechanics and Materials 625: 223-228.

Limitation on current Hg removal techniques.

Mercury Removal Systems	Limitations	Reference
Adsorption on activated carbon	Low saturation loading but cheap, disposal problem and cant remove all types of mercury in one step.	Si Hyun Lee et al, 2006
Adsorption on molecular sieve	Common sieves have low capacity; big beds.	Canute Hudson 2010, Terrigeol, A 2013
Adsorption on metal sulphide impregnated on mesoporous Alumina	Low resistance to both capillary condensation and liquid carry-over, only capable to remove elemental mercury	Shafawi A et al, 2000
Adsorption by amalgamation with a metal such as aluminium, silver Zeolites etc.	High investment cost, lower capacities	Mahmoud et al, 2006
Oxidising solutions- potassium permanganate, sodium hypochlorite, and sodium vanadate	Regeneration problems; system contamination	
Acid absorption of mercury-acidic permanganate and chromic acid	Increased corrosion; low saturations; system contamination	April Heffart et al 2006
Condensation and separation	Poor removal efficiency: liquid contamination	Abu El Ela et al, 2007
Stripping through liquid hydrocarbon medium	Poor removal efficiency: liquid contamination	Hadden and Moss 2010

Provides further motivation for the research...

20

Why Ionic Liquids to remove Hg ?

- **Ionic Liquids**
  - IL's have the ability to capture all types of Hg species
  - IL incorporates an oxidizing moiety

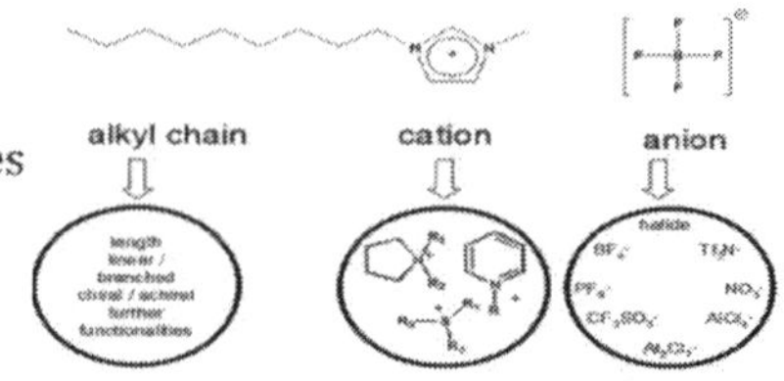

- **Solid Supported ionic Liquids (SSILs)**
  - Prove practicality of ionic liquid reactive extraction processes at scale.
  - Stability of the mercury capture within the ionic liquids enabled safe disposal of the spent adsorbent.

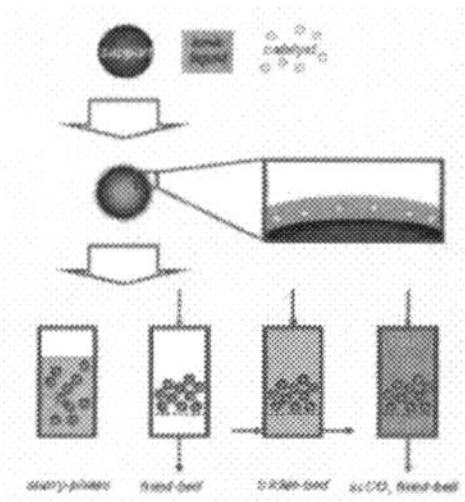

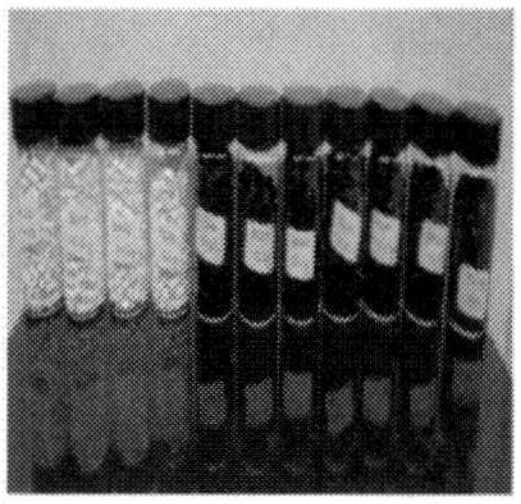

21

Key findings from the study....

- Halogen Based or sulfide based Ionic Liquids with different cationic core shows potential for high adsorption capacity of Hg – up to 100% - **provide direct cost competitive for application.**
- Oxidizing agents such as $KMnO_4$ could also be added to enhance performance
- There are different range of microporous to mesoporous supports possible and it has effect on the performance of the Ionic Liquids adsorbent.
- The removal efficiency was found to be up to 5 times better than carbon sorbents in a high concentration of mercury flow.

Element	Weight %
C	44.51
O	23.43
Al	1.21
Si	2.30
Cl	4.74
K	4.94
Ca	1.30
Mn	15.10
Hg	2.45

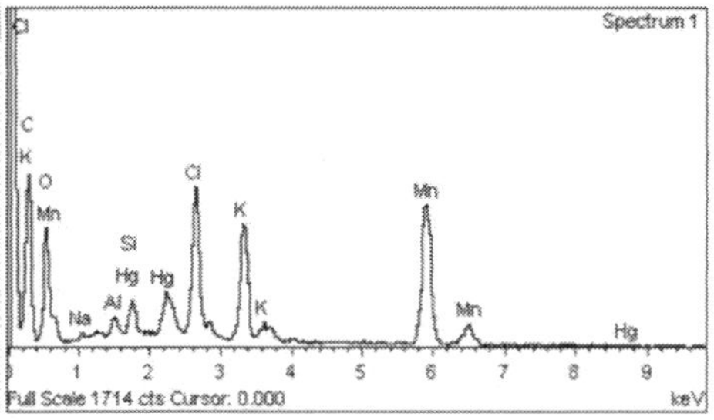

EDX Analysis of Spent Adsorbent

22

Toxicity Modelling of Ionic Liquids – QSAR Model

- There are significant number of potential Ionic Liquids pairing and determining the toxicity of each will be an impossible task. The numbers of commercially available ILs are increasing significantly.
- Even, if ILs do not significantly contribute to air pollution, but they may accumulate and harm the living aquatic organisms by accidental spills or effluents because of their considerable solubility in water.

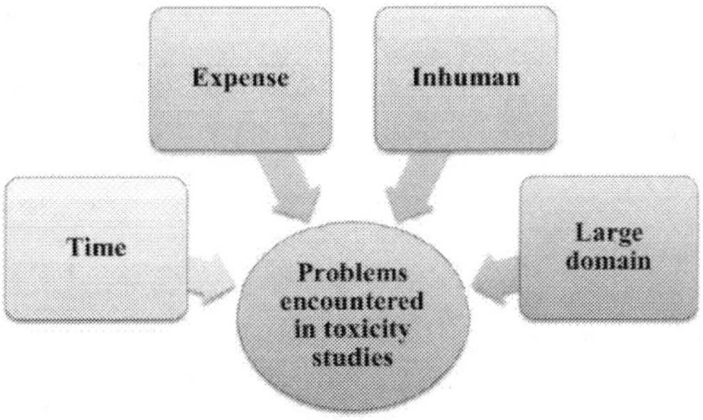

The approach we took is to relate the individual impact of each component in the ionic liquids to the toxicity effect.

- **Relate the structure of Ionic Liquids to antimicrobial activity**

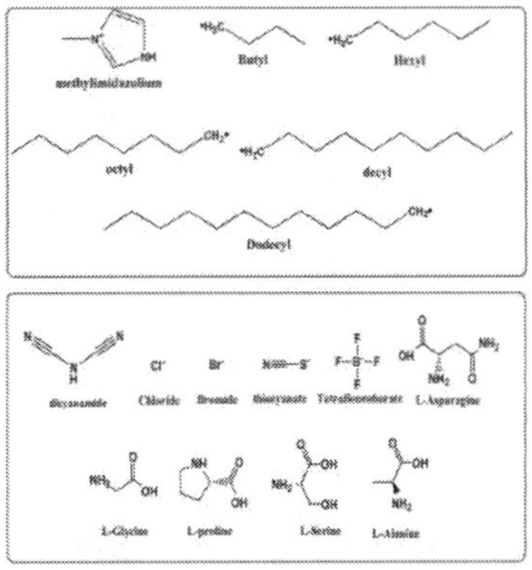

- bacteria is good starting point to examine ILs toxicity as *they have short generation time compared to other organisms.)*
- ***Test Microorganisms***
- ***Gram Positive***
  - *Listeria monocytogenes L 49 (LM)*
  - *Staphylococcus aureus S 1426 (SA)*
- ***Gram Negative***
  - *Escherichia coli (E.Coli)*
  - *Aeromonas hydrophila (AH)*

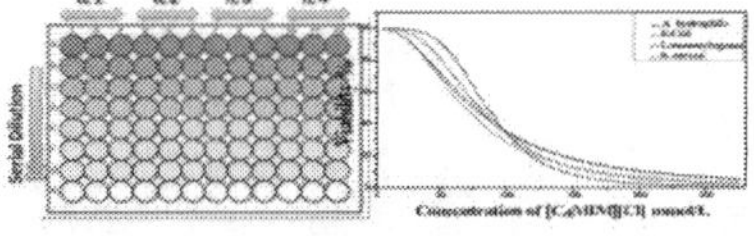

- The 50% effective concentration (EC_{50}) {micro broth dilution test }

Perform experimental measurement to investigate the impact of type of cations and anions and the alkyl chain length on the EC_{50} concentration.

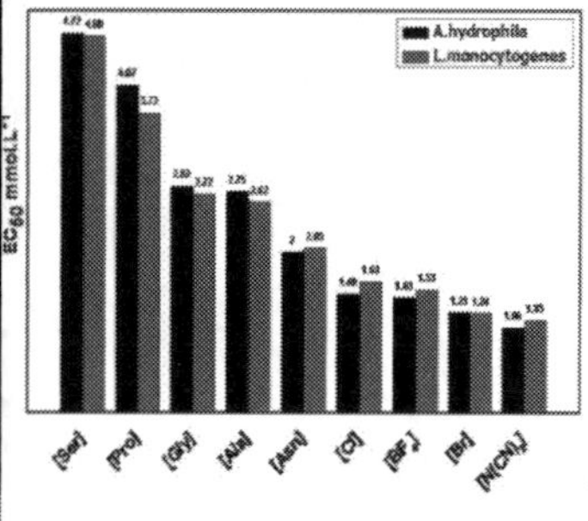

Influence of anions on EC_{50} of $[C_8mim]$ based ILs

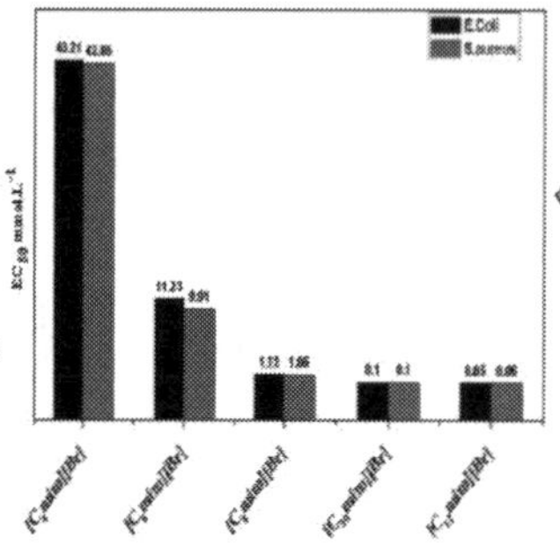

Influence of alkyl chain length on EC_{50} of $[C_nmim][Br]$ based ILs

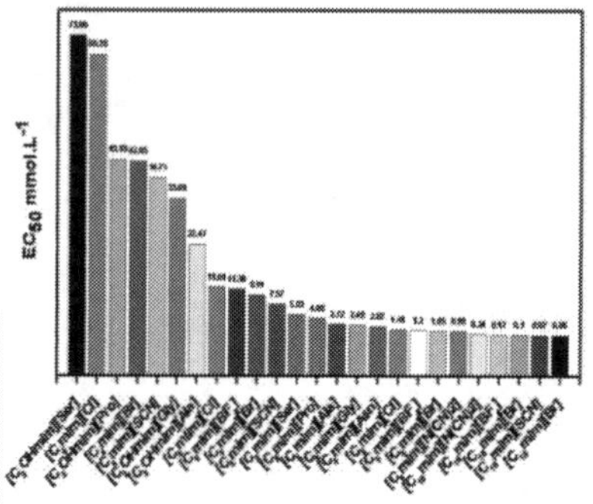

Close observation on the antimicrobial activity shows that some ionic liquids can be toxic and a particular impact on the overall toxicity could be observed.

Based on the extensive experimental data collected, several QSAR models were developed by correlating the experimental log (EC_{50}) values with the various cations (represented by descriptors) to form sub-$(QSAR)_{cation}$ model and with the various anions (represented by descriptors) to form sub-$(QSAR)_{anion}$ model separately.

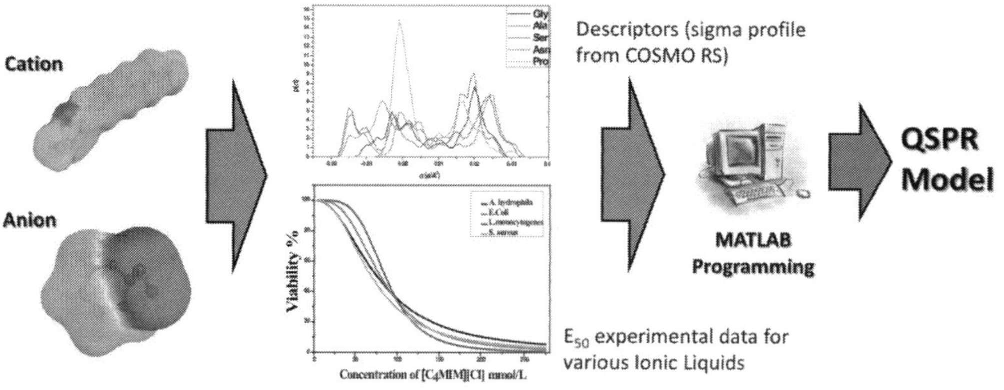

Chemical Engineering
Inspiring Potential·Generating Futures

Molecular descriptor

- COSMO-RS (σ-profile) were used as molecular descriptors for the ILs, which indicate their electronic structure and the molecular size.
- The area under the σ-profile curve for each cation and anion were divided to four regions. Eight descriptors (S_1- S_8) were used to represent the ILs structures.

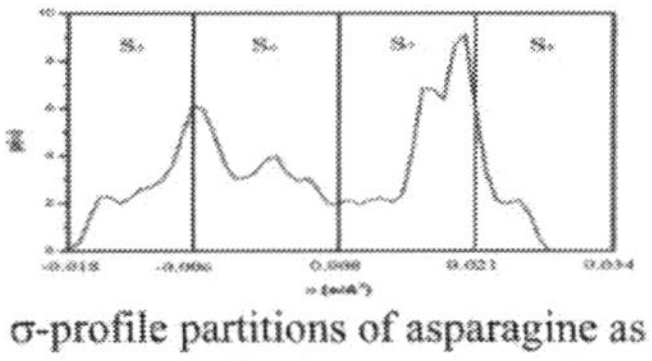

σ-profile partitions of asparagine as an example for dataset preparation

ILs		Cations molecular descriptors				Anions molecular descriptors			
		S_1	S_2	S_3	S_4	S_5	S_6	S_7	S_8
1	[C_2mim][SCN]	0.0180835	0.064439	0.1024925	0.050613	0	0.053582	0.0428775	0
2	[C_2mim][Br]	0.0180835	0.064439	0.1024925	0.050813	0	0	0.058629	0
3	[C_2mim][Cl]	0.0180835	0.064439	0.1024925	0.050813	0	0	0.05281	0
4	[C_4mim][BF_4]	0.018162	0.064328	0.1183245	0.081717	0	0.083755	0.0250115	0
5	[C_4mim][Br]	0.018162	0.064328	0.1183245	0.081717	0	0	0.058629	0
6	[C_4mim][Cl]	0.018162	0.064328	0.1183245	0.081717	0	0	0.05281	0
7	[C_4mim][SCN]	0.018162	0.064328	0.1183245	0.081717	0	0.053582	0.0428775	0
8	[C_6mim][Cl]	0.0181365	0.0642015	0.133408	0.11322	0	0	0.05281	0
9	[C_6mim][Br]	0.0181365	0.0642015	0.133408	0.11322	0	0	0.058629	0
10	[C_6mim][$N(CN)_2$]	0.0181365	0.0642015	0.133408	0.11322	0.0024005	0.060037	0.047234	0
11	[C_6mim][BF_4]	0.0181365	0.0642015	0.133408	0.11322	0	0.083756	0.0250115	0
12	[C_6mim][Ala]	0.0181365	0.0642015	0.133408	0.11322	0.026491	0.0383145	0.0286225	0.040365
13	[C_6mim][Ser]	0.0181365	0.0642015	0.133408	0.11322	0.033789	0.0320265	0.03489	0.038397
14	[C_6mim][Asn]	0.0181365	0.0642015	0.133408	0.11322	0.036344	0.0519305	0.0635675	0.0146825
15	[C_6mim][Pro]	0.0181365	0.0642015	0.133408	0.11322	0.001099	0.084943	0.04882	0.011961
16	[C_6mim][Gly]	0.0181365	0.0642015	0.133408	0.11322	0.0105355	0.0596305	0.0422955	0.0194415
17	[C_{10}mim][Br]	0.017982	0.0641575	0.1481335	0.1450045	0	0	0.058629	0
18	[C_{10}mim][BF_4]	0.017982	0.0641575	0.1481335	0.1450045	0	0.083756	0.0250115	0
19	[C_{10}mim][$N(CN)_2$]	0.017982	0.0641575	0.1481335	0.1450045	0.0024005	0.060037	0.047234	0
20	[C_{10}mim][SCN]	0.017982	0.0641575	0.1481335	0.1450045	0	0.053582	0.0428775	0
21	[C_{12}mim][Br]	0.0179585	0.0640975	0.1625145	0.1767695	0	0	0.058629	0
22	[C_2OHmim][Gly]	0.070256	0.072425	0.023864	0.005958	0.0105355	0.0596305	0.0422955	0.0194415
23	[C_2OHmim][Ala]	0.070256	0.072425	0.023864	0.005958	0.026491	0.0383145	0.0286225	0.040365
24	[C_2OHmim][Ser]	0.070256	0.072425	0.023864	0.005958	0.033789	0.0320265	0.03489	0.038397
25	[C_2OHmim][Pro]	0.070256	0.072425	0.023864	0.005958	0.001099	0.084943	0.04882	0.011961

Input Dataset for the Matlab Prog.

Chemical Engineering
Inspiring Potential·Generating Futures

Multiple Linear Regression result for the QSAR models, R^2, mean absolute error (MAE) and $R^2_{average}$.

	A. hydrophila	*E. Coli*	*L. monocytogenes*	*S. aureus*
R^2	0.972	0.966	0.963	0.966
MAE	0.127	0.141	0.147	0.139
Cross validation, $R^2_{average}$	0.978	0.970	0.969	0.970

	A. hydrophila	*E. coli*	*L. monocytogenes*	*S. aureus*
R^2 - $(QSAR)_{cation}$	0.958	0.953	0.951	0.949
MAE - $(QSAR)_{cation}$	0.166	0.141	0.178	0.178
R^2 - $(QSAR)_{anion}$	0.138	0.132	0.127	0.144
MAE - $(QSAR)_{anion}$	0.733	0.738	0.732	0.733

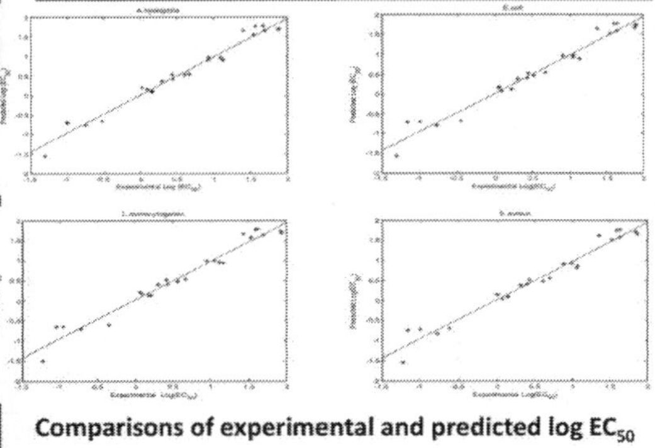

Comparisons of experimental and predicted log EC_{50}

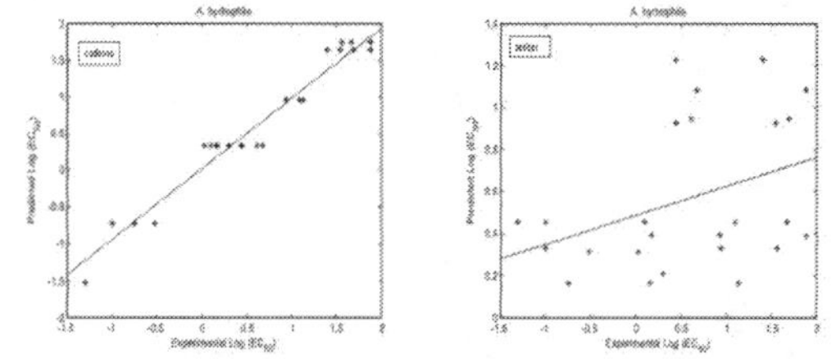

An example for sub-QSAR : comparison between $(QSAR)_{cation}$ and $(QSAR)_{anion}$ models for *A.hydrophila*

- Poor correlation for anions

So far the studies indicate ;

- The antimicrobial activity of the ILs relies mainly on the alkyl chain length and, to a lesser extent, on the nature of the anions.
- The ILs based on hydroxy-ethyl and butyl alkyl chains show more eco-friendly behavior than the longer alkyl chain analogues (imidazolium based)
- The introduction of amino acid anions leads to the reduction in toxicity compared to the common anions used such as [[$N(CN)_2$], [Br], [Cl], and [BF_4]).
- The sub-QSAR models show the importance of the cations descriptors set on the main predictive models performance.

There are still much more to discover as the area is some what under explored ....

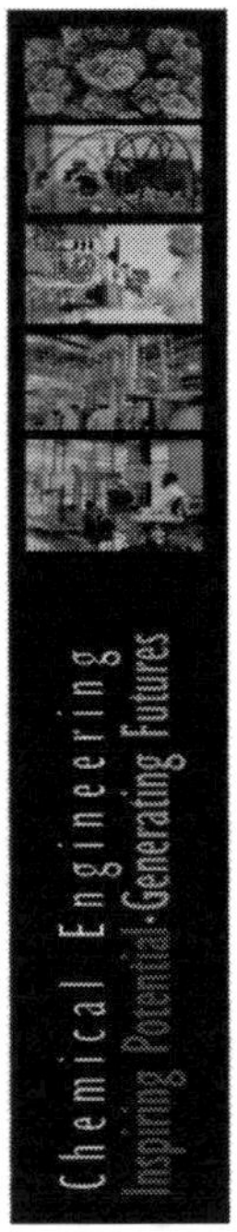

CONCLUSION

- Ionic Liquids presents one of the potential green solutions to industry due to their tunable properties with the potential to out perform the conventional organic solvents (and in selected cases - heavy metal based catalyst)
- Although Ionic Liquids have several constraints such as high viscosity and material cost, it can still be considered for feasible industrial applications due to its ease of recovery and recyclability and direct cost competitive in certain process applications such as adsorption.
- Despite being known as greener solvent, toxicity of Ionic Liquids should not be overlooked but there are many types of cations and anions which can be used to reduce the toxicity level to a much lower values.
- In view of the tremendous number of ionic Liquids that could be developed from the matching of the various types of cations and anions, predictive and generic method for screening its performance as well as determining its physical properties and toxicity are needed especially for process design purpose.

Chemical Engineering
Inspiring Potential·Generating Futures

ACKNOWLEDGEMENT

Our special thanks goes to the ;

- PETRONAS Petroleum Research Fund for providing the funding in undertaking the fundamental and applied research that has enabled us to build the capacity and capability at the Ionic Liquids research center at UTP.
- Our partner PETRONAS Research for their cooperation without which we would not have able to direct our research for the benefit of the company.
- The ICOSSE 2015 organiser – University of Pannonia , AIChE and EFCE for giving me the opportunity to be here to share our work with you all.
- To all the participant for the warmth companionship and sharing your knowledge which has given me a fresh perspective on the concept of Sustainability.

Chemical Engineering
Inspiring Potential·Generating Futures

THANK YOU FOR YOUR KIND ATTENTION

Incorporating the design of sustainable systems into engineering education

Monday, November 9, 2015: 1:20 PM
258 (Salt Palace Convention Center)

David Allen, Department of Chemical Engineering, and Center for Energy and Environmental Resources, University of Texas at Austin, Austin, TX

What does it mean for an engineering design to be sustainable? What is the body of knowledge needed for the design of sustainable engineered systems? Design tools that allow engineers to choose inherently safer chemicals, improve energy efficiency, improve mass efficiency, and reduce emissions and exposures are certainly part of an emerging tool set engineers will need for sustainable designs. However, these are not entirely new tools for engineers. Energy and mass efficiency are objectives that have always been included in engineering design. What is new is the need to systematically and simultaneously incorporate economic, environmental and social objectives into engineering designs, at multiple scales. These tools and skills will be increasingly important, and challenging to apply, as engineers operate in diverse global environments, with varying social perspectives and environmental priorities.

Over the past decade, sustainability concepts, and tools for sustainable engineering, have begun appearing in engineering curricula in the United States. In this presentation, current practices for incorporating sustainability into engineering education will be reviewed, and categorized into an emergent body of knowledge. Best practices and tools for presenting this body of knowledge will be summarized and new AIChE initiatives will be described.

The Challenges of Water Availability

Can we Eat, Drink AND Turn on the Lights?

Danny Reible
Civil and Environmental Engineering
Chemical Engineering
Texas Tech University

05/27/15

Food/Energy/Water Nexus

- **Water/Energy required to produce food**
  - ✓ Irrigation central to increased agricultural productivity
- **Water required to produce energy**
  - ✓ High withdrawals, limited consumption
  - ✓ Much more substantial energy/water link with biofuels
- **Energy required to treat, transport water**
  - ✓ We don't pay for water, we pay to move it around
  - ✓ Advanced treatment largely defined by energy requirements
  - ✓ Substantial energy demand to heat water

Food Energy Water Nexus

- **Water for Energy and Economy**
  - ✓ Oil and gas production requires 1-10 L of water per GJ of energy produced,
  - ✓ Coal 40 L/GJ
  - ✓ Bofuels may require 10,000 L/GJ or more.
  - ✓ Manufactured products requiring an average of 80 L of water/$ of product value
- **Energy for Water**
  - ✓ 0.4 kWh/m3 of water required to provide conventional surface water supplies
  - ✓ 1 kWh/m3 required for reclaimed water
  - ✓ 3-4 kWh/m3 required to desalinate seawater.
- **Agricultural irrigation water is a relatively low value use of water but positively impacts rural communities and and US exports**
  - ✓ 24% of Great Plains agricultural output exported from the U.S

The Water/Energy/Food Nexus

- URGENCY

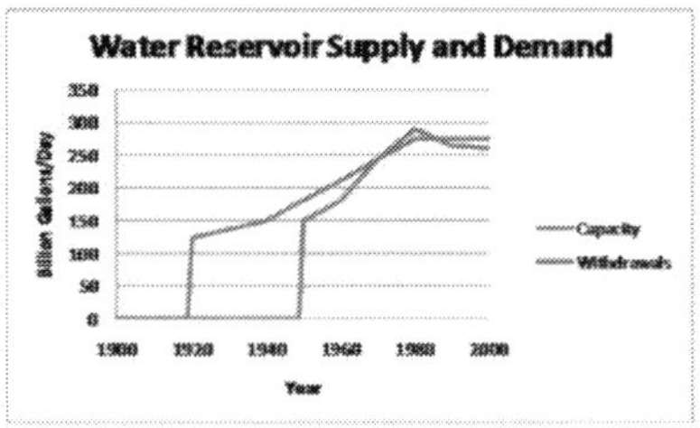

Water use and availability *will* limit our energy, economic and food choices without substantial efficiency improvements.

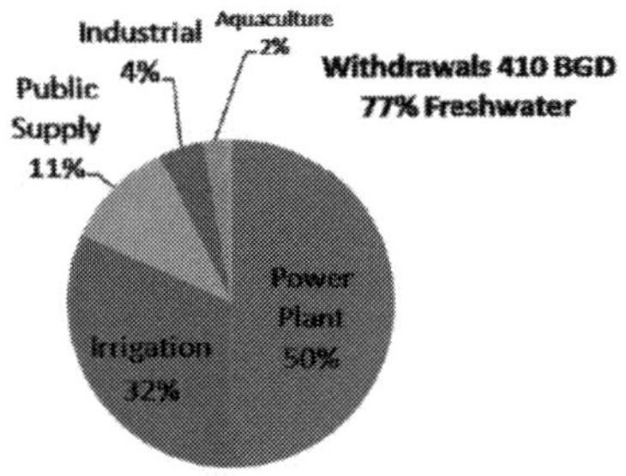

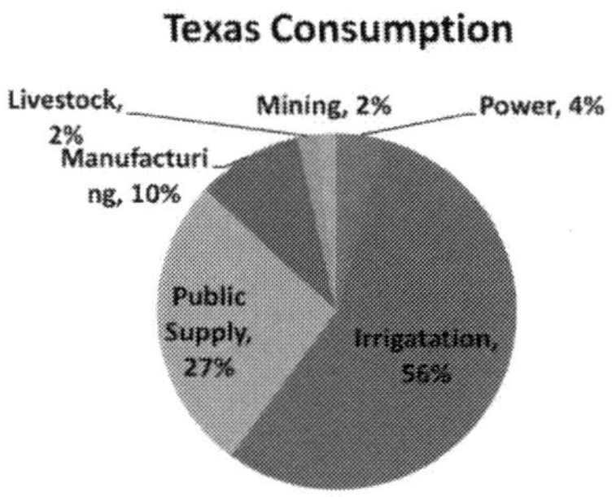

Challenges

- **Water is not valued**
  - ✓ Value added by 1 acre-ft of water in agriculture <$100
  - ✓ Municipal value of water $1000-2000/acre-ft
  - ✓ Value of acre-ft of water used in hydraulic fracturing >$100,000
  - ✓ Compare these water values to oil at ~$500,000 acre-ft (~$60/bbl)
- **All water problems and solutions are local**
  - ✓ Economics deter any trans-watershed solutions
  - ✓ Legal- social impediments pose serious challenges to trans-watershed solution
- **Disposal of water is cheaper than treating/recycling**
  - ✓ Social/economic resistance to "toilet to tap"
  - ✓ Owned disposal wells for produced water in Texas $0.10/bbl
  - ✓ Commercial disposal well $2-3/bbl

Land Use - Southern Great Plains

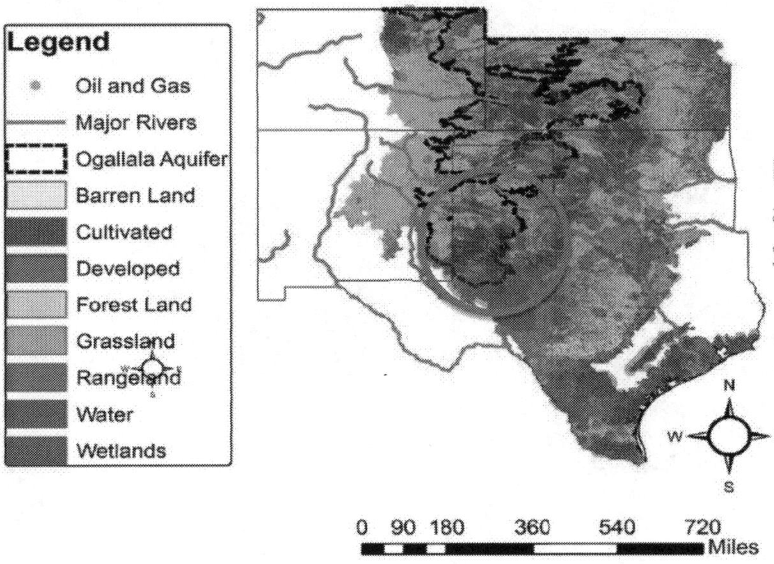

Permian Basin and Surroundings

Texas Water Demand and Value

Sources:
Texas Water Development Board
Office of State Comptroller

Irrigated agriculture 56% of consumptive water demand but 0.6-0.8% of economy

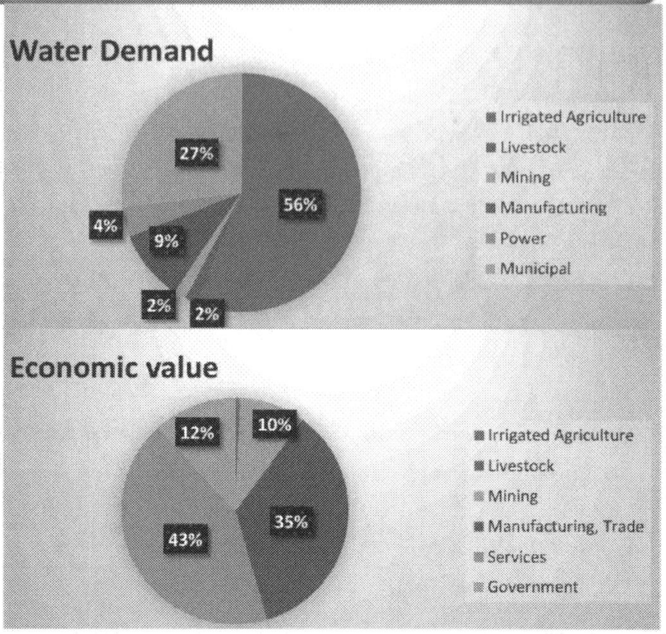

Cumulative Water Management Strategies by 2060- Texas

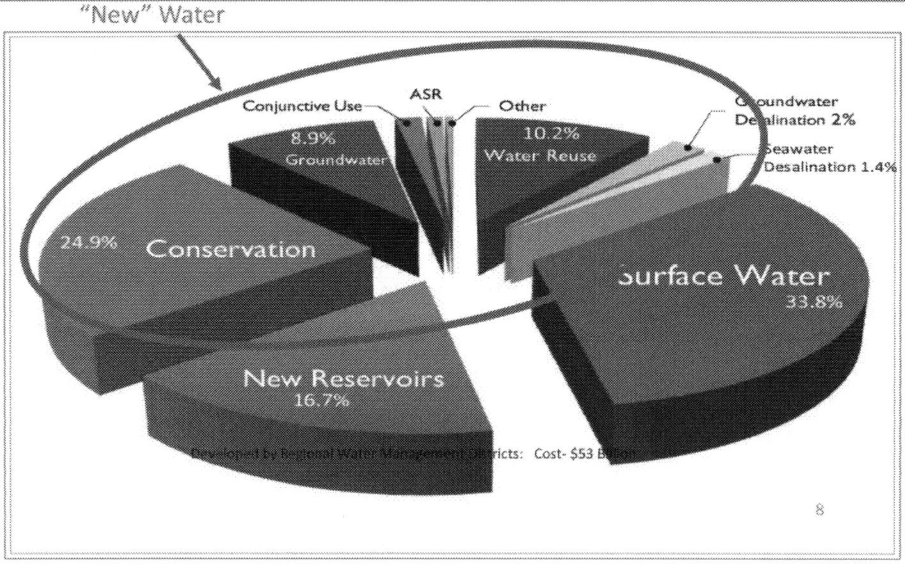

Agricultural Irrigation Conservation

- **Approaches**
  - ✓ Appropriate crop selection
  - ✓ Efficient hybrids
  - ✓ Efficient Irrigation Systems
    - **Drip irrigation**
  - ✓ Efficient scheduling
    - **Canopy Temperature Control**
    - **Satellite Soil Moisture Sensing**
  - ✓ Target ~80% of crop ET needs

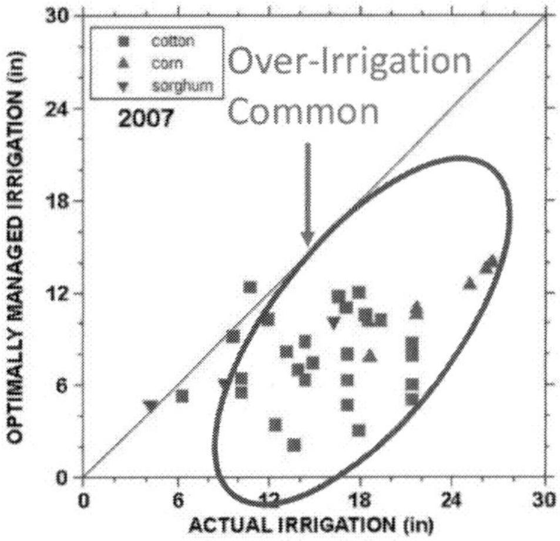

West, 2014

Municipal Conservation

San Antonio 1984-2009

Customers ↑ 67%

Water ↑ 0%

Wide variety of approaches employed to alleviate needs for capital investment in water and waster treatment and distribution facilities

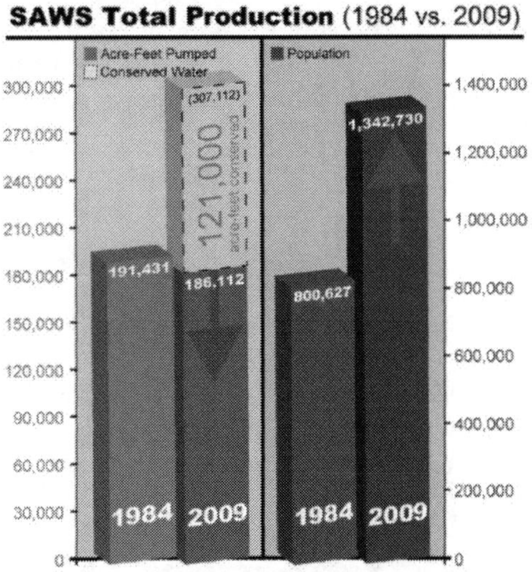

Puente, 2012

Making new Water Desalination in Texas

- **40 brackish water desalination facility in Texas**
  - ✓ Cost using brackish groundwater $357 to $782 per acre foot of water
  - ✓ Cost using seawater $800 to about $1,400 per acre foot of water
- **Municipal Desalination capacity**
  - ✓ 123 M gal/day total
  - ✓ 73 M gal/day brackish groundwater
  - ✓ 50 M gal/day brackish surface water
- **Industrial desalination capacity 60 to 100 MGD mainly in the power and semi-conductor industries.**
- **Is there opportunity for 500 MGD coast desalination providing water to central and southeast Texas?**
  - ✓ ~20% of expected water needed in 2060

Water Consumption for Power

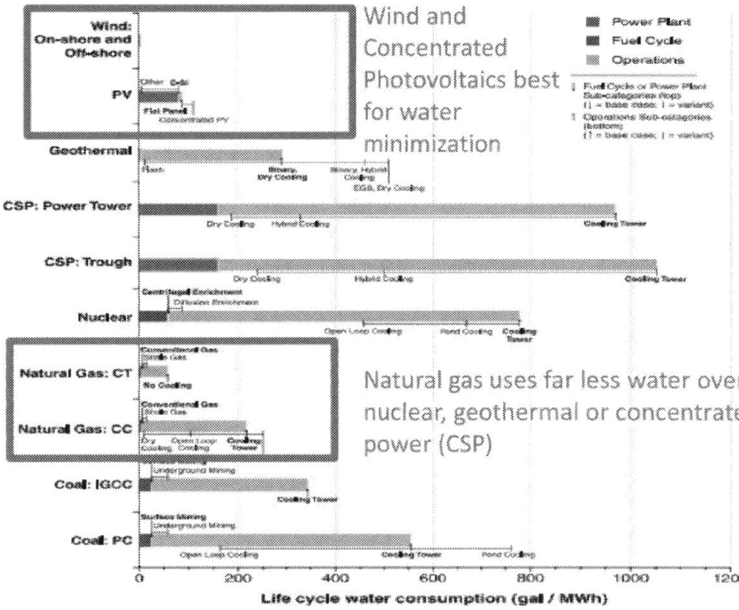

Wind and Concentrated Photovoltaics best for water minimization

Natural gas uses far less water overall than coal, nuclear, geothermal or concentrated solar power (CSP)

Meldrum et al. 2014

Texas Energy/Water Mix

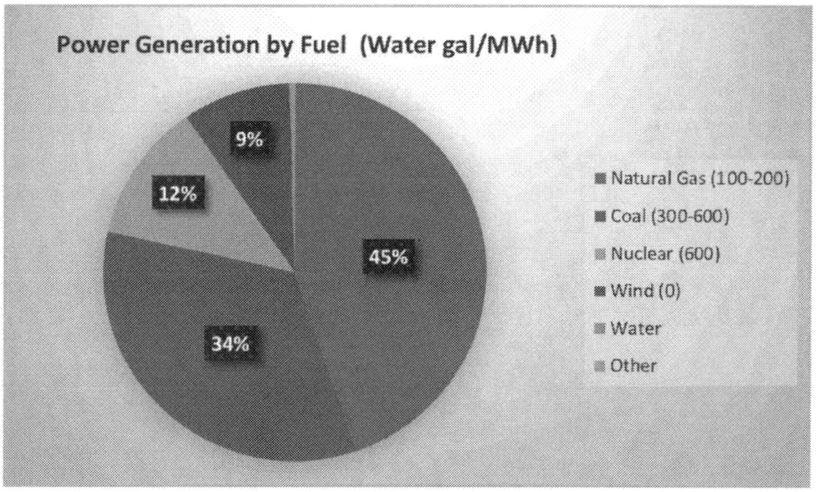

Dry Cooling viable if water costs 1000-2000 $/acre-ft

Water for Energy: Hydraulic Fracturing and Water Availability

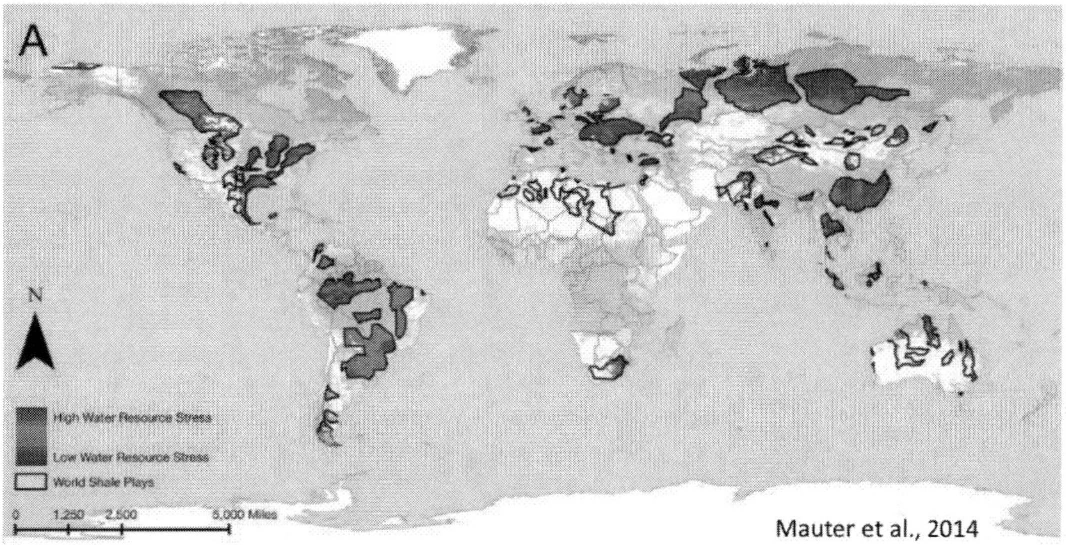

US Shale Plays and Availability

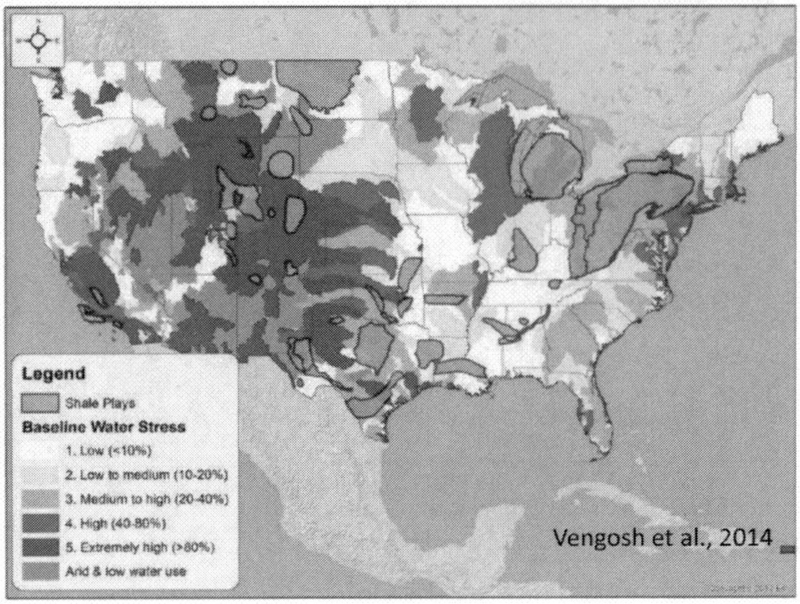

Water Availability Eagle Ford

- **Water demand- 5-6.7% of total (Jester, 2011)**
- **Local dislocations possible (Nicot & Scanlon, 2012)**
  - ✓ Projected water needs as % of total water use by county in Eagle Ford
    - **Webb – 5.2%**
    - **De Witt – 35%**
    - **Karnes – 39%**
    - **Live Oak – 12%**
    - **Dimmit – 55%**
    - **La Salle – 89%**

  ↓ Increasingly rural and lower overall water use
- **State total water needs for hydraulic fracturing**
  - ✓ ~125,000 acre-ft/yr (~ 0.5% of state total use)
  - ✓ Approximately equivalent to evaporation from Texas Highland Lakes (water source for Austin)

What are our Alternatives to Freshwater for Extractive Mining?

- **Reuse Flowback and Produced Water**
- **Reuse Municipal Effluents**
  - ✓ Treated wastewater
  - ✓ RO reject water
- **Employ brackish waters**
- **Disincentives common to each of these**
  - ✓ Lack of a strong driver except in Marcellus
  - ✓ Transportation to point of use
  - ✓ Appropriate chemistry (precipitation and fouling, corrosion)
  - ✓ Treatment to achieve appropriate chemistry

Practical brine thermodynamic modeling

- **Electrolytes NRTL (eNRTL) model for describing thermodynamics of the ionic solution**
  - ✓ Potentially more robust model and more firmly grounded in thermodynamic theory than simpler models (Debye-Huckel)
  - ✓ More feasible than complicated Pritzer model
  - ✓ Applicable to broader range of conditions (subsurface and surface processing and treatment of produced water)
  - ✓ Directly useful to process simulators such as ASPEN
  - ✓ Substantial development required for specific ions
- **Initial efforts focused on barium, strontium, calcium, and magnesium sulfate and carbonate scale**

Gel Fracturing Water Requirements

- **Not limited by water chemistry but by economics of achieving desired gel characteristics with that chemistry**
- pH: 6-8
- Ca: <1200ppm preferred less than 600
- Mg: <400ppm preferred less than 200
- Cl: <4000ppm
- Si: <20ppm 0 is preferred
- Fe: <10ppm 0 is preferred
- Sulfates: <2500ppm 100 is preferred
- Bicarbonates: <500ppm 200 is preferred
- TDS < 40,000 ppm preferred

Slick Water Fracturing Requirements

- **Compatibility with formation waters**
  - ✓ Avoidance of subsurface scale and plugging
  - ✓ Produced water is by definition compatible with same formation
- **Potential concerns**
  - ✓ Oil produced with the water
  - ✓ Iron
  - ✓ Potential trace ions (Ba, St, and SO_4, CO_3 scaling)
  - ✓ TDS not a significant concern
- **Limitations**
  - ✓ Cost
  - ✓ Complexity
  - ✓ Availability of alternative water sources

Reuse Produced Water?

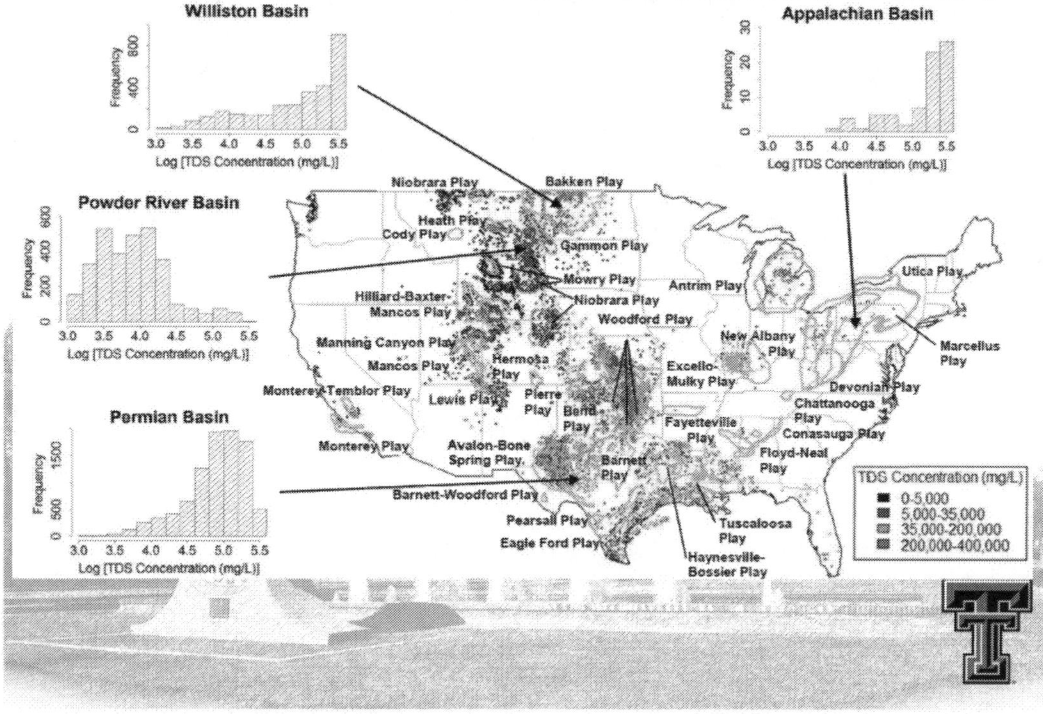

Recycling Example in Region of High Well Density

Conditions Conducive to Reuse

- **Ongoing fracturing**
  - ✓ 30,000-60,000 bbls per day needed
- **Close proximity of producing and fracturing wells and volume sufficient to justify centralized collection & treatment**
  - ✓ 30,000 bbls per day of produced water
  - ✓ 30,000 bbls per day of brackish water available
    - ~5,000 mg/L TDS
- **Minimal treatment requirements**
  - ✓ ClO_2 iron oxidation
  - ✓ Oil-water separation for produced water
  - ✓ Residual ClO_2 for biocide
- **Recovered oil from produced water aids economics**
- **Lack of landowner/regulatory resistance**

Reuse Municipal Effluents

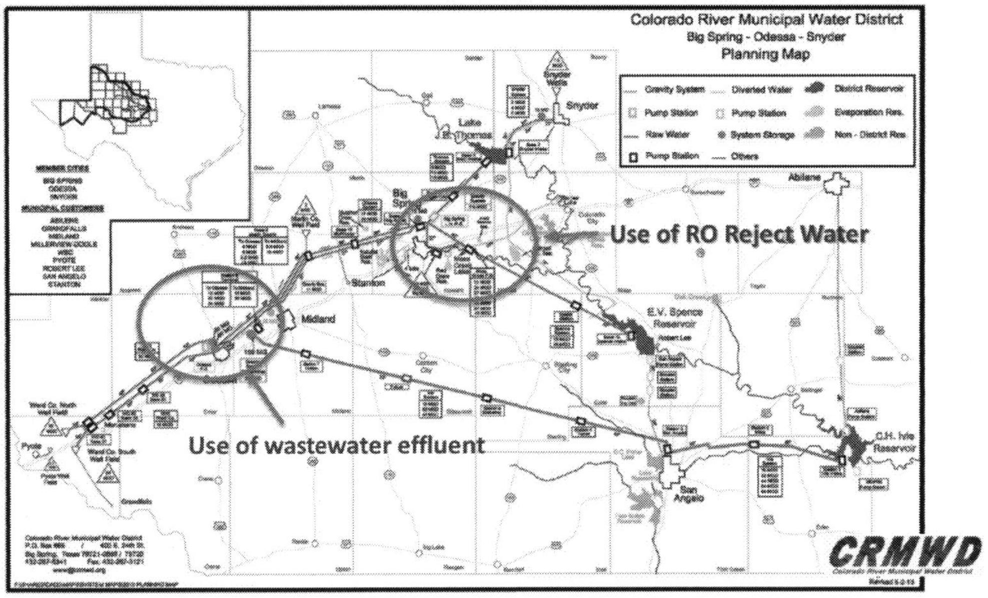

Saline Groundwater (Brackish Water)?

Mauter et al, 2014

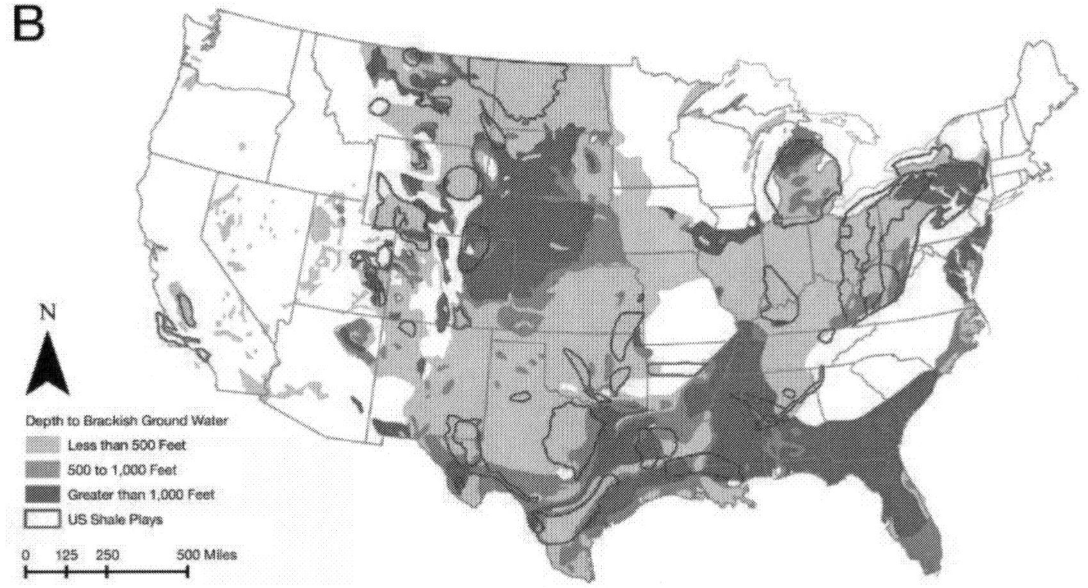

Research Needs to Expand Use of Brackish Groundwater

- **Depth and areal extent largely known**
- **Quality partially known but resolution marginal**
- **Substantial gaps**
  - ✓ Aquifer Characteristics
    - **Transmissivity, Storage coefficient of brackish aquifers**
    - **T = 5-10 ft/day (60+ft/day in Ogallala)**
    - **S = 10^{-4} (0.1 in Ogallala)**
  - ✓ Aqueous Chemistry
    - **Aqueous chemistry effects on plugging, fouling, operations**
    - **Limit of reuse requirements and treatment**
    - **Optimal (minimal) treatment requirements, if needed**

Brackish Water Characteristics

Strontium WVUDOE Limit 10 mg/L

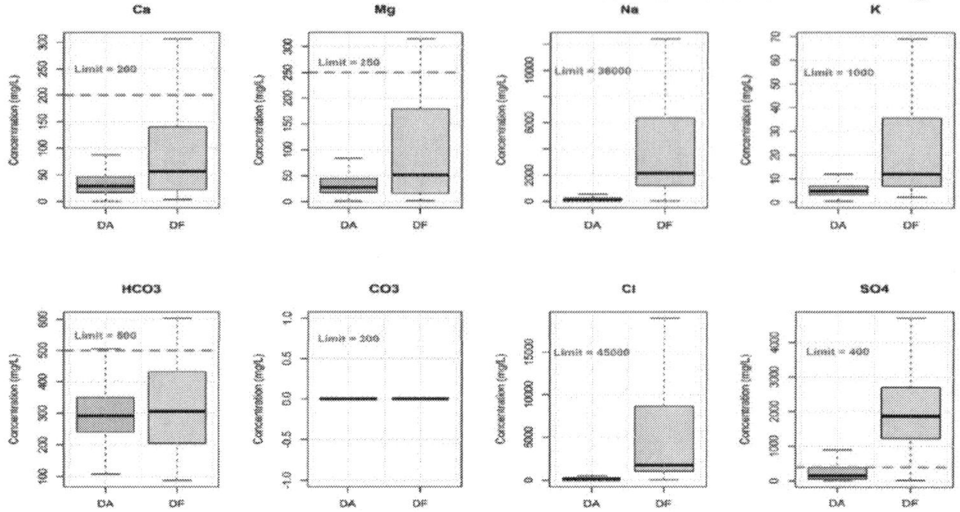

Can we Eat, Drink AND Turn on the Lights?

Turn on the Lights? √

Water consumption is not all that great and improving with natural gas & wind/solar power

Drink? √

But it will likely cost us more!

Eat? ?

Agricultural cannot easily support investments necessary to achieve maximum efficiency and there are other high value needs for the water

TÁMOP-4.1.1.C-12/1/KONV-2012-0017
„Green Energy" – Cooperation of the higher education sector for the development of green economy in the area of energetics

GENO- AND CYTOTOXICOLOGIC ASSESSMENT OF WASTEWATER EFFLUENTS WITH A NOVEL FLOW CYTOMETRIC SPERM TOXICITY ASSAY

Balázs Kakasi[1], Szabolcs Tamás Nagy[2], Nóra Kováts[3], Endre Nagy[1]

University of Pannonia, Veszprém, Hungary
[1]Research Institute of Chemical and Process Engineering,
[2]Department of Animal Sciences and Animal Husbandry
[3]Institute of Environmental Sciences

ICOSSE conference 2015

SZÉCHENYI 2020

HUNGARIAN GOVERNMENT

European Union
European Social Fund
INVESTING IN YOUR FUTURE

OVERVIEW

TÁMOP-4.1.1.C-12/1/KONV-2012-0017
„Green Energy" – Cooperation of the higher education sector for the development of green economy in the area of energetics

Introduction
- problems of waste water treatment

Aims
- Defining the aims of the study

Boar spem toxicity assay
- Introduction of the test
- Experimental design

Results
- Mitochondrial transmembrane potential changes
- Oxidative DNA lesions
- DNA fragmentation

Conclusions
- Conclusions and summary

INTRODUCTION

TÁMOP-4.1.1.C-12/1/KONV-2012-0017
„Green Energy" – Cooperation of the higher education sector for the development of green economy in the area of energetics

Diminution of freshwater resources makes wastewater treatment important

primary treatment

- including preliminary treatment
- coarse solids are removed
- suspended matter settles out
- grease, oil, fat removal

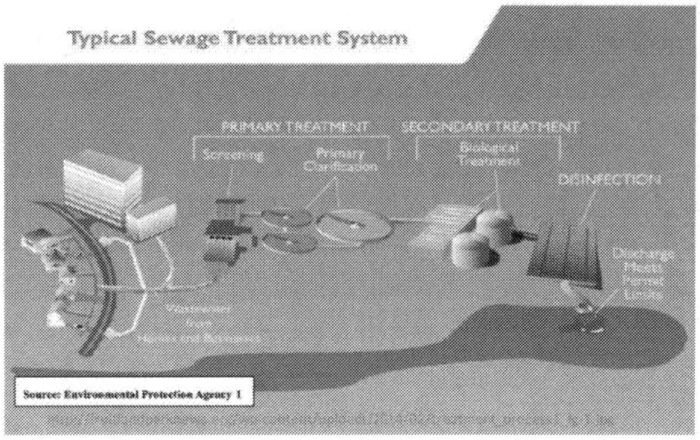

secondary treatment

biological treatment processes

- organic matter removal
- nitrification/denitrification
- biological phosphorus removal
- nutrient concentration reduction.

combined primary and secondary stages

tertiary treatment

- Disinfection – removal of pathogen microorganisms (UV, chlorination)
- further reduce parameter values below the standards

(EPA - Wastewater Treatment Manuals – Primary Secondary and Tertiary treatments 1997)

INTRODUCTION

TÁMOP-4.1.1.C-12/1/KONV-2012-0017
„Green Energy" – Cooperation of the higher education sector for the development of green economy in the area of energetics

Wastewater treatment

physical, chemical, biological processes

Remaining dangerous components

- Pharmaceuticals, PCBs, PAHs, EDCs

disposal into receiving water bodies

concentrations reach background levels kms away from the effluent outfall

The cleaning efficiency are growing

harmful components in the effluents

Biological effects not monitored

AIMS

TÁMOP-4.1.1.C-12/1/KONV-2012-0017
„Green Energy" – Cooperation of the higher education sector for the development of green economy in the area of energetics

Major parameters measured

- Dissolved solids (TDS)
- Suspended solids
- Nitrogen
- Phosphorus
- Alkalinity
- Biological Oxigen Demand (BOD_5)
- Chemical Oxigen Demand (COD)
- Total Organic Carbon (TOC)
- Microbiological characteristics

Biological effects not monitored

Aims

- **cyto- and genotoxicological characteristics**
- **boar sperm test** with **flow cytometry**

BOAR SPEM TOXICITY ASSAY

TÁMOP-4.1.1.C-12/1/KONV-2012-0017
„Green Energy" – Cooperation of the higher education sector for the development of green economy in the area of energetics

Advantages

- Mammal cells
- Fast, easy and ethical
- Wide range of cellular parameters

Boar sperm toxicity assay

- 2ml sperm exposed to 20µl sewage sample
- 30 minutes at room temperature
- triplicate
- cellular parameters analyzed

(Andersson et al., 2004)

Flow cytometry

- Beckman Coulter FC500,
- acquisition settings 20000 events or 5 minutes,
- stored as List Mode Files (LMD)

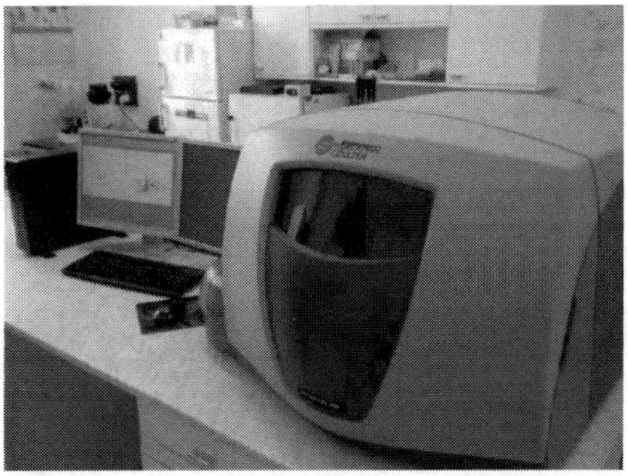

EXPERIMENTAL DESIGN

TÁMOP-4.1.1.C-12/1/KONV-2012-0017
„Green Energy" – Cooperation of the higher education sector for the development of green economy in the area of energetics

Samples

- Domestic wastewater treatment plant (accepts pharmaceutical sewage)
- two wastewater samples collected at effluents

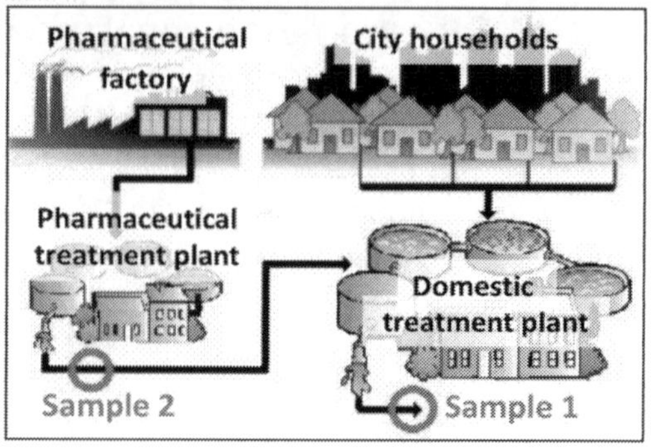

Pretreatment

- PBS tablet (Sigma) was dissolved in 200ml deionized water (used as control)
- PBS tablet (Sigma) and in 200 ml of each samples.

Porcine sperm

- from a commercial supplier (Véghfarm Kft., Pakod, Hungary)

EXPERIMENTAL DESIGN

TÁMOP-4.1.1.C-12/1/KONV-2012-0017
„Green Energy" – Cooperation of the higher education sector for the development of green economy in the area of energetics

Parameters

Mitochondrial transmembrane potential changes

- 5-5ul MitoTracker Green FM, MitoTracker Deep Red FM (Molecular Probes)

Oxidative DNA lesions

- OxyDNA kit (Argutus Medical BIO81DNA)

DNA fragmentation

- Nicoletti assay (Riccardi and Nicoletti, 2006)

Data analysis

Flowing data analysis software (v. 2.5.1., www.flowingsoftware.com)

- cells were identified by FSC-SSC properties
- non-sperm events were gated out

RESULTS

TÁMOP-4.1.1.C-12/1/KONV-2012-0017
„Green Energy" – Cooperation of the higher education sector for the development of green economy in the area of energetics

Mitochondrial transmembrane potential changes

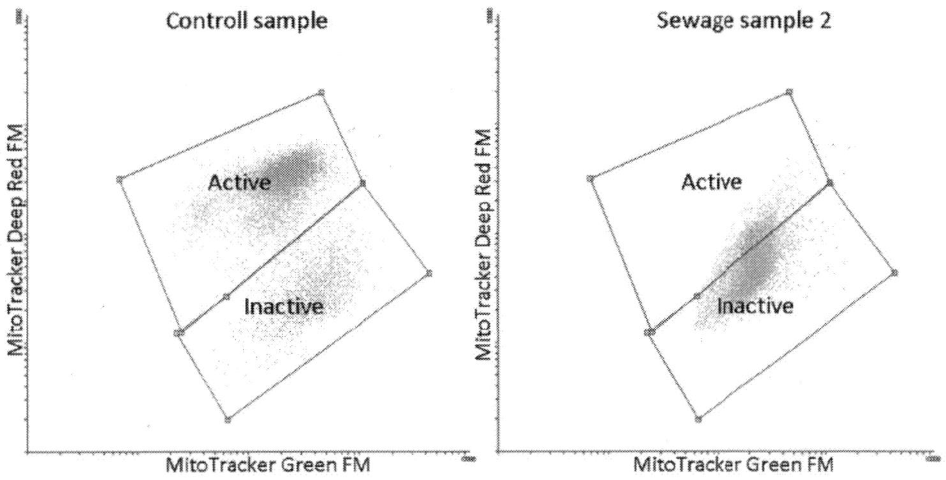

RESULTS

TÁMOP-4.1.1.C-12/1/KONV-2012-0017
„Green Energy" – Cooperation of the higher education sector for the development of green economy in the area of energetics

Oxidative DNA lesions

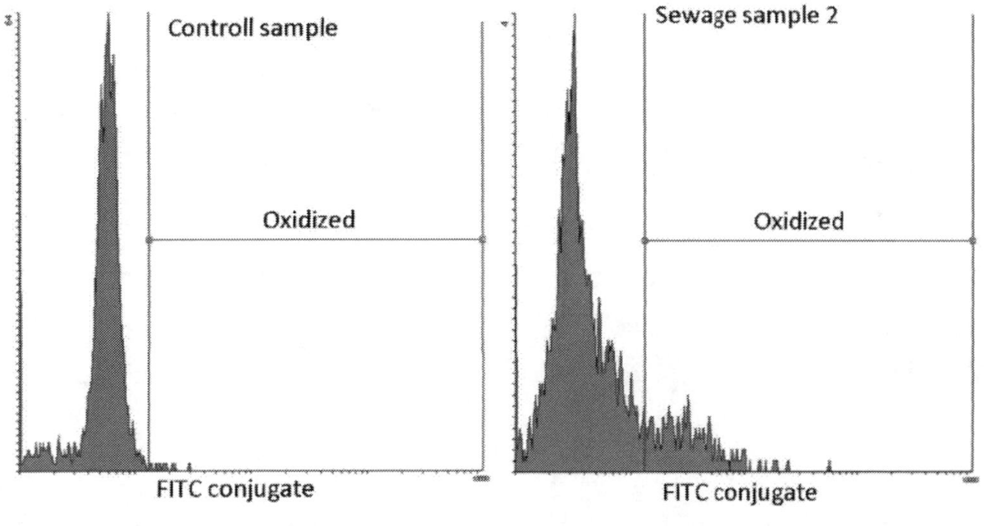

RESULTS

TÁMOP-4.1.1.C-12/1/KONV-2012-0017
„Green Energy" – Cooperation of the higher education sector for the development of green economy in the area of energetics

DNA fragmentation

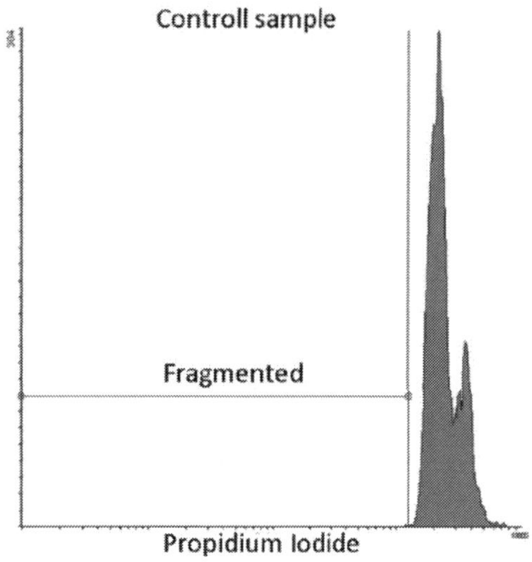

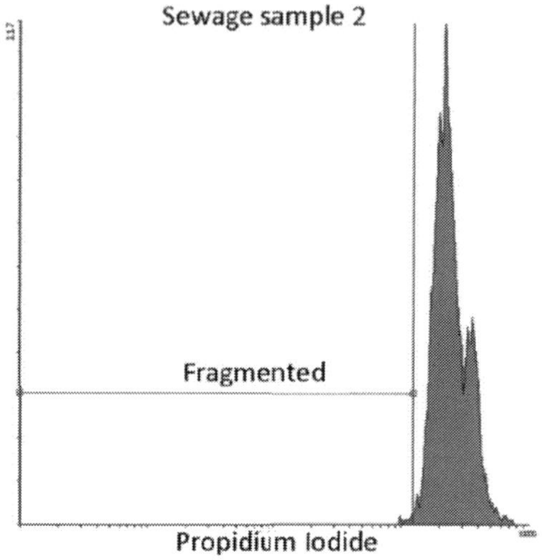

SUMMARY OF RESULTS

TÁMOP-4.1.1.C-12/1/KONV-2012-0017
„Green Energy" – Cooperation of the higher education sector for the development of green economy in the area of energetics

Sample (replicate)	Active mitochondria (%)	Oxidized DNA (%)	Fragmented DNA (%)
Control (1)	86.16	1.63	0.04
Control (2)	87.45	4.85	0.01
Control (3)	87.92	3.10	0.01
Sewage-1 (1)	83.30	3.41	0.07
Sewage-1 (2)	82.99	3.22	0.14
Sewage-1 (3)	84.57	2.27	0.09
Sewage-2 (1)	38.69	16.68	0.75
Sewage-2 (2)	36.86	11.87	0.59
Sewage-2 (3)	28.22	15.16	0.46

CONCLUSIONS

TÁMOP-4.1.1.C-12/1/KONV-2012-0017
„Green Energy" – Cooperation of the higher education sector for the development of green economy in the area of energetics

two Hungarian **wastewater effluent samples**
cellular parameters of boar spermatozoa **with flow cytometry**

changes in all measured parameters compared to controls

- mitochondrial transmembrane potential
- oxidative DNA lesions
- DNA fragmentation

treated wastewater may still contain **harmful components**
concentrations might **reach background levels** kms away from outfall

boar sperm assay
easy and ethical alternative in toxicology monitoring

TÁMOP-4.1.1.C-12/1/KONV-2012-0017
„Green Energy" – Cooperation of the higher education sector for the development of green economy in the area of energetics

THANK YOU FOR YOUR ATTENTION!

SZÉCHENYI 2020

HUNGARIAN GOVERNMENT

European Union
European Social Fund

INVESTING IN YOUR FUTURE

REFERENCES

TÁMOP-4.1.1.C-12/1/KONV-2012-0017

„Green Energy" – Cooperation of the higher education sector for the development of green economy in the area of energetics

- Andersson, M.A., Jaaskelainen, E.L., Shaheen, R., Pirhonen, T., Wijnands, L.M., Salkinoja-Salonen, M.S., 2004. Sperm bioassay for rapid detection of cereulide-producing Bacillus cereus in food and related environments. International journal of food microbiology 94, 175-183.
- EPA - Wastewater Treatment Manuals – Primary Secondary and Tertiary treatments 1997
- Riccardi, C., Nicoletti, I., 2006. Analysis of apoptosis by propidium iodide staining and flow cytometry. Nature protocols 1, 1458-1461.
- www.flowingsoftware.com

Figures

- http://fruitlandparknews.org/wp-content/uploads/2014/02/treatment_process1_lg-1.jpg
- https://www.ufz.de/export/data/1/27320_probennahme_fluss.jpg

Anaerobic Digestion of Composite Organic Wastes: An Ecologically Balanced Approach for Improvement of Health, Hygiene, and Environment of Rural Communities

- **Conference:** International Congress on Sustainability Science Engineering ICOSSE
- **Year:** 2015
- **Proceeding:** ICOSSE 2015
- **Group:** Abstract Submissions
- **Session:**

Session 1: Water Resource and Availability Issues I

- **Time:**

Wednesday, May 27, 2015 - 11:00am-11:20am

Authors:

Banerjee, G., The University of The West Indies, St. Augustine, Trinidad & Tobago
Banerjee, S., The University of The West Indies, St. Augustine, Trinidad & Tobago

Anaerobic digestion of composite organic wastes comprising human excreta (HE), partially-digested market waste (PMW), and partially-digested water hyacinth (PWH) in community latrine-linked field digester as well as in laboratory-scale anaerobic digester using sewage treatment plant primary sludge (PS), liquefied market wastes (LMW), and liquefied water hyacinth (LWH) was studied with an aim to curtail huge costs associated with transportation and ultimate disposal of market wastes into landfill.

While the optimum proportions of waste feed components viz, PS, LMW, and LWH of 2:1.5:1 and HE: PMW: PWH of 11.8:6:1 were observed, the maximum bio-gas productions were 0.015 sm^3 and 0.042 sm^3 per kg of composite wastes for laboratory and field digesters, respectively. The total coliform removals were 98.6% and 99.22%, whereas the faecal coliform removals were 98.45% and 99.62% in the laboratory and field digesters, respectively. The digester effluent was beneficially used for irrigation and pisciculture through post-treatment in WSP system.

Uses Of Nano-Water For Living, Soil-Water Remediation and Agricultural Irrigation

- **Conference:** International Congress on Sustainability Science Engineering ICOSSE
- **Year:** 2015
- **Proceeding:** ICOSSE 2015
- **Group:** Abstract Submissions
- **Session:**

Session 1: Water Resource and Availability Issues I

- **Time:**

Wednesday, May 27, 2015 - 11:20am-11:40am

Authors:

Wang, I. S.
Loo, W., Enviro Best Ltd.

The uses of nano-water have basically open new pages in the field of water sciences and engineering. Nanowater is microscopic nano-micron size water bubbles which will not easily gas out. It will create super saturated oxygen, carbon dioxide, and other gases in water. Nano-water provides perfect water for plant growth irrigation and also for human and animal consumption. Also, high dissolved oxygenated water with nano-bubbles is perfect for remediation of all toxic organic chemicals in soil and water.

Urban Food Systems: Closing the Food Life Cycle

Thomas L. Theis
Institute for Environmental Science and Policy
University of Illinois at Chicago
(www.iesp.uic.edu)

Sustainability Science and Urban Food Systems

SS is:

- interdisciplinary
- requires an integrative methodology
- aimed at contributing to the solution of complex dynamic problems embedded in human-natural systems
- requires a partnership between researchers and practitioners for transmitting knowledge and implementing the insights of sound science

UFSs have the potential to:

- become an important part of the national food system
- transform urban spaces
- improve overall accessibility to healthy food
- serve as a source of social cohesion
- generate new economic opportunities
- serve as a source of insight for new more sustainable agricultural practices and policies

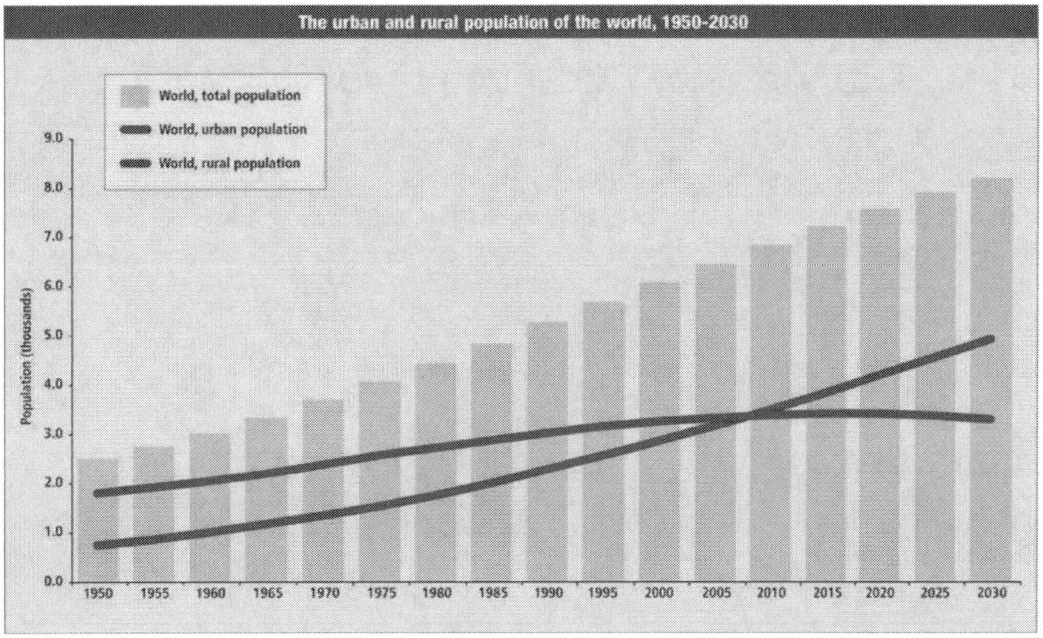

www.un.org/esa/population

Urban growth in the Chicago metropolitan region

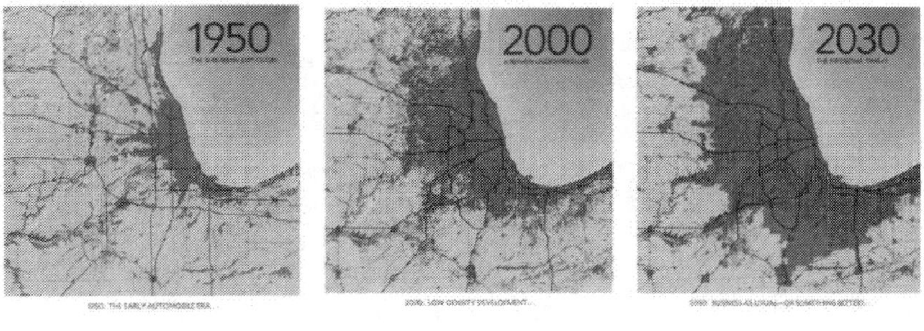

Two Views of CO2 Footprints (Chicago)

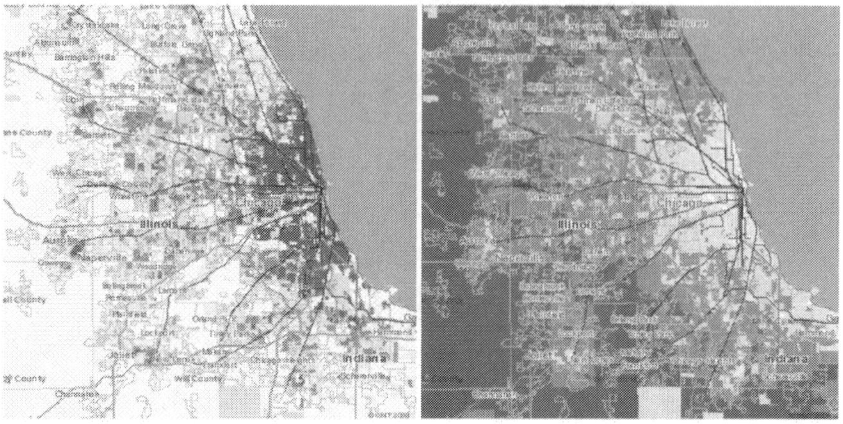

CO2/area **CO2/household**

Network Diagram of the Life Cycle of Food Systems

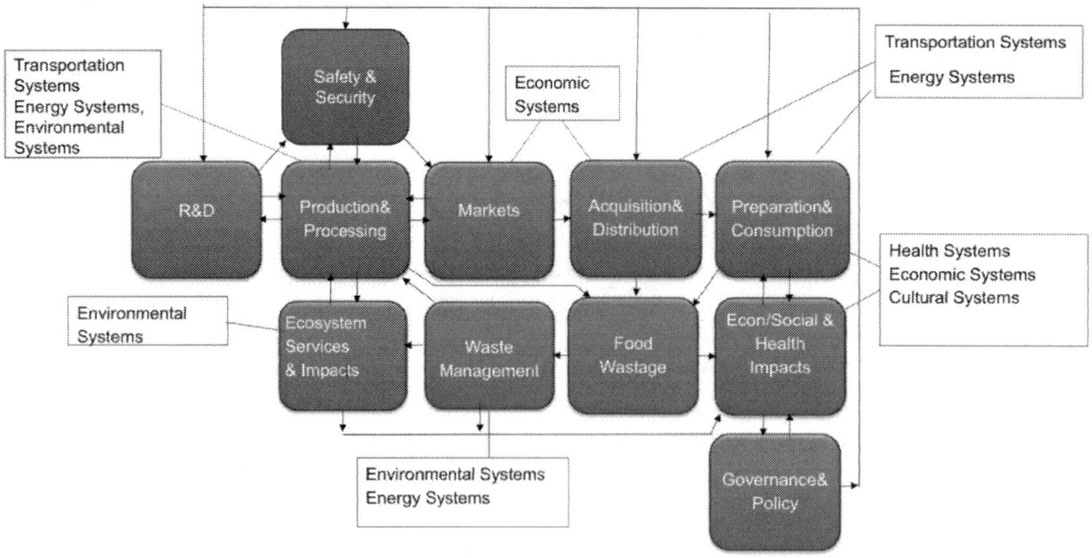

Urban agriculture

The practice of cultivating, processing, and distributing food in or around a village, town, or city. Urban agriculture can also involve animal husbandry, aquaculture, aquaponics, agroforestry, beekeeping, and horticulture. These activities occur in peri-urban areas (within 20km of the urban fringe) as well.

Modified after Hampwaye, G.; Nel, E. and Ingombe, L. "The role of urban agriculture in addressing household poverty and food security: the case of Zambia". Gdnet.org

Multifunctional Co-benefits of Urban Agriculture

Function	Benefits	Barriers
Production	Produces fruits, vegetables, herbs, and other products in proximity to residents	Requires spaces and nutrients, unit costs can be higher; competition from national ag markets; hampered by air pollution
Water conservation	Can require less water than rural farms	A burden on municipal supplies; must tap into existing lines and pay going rates
Energy conservation	Can reduce the embodied energy of material inputs, transport, and packaging	Material supply and crop distribution can contribute to local congestion
Human health	Helps to alleviate "food deserts"; better nutrition; requires physical activity	Accumulated toxins (e.g. lead) must be managed
Waste management	Closes the life cycle loop, organic waste products composted, nutrients recycled	Requires a local collection and composting system
Biodiversity	Can support a wide range of species, including native plants	Requires robust ecosystem services (e.g. pollination)
Urban greening	Convert vacant and derelict lands into productive green spaces	Competition for other land and space uses
Economic revitalization	Creates new jobs for residents	Profit margins can be low without closed loop operation
Education and socialization	Learn about foods, nutrition, cooking, environment; a shared social activity	Requires suitable oversight and programming

A Question of Land Use?

Google Earth Reference Images (Chicago)

(A) Community garden, (B) Residential garden

(Taylor, J. R., S. T. Lovell (2012) "Mapping public and private spaces of urban agriculture in Chicago through the analysis of high-resolution aerial images in Google Earth". *Landscape and Urban Planning*, 108 (1): 57-70)

Urban Farms and Gardens in Chicago

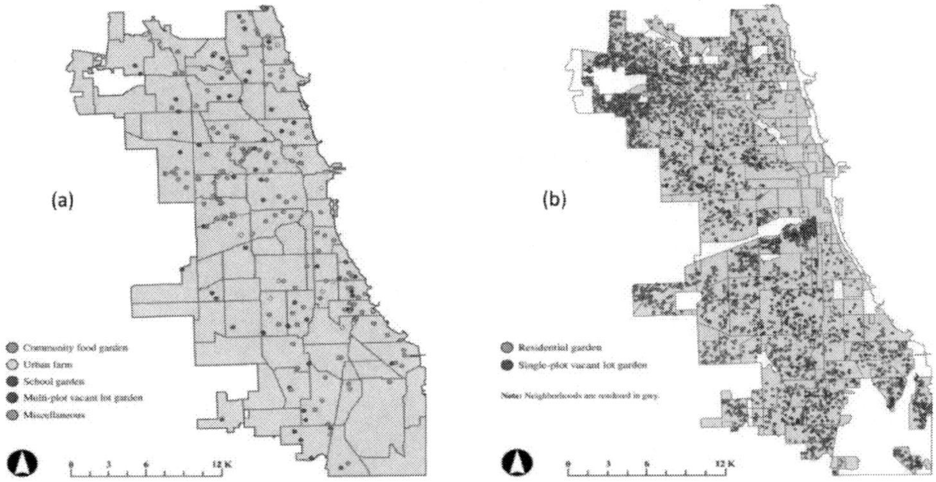

Mapping of Chicago urban farms and gardens, (a) identified through visual analysis of high-resolution aerial imagery in Google Earth superimposed on Chicago's 77 community areas, and (b) identified through manual interpretation of high-resolution aerial imagery in Google Earth superimposed on the city's 228 neighborhoods (Taylor and Lovell, 2012).

Urban and non-urban irrigated and rainfed croplands by region

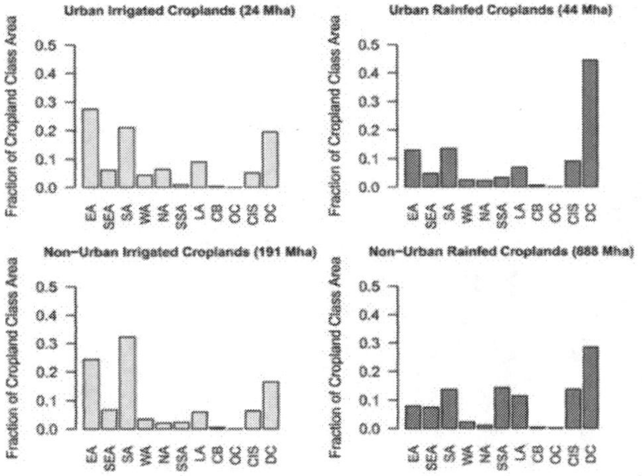

Regional groupings and abbreviations are: Commonwealth of Independent States (CIS), Caribbean (CB), Latin America (LA), Oceania (OC), East Asia (EA), Southeast Asia (SEA), South Asia (SA), West Asia (WA), North Africa (NA), Sub-Saharan Africa (SSA), and Developed Countries (DC).

Thebo, A.L., P. Dreschel, and E. Lambin (2014). "Global assessment of urban and peri-urban agriculture: irrigated and rainfed croplands", Environ. Res. Lett. 9 114002.

Food Deserts and Food Insecurity

Urban neighborhoods and rural towns without ready access to fresh, healthy, and affordable food

Ready access in an urban area is defined as more than one mile (1.6 km) to a supermarket or market selling fresh food (10 miles in rural areas), or lack of adequate transportation to reach a market

USDA Agricultural Marketing Service

Could Cities Meet Their Vegetable Demand Through Urban Agriculture?

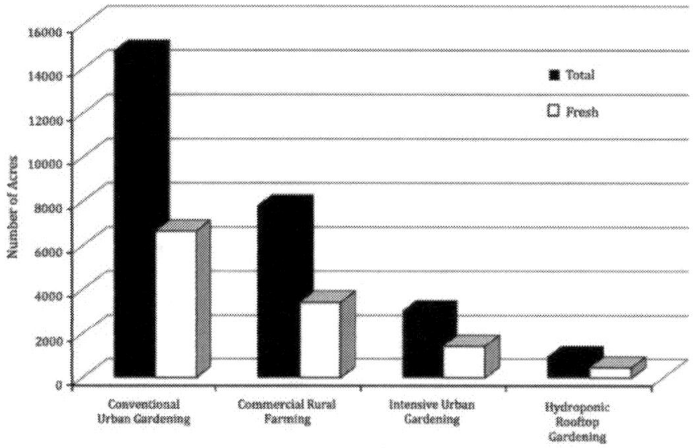

Number of acres required to meet 100% of Cleveland's fresh and total (fresh and processed) vegetable demand using various production practices and average yield obtained from different published sources

(Grewal, S., and P. Grewal (2012). "Can cities become self-reliant in food?" *Cities* 29(1): 1-11.)

Lufa Farms rooftop hydroponic greenhouses, Montreal

Greenhouses in the Westland Region of The Netherlands

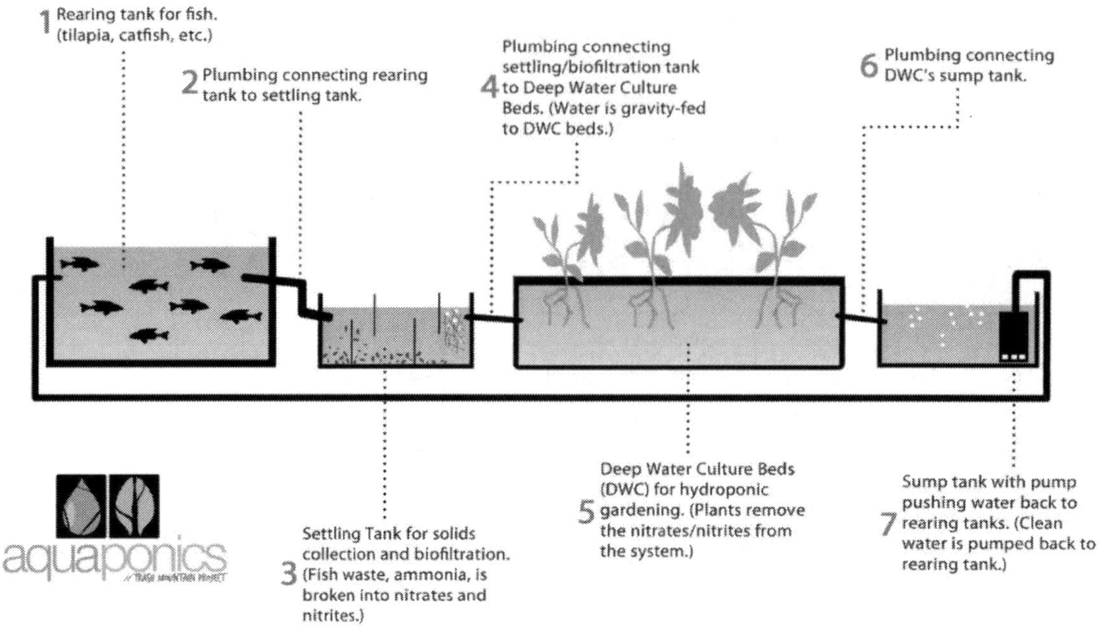

Closing the Loop

Large-scale agriculture operates as an "open" system in which large quantities of chemicals (imported) and wastes (exported) are exchanged with the environment

In contrast, urban agricultural systems must, to maintain profitability, focus on closing the life cycle loop, recycling and reusing wastes and nutrients

Sources of reactive nitrogen introduced into the US environment (Tg N/yr)

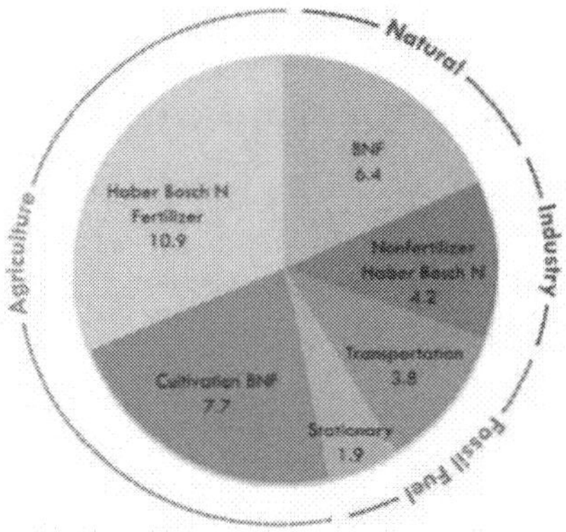

Agricultural Drainage

Soil Losses

Erosion and loss of soils commonly occur following conversion of natural vegetation to agricultural land – carrying away fertile soil as well as fertilizers, pesticides and other agrochemicals.

Cultivation erosion (WWF)

Water erosion

Rill erosion

Gully erosion

http://www.omafra.gov.on.ca/english/engineer/facts/12-053.htm

Comparison of rates of soil erosion from agricultural fields under conventional agriculture (n = 448) and geologic erosion rates from low-gradient continental cratons (n = 218), soil-mantled landscapes (n = 663), and alpine terrain (n = 44) (sources are lis...

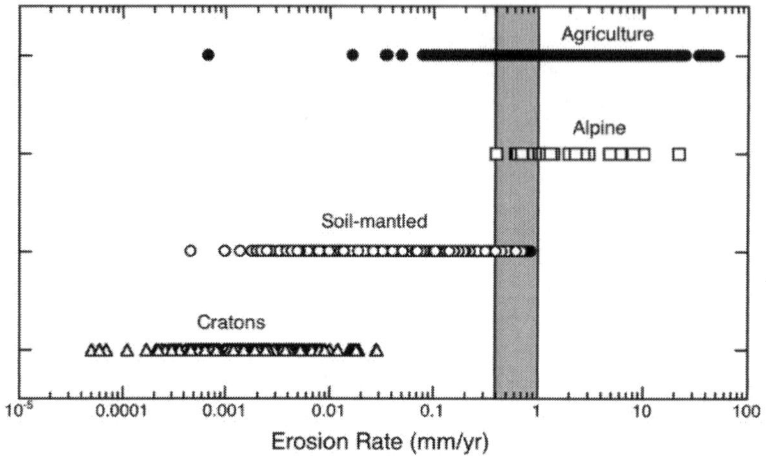

David R. Montgomery PNAS 2007;104:13268-13272

PNAS

"The Plant" Vertical Aquaponic Farm and Industrial Ecology Site, Chicago

"The Plant": Vertical Farm Schematic

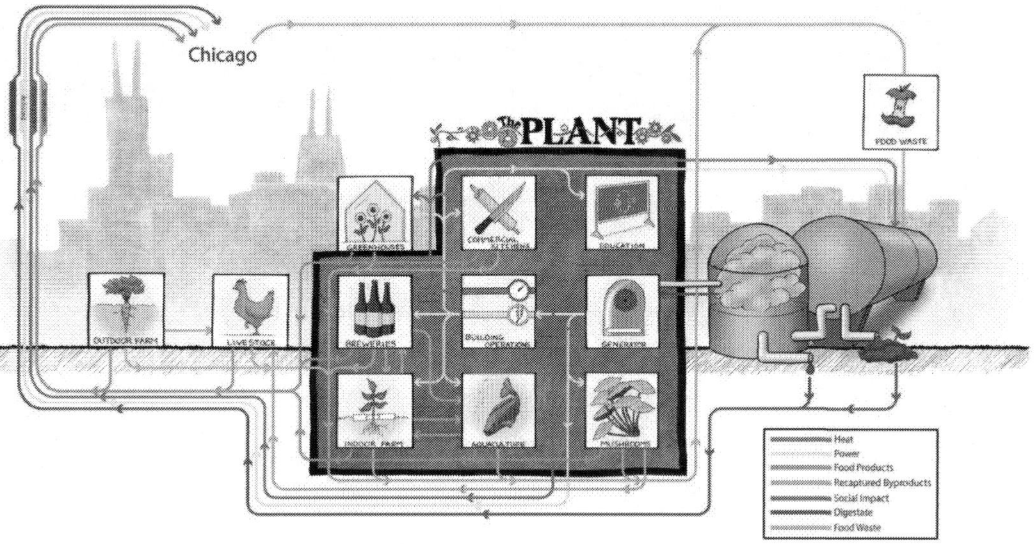

Research Needs

- Mapping and scaling up urban and peri-urban food production systems to enhance sustainable local food systems
- Gathering data on urban/peri-urban agriculture resource and energy needs and impacts for different farming designs and methods
- Designing urban/peri-urban food waste management systems
- Understanding the interplay of urban/peri-urban agriculture, the urban biodiversity matrix, and supply of ecosystem services
- Investigating food insecurity, safety, and distribution in urban systems
- Understanding social and land-use drivers and impacts of urban/peri-urban food production
- Establishing the economic viability of urban agriculture within local and national market systems
- Measuring the efficacy of policy interventions on urban agriculture and the local food retail environment, especially in low income and minority communities
- Applications of sustainability science in support of food policies

RAFT
Regional Algal Feedstock Testbed

A Perspective on How Microalgae can Address the Water, Energy Food Nexus

ARIZONA
NM STATE UNIVERSITY
Pacific Northwest NATIONAL LABORATORY
TEXAS A&M AGRILIFE RESEARCH

RAFT Regional Algal Feedstock Testbed

Overview

- Algal Cultivation in Open Systems
  - Rotating Crop Strategy
  - Measurement and Control
  - Contamination
- Wastewater
- Harvesting/Extraction – Conversion to Fuel
- Food
- Conclusion

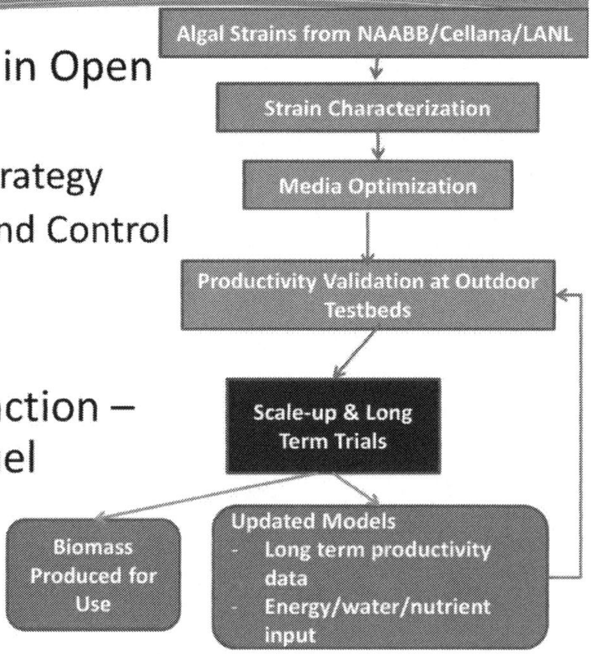

Characterization

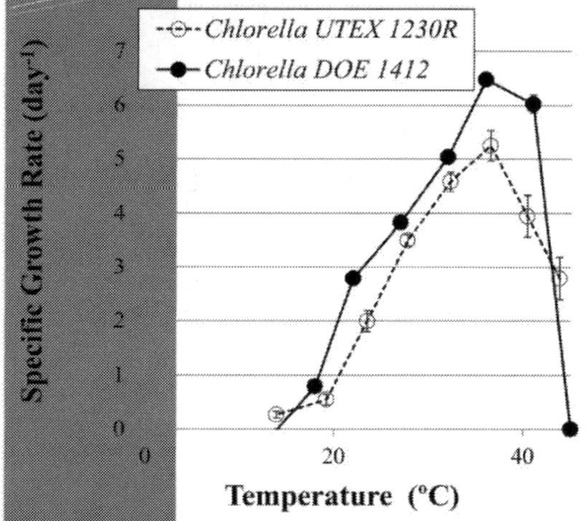

High growth rate even up to low 40s
Summer strain

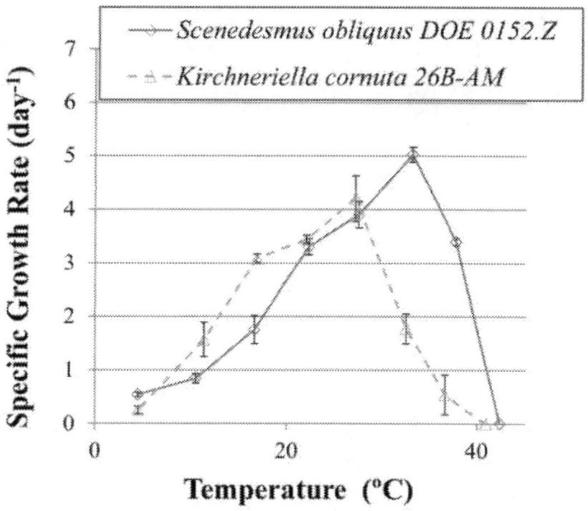

Growth even at 10 C
Two winter strains

Media Optimization

Chemical	Common name	g/L
Na_2CO_3	Soda Ash	0.02
NaCl	TruSoft	5
(NH2)2CO	Urea	0.1
$MgSO_4 * 7H_2O$	Magnesium sulphate	0.012
$NH_4H_2PO_4$	MAP	0.035
KCl	Potash	0.175
FeCl	Iron Chloride	0.0035
Di Sodium EDTA		0.00436
Trace metals solution		1
Vitamin solution (100 X)	Vitamin solution	0.005

- Complete Media Optimization
  - *Chlorella sorokinia (DOE 1412)*
  - *Scenedesmus obliquus*
  - *Kirchneriella* sp.
- **Cost is about 1/10 that of BG11 with same growth rate**

Testbed Facilities

ARID Raceway US Patent # 8,245,440

Ability to Cultivate GMO Algae in Climate Controlled Green Houses

Chlorella sorokiniana

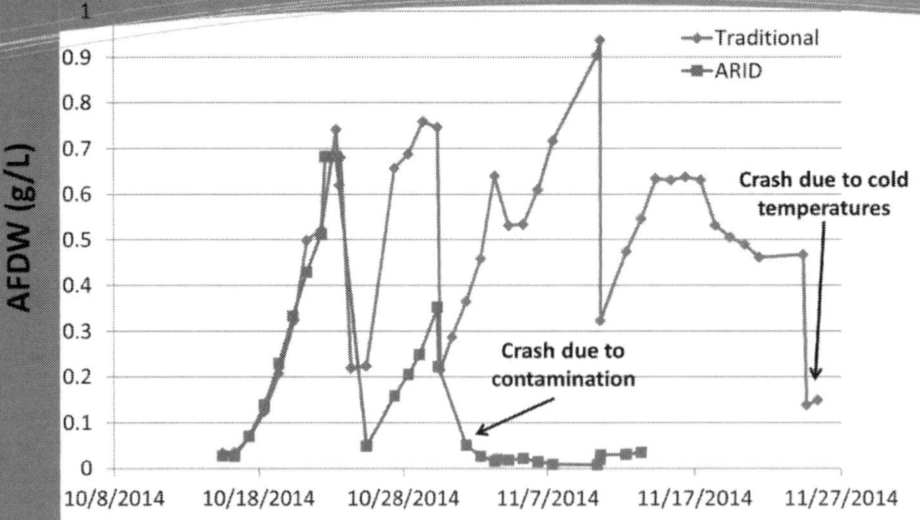

Productivity (9 to 10 g/m^2/day fall 21 g/m^2/day summer) – but culture crashes requiring monitoring of culture health

Early Indication – Dissolved Oxygen

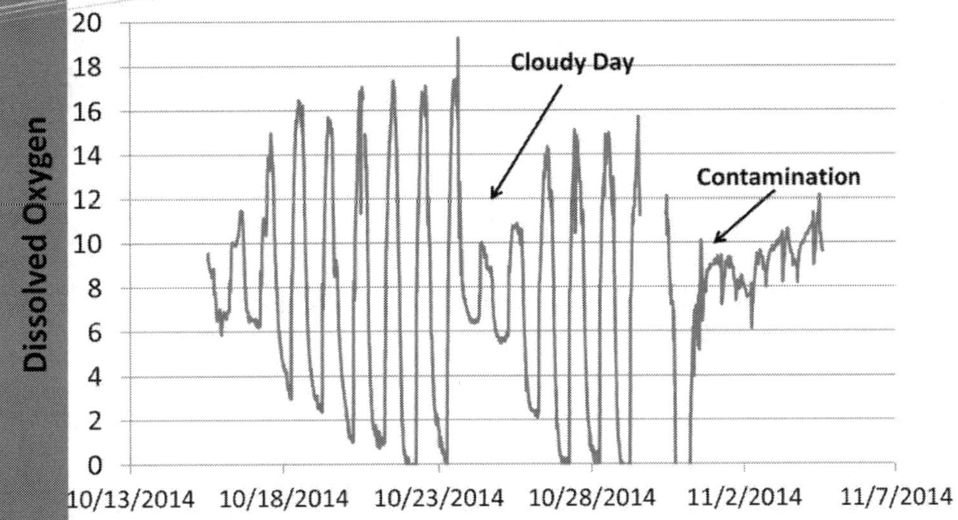

Vampirovibrio chlorellavorus

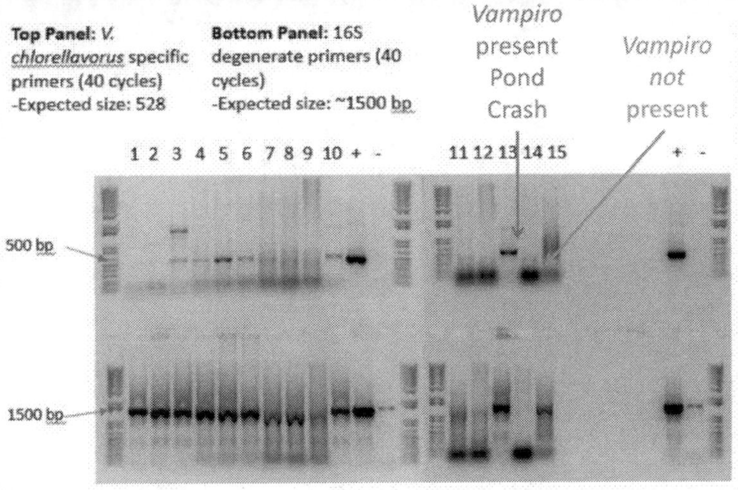

Samples From RAFT05

1. ARID 10/15/14
2. BT1 10/15/14
3. BT2 10/15/14
4. ARID 10/19/14
5. BT1 10/19/14
6. BT210/19/14
7. ARID 10/31/14
8. BT1 10/31/14
9. BT210/31/14
10. ARID 11/3/14
11. BT1 11/3/14
12. BT211/3/14
13. ARID 11/10/14
14. BT1 11/10/14
15. BT211/10/14

+. *V. chlorellavorous* 16S
-. ddH2O

Molecular monitoring enabled the 'association' of *Vampirovibrio* with production 'crashes' of *Chlorella* spp. in AZ and NM

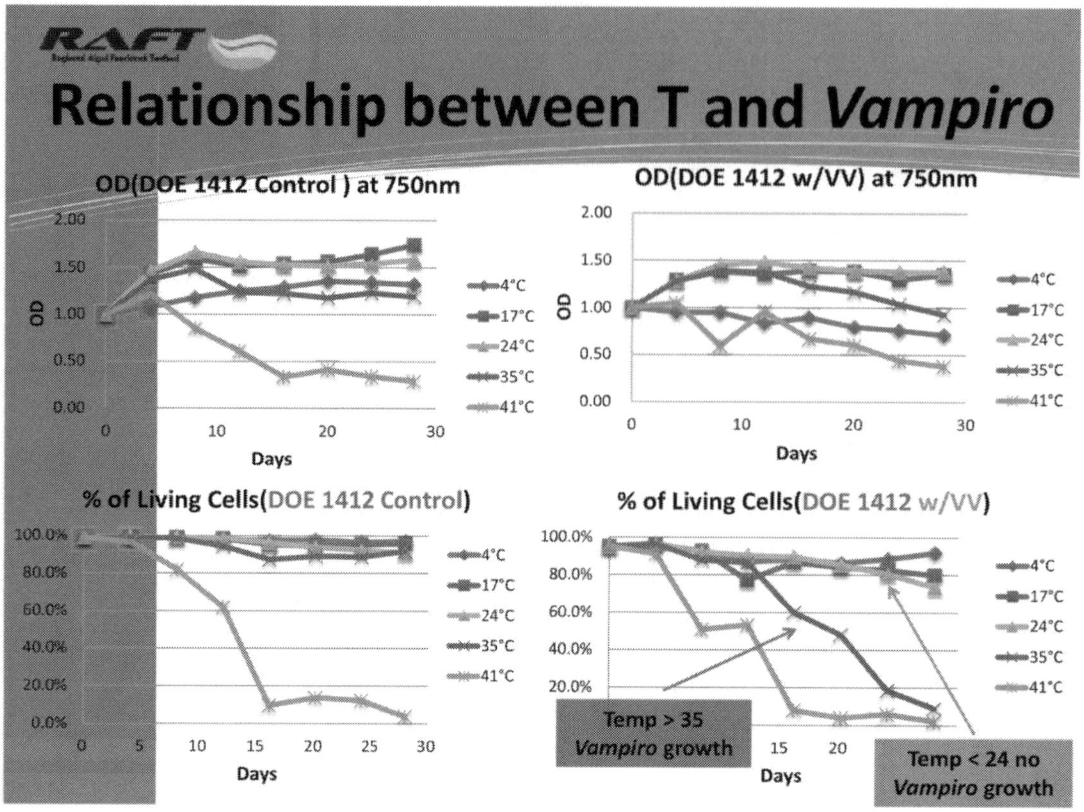
RAFT
Relationship between T and Vampiro
OD(DOE 1412 Control) at 750nm
OD(DOE 1412 w/VV) at 750nm
% of Living Cells(DOE 1412 Control)
% of Living Cells(DOE 1412 w/VV)
OD
Days
4°C
17°C
24°C
35°C
41°C
Temp > 35 Vampiro growth
Temp < 24 no Vampiro growth

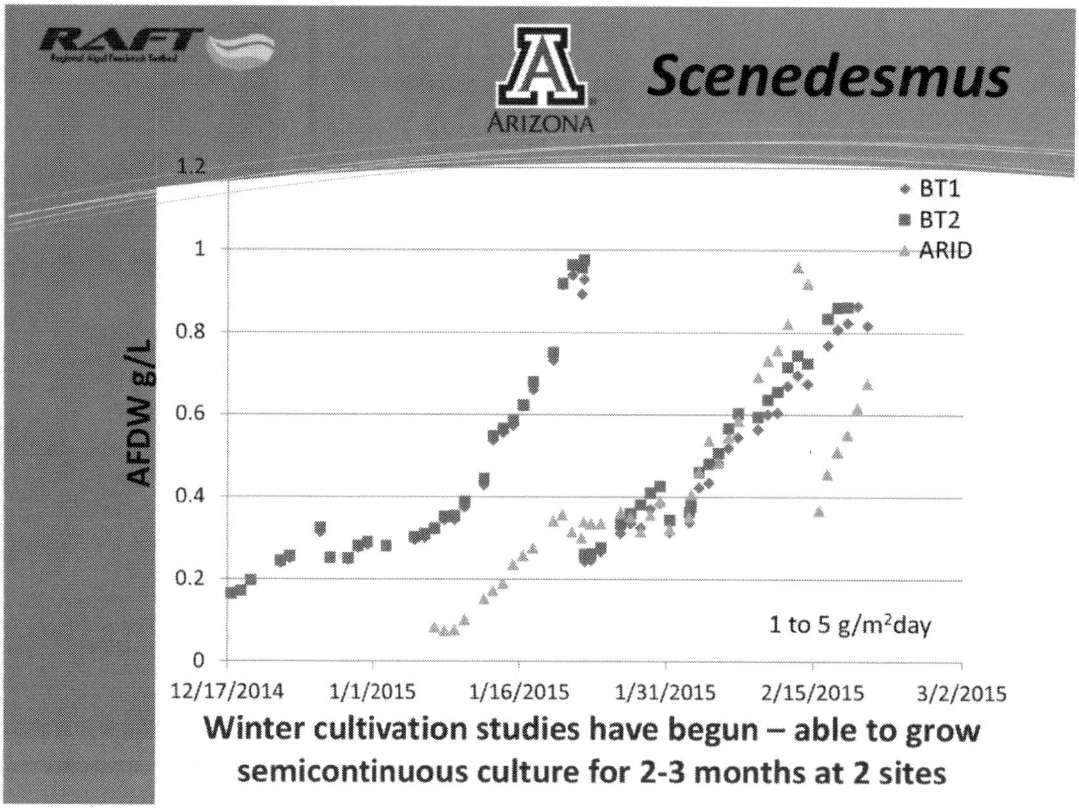
RAFT
ARIZONA
Scenedesmus
BT1
BT2
ARID
AFDW g/L
1 to 5 g/m²day
12/17/2014
1/1/2015
1/16/2015
1/31/2015
2/15/2015
3/2/2015
Winter cultivation studies have begun – able to grow semicontinuous culture for 2-3 months at 2 sites

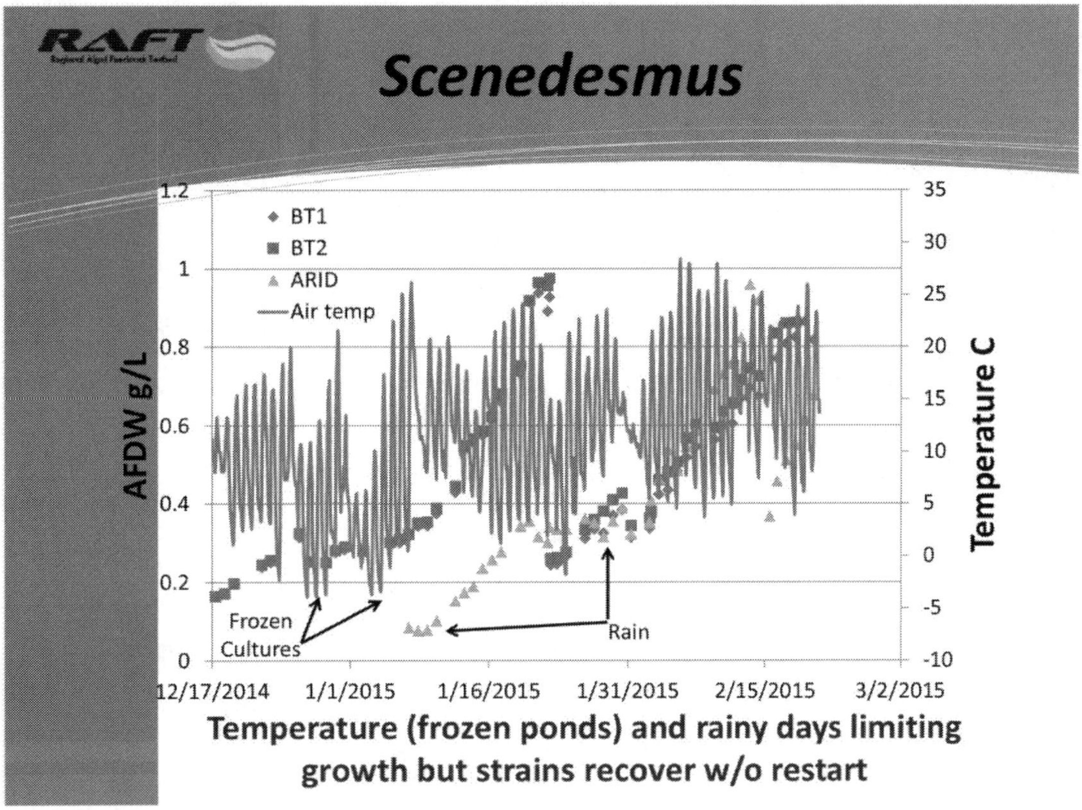

RAFT
Scenedesmus
BT1
BT2
ARID
Air temp
AFDW g/L
Temperature C
Frozen Cultures
Rain
12/17/2014
1/1/2015
1/16/2015
1/31/2015
2/15/2015
3/2/2015
Temperature (frozen ponds) and rainy days limiting growth but strains recover w/o restart

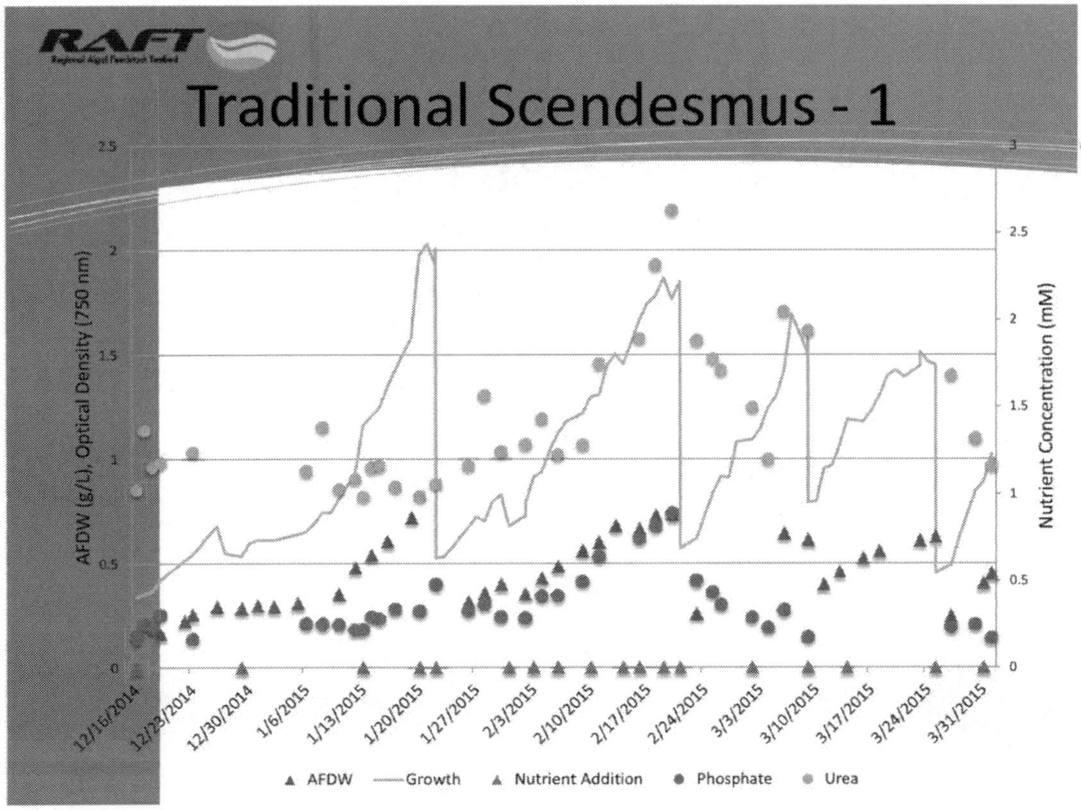

RAFT
Traditional Scendesmus - 1
AFDW (g/L), Optical Density (750 nm)
Nutrient Concentration (mM)
AFDW
Growth
Nutrient Addition
Phosphate
Urea

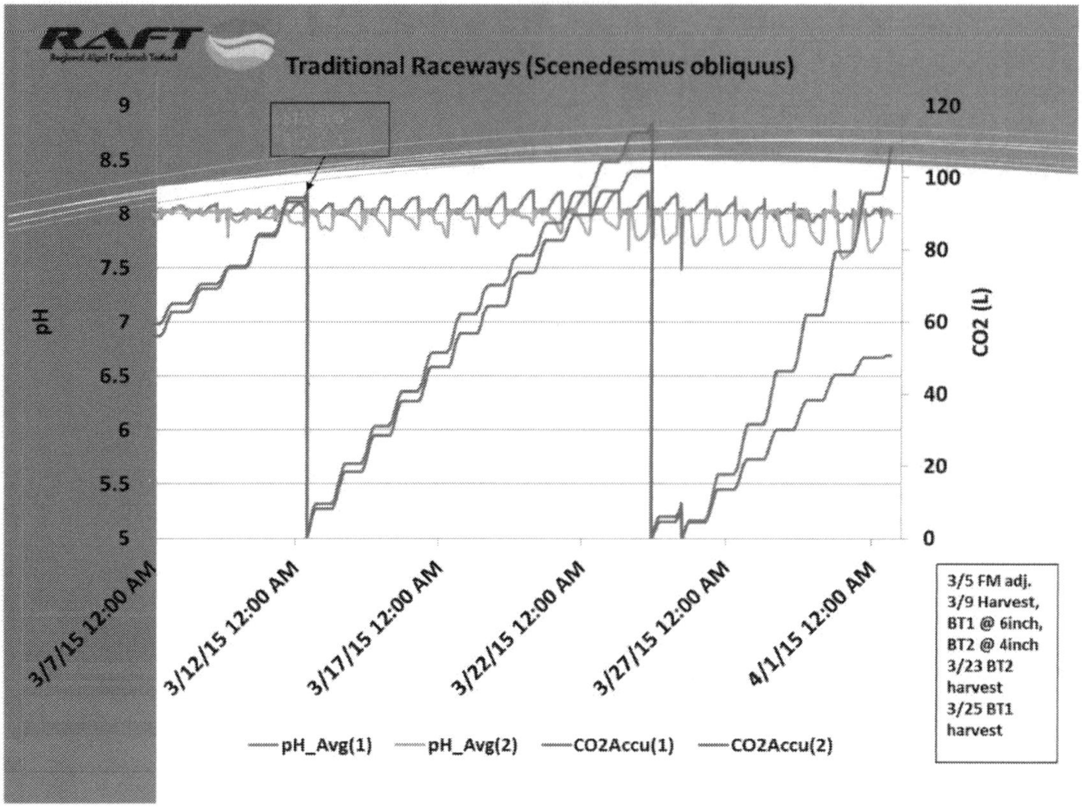
RAFT
Traditional Raceways (Scenedesmus obliquus)
pH
CO2 (L)
9
8.5
8
7.5
7
6.5
6
5.5
5
120
100
80
60
40
20
0
3/7/15 12:00 AM
3/12/15 12:00 AM
3/17/15 12:00 AM
3/22/15 12:00 AM
3/27/15 12:00 AM
4/1/15 12:00 AM
pH_Avg(1)
pH_Avg(2)
CO2Accu(1)
CO2Accu(2)
3/5 FM adj.
3/9 Harvest,
BT1 @ 6inch,
BT2 @ 4inch
3/23 BT2
harvest
3/25 BT1
harvest

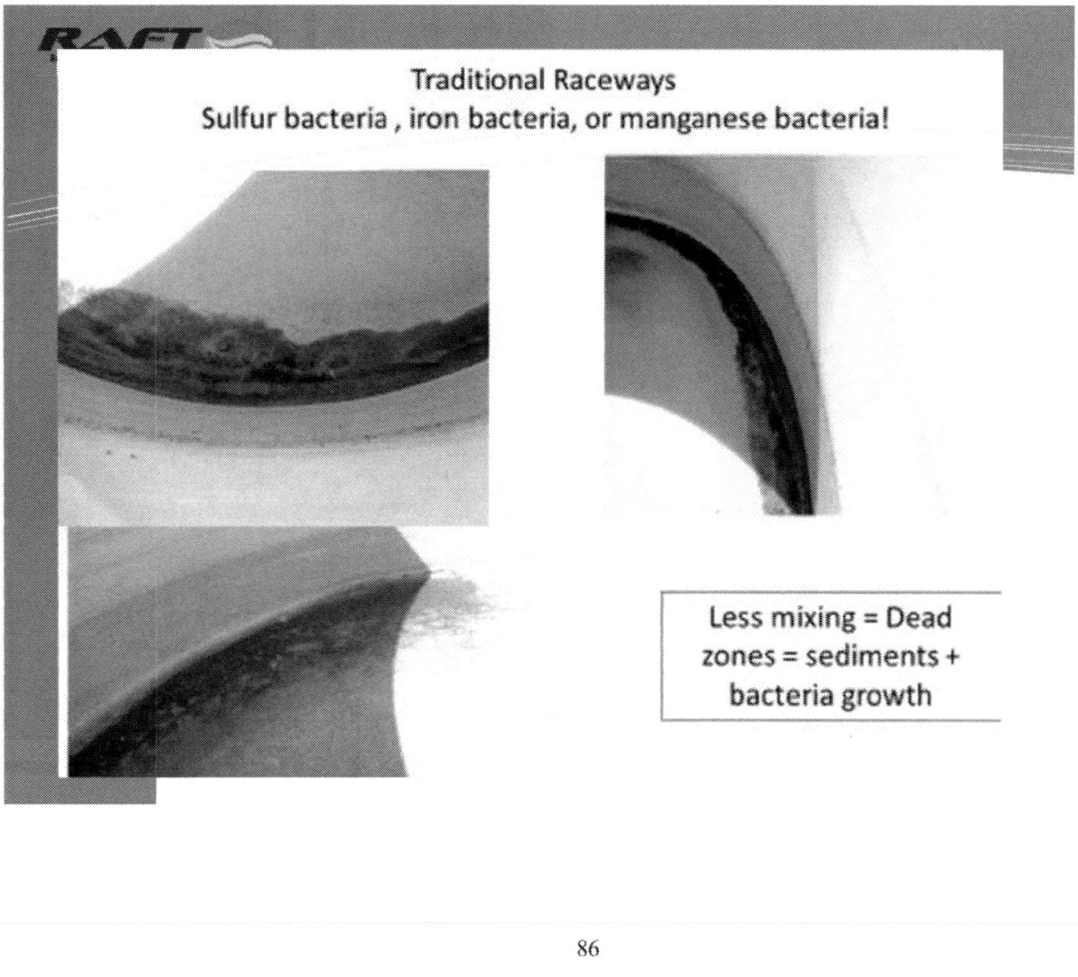
Traditional Raceways
Sulfur bacteria , iron bacteria, or manganese bacteria!
Less mixing = Dead zones = sediments + bacteria growth

Healthy vs Stressed

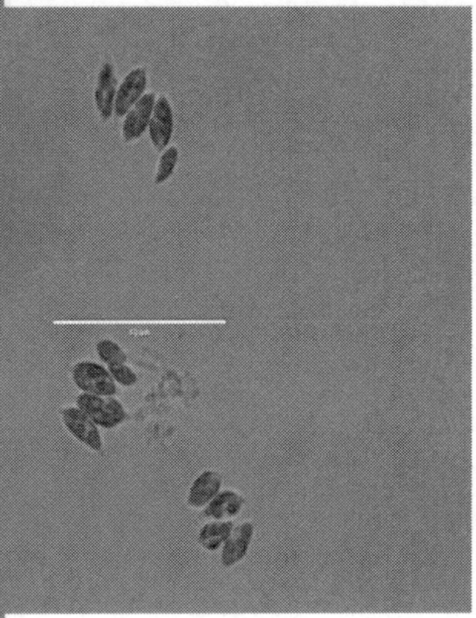

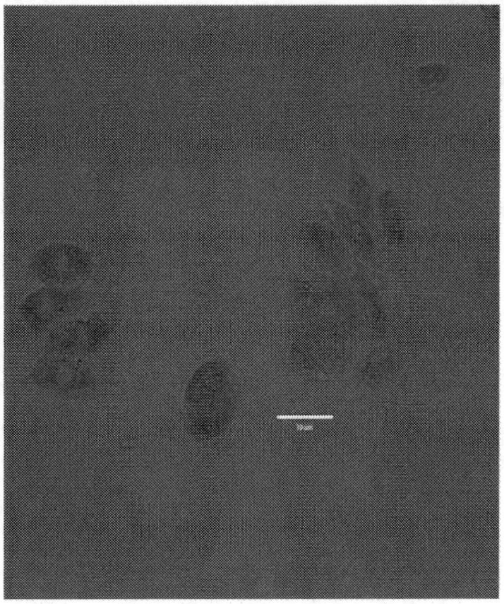

Chlorella on Wastewater

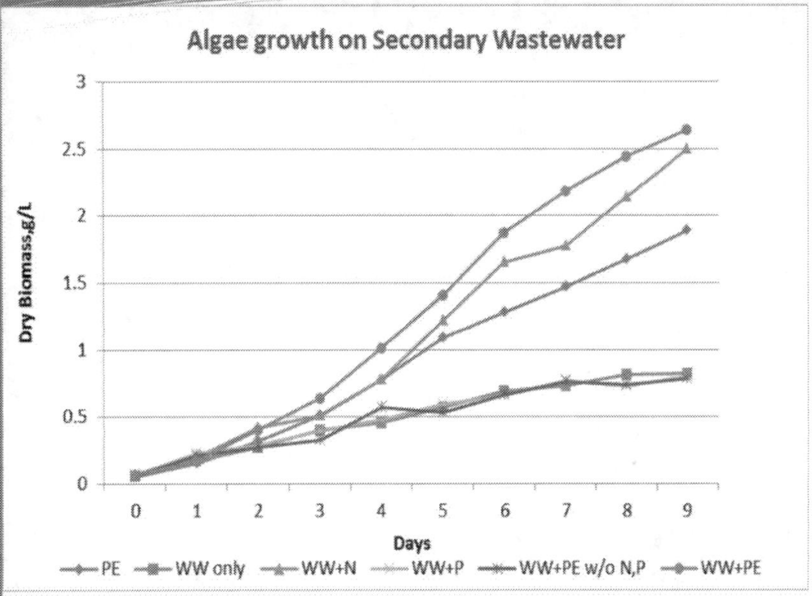

16

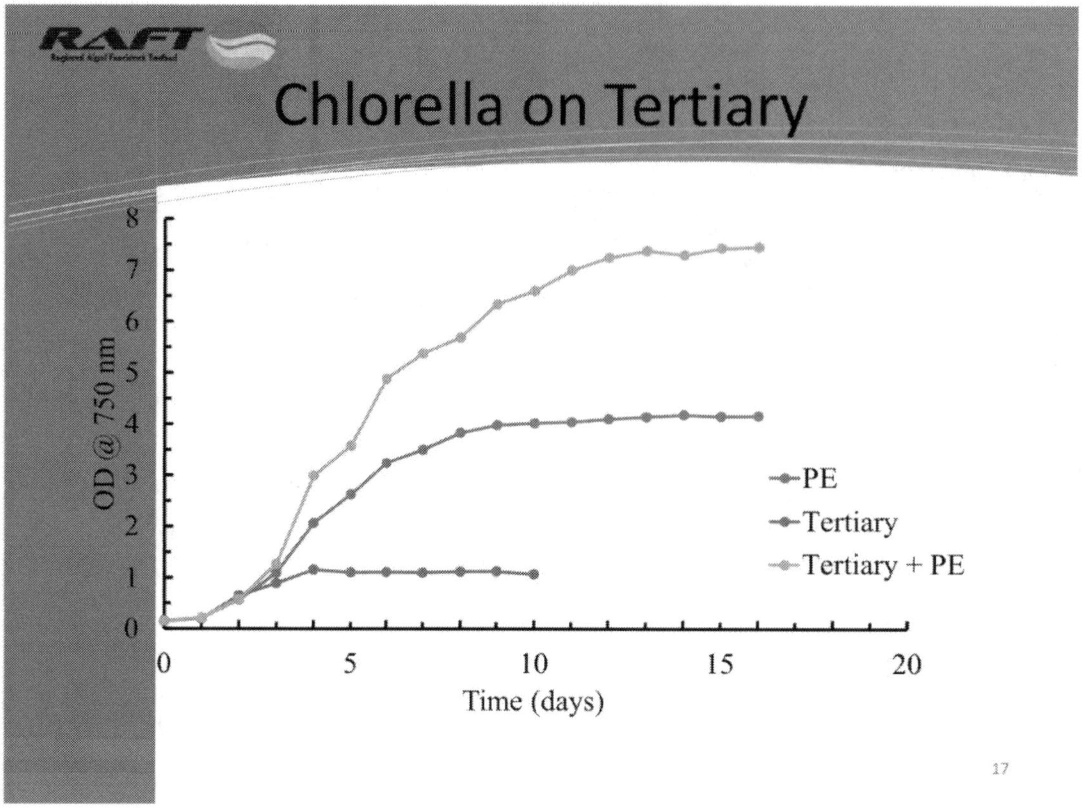
RAFT
Regional Algal Feedstock Testbed
Chlorella on Tertiary
OD @ 750 nm
Time (days)
PE
Tertiary
Tertiary + PE
0
5
10
15
20
1
2
3
4
5
6
7
8
17

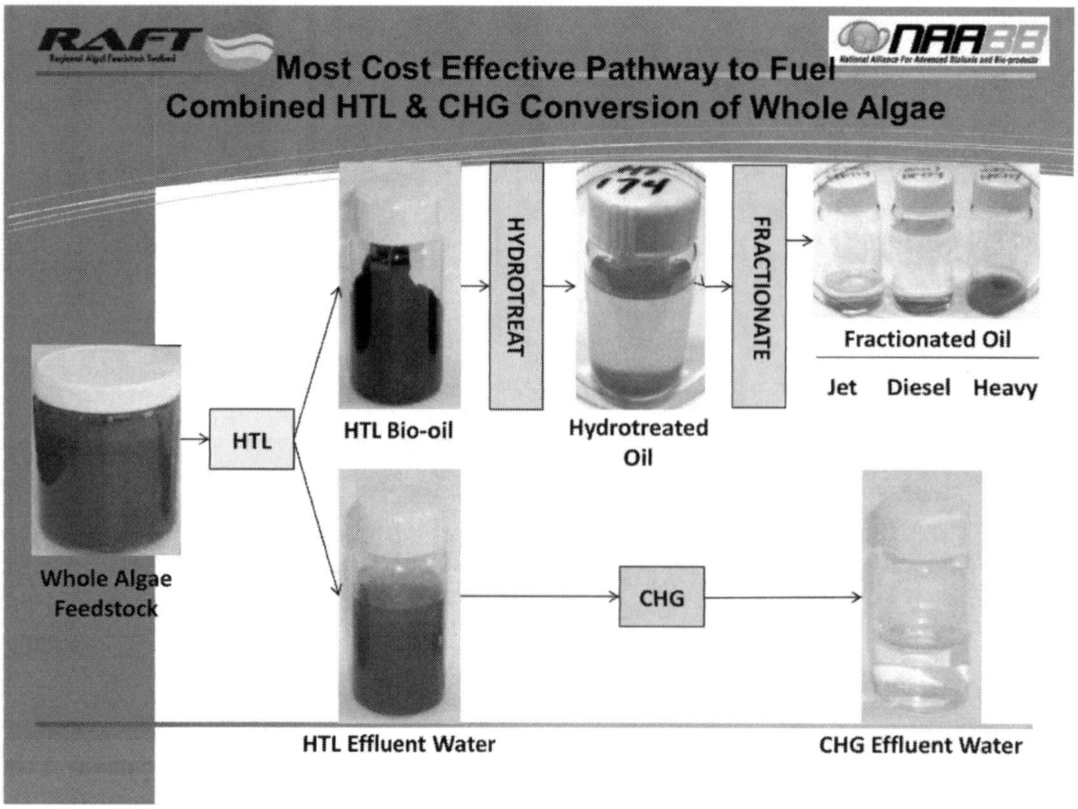
RAFT
Regional Algal Feedstock Testbed
NAABB
National Alliance For Advanced Biofuels and Bio-products
Most Cost Effective Pathway to Fuel
Combined HTL & CHG Conversion of Whole Algae
Whole Algae Feedstock
HTL
HTL Bio-oil
HYDROTREAT
Hydrotreated Oil
FRACTIONATE
Fractionated Oil
Jet
Diesel
Heavy
CHG
HTL Effluent Water
CHG Effluent Water

Chemical/Physical Characterization of Jet Fuel from HTL Conversion

Parameter	Jet A	Jet A1	Jet Fuel From HTL Algal Bio-Oil #1	Jet Fuel From HTL Algal Bio-Oil #2
Density (g/L)	775 - 840	775 - 840	786.3	780.2
Freeze point (°C) max	-40	-47	-45.8	-57
Flash Point (°C) min	38	38	61.2	59.6
Distillation				
10% Recovered Temp (T10) °C max	205	205	167.8	167
50% Recovered Temp (T50) °C	Report	Report	207.4	203.6
90% Recovered Temp (T90) °C	Report	Report	244.6	242.2
Final Boiling Point (°C) max	300	300	273.2	272

Algae for Feed

Fish Feed

Ingredients	%
Soymeal	38
Wheat	16
Wheat bran	6
Scenedesmus	22
Meat and bone meal	10
Canola oil	4.5
Soy lecithin	2
Trace mineral premix	1
Fish oil	0.5
Fish meal	0

21

Composition	Algae (calculated)	Control (Guaranteed analysis)
Protein %	32.7	32
Fat %	11.3	7
Fiber %	3.9	4
Ash %	7.9	5
Omega 3 in fish feed %	0.7	To be determined
Disposable energy (Mj/kg)	11.4	To be determined

22

RAFT Regional Algal Feedstock Testbed

Preliminary feeding study results

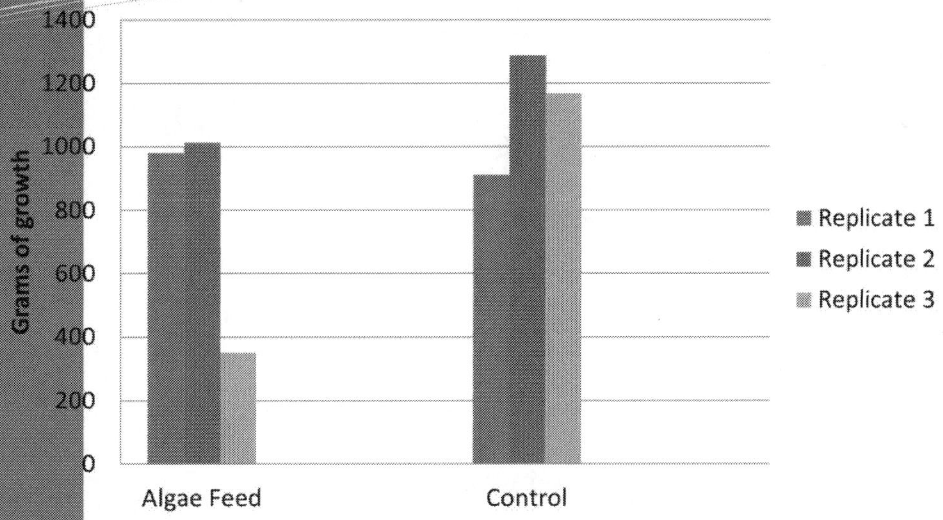

23

RAFT Regional Algal Feedstock Testbed

Value of Algae as Fish Feed

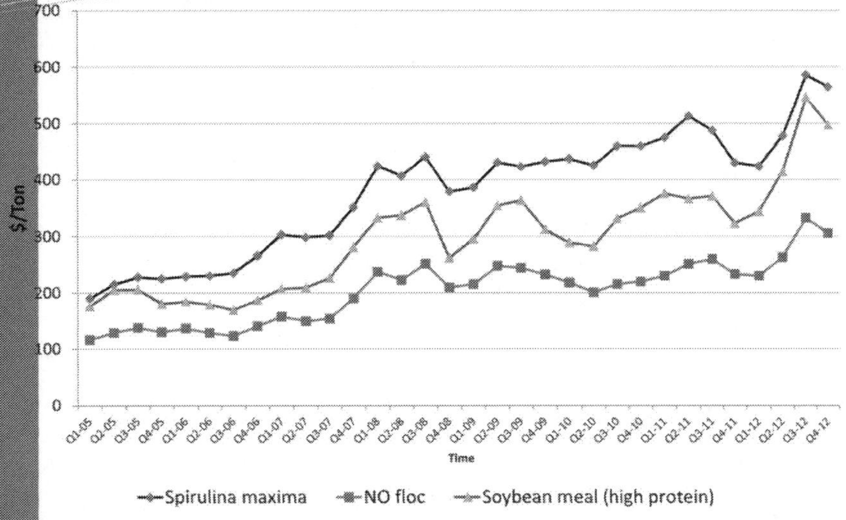

24

Summary

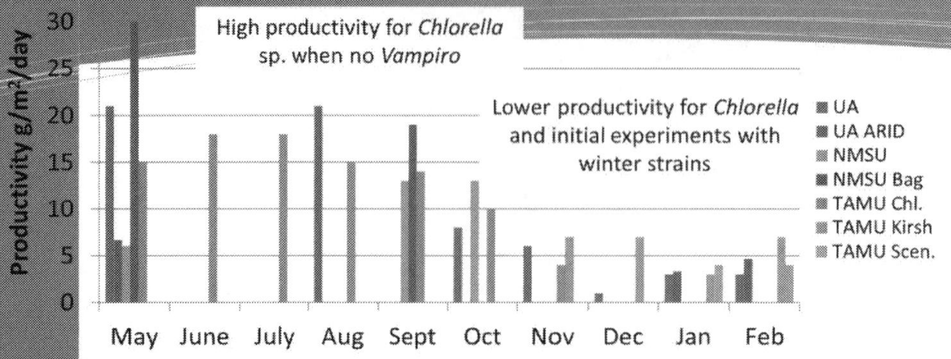

1. **Overview** - Long term cultivation at 3 testbeds using multiple strains
2. **Approach** - Screen strains for optimal temperature and light; Semi continuous growth over all seasons in outdoor systems; Provide data to modelers – receive feedback
3. **Conclusion** - Algae cost effective for fish feed at this time.
4. **Future work** - Impaired waters; Larger scale; Process control

25

Acknowledgements

- University of Arizona
  - Pete Waller
  - Xuehue Li
  - Renhe Qiu
  - Margarita Acedo
  - Said Attalah
  - Kai Lepley
  - Judy Brown
  - Qi Li
  - Randy Ryan
  - Kevin Fitzsimmons
  - Manual Vasquez
  - Juan Sandoval
  - Leanne Strauss
  - Yu-Shien Chu
  - Greg Ogden
  - Cassandra Galvez
- Pacific Northwest Laboratory
  - Michael Huessman
  - Dan Anderson
  - Rick Skaggs
  - Richard Halen
- Texas A&M University
  - Lou Brown
  - Jola Brown
  - James Richardson
- New Mexico State University
  - Meghan Downes
- UOP
  - Steve Lupton

U.S. DEPARTMENT OF ENERGY | Energy Efficiency & Renewable Energy

Energy and Water Interactions
Implications for Industry within the European Context

Jiří Jaromír Klemeš[1], Petar Sabev Varbanov[1], Alajos Mészáros[2]

[1]Centre for Process Integration and Intensification – CPI[2], Research Institute of Chemical and Process Engineering, Faculty of Information Technology, University of Pannonia, Egyetem u. 10, H-8200 Veszprém, Hungary

[2]Faculty of Chemical and Food Technology, Institute of Information Engineering, Automation, and Mathematics, Slovak University of Technology in Bratislava, Radlinského 9, 812 37 Bratislava, Slovak Republic

ICOSSE
International Congress on Sustainability Science & Engineering

Motto / Motivation

- The increase in the world population has placed increasing pressure on the demands of world society, and especially that of industrial and agricultural production.
- The accelerating development of countries with large populations has resulted in increased demands on agricultural production and processing, which resulted in further increases in energy and water demands.
- The supplies sharply increase in cost and many cases of shortages of all forms of energy and water are witnessed
- EU policy and legislation have been delayed in energy-water nexus context (Energy Strategy 20-20-20)

Outline

- The motivating problems - energy and water
- Global energy and water flows
- Key Overall Implications of the flows
- Water and Virtual Water
- Inherent Limitations and Industry Implications
- Important Trends
- Recent Research and Demonstration
- Summary and Suggestions

Energy

- **Fossil fuel** resources and delivery has been of **paramount importance** to industrialized countries and **cause considerable struggle** resulting in many conflicts.
- **Nuclear-origin** energy can be generated in the region of the needed consumption.

  However, it raises **safety and environmental issues**.
- **Renewable sources** can be harvested inside or close to the **region of consumption**.

  However, they have problems **with variable availability** in time, intensity and handling their supply flows.

Water

- A similar situation has been developing in water and fresh water supply and waste water minimization.
- This has been also related to **industrial and agricultural development** as well as increasing civic demand and consumption.
- **Water** is very often used hot or chilled which **requires energy**. What solutions improving the state-of-the-art can be offered by advanced engineering?
- Evaluating the **interactions** of energy and water flows on the global scale **demands enormously extensive models** supplied with the large volume of data.

ICOSSE 2015, 26 – 29 May, Balatonfüred, Hungary 5

Global Energy and Water Flows

ICOSSE 2015, 26 – 29 May, Balatonfüred, Hungary 6

Carbon Cycle

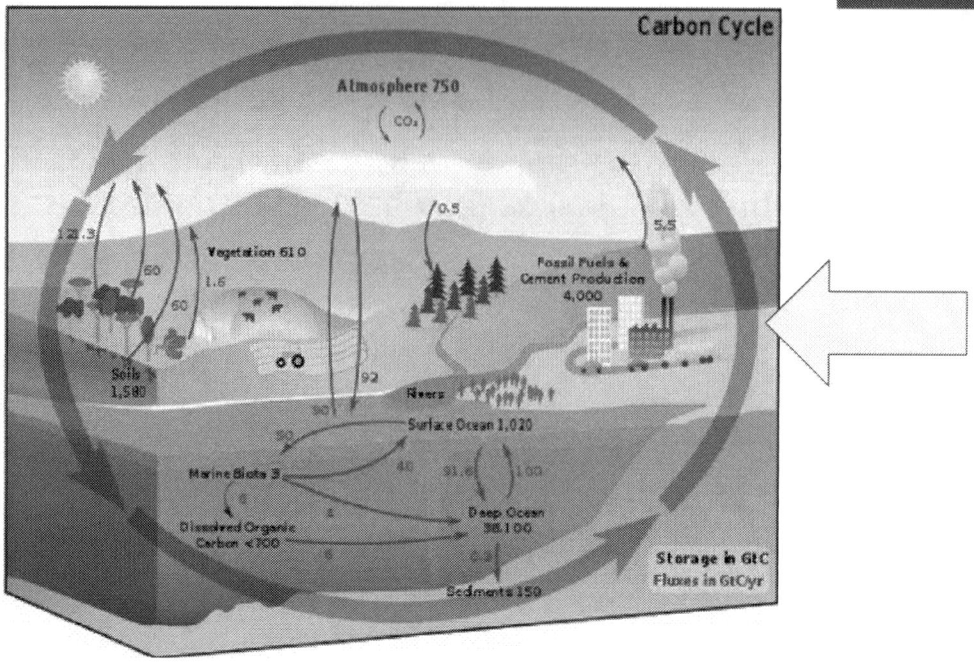

<www.worldculturepictorial.com/blog/content/earth-does-not-bargain-reacts-major-emitters-job-coo>

Energy Flow Losses

Among the most substantial losses are:

- **Conversion**
- **Transmissions / transport**

US Balance: Economy Wide Losses

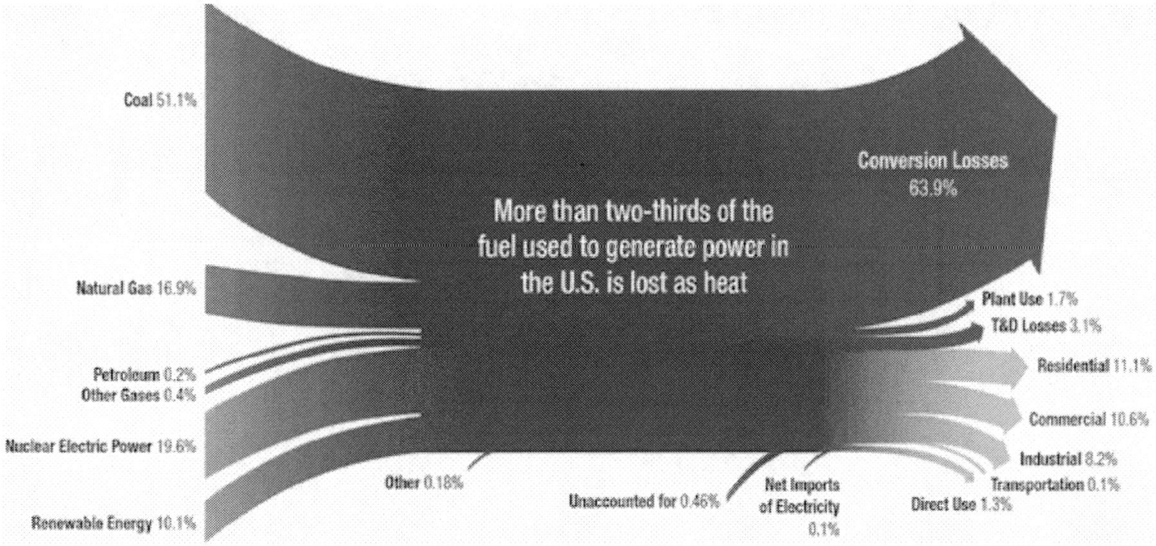

<www.gulfcoastcleanenergy.org/CLEANENERGY/CombinedHeatandPower/tabid/1698/Default.aspx>

Losses for Electricity Conversion and Transmission

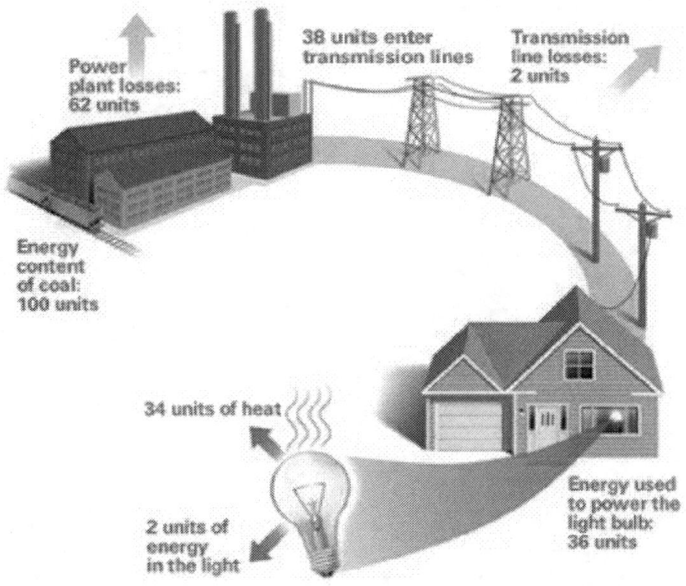

<www.energyliteracy.com/?p=235>

STU

Daily Energy Losses

"More than 90 % of energy extracted from the ground is wasted before it becomes useful work at a water pump."

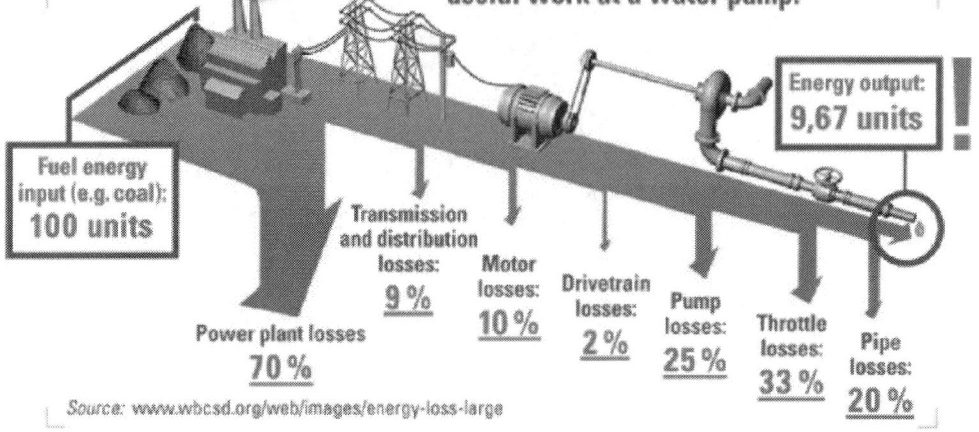

<www.hydrogenambassadors.com/background/daily-energy-losses.php>

Transport: Petrol powered vehicle losses

STU

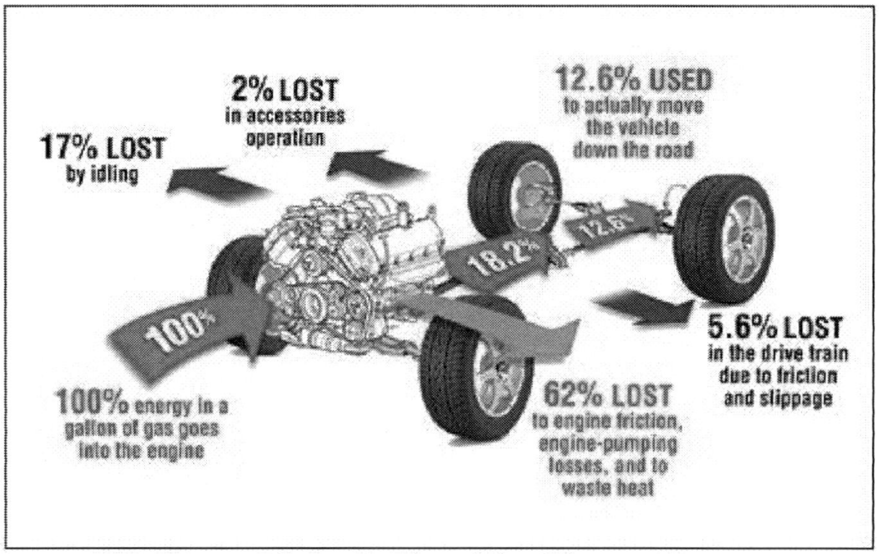

This diagram illustrates the paths of energy through a typical gas-powered vehicle in city driving.

<www.microlon.com/>

Key Overall Implications

Huge investments are needed for transporting the fuels

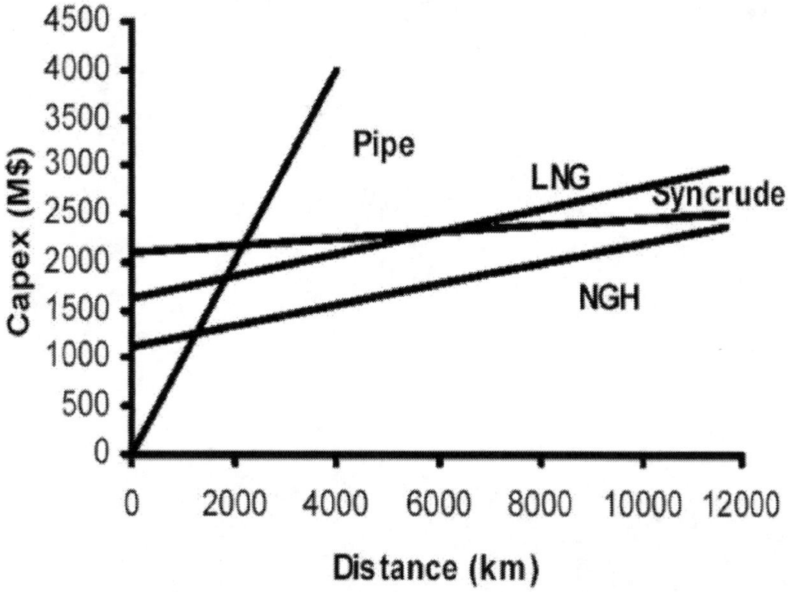

<climatetechwiki.org/technology/ngcc>

Germany after Nuclear Energy

STU

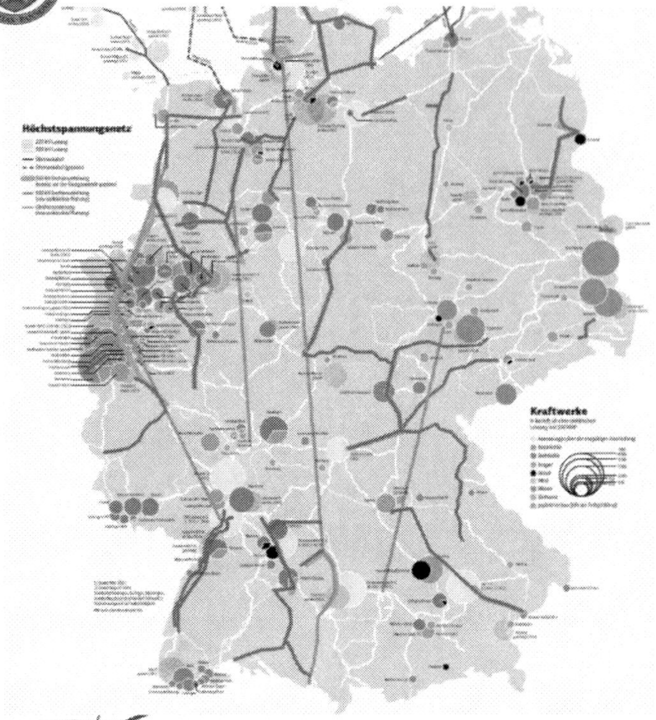

- Coal and natural gas demand is expected to rise
- Security of supply may be compromised
- A chance for renewables

<prezi.com/uss0j8upexiw /deutschland-unter-strom/>

Oil Exporting Regions

STU

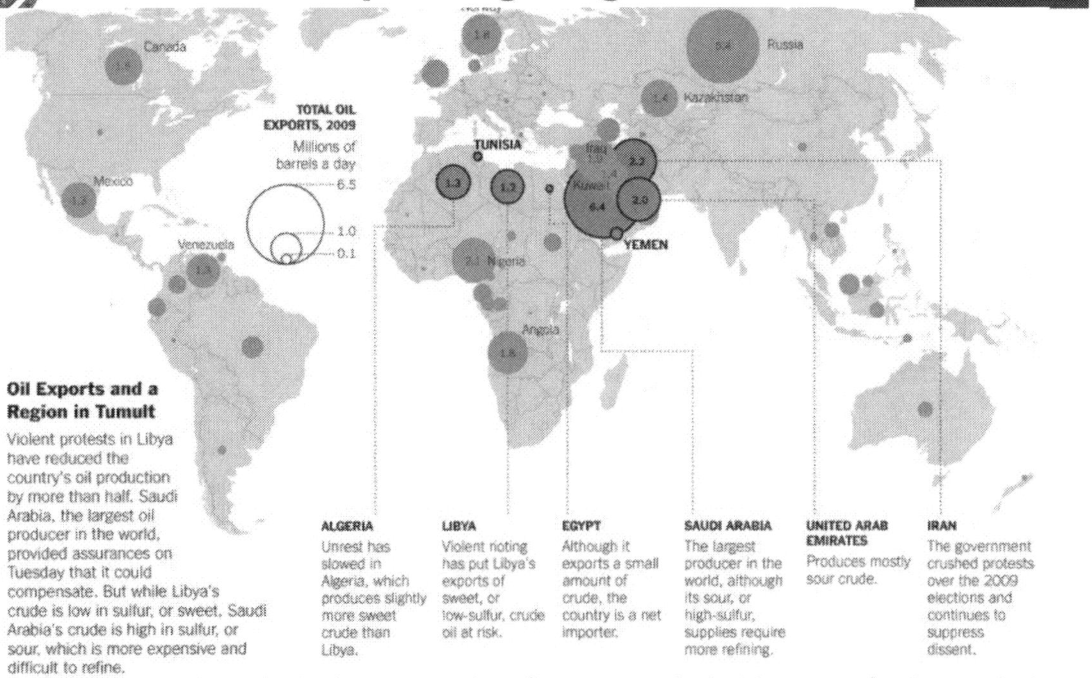

<blog.surveytool.com/2011/04/21/why-gas-prices-fluctuate-and-what-it-means-for-the-market>

STU

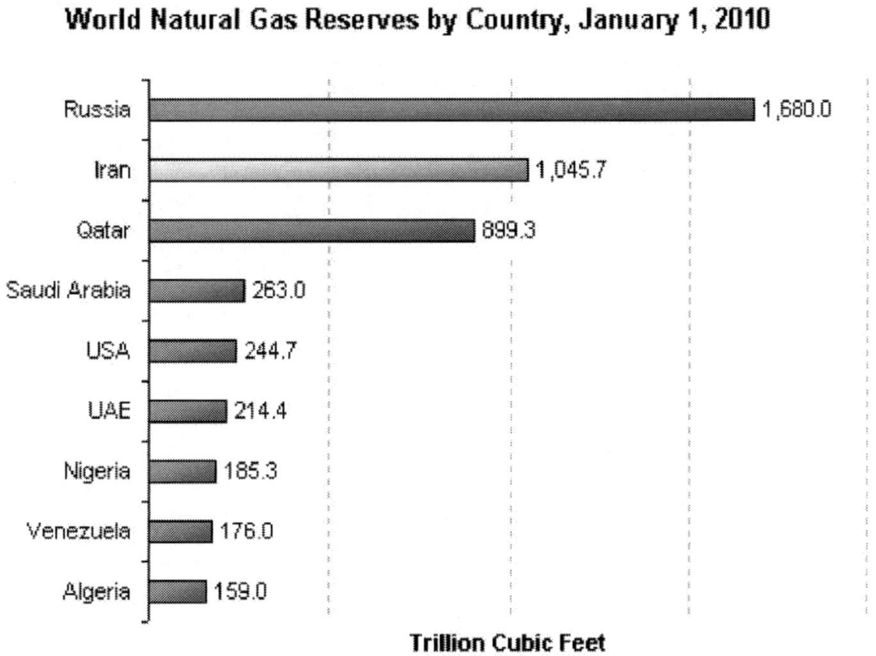

Source: Oil & Gas Journal, Jan. 1, 2010

<www.eoearth.org/article/Energy_profile_of_Iran>

STU

Inherent Limitations

Industry Implications

Meeting point of energy and water demands
Hot-and-cold-utilities

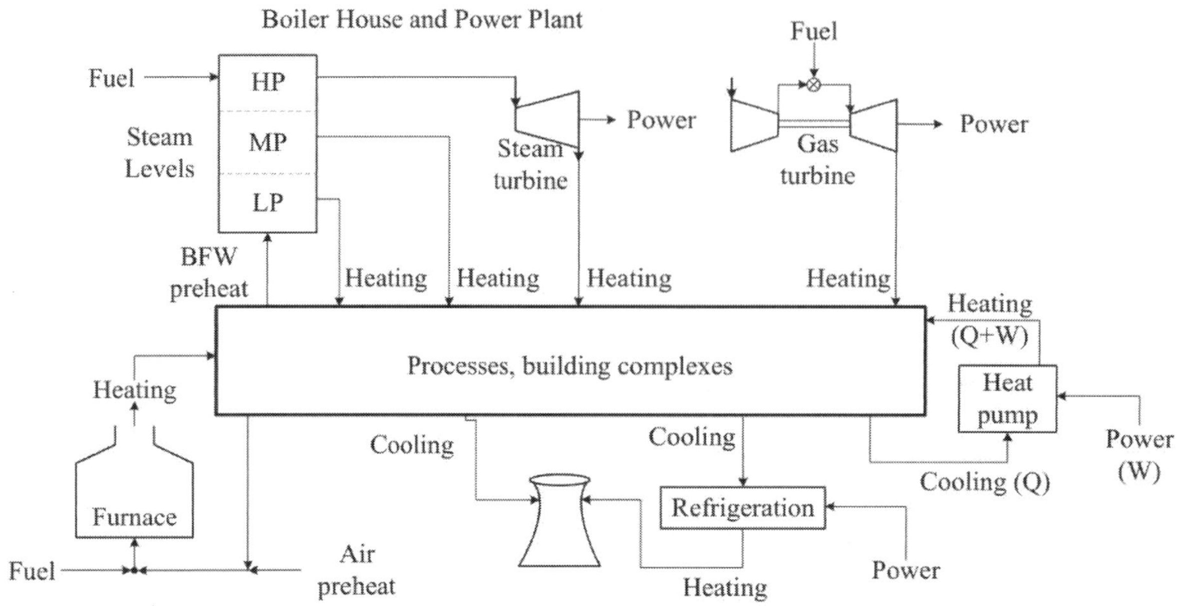

Example: Oil from Sands

www.pembina.org/oil-sands/os101/water

www.pembina.org/oil-sands/os101/water

Water Impacts

Click each heading for more detailed information.

- Oilsands operations return almost none of the water they use to the natural cycle.1
  - Oilsands mining operations are subject to a zero-discharge policy because of the toxins in the waste water. Waste water that is not recycled is stored in tailings lakes.2,3
  - Waste water from in-situ processes is often re-injected deep underground.4,5
- Oilsands operators used approximately 170 million cubic meters of water in 2011, equivalent to the residential water use of 1.7 million Canadians.6
- The ecosystem of the Athabasca River, which flows into one of the world's largest freshwater deltas, is at risk from current water withdrawals.
- The cumulative impacts of oilsands development on water are largely unknown due to inadequate monitoring.

Potential Solution: Technology Switch

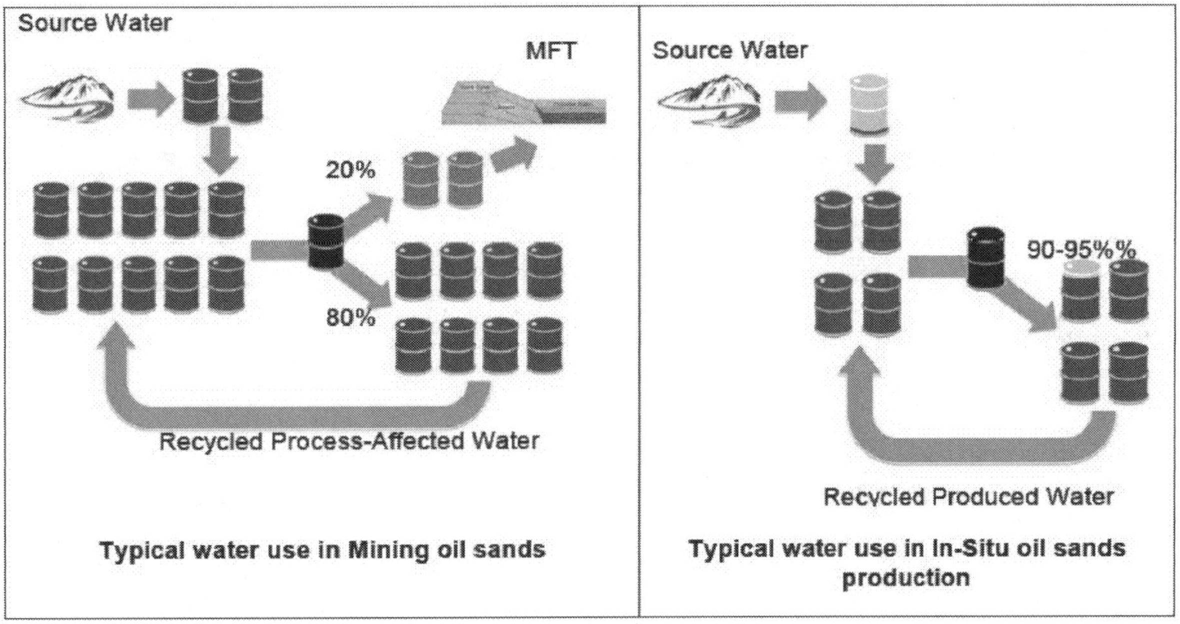

<www.bluetechresearch.com/author/tyler/>

Summary and Suggestions

A brief summary

STU

- Huge and growing energy and water flows are the reality
- These flows are subject to considerable losses and a potential danger for the environment
  - Combination of large volumes and large loss rates
- The energy–water nexus is pervasive and almost universal for human society
- There are implications and opportunities at several levels ranging from plant/site level up to regional and country level

How to tackle the problem?

STU

- Energy and water flows consume huge amounts of resources
- Individual plants have been frequently optimized
- The optimization has been extended to larger complexes – Total Sites
- However, combined energy and water flows have not been satisfactorily covered so far at site and higher level
- Based on the analysis of present situation a well rehearsed engineering bottom to top approach is suggested.

Energy-Water Nexus: Solution Side

- The nexus works **both ways** – not only in the direction of causing problems
  - In the same way as energy use and water use amplify each other, also energy and water savings should amplify each other.
- It can be used to generate solutions **using the nexus as a synergy mechanism**
- So far applied mainly at process level by the discussed methodologies
- Extend the scope of energy-water integration to site and supply chain level

More Dimensions?

- Water and energy are only part of the picture
- Food supply and food security?
- Environmental performance in terms of pollution
- Multi-criteria problems

Extend the Synergies

- Process Intensification – combine with Process Integration
  - to make them economically competitive
- Finding more synergy routes for improved energy, water and emission performance – e.g. chemical energy recovery in addition to direct heat or mass exchange.
- Combine the renewables with integration and intensification
- Minimize energy/water overheads resulting from logistics
  - Preference to trade virtual water over real water
  - Source (not just generation!) water and energy as close to the users as possible
  - Reuse any waste heat and water that can be cleaned/recycled

Benefits of Sourcing Regionally

- Short-medium term: direct energy savings from transportation
  - Substantial
  - Easily quantifiable
- Long-term and strategic
  - Improved supply security (quantifiable, but difficult)
  - Eventual higher employment locally (quantifiable, but difficult)
  - Avoiding/mitigating conflicts – not quantifiable, expected to improve safety and save costs of destruction and costs for military spending

Computational and Modeling Paradigm Issues

- Need to manage the existing and newly acquired knowledge more efficiently – options:
  - Uniform or compatible modeling languages
  - Powerful and efficient solvers
  - Proper business processes to maximize the use of human expertise and tacit knowledge
  - Knowledge Management systems to integrate the overall knowledge life cycle
- Novel modeling concepts inherently linked with efficient and effective visualisations

A Recent Publication

Available online at www.sciencedirect.com

ScienceDirect

ELSEVIER

Current Opinion in Chemical Engineering

Energy and water interactions: implications for industry

Petar Sabev Varbanov

The constant increase of the world-wide demand for water and energy makes it necessary to make their use more efficient. For supplying energy, so far mainly fossil fuels are used. The apparent link of energy and water supply increases the cost of delivery and, in many cases, causes shortages of all forms of energy and water. This paper reviews the main trends in the global flows of energy and water supply, identifies the inherent limitations and pays attention also to the concept of virtual water. The implications for industry and of some notable recent research efforts are also reviewed, and conclusions are drawn for the directions of possibly promisi development.

Addresses
Centre for Process Integration and Intensi
Institute of Chemical and Process Engine
Information Technology, University of Pan
Veszprém, Hungary

countries and causes considerable struggle and conflicts. Nuclear energy can be generated close to the demands, but raises safety and environmental concerns. Renewable energy resources can be harvested close to the demands too. However, they have problems with variable availability in time, intensity and handling their supply flows.

The energy-water nexus has been addressed at the levels of economies and regions — for example, the US Congress' analysis by Carter [1], the regional modelling study

Current Opinion in Chemical Engineering 2014, **5**:15–21

This review comes from a themed issue on **Energy and environmental engineering**

Edited by **Rakesh Agrawal** and **Subhas K Sikdar**

Thank you for your attention!

STU

Acknowledgement

The authors gratefully acknowledge the financial support from the Hungarian State and the European Union under project TÁMOP-4.1.1.C-12/1/KONV-2012-0017

"Green Energy" – Cooperation of the higher education sector for the development of green economy in the area of energetics

The authors gratefully acknowledge the contribution of the Scientific Grant Agency of the Slovak Republic under the grant 1/0973/12.

CPI²

4th International Congress on Sustainability Science & Engineering
26-29 May 2015

UNIVERSITEIT GENT

Environmental Sustainability Assessment of a Microalgae Raceway Pond Treating Wastewater from a Recirculating Aquaculture System

From Upscaling to System Integration

Sophie Sfez[(a)], Sofie Van Den Hende[(b)], Sue Ellen Taelman[(a)], Steven De Meester[(a)], Jo Dewulf[(a)]

(a) Department of Sustainable Organic Chemistry and Technology, Ghent University, Coupure Links 653, B-9000 Ghent, Belgium

(b) Laboratory for Industrial Water and Eco-Technology (LIWET), Faculty of Bioscience Engineering, Ghent University, Graaf Karel de Goedelaan 5, B-8500 Kortrijk, Belgium

Vlaamse overheid

west-vlaanderen de gedreven provincie

EnAlgae

Sustainable Pathways for Algal Bioenergy

Introduction

- EnAlgae: INTERREG IVB North West Strategic Initiative (03/2011 – 06/2015)
- 9 pilot scale algae cultivation sites (micro- and macroalgae)
- In Roeselare, Belgium: Algae-based wastewater treatment plant, treating wastewater from a pikeperch recirculating aquaculture systems (RAS)

EnAlgae

Introduction

- Aquaculture: fast growing sector competing for freshwater resources
- RASs: promising option to mitigate the environmental footprint of aquaculture systems

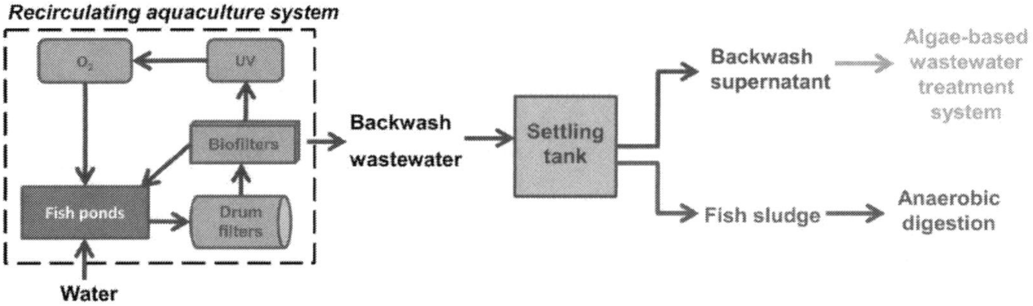

- The MaB-floc technology tested in 2013 in Belgium at pilot scale to treat pikeperch aquaculture wastewater from the Aquaculture Research Center of Inagro (Belgium)

Introduction

- MaB-flocs: bioflocculating consortium of bacteria and microalgae

- As they grow, MaB-flocs need to be harvested, delivering a new source of biomass: valorisation as *shrimp feed* and *anaerobic digestion* were tested at pilot scale

→ Industry needs insights to know which direction to take

Goal of the study

Goal 1: Assess the environmental footprint of a pilot MaB-floc SBR treating pikeperch culture WW and identify its improvement potential

Goal 2: Forecast the most sustainable valorisation pathway for MaB-flocs in the framework of an integrated aquaculture waste treatment system at industrial scale

Studied MaB-floc based WWT plants

- Pilot MaB-floc SBR treating pikeperch wastewater (real case)

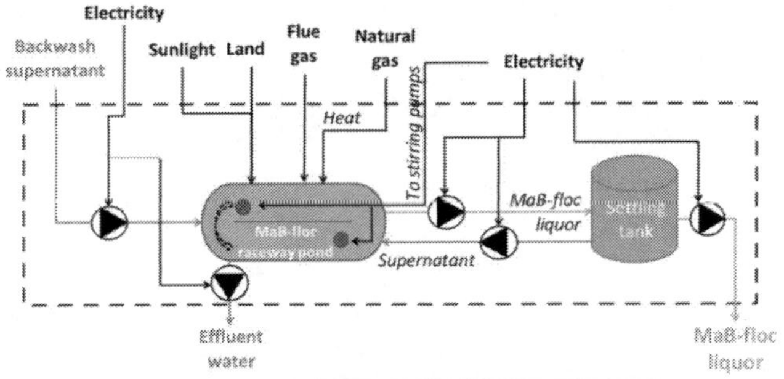

1 pond
Area: 12 m^2
Volume: 28 m^3
Flow: 2.59 m^3 day^{-1}

Van Den Hende 2014

EnAlgae

Studied MaB-floc based WWT plants

- Pilot MaB-floc SBR treating pikeperch wastewater (real case)
- Hypothetical up-scaled cases (1000 m^3 of WW treated per day):
  - L: linearly up-scaled MaB-floc plant

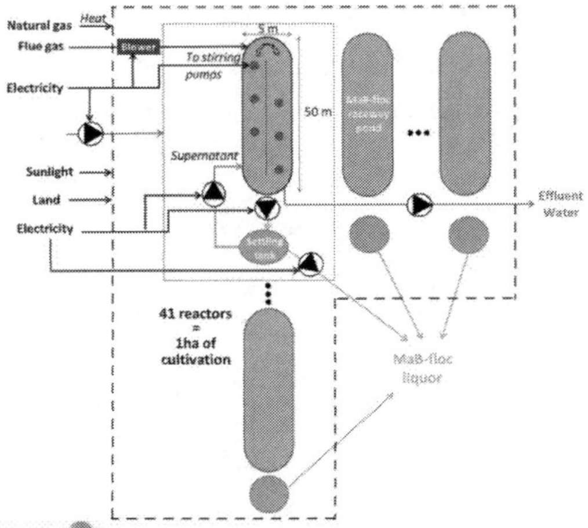

41 ponds
Area: 245 m^2 $pond^{-1}$
Volume: 98 m^3 $pond^{-1}$
Flow: 24.5 m^3 day^{-1} $pond^{-1}$

EnAlgae

Studied MaB-floc based WWT plants

- Pilot MaB-floc SBR treating pikeperch wastewater (real case)
- Hypothetical up-scaled cases (1000 m^3 of WW treated per day):
  - L: linearly up-scaled MaB-floc plant
  - S: linearly up-scaled MaB-floc plant with improved stirring system

Propeller pump 22 W m^{-2} → Paddle wheel 5.1 W m^{-2}

EnAlgae

Studied MaB-floc based WWT plants

- Pilot MaB-floc SBR treating pikeperch wastewater (real case)
- Hypothetical up-scaled cases (1000 m^3 of WW treated per day):
  - L: linearly up-scaled MaB-floc plant
  - S: linearly up-scaled MaB-floc plant with improved stirring system
  - E: linearly up-scaled MaB-floc plant with Belgian electricity mix replaced by 100% wind energy

EnAlgae

Studied MaB-floc based WWT plants

- Pilot MaB-floc SBR treating pikeperch wastewater (real case)
- Hypothetical up-scaled cases (1000 m^3 of WW treated per day):
  - L: linearly up-scaled MaB-floc plant
  - S: linearly up-scaled MaB-floc plant with improved stirring system
  - E: linearly up-scaled MaB-floc plant with Belgian electricity mix replaced by 100% wind energy
  - M: linearly up-scaled MaB-floc plant with MaB-floc productivity improved by 30%

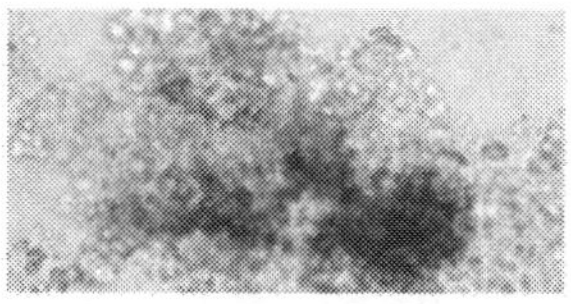

EnAlgae

Studied integrated system

- Three scenarios are compared:
  - Valorisation of MaB-flocs as shrimp feed

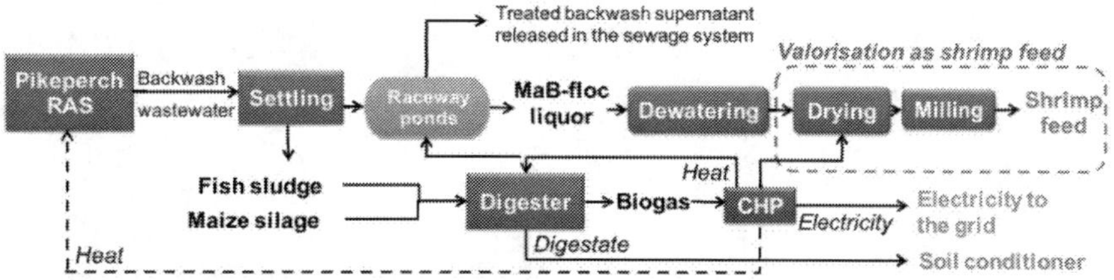

  - Valorisation of MaB-flocs as biogas

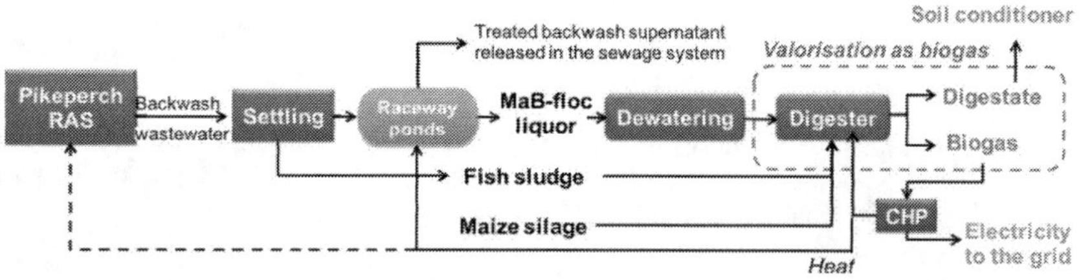

EnAlgae

Studied integrated system

- Three scenarios are compared:
  - Valorisation of MaB-flocs as shrimp feed
  - Valorisation of MaB-flocs as biogas
  - Baseline scenario

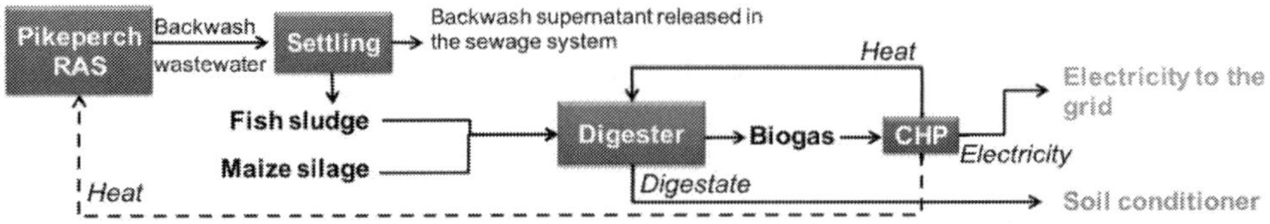

- 2 MaB-flocs plants are integrated:
  → Plant L (linearly up-scaled plant)

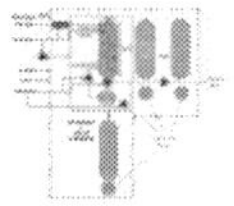

  → Plant SEM (plant L with the 3 improvements implemented

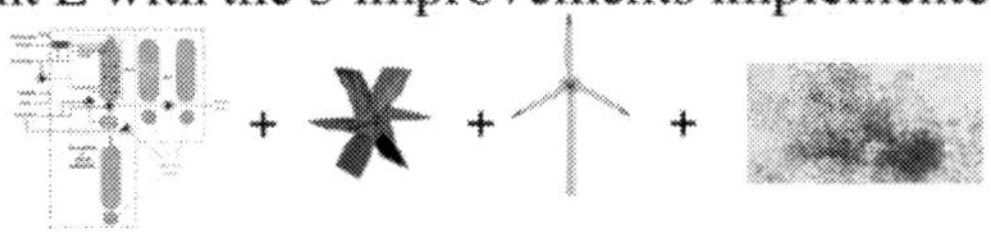

EnAlgae

Env. Sustainability Analysis

- Life Cycle Assessment (LCA), ISO standards 14040 & 14044

	Goal 1: comparison of the 4 MaB-floc based WWTP	Goal 2: SA of the integration of MaB-floc based WWTP in an aquaculture system
Functional unit	Production of 1 kg TSS MaB-floc liquor	Treatment of 1 m^3 of wastewater
Syst. boundaries	Cradle-to-gate	
Foreground system	Pilot: site data Up-scaled: pilot data + literature	Data from up-scaled plant + ecoinvent v 2.2 + literature
Background system	ecoinvent v 2.2 + literature	

Resource consumption (CEENE 2013)
→ *resource efficiency analysis*
Global warming potential (IPCC 2007)
→ *air emission efficiency analysis*
Marine and freshwater eutrophication (ReCiPe 2013)
→ *water emission efficiency analysis*

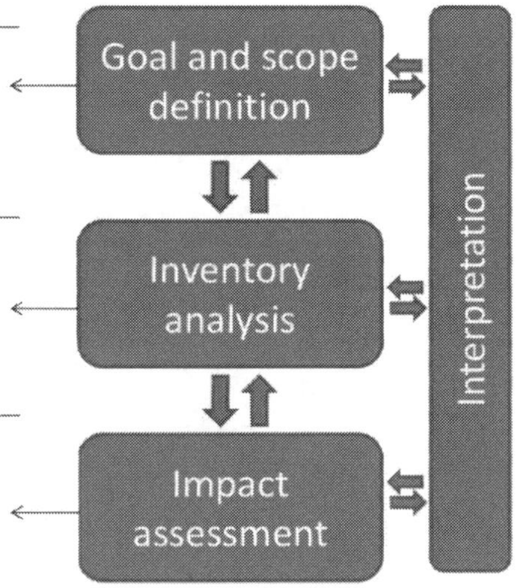

EnAlgae

LCA results: environmental sustainability of the MaB-floc based WWTP

Resource footprint (CEENE results)

Total CEENE: 848 MJ kg^{-1} MaB-floc TSS

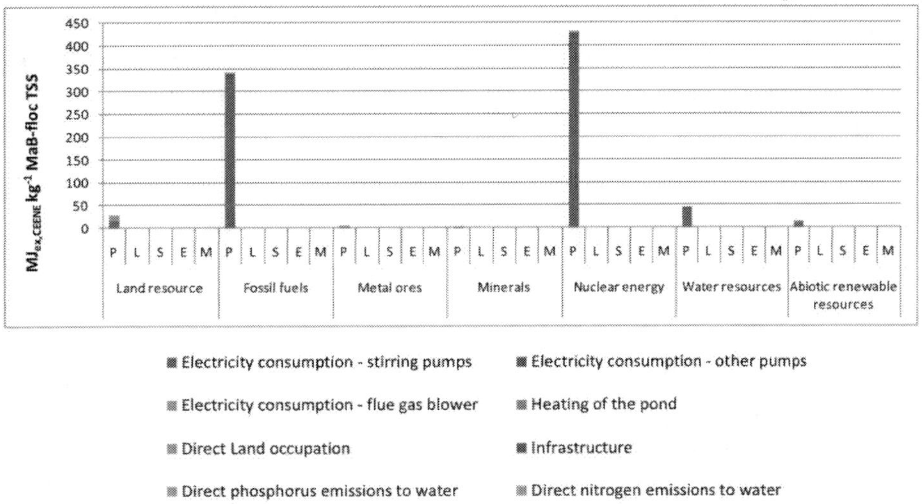

LCA results: environmental sustainability of the MaB-floc based WWTP

Resource footprint (CEENE results)

Total CEENE plant L: 278 MJ kg^{-1} MaB-floc TSS

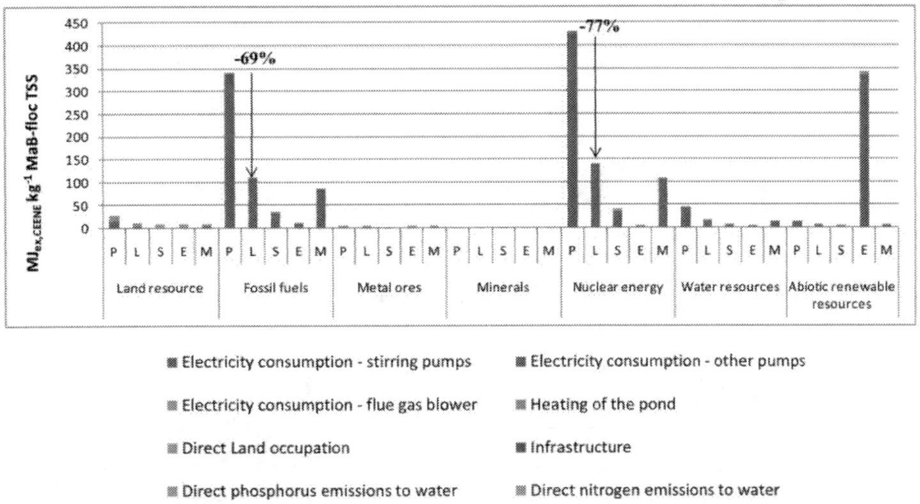

LCA results: environmental sustainability of the MaB-floc based WWTP

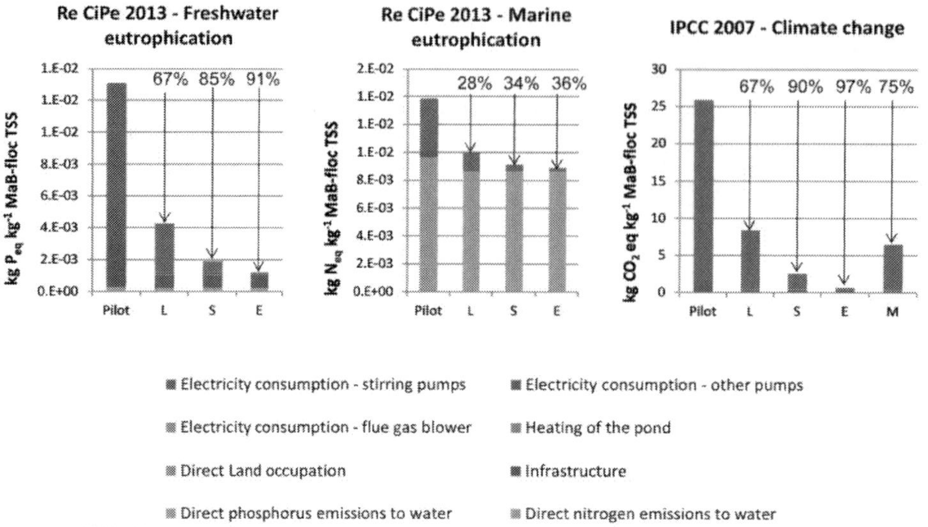

EnAlgae

LCA results: environmental sustainability of the Integrated systems

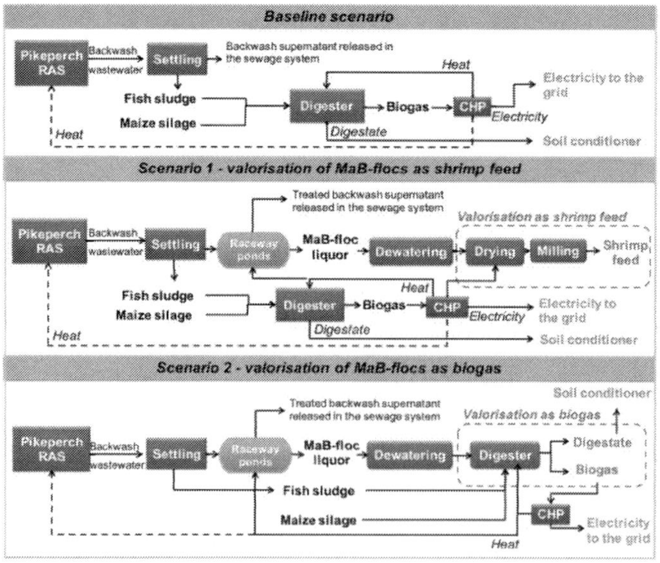

EnAlgae

LCA results: environmental sustainability of the Integrated systems

Resource footprint[1]

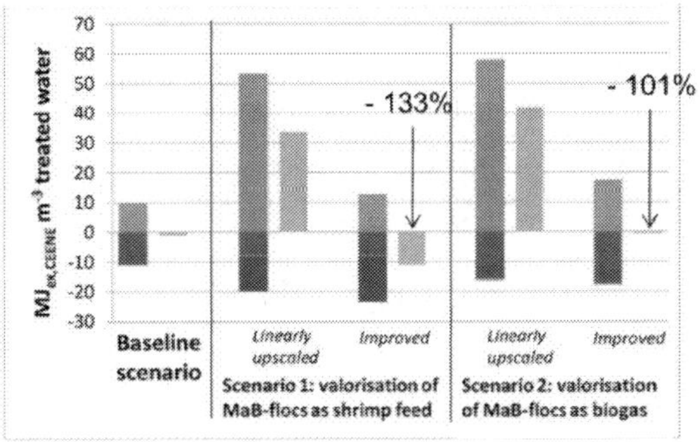

Left bar:
- Phosphorus emission
- Nitrogen emission
- Anaerobic digestion
- Shrimp feed production
- Algae-based wastewater treatment plant
- MaB-flocs dewatering

- Wheat flour (shrimp feed production)
- Compost production (AD)
- Heat production from boiler (AD)
- Electricity from the grid (AD)

Avoided processes

Right bar: Net impact

[1] *CEENE results* *__without__* *abiotic renewable resources*

EnAlgae

LCA results: environmental sustainability of the Integrated systems

Freshwater eutrophication
(ReCiPe 2013)

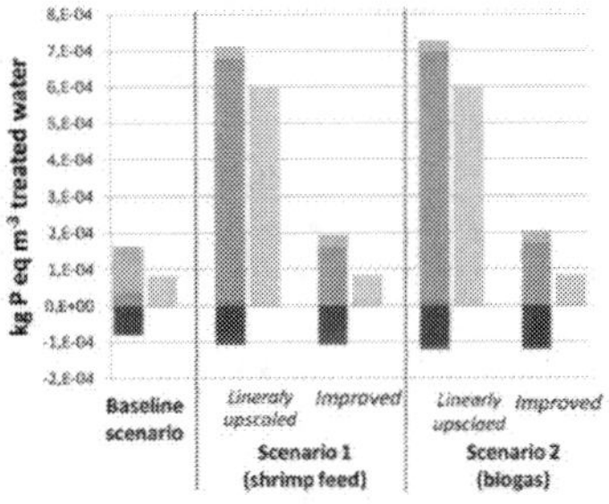

Marine eutrophication
(ReCiPe 2013)

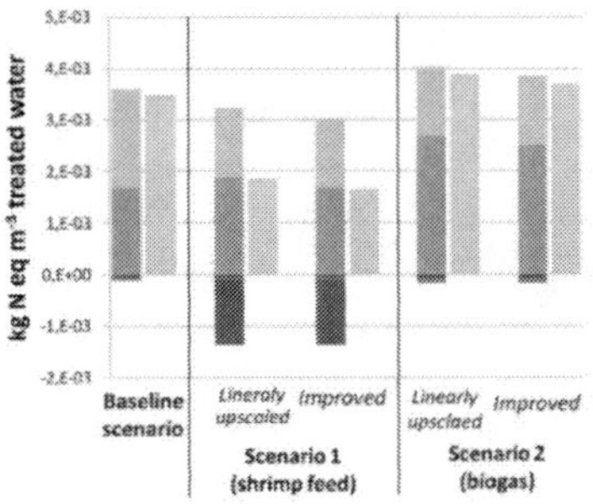

Carbon footprint
(IPCC 2007)

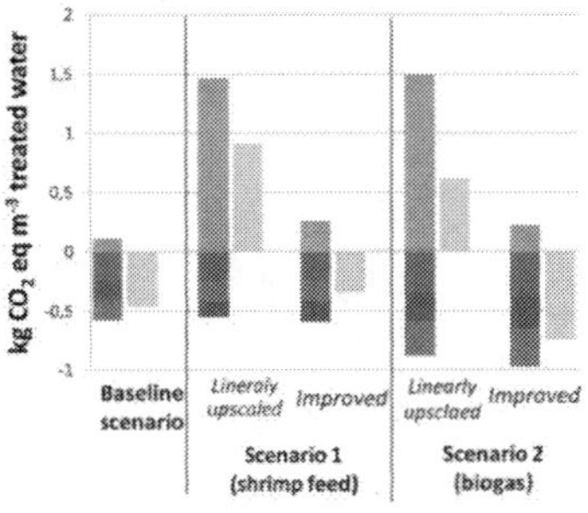

Left bar:
- Phosphorus emission
- Nitrogen emission
- Anaerobic digestion
- Shrimp feed production
- Algae-based wastewater treatment plant
- MaB-flocs dewatering

- Wheat flour (shrimp feed production)
- Compost production (AD)
- Heat production from boiler (AD)
- Electricity from the grid (AD)

Avoided processes

Right bar: Net impact

EnAlgae

Conclusion

- MaB-floc technology: stirring has the highest contribution to most impact categories
- Integrated aquaculture waste treatment system:
  - Potential to compete with the baseline scenario and contribute to a sustainable connection of the water-food-energy nexus in the aquaculture sector
  - Valorizing MaB-flocs into shrimp feed: overall more sustainable than into biogas

    Bottleneck: EU legislation
- Future research:
  - Improvement of LCA with more complete data on nutrient cycle (measurements needed)
  - Focus on the improvement of the energy efficiency of the system, rather than of MaB-flocs productivity

EnAlgae
collaborate innovate communicate

UNIVERSITEIT GENT

collaborate
innovate
communicate

Thank you!

Sophie.Sfez@UGent.be
+32 (0) 9 264 99 27

Vlaamse overheid

EnAlgae
collaborate innovate communicate

west-vlaanderen
de gedreven provincie

www.enalgae.eu info@enalgae.eu

Enalgae is a Strategic Initiative of the INTERREG IVB North West Europe (NWE) Programme

Early Stage Synthesis and Design of Integrated Process and Wastewater Treatment Networks

Zainatul Bahiyah Handani, Rafiqul Gani

Department of Chemical and Biochemical Engineering,

Technical University of Denmark,

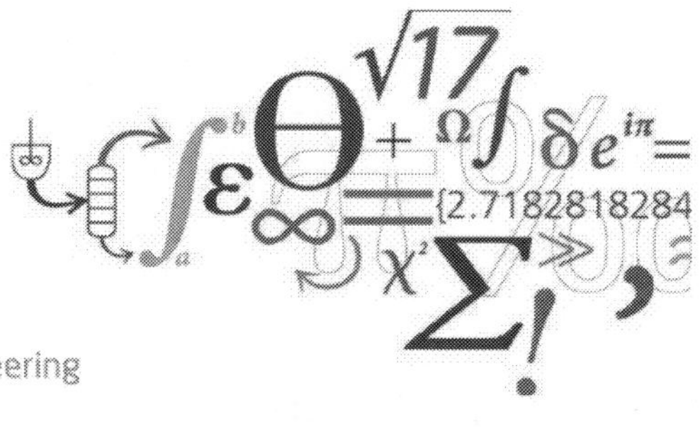

DTU Chemical Engineering
Department of Chemical and Biochemical Engineering

Presentation Outline

- Introduction
  - ✓ Motivation and Objective
- Methodology
  - ✓ Superstructure representation
  - ✓ Generic interval model
- Case study: A bioethanol production
- Conclusions and Future Works

Motivation

DTU

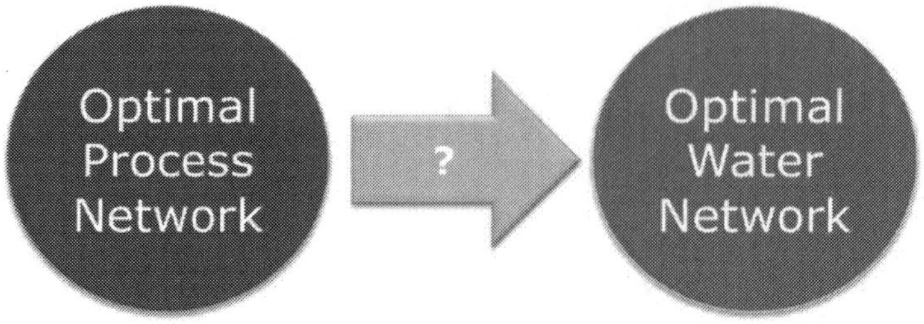

Drawbacks:

Sequential approach may not fully optimize the whole process and may overestimate the utility cost.

Motivation

DTU

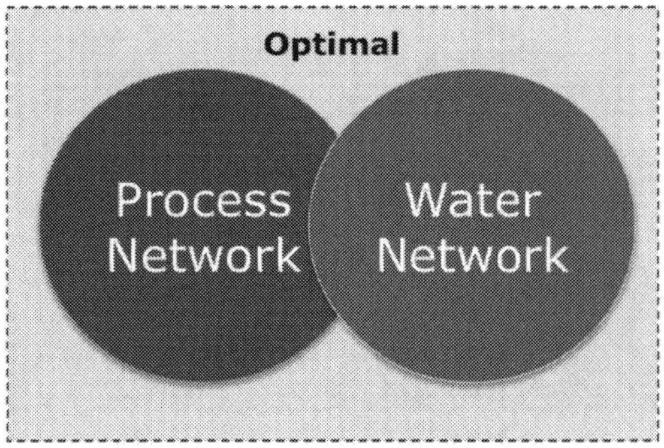

Advantages:

- ✓ All interactions and economic trade-offs would be taken explicitly into account (Duran and Grossmann, 1986; Yee et al., 1990; Yang and Grossmann; 2013)
- ✓ Result obtained from simultaneous optimization is better than sequential optimization

Decision-making at early stage of design

Process network
- ✓ raw materials
- ✓ processing technologies
- ✓ product portfolio

Wastewater network
- ✓ Wastewater sources & characterization
- ✓ treatment technologies
- ✓ Treatment objectives

Objective

To develop a **systematic framework** for **integrated** synthesis and design of **process and water networks** with respect to **environment, resource consumption, economics and sustainability** using **superstructure-based** optimization approach

5 **DTU Chemical Engineering, Technical University of Denmark**

Methodology

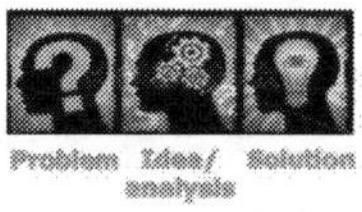

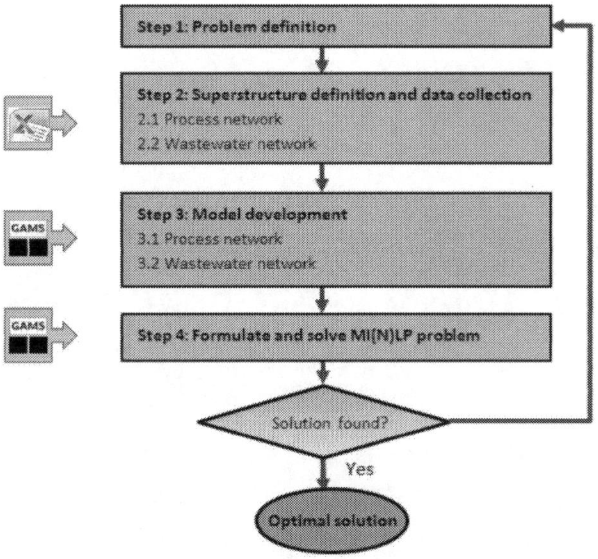

- Superstructure
- Generic model development

6 **DTU Chemical Engineering, Technical University of Denmark**

Superstructure representation

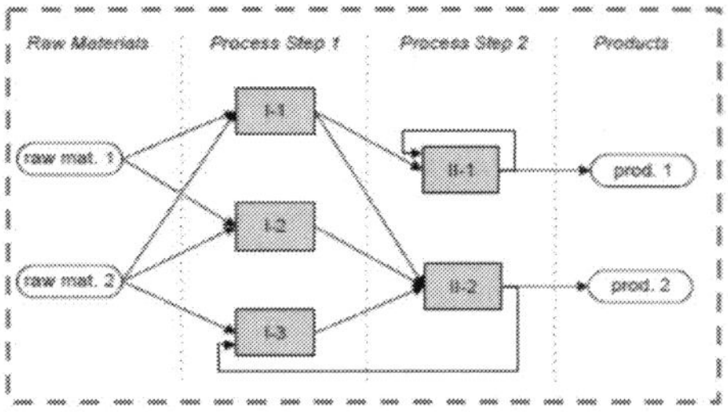

+

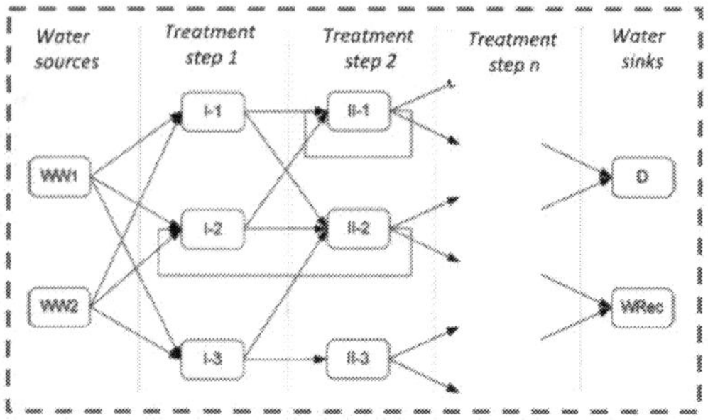

Process Network
(Quaglia et al. [1])

Water Treatment Network
(Quaglia et al. [2])

[1] A. Quaglia, B. Sarup, G. Sin, R. Gani. Computers and Chemical Engineering. 38 (2012). 213-223
[2] A. Quaglia, A. Pennati, M. Bogataj, Z. Kravanja, G. Sin, R. Gani. Ind. Eng. Chem. Res., 53 (13) (2014). 5160–5171
[3] Čuček, L. , Martín, M., Grossmann, I. E., Kravanja, Z. Computers and Chemical Engineering. 35 (2011). 1547-1557

Combine process and water treatment network superstructure

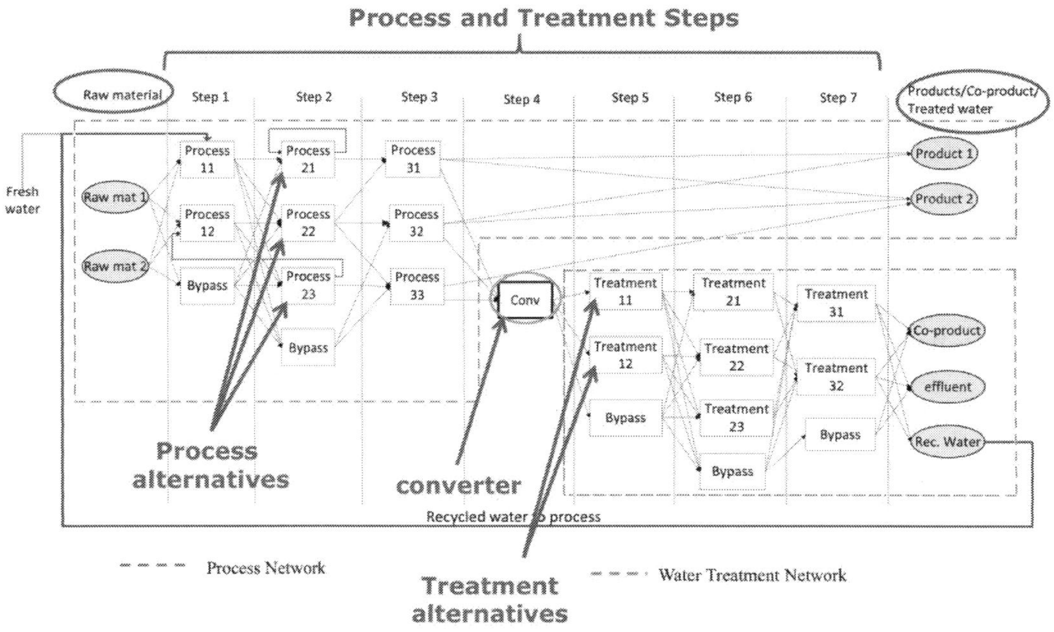

Model development

Generic process interval block and model [1]

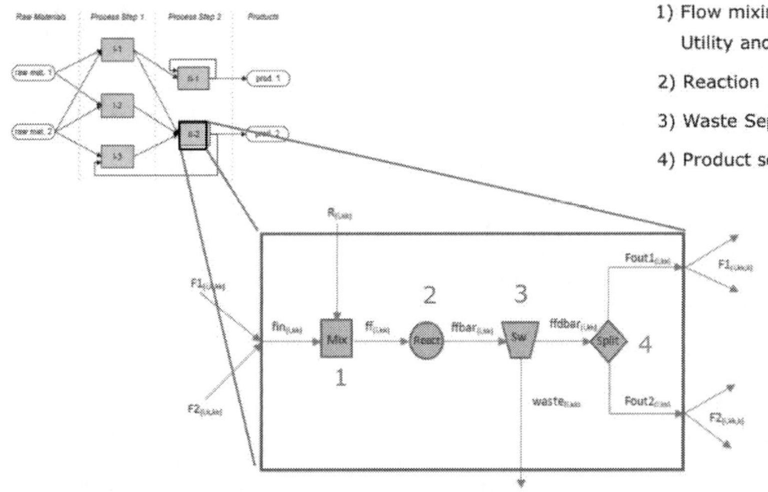

1) Flow mixing
   Utility and chemical quantity
2) Reaction
3) Waste Separation
4) Product separation

General block – a block representing activities of the process technology (unit operations, several processes or a plant) using simple short-cut model

[1] A. Quaglia, B. Sarup, G. Sin, R. Gani. Computers and Chemical Engineering. 38 (2012). 213-223

Case Study: Conceptual example

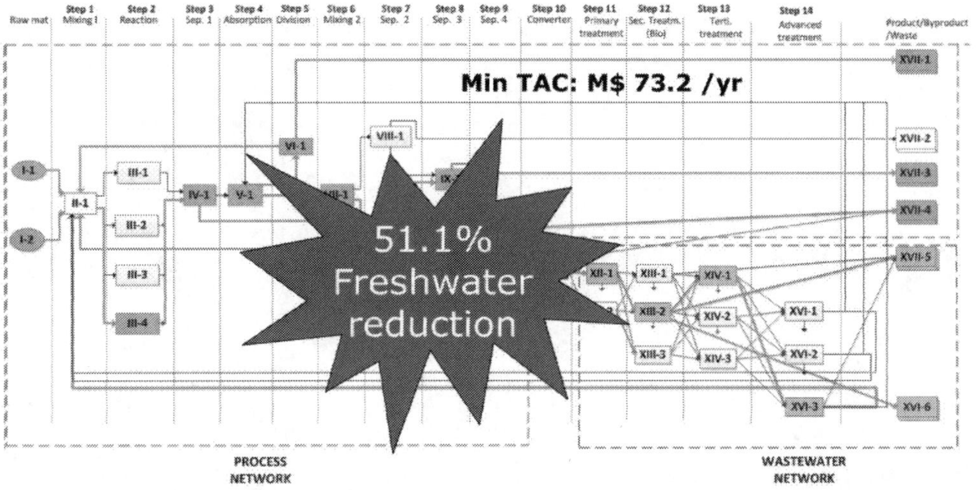

Optimum topology: RM1, RM2, Mix1, React4, Sep11, Flash, Divider, Sep22, Sep31, Sep41, Conv, Sed, UASB, UF, BP4, Purge, Product C, Byprod. D, Effluent, Methane, Recycle as raw material

Case Study: Bioethanol production

Biomass

Biochemical process

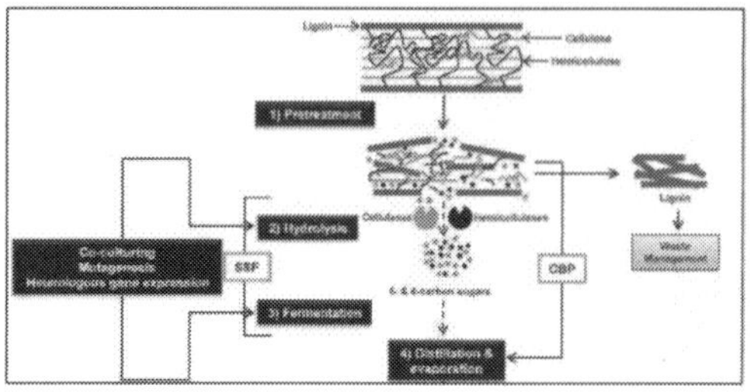

Bioethanol

- One of the main concerns in bioethanol production plant is consumptive use of water
- Water mainly used in pretreatment, detoxification and conditioning process steps
- Wastewater produced from the process contains of high organic content, solid waste and other contaminants and need to be treated in wastewater treatment facility

Case Study: Bioethanol production

Step 1: Problem definition

GOAL

Synthesis and design of optimal process and wastewater networks

SCOPE

-Effluent discharge, solids treatment, recycle water is not considered

OPTIMALITY CONDITION

Minimize Total Annual Cost

Case Study:

Step 2: Superstructure definition

Process Network:

1 raw material, 15 process intervals in 14 processing steps, 1 product & 6 wastewater

Wastewater Treatment Network:

6 wastewater sources, 23 treatment intervals in 8 treatment steps, 4 products and co-products

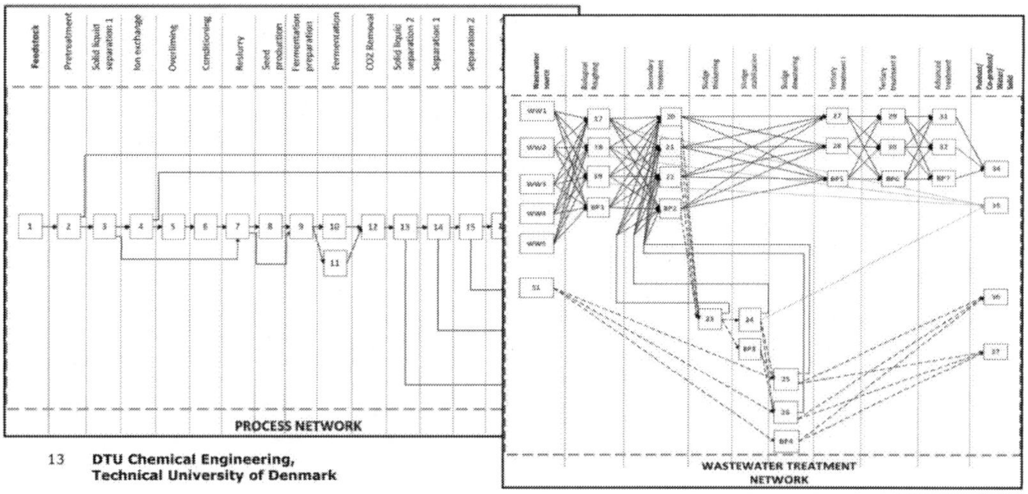

Case Study: Bioethanol production

Superstructure definition

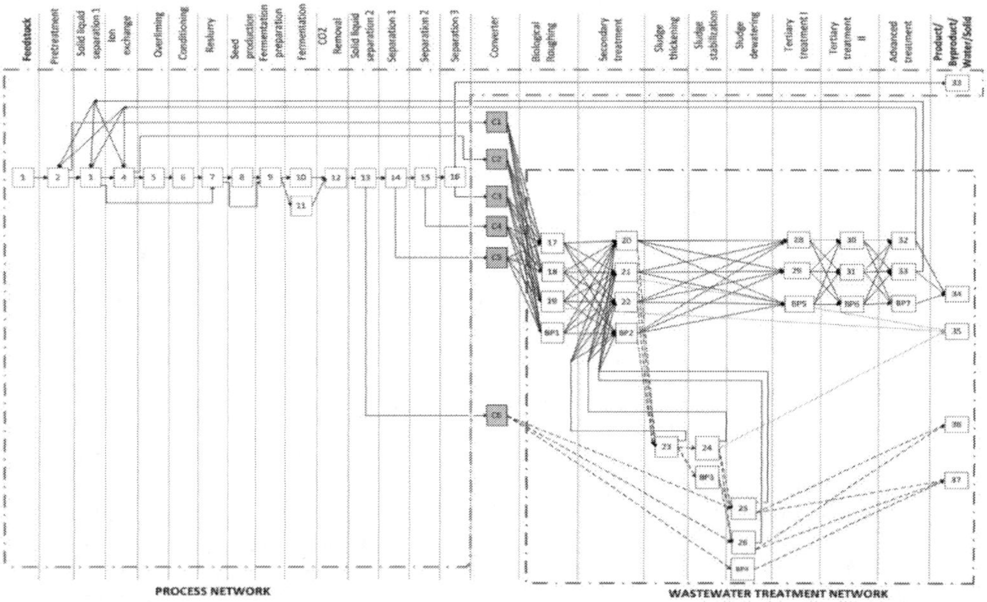

Case Study: Bioethanol production

Step 3 : Model development

Generic Interval Model

1) Flow mixing:

$$ff_{i,kk} = fin_{i,kk} + \alpha_{i,kk} R_{i,kk}$$

Utility and chemical quantity:

$$R_{i,kk} = \sum_{ii} (\mu_{i,ii,kk} \cdot fin_{ii,kk})$$

2) Reaction:

$$ffbar_{i,kk} = ff_{i,kk} + \sum_{rr,react} (\gamma_{i,kk,rr} \cdot \theta_{react,kk,rr} \cdot ff_{react,kk})$$

3) Waste Separation:

$$waste_{i,kk} = ffbar_{i,kk} \cdot SW_{i,kk}$$

$$ffdbar_{i,kk} = ffbar_{i,kk} \cdot (1 - SW_{i,kk})$$

4) Product separation:

$$Fout1_{i,kk} = ffdbar_{i,kk} \cdot SPr_{i,kk}$$

$$Fout2_{i,kk} = ffdbar_{i,kk} \cdot (1 - SPr_{i,kk})$$

Explanations

i,ii: component index
k,kk: process interval index
$fin_{i,kk}$: inflow to the process interval
$ff_{i,kk}$: component flow after mixing
$R_{i,kk}$: utility flow
$\alpha_{i,kk}$: fraction of utility consumed
$\mu_{i,ii,kk}$: specific utility consumption
$ffbar_{i,kk}$: component flow after reaction
$\gamma_{i,kk,rr}$: reaction stoichiometry
$\theta_{react,kk,rr}$: conversion of key reactant
$waste_{i,kk}$: component flow of waste stream
$ffdbar_{i,kk}$: component flow after waste separation
$SW_{i,kk}$: waste fraction
$Fout1_{i,kk}$: component flow of primary outlet
$Fout2_{i,kk}$: component flow of secondary outlet
$SPr_{i,kk}$: product split fraction

15 **DTU Chemical Engineering, Technical University of Denmark**

Case Study: Bioethanol production

Step 4 : Formulate and solve MI(N)LP problem

Problem formulates as MINLP

OBJECTIVE FUNCTION: TOTAL ANNUAL COST (TAC)

$$TAC = OPEX + CAPEX \cdot \left[\frac{i(1+i)^n}{(1+i)^n - 1}\right]$$

Assume i, = 5%, n= 10 years

s.t. $g(x,y) \geq 0$ (process constraints, environmental limit, product specification etc.)

$h(x,y) = 0$ (process model, performance indicator)

$x \epsilon X$ (flow rates, split matrix)

$x^{lo} \leq x \leq x^{up}$

$y \epsilon \{0;1\}^n$ (selection of intervals)

MODEL STATISTICS

Continuous Variables	*182,790*
Binary variables	*24*
Objective Function	*1*
Number of equations	*143,492*

16 **DTU Chemical Engineering, Technical University of Denmark**

Case Study: Bioethanol production

Result : Optimal solution (topology)

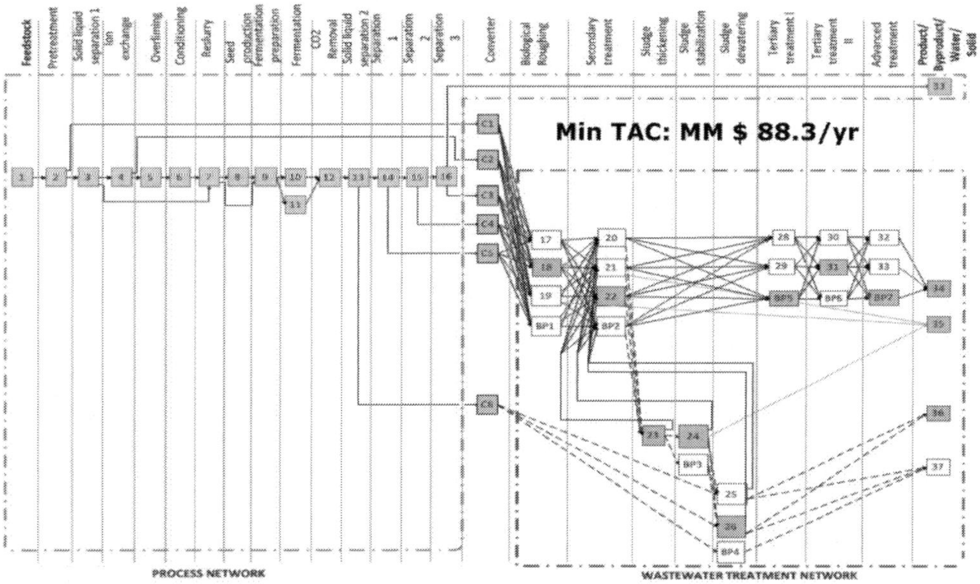

Conclusion:

- A **systematic framework** for an early stage synthesis and design of integrated process and wastewater treatment network has been demonstrated on a bioethanol production case study.
- The superstructure-based optimization methodology has shown to be a useful **decision support tool for early stage** synthesis and design networks by screening a large number of design alternatives .
- The **interaction between two networks** is visualized via selection appropriate raw materials, technologies and alternatives for process and water treatment as well as products.

Current and Future works:

- Apply the methodology with **different optimization scenarios** in order to develop a flexible and versatile methodology.
- Extend the superstructure by **integrating heat exchanger network with process and wastewater networks**

19 **DTU Chemical Engineering, Technical University of Denmark**

Environmental and Economic Sustainability Evaluation for Design Improvement and Optimization of Ion Exchange Drinking Water Treatment

- **Conference:** International Congress on Sustainability Science Engineering ICOSSE
- **Year:** 2015
- **Proceeding:** ICOSSE 2015
- **Group:** Abstract Submissions
- **Session:**

Session 3: Water Resource and Availability Issues II

- **Time:**

Wednesday, May 27, 2015 - 1:50pm-2:10pm

Authors:

Amini, A.
Payne, K., University of South Florida
Zhang, J., University of South Florida
Zhang, Q.,

Water treatment infrastructure faces numerous operational, financial, and environmental challenges. Implications of design choices on environmental impacts and costs at full-scale implementation are often poorly understood or evaluated. Furthermore, as new facilities are constructed and designs adopted, the implications of design choices gain importance. Ion Exchange (IX) is a robust technology that can be used to remove various types of contaminants in drinking water and has shown increased adoption for applications such as disinfection byproduct precursor removal. Through this research the sustainability of IX technology has been evaluated and utilized to target design improvements. This evaluation is being used to develop a framework for technology improvement and optimization of water technologies by dynamically linking a mathematical process model with environmental and economic assessment, using a case study of

IX. This model and framework allow for multi-objective optimization of design parameters to minimize environmental impacts and costs of water treatment technology.

Waste Water Treatment Plants As Energy Centres

- **Conference:** International Congress on Sustainability Science Engineering ICOSSE
- **Year:** 2015
- **Proceeding:** ICOSSE 2015
- **Group:** Abstract Submissions
- **Session:**

Session 3: Water Resource and Availability Issues II

- **Time:**

Wednesday, May 27, 2015 - 2:10pm-2:30pm

Authors:

Narodoslawsky, M., Institute for Resource Efficient and Sustainable Systems
Kindermann, H.
Truger, B., TU Graz
Kollmann, R., TU Graz

Waste water treatment plants handle large amounts of water and generate biomass. Using heat pumps the energy gained by cooling waste water stream a few degrees and digesting (or burning) the sludge generated by the plant can provide considerable energy.

The presentation will report about a method to design optimal energy recovery and utilisation schemes that are adapted to local context. Case studies at two treatment plants will show the application of this method and provide insight into how energy from waste flows of society may be put to use and waste water treatment plants can be converted to energy centres.

Implications of Implementation Scale on the Environmental Sustainability of Wastewater Treatment with Resource Recovery

- **Conference:** International Congress on Sustainability Science Engineering ICOSSE
- **Year:** 2015
- **Proceeding:** ICOSSE 2015
- **Group:** Abstract Submissions
- **Session:**

Session 3: Water Resource and Availability Issues II

- **Time:**

Wednesday, May 27, 2015 - 2:50pm-3:10pm

Authors:

Zhang, Q.,
Cornejo-Warner, P.
Mihelcic, J.

Many wastewater treatment plants (WWTPs) require high levels of resource consumption to transport and treat wastewater. Resource recovery is beneficial to reducing the environmental impacts of WWTPs; however, few studies have investigated how scale of implementation impacts the sustainability of WWTPs with resource recovery. This study investigated the impact of scale on the environmental sustainability of WWTPs with resource recovery using lifecycle assessment (LCA). Three systems in Florida were selected: septic tank with aerobic treatment at household scale, advanced water reclamation facility at community scale, and advanced water reclamation facility at city scale. Three sustainability indicators were considered: embodied energy, carbon footprint, and eutrophication potential. This study finds that there are benefits to centralization of WWTPs with resource recovery in terms of embodied energy and carbon footprint; however, the community scale results in the lowest eutrophication potential. Additionally, technology selection may have a larger impact on environmental sustainability than scale.

Rethinking Future Energy Systems

- **Conference:** International Congress on Sustainability Science Engineering ICOSSE
- **Year:** 2015
- **Proceeding:** ICOSSE 2015
- **Group:** Abstract Submissions
- **Session:**

Session 4: Energy-Water-Food Nexus: Energy

- **Time:**

Wednesday, May 27, 2015 - 1:30pm-2:00pm

...Given that society requires electricity, heating, and transportation fuels, the best way to develop the most economic and environmentally friendly solution is to consider a holistic approach in which the production of heat, electricity and transportation fuels are considered simultaneously. A range of sources of primary energy need to be considered. If biomass is used for the production of electricity or transportation fuels, this can compete with food production and create additional demands on water sources. Systems need to be flexible and to integrate storage of both power and heat. The design of such systems requires a framework that will allow the production of electricity, heat and transportation fuels to be optimized simultaneously.

ETH zürich

Reduction of the Environmental Impact While Improving the Economic Benefits:
A Win-Win Situation in Industrial Waste Incineration through Optimized Treatment Schedules

Matteo L. Abächerli, Elisabet Capón-García & Konrad Hungerbühler
matteo.abaecherli@chem.ethz.ch

Safety and Environmental Technology Group
Institute for Chemical and Bioengineering
Department of Chemistry and Applied Bioscience
ETH Zürich

ETH zürich

Contents

- Waste Incineration
- Model
- Case Study
- Results
- Conclusion

ETH zürich

Waste Incineration

Safety and Environmental Technology Group
Institute for Chemical and Bioengineering
ETH Zürich

Matteo Abächerli | 19.10.2015 | 3

ETH zürich

Schematic Overview

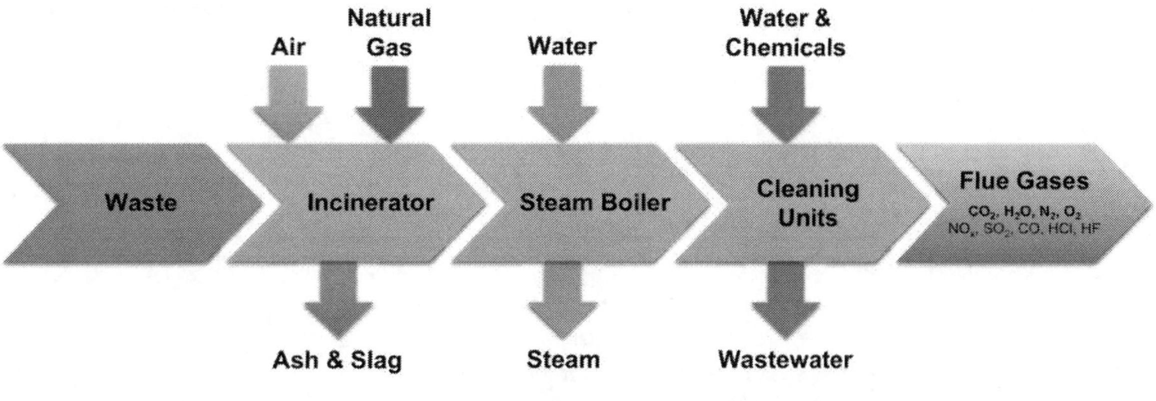

Safety and Environmental Technology Group
Institute for Chemical and Bioengineering
ETH Zürich

Matteo Abächerli | 19.10.2015 | 4

ETH zürich

Challenges in Industrial Waste Incineration Management

- Large and complex problem
  - Many different waste streams
  - Many storage & mixing possibilities
  - Different technical, environmental and regulatory constraints
- Good performance is crucial to avoid holdups in production
- Treatment costs affect product prices and profit
- Decision making process often based on purely empirical basis

Safety and Environmental Technology Group
Institute for Chemical and Bioengineering
ETH Zürich

Matteo Abächerli | 19.10.2015 | 5

ETH zürich

Targets and Solution Strategy

- Provide the industry with a short-term optimization schedule in order to:
  - Reduce environmental burden & operational costs
  - Support the daily decision making process
  - Avoid bottlenecks in incineration and therefore holdups in the production
  - Respect all technical, environmental and regulatory constraints
- How to reduce environmental burden & costs?
  - Reduce natural gas consumption
- How to reduce natural gas consumption?
  - Better utilization of the energy contained in waste
    - by improving the storage, mixing & incineration strategy

Safety and Environmental Technology Group
Institute for Chemical and Bioengineering
ETH Zürich

Matteo Abächerli | 19.10.2015 | 6

ETH zürich

The Energy Challenge

- Avoid situation of energy excess or shortage

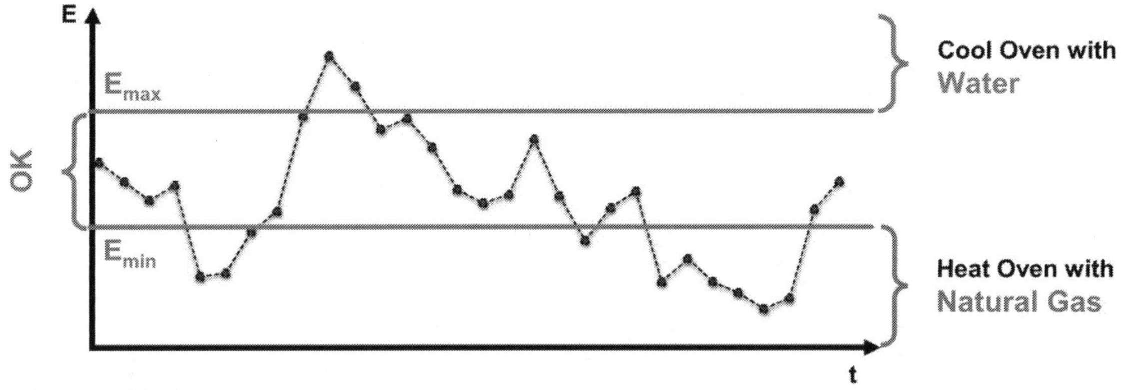

ETH zürich

Model

ETH zürich

Model Overview

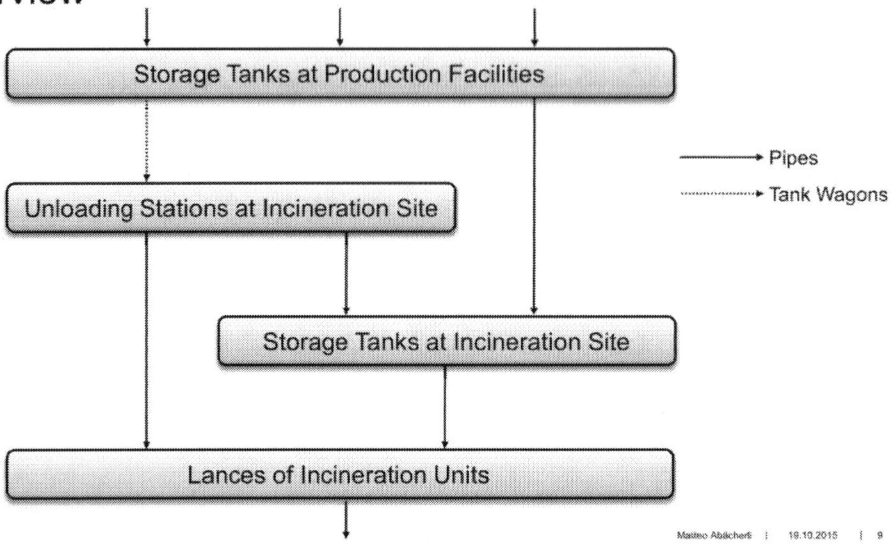

Safety and Environmental Technology Group
Institute for Chemical and Bioengineering
ETH Zürich

Matteo Abächerli | 19.10.2015 | 9

ETH zürich

Model Classification

- Process scheduling problem
  - Logistics: Transfer, storage & mixing issues
  - Operational: Incineration process
- MILP based formulation
  - Original formulation MINLP $\rightarrow$ too resource intensive to solve
    - Linearization
- Discrete time formulation
  - Single, uniform time-grid

Safety and Environmental Technology Group
Institute for Chemical and Bioengineering
ETH Zürich

Matteo Abächerli | 19.10.2015 | 10

ETH zürich

Core Mathematical Formulation

Sets
i Tanks
k Incineration units
t Time periods
w Waste streams
y Flue gases

- Mass balances $M_{w,i,t}^{stored} = M_{w,i}^{0}\Big|_{t=1} + M_{w,i,t-1}^{stored}\Big|_{t>1} + M_{w,i,t}^{in} - M_{w,i,t}^{out} \quad \forall w,i,t$

- Mixing $\dfrac{M_{w,i,t}^{stored}}{\sum_{w} M_{w',i,t}^{stored}} = \dfrac{M_{w,i,t}^{out}}{\sum_{w} M_{w',i,t}^{out}} \quad \forall w,i,t \,|\, (w \in w^{mix})$

- Energy balance

$$\sum_{w}\sum_{l\in l^k}\left(M_{w,l,t}\cdot H_w\right) + M_{k,t}^{G}\cdot H^{G} = \sum_{y\notin y^{H_2O}}\left\{M_{k,y,t}^{Fumes}\cdot\int_{T^0}^{T_k} Cp_{k,y,t}\partial T\cdot\left(T_k - T^0\right)\right\} +$$

$$\sum_{y\notin y^{H_2O}}\left\{M_{k,y,t}^{Fumes}\cdot\left[\int_{T^0}^{T_k} Cp_{k,y,t}\partial T\cdot\left(T_k - T^0\right) + Cp^{H_2O}\cdot\left(100 - T^0\right) + \Delta H_{vap}^{H_2O}\right]\right\} \quad \forall k,t$$

Safety and Environmental Technology Group
Institute for Chemical and Bioengineering
ETH Zürich

Matteo Abächerli | 19.10.2015 | 11

ETH zürich

Case Study

Safety and Environmental Technology Group
Institute for Chemical and Bioengineering
ETH Zürich

Matteo Abächerli | 19.10.2015 | 12

ETH zürich

Industrial Partner

Lonza

- Lonza
  - One of the world's leading suppliers to the Pharma, Biotech and Specialty Ingredient markets
- Chemical & Biotechnological Site
  - Location: Visp (CH)
  - Size: 0.9 km^2
  - Employees: ~ 2890
  - Production facilities: > 30
    - Continuous and batch production
  - Incinerated waste: ~ 60'000 t/year
    - > 1500 different waste streams per year

Safety and Environmental Technology Group
Institute for Chemical and Bioengineering
ETH Zürich

Matteo Abächerli | 19.10.2015 | 13

ETH zürich

Industrial Partner - System Overview

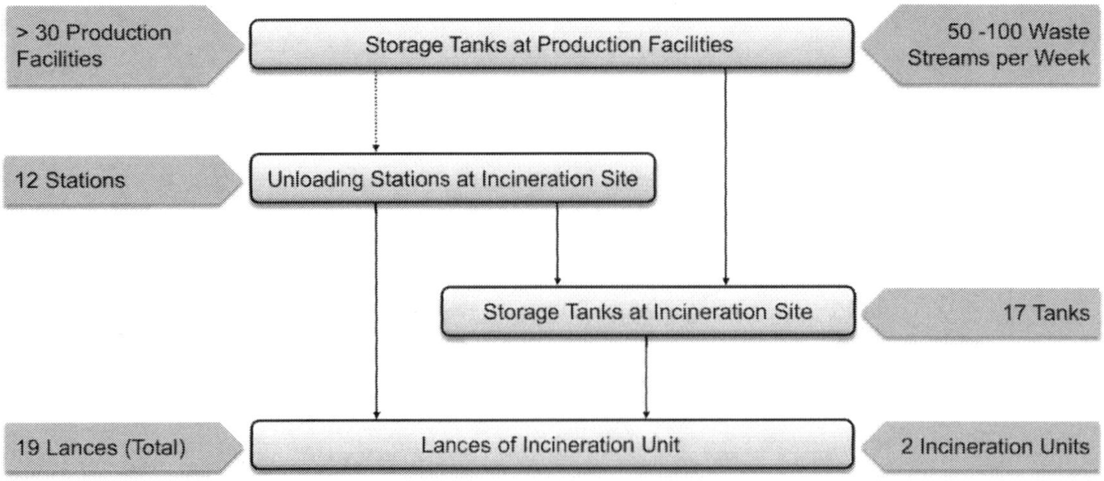

Safety and Environmental Technology Group
Institute for Chemical and Bioengineering
ETH Zürich

Matteo Abächerli | 19.10.2015 | 14

ETH zürich

Average Energy Requirement per ton of Waste Incinerated

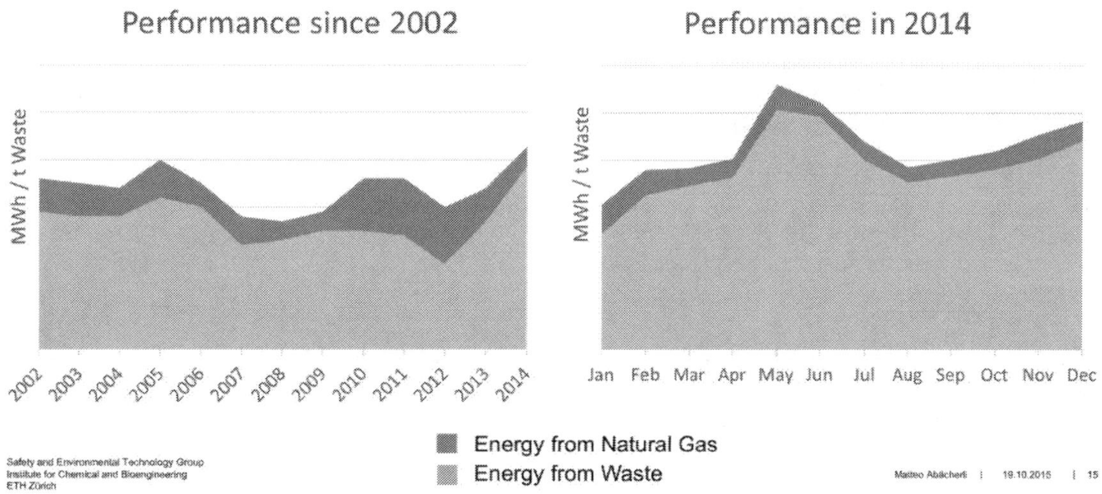

ETH zürich

Procedure

- Performance comparison for 2014
  - Actual vs. theoretically possible performance (simulated)
    - Empirical vs. optimization supported waste management
- Annual performance simulated in 4 week blocks
  - Rolling-Horizon approach
- Constraint allowing comparison: $M_{Inc}^{Model} = M_{Inc}^{Actual} \pm \varepsilon$
- 4 Weeks Shutdown of one incineration unit
  - August 27th - September 23th

ETH zürich

Rolling-Horizon Approach

- Integration of models with different horizon length, resolution and level of detail
- Planning Model
  - 4 Weeks horizon in 1 day steps

  $$\min z = Costs - Revenue$$
- Scheduling Model
  - 1 Week horizon in 4 h steps

  $$\min z = abs\left(M_{Inc}^{Plan} - M_{Inc}^{Sched}\right)$$

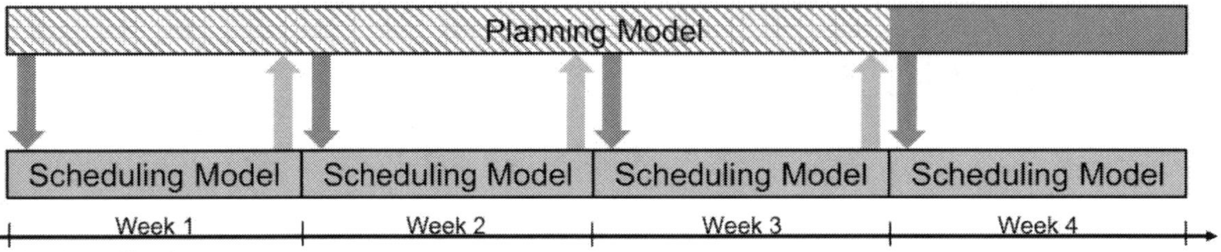

Safety and Environmental Technology Group
Institute for Chemical and Bioengineering
ETH Zürich

Matteo Abächerli | 19.10.2015 | 17

ETH zürich

Results

Safety and Environmental Technology Group
Institute for Chemical and Bioengineering
ETH Zürich

Matteo Abächerli | 19.10.2015 | 18

Average Energy Requirement per ton of Waste Treated

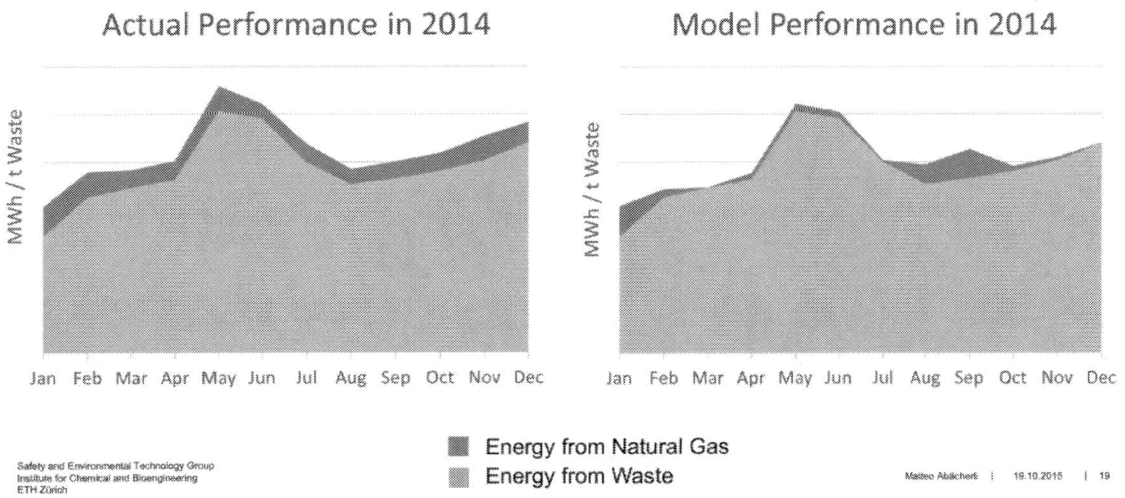

Safety and Environmental Technology Group
Institute for Chemical and Bioengineering
ETH Zürich

Matteo Abächerli | 19.10.2015 | 19

Average Natural Gas Consumption per ton of Waste Treated

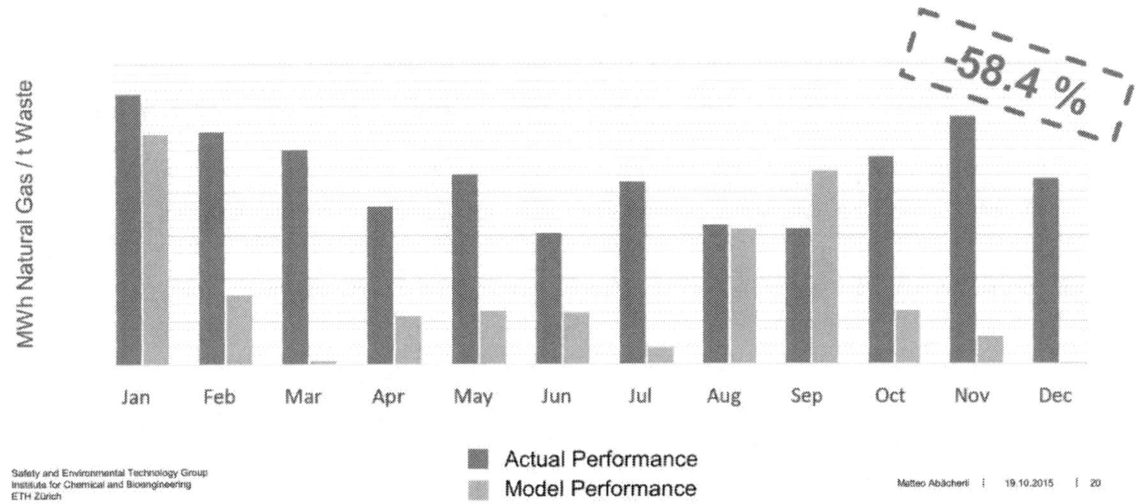

Safety and Environmental Technology Group
Institute for Chemical and Bioengineering
ETH Zürich

Matteo Abächerli | 19.10.2015 | 20

ETH zürich

Average CO_2 Emissions per ton of Waste Treated

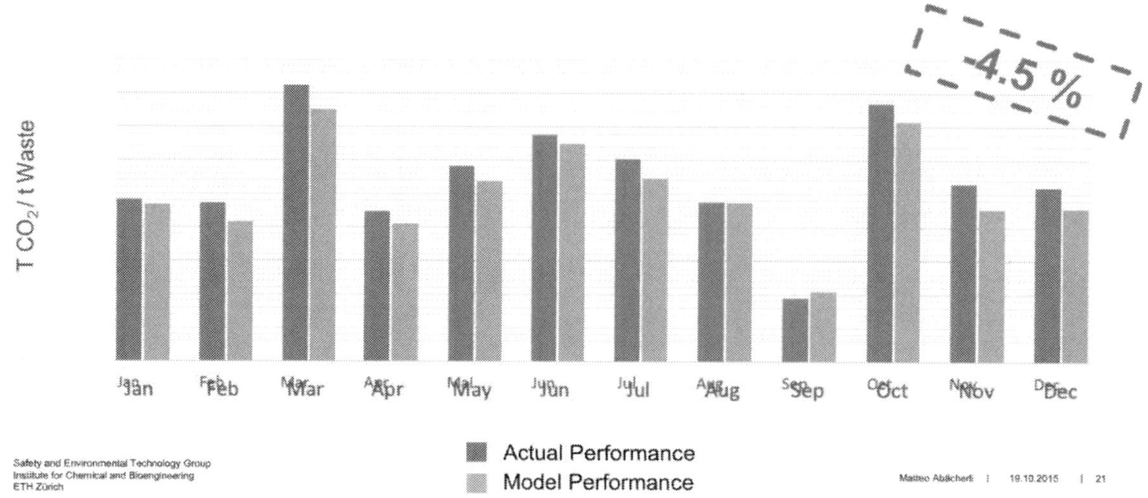

ETH zürich

Computational Performance

- DELL Precision T3500
  - Intel 3.33GHz CPU
  - 18 GB RAM
- Implementation in GAMS
  - Solver: CPLEX 12.5 (for MILP problems)
- Stopping criteria for each week
  - Optimality gap: 1 %

No. of Equations (in millions)	6 - 13
No. of Variables (in millions)	6 - 10
No. of discrete Variables (in thousands)	120 - 150
Solving Time [min]	2 - 20

ETH zürich

Conclusion

ETH zürich

Implementation

- Straightforward Handling
  - Simple handling of in and output data
  - Implementation of changes easy and fast
- Almost no additional investment costs
- Fast solution provision
  - Possibility of automation
- Possibility of scenario simulation
  - Answering "What if?" questions

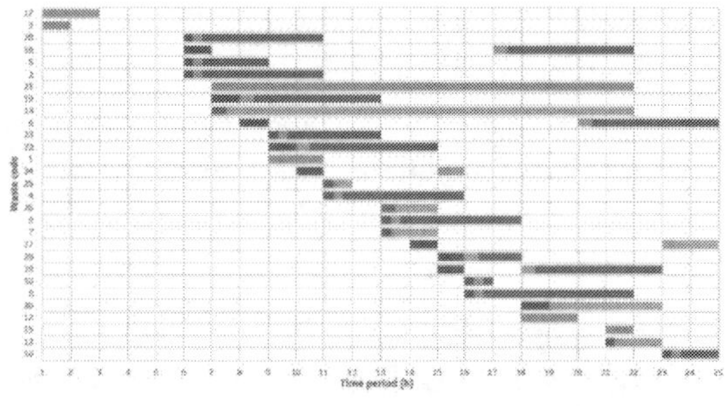

ETH zürich

Recapitulation

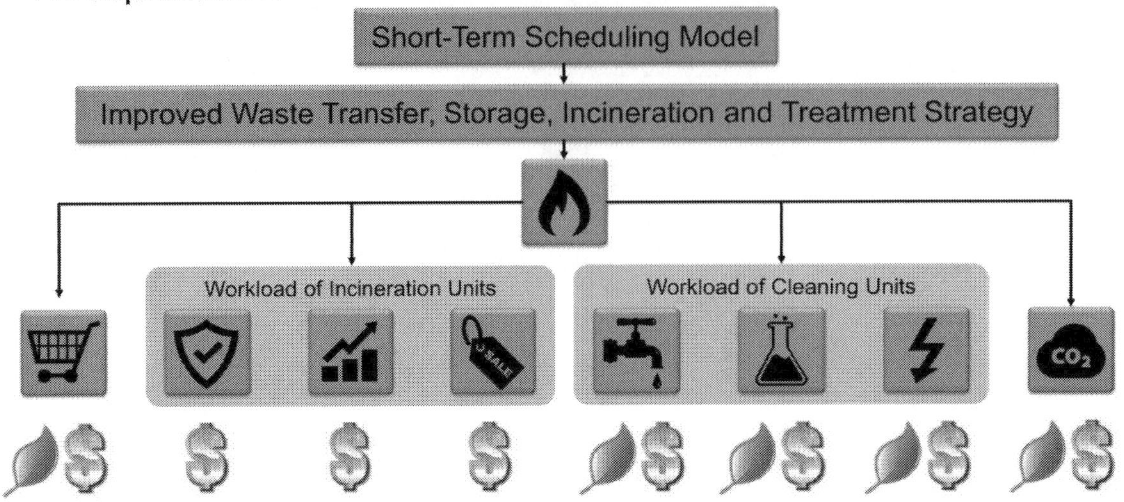

Safety and Environmental Technology Group
Institute for Chemical and Bioengineering
ETH Zürich

Matteo Abächerli | 19.10.2015 | 25

ETH zürich

Acknowledgements

- Safety & Environmental Technology Group, ETH Zürich
  - Prof. Dr. Konrad Hungerbühler, Dr. Elisabet Capón-García Oliver Weder, Philipp Steinleitner, Silvia Bally & Daniel Santos

Safety and Environmental Technology Group

- Energy & Waste Management, Lonza AG Visp
  - Dr. Andrej Szijjarto, Beat Steinmann, Andreas Imstepf Patrick Blumenthal, Veronika Scholz & Thomas Kahoun

Lonza

- Financial Support
  - Lonza AG
  - Commission for Technology and Innovation CTI
    - Federal Department of Economic Affairs, Education and Research

Schweizerische Eidgenossenschaft
Confédération suisse
Confederazione Svizzera
Confederaziun svizra

Safety and Environmental Technology Group
Institute for Chemical and Bioengineering
ETH Zürich

Matteo Abächerli | 19.10.2015 | 26

ETH zürich
Thank you for your kind attention!

ICOSSE

International Congress on Sustainability Science & Engineering

May 26-29, 2015 Balatonfüred, Hungary

Session 4: Energy-Water-Food Nexus: Energy, May 27, 2015

"Ecoefficiency of Buildings Including Reduction of CO_2 Emission in Life Cycle of Building Materials, Elements and Technologies"

GŁÓWNY INSTYTUT GÓRNICTWA

Włodzimierz A. Sokół (D.Sc.Eng.)
Central Mining Institute, Katowice (PL)

Longlife Invest

Baltic Sea Region Programme 2007-2013

Part-financed by the European Union (European Regional Development Fund)

ICOSSE

International Congress on Sustainability Science & Engineering
May 26-29, 2015 Balatonfüred, Hungary

ABSTRACT

In the European Union, the building sector is responsible for the consumption of 40 % of the energy and relative emissions of CO_{2e} in a lifecycle.

The methodology for the evaluation of building ecoefficiency that includes resource and energy efficiency and efficiency in the reduction of CO_{2e} emission, including CO_{2e} emission in life cycle of building materials, elements and technologies is presented in the example of a new student dormitory in TU-Klaipeda (Lithuania). It is implemented within a framework of the international project LONGLIFE-INVEST being part financed by the European Regional Development Fund via Baltic Sea Region Programme 2007-2013.

The most important standards being developed by their international staff i.e.: primary energy demand, energy sources from renewable energy, reduction of CO_{2e} emissions, highly minimized lifecycle costs and others are discussed.

Evaluation and selection of ecoefficient technologies as task of company, sectorial and regional environmental management system is presented including involvement of all significant actors.

2

CONTENT

ICOSSE
International Congress on Sustainability Science & Engineering
May 26-29, 2015 Balatonfüred, Hungary

INTRODUCTION

In the **European Union**, the **building sector** is responsible for the **consumption** of **40 % of the energy** [1] and relative emissions of CO_2 in a lifecycle.

That is why a new building concept has been developed by international staff of LONGLIFE project [2] that offers a new approach for building investment process.

Suggested principles are based on evaluation of energy resources' needs for the **entire lifecycle of a building**.

Although, the **cost of a building** based on Longlife construction and engineering principles is on average about **20% higher** than the conventional construction costs, in a long term the total Longlife **building maintenance costs** are up to **40%, lower**. Life cycle CO_2 emission is evaluated and respectively reduced.

That is why LONGLIFE standards are **implemented and checked** in practice now by international team in framework of LONGLIFE-INVEST [3] extension stage project while design, construction and exploitation of new a **student dormitory for Klaipeda University in Lithuania.**

LONGLIFE INVEST partnership:

1. STANDARD & LONGLIFE EVALUATION OF CO_2 EMISSIONS FORM THE BUILDING

If standard approach had been applied the consideration of **energy efficiency and CO_{2e} emission** include the **direct CO_{2e} emission** and **indirect CO_{2e} emission** only based on the results of evaluation of energy demand for the building

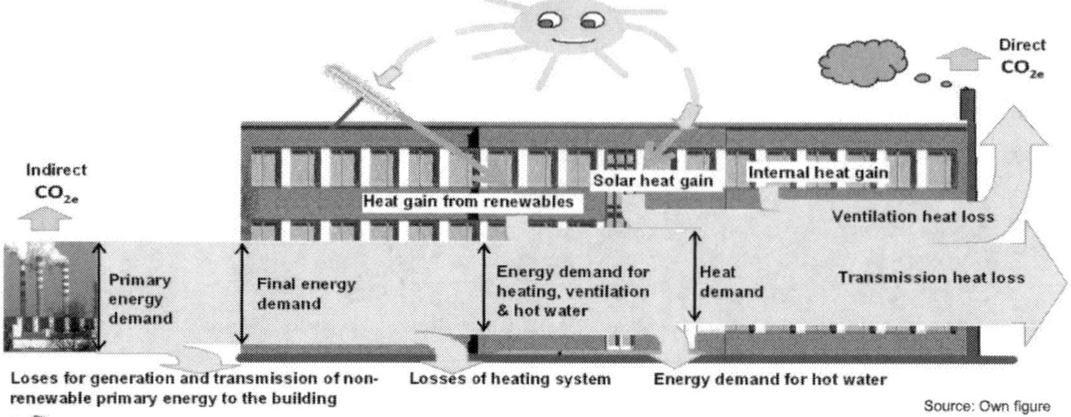

Source: Own figure

5

1. STANDARD & LONGLIFE EVALUATION OF CO_2 EMISSIONS FORM THE BUILDING

In case of LONGLIFE approach beside of standard **direct CO_{2e} emission** and **indirect CO_{2e} emission** some **other indirect CO_{2e} emissions** are taking into account

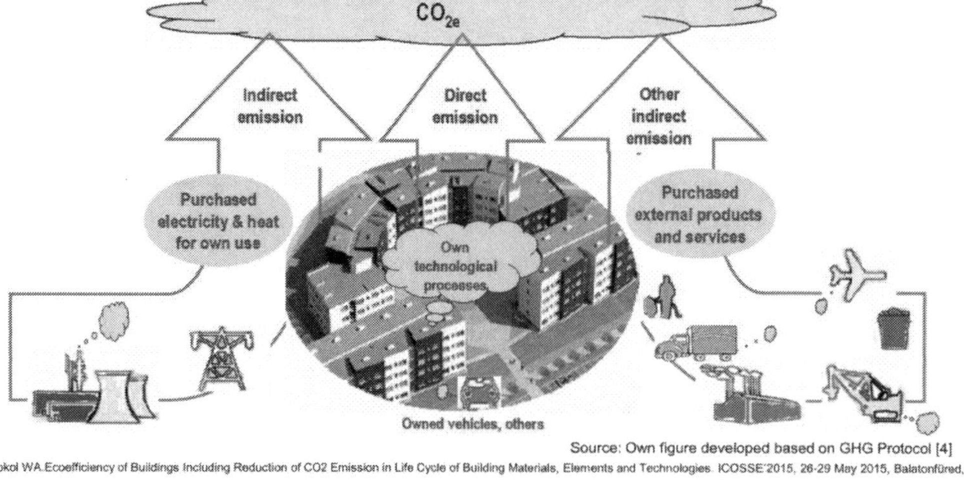

Source: Own figure developed based on GHG Protocol [4]

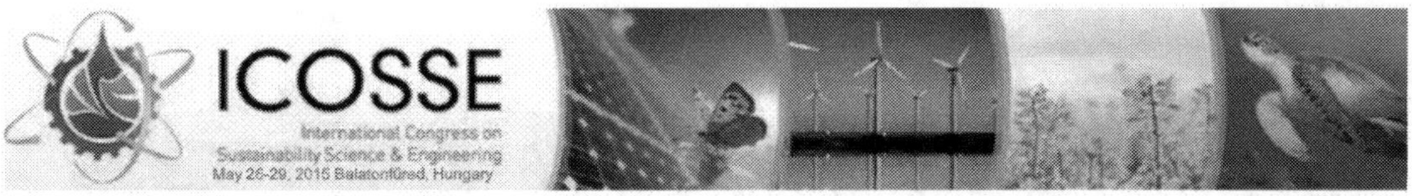

1. STANDARD & LONGLIFE EVALUATION OF CO_2 EMISSIONS FORM THE BUILDING

Moreover the building **energy efficiency and reduction of total CO_{2e} emission** are considered in whole **lifecycle.** The **other indirect CO_{2e} emissions** are evaluated for a lifecycle of **building materials, elements and technologies**

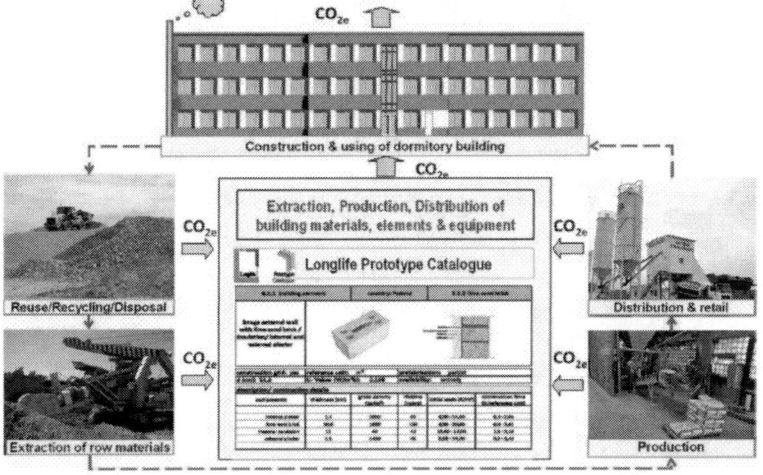

Source of photos :www.kruszgeo.com.pl

7

Sokol WA.Ecoefficiency of Buildings Including Reduction of CO2 Emission in Life Cycle of Building Materials, Elements and Technologies. ICOSSE'2015, 26-29 May 2015, Balatonfüred, HU

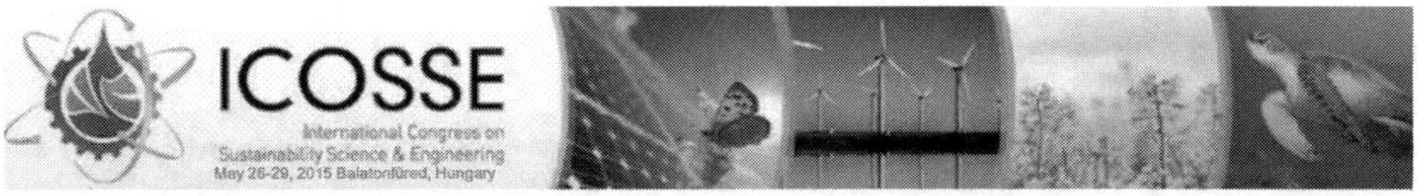

1. STANDARD & LONGLIFE EVALUATION OF CO_2 EMISSIONS FORM THE BUILDING

The boundaries of the building and building materials, elements and technologies are considered according ISO 14072 [5] approach

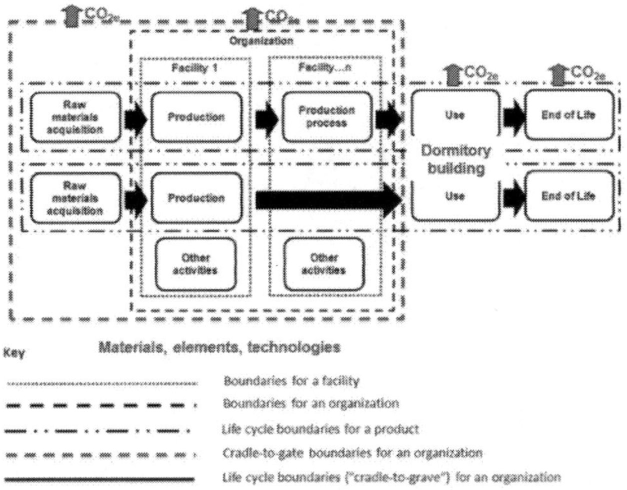

Source: adopted ISO/TS 14072:2014(E)

8

Sokol WA.Ecoefficiency of Buildings Including Reduction of CO2 Emission in Life Cycle of Building Materials, Elements and Technologies. ICOSSE'2015, 26-29 May 2015, Balatonfüred, HU

2. METHODOLOGY FOR MINIMIZING CO_2 EMISSION FROM THE BUILDING

The recommended **methodology aims to select** for the implementation **the best technology variant** from other being considered i.e. **most ecoefficient** including minimization of CO_{2e} emission in building life cycle. It includes three basic steps

2.1. STEP 1 – BUILDING DESIGN PHASE:

This step relates a design phase **while the suppliers** of building materials, elements and technologies as well as construction company (investor) **are unknown** but the **best technology option has to be selected** that is why:

- **Direct CO_{2e} emissions** resulting from energy demand for heat and domestic hot water and **indirect CO_{2e} emissions** caused by electricity that propels the auxiliary equipment being used for production of heat, hot water and ventilation are calculated based on designed energy demand
- **Other indirect CO_{2e} emissions** connected with **building materials, elements and technologies** are **preliminary** calculated taking GWP data from the **Longlife Prototype Catalogue** [6] (according ÖCOBAUDAT [7])
- **Summed direct, indirect and other indirect emissions give a total CO_{2e} emission** during the building lifecycle and data for **Longlife Performance Pass** [2]

ICOSSE
International Congress on Sustainability Science & Engineering
May 26-29, 2015 Balatonfüred, Hungary

2. METHODOLOGY FOR MINIMIZING CO_2 EMISSION FROM THE BUILDING

- For evaluation the **ecoefficiency** of each technology variants and then selecting the best solution, **the methodology for calculation of aggregated indicator - ecoefficiency enhancement ΔE** [8,9] has been adopted.

Ecoefficiency of each technology option is evaluated by comparison to selected **reference technology** and taking into account:

- **efficient use of resources** (degree of reduction in consumption of raw materials, water, natural fuels etc.),
- **efficient use of energy** (degree of reduction in consumption of energy and using of renewable sources of energy) and,
- **efficiency in reduction of emissions** (degree of reduction of total CO_{2e} emission but as well other emissions to the air, water and soil and content of environmentally harmful substances if any).

- **Evaluation of risk** is carried out in case of lacking or **uncertainty** of **lifecycle data** for the building materials, elements and technologies **by adopting the methodology being described in [9].**

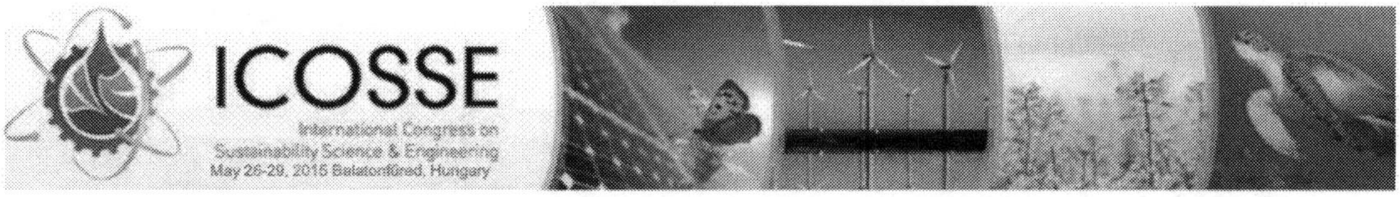

2. METHODOLOGY FOR MINIMIZING CO_2 EMISSION FROM THE BUILDING

Ecoefficiency enhancement ΔE [8,9] is calculated as following:

$$\Delta E[\%] = 100 \cdot (E_{new} - E_{old}) / E_{old} = 100\left\{ \sqrt{3\left(\sum_{i=1}^{3} I_i^2\right)^{-1}} - 1 \right\}$$

where: $I_i = \sqrt{\frac{1}{n_i}\sum_{j=1}^{n_i} I_{ij}^2}$, $i = 1 \div 3$, $j = 1 \div n_i$

I_i - **means the environmental impact** of considered (new) technology (improved) with comparison to the reference (old) solution i.e. :

- **I_1 – Efficiency of resources**
- **I_2 – Efficient use of energy**
- **I_3 – Emissions released**

n_i – quantity of the parameters I_{ij} in framework of the environmental impacts I_i,

E_{new}, E_{old} – the ecoefficiency of new and old (reference) technology respectively

Sokol WA.Ecoefficiency of Buildings Including Reduction of CO2 Emission in Life Cycle of Building Materials, Elements and Technologies. ICOSSE'2015, 26-29 May 2015, Balatonfüred, HU

2. METHODOLOGY FOR MINIMIZING CO_2 EMISSION FROM THE BUILDING

RESULTS OF THE 1st STEP:

1. **Total CO_{2e} emission** from the building during a lifecycle for all technology variants and **share of** direct, indirect and other indirect CO_{2e} emissions

2. **Recommendation** the **most ecoefficient technology** for implementation however **lifecycle costs** and **legal framework** play a decisive role

2.2. STEP 2 – BUILDING CONSTRUCTION PHASE

During the construction phase the **suppliers** of building materials, elements and technologies **are known** as well as construction company.

Evaluation of building **ecoefficiency, including evaluation of total CO_{2e} emission** in lifecycle is recommended like in step 1 but for **real data** of materials, elements and technologies being **submitted by the suppliers** and **construction company**

12

Sokol WA.Ecoefficiency of Buildings Including Reduction of CO2 Emission in Life Cycle of Building Materials, Elements and Technologies. ICOSSE'2015, 26-29 May 2015, Balatonfüred, HU

2. METHODOLOGY FOR MINIMIZING CO_2 EMISSION FROM THE BUILDING

The reliable sources of this data on CO_{2e} emission are respectively:

- **Type III of Environmental Product Declaration** according ISO EN 15804 (**EPD**) [10] that include life cycle A1-A3 if are submitted by the supplier
- **Environmental statements** if the suppliers participate in **EMAS** [11]
- **Annual reports on GHG emission** if the suppliers reports to **EU ETS** [12]

Otherwise the **relevant Data** related energy efficiency have to be submitted by companies for evaluation of basic lifecycle A1-A3 and A4-A5 and other stages

Product			Construction		Use stage							End-of-life				Benefits and loads beyond the system boundary
					Related to the building fabric					Related to the building operation						
A1	A2	A3	A4	A5	B1	B2	B3	B4	B5	B6	B7	C1	C2	C3	C4	D
Raw material supply	Transport	Manufacturing	Transport	Construction	Use	Maintenance	Repair	Replacement	Refurbishment	Operational energy use	Operational water use	Demolition	Transport	Waste processing	Disposal	Reuse / Recovery / Recycling potential

According ISO EN 15804 (PN-EN 15804) life cycle of building products from cradle to grave considers 4 cycles and 16 activities in the system boundaries

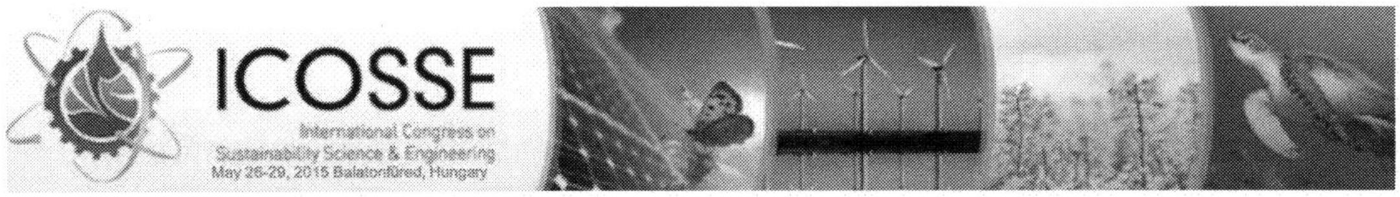

2. METHODOLOGY FOR MINIMIZING CO_2 EMISSION FROM THE BUILDING

RESULTS OF 2nd STEP:

1. **Total CO_{2e} emission** from the building during the lifecycle and share of CO_{2e} emission evaluated for real supply of materials, elements and technologies

2.3. STEP 3 - USING THE BUILDING

1. During exploitation of the building the consumption of all kind of energy (heat, electricity, renewable energy) that were considered as data for the project have to be collected **to prove achievement of energy efficiency targets**

2. **Intelligent energy management** system would be a useful tool for monitoring and minimizing energy consumption of the dormitory building and CO_{2e} emission.

RESULTS OF 3rd STEP

Proving achievement of LONGLIFE targets related energy efficiency of the building and minimization of total CO_{2e} emission

3. EXAMPLE OF EVALUATION OF THE DORMITORY BUILDING IN TU-KLAIPED

3.1. Evaluation of direct and indirect CO_{2e} emissions and ecoefficiency

During a design phase (step1) the **direct CO_{2e} emissions** resulting from energy demand for heat and domestic hot water and **indirect CO_{2e} emissions** are caused by electricity that propels the auxiliary equipment being used for production of heat, hot water and ventilation were calculated based on designed energy demand

Over 20 technology options (variants) have been considered by Longlife-Invest team but only results for selected **technology variants** are presented here

The schemes of selected variants are described bellow [13] that have been developed by Polish team however the results of evaluation of seven technology variants that relate using of geothermal were considered as well.

Because of reduced volume of the presentation the results of evaluation of CO_{2e} and ecoefficiency enhancement are presented only

Variant 1:natural gas boiler + convection heaters

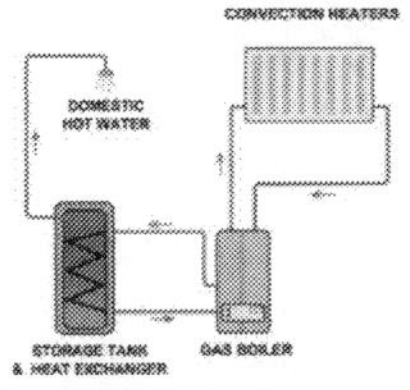

Variant 2: coal boiler + convection heaters
Variant 3: wood-pellet boiler + convection heaters

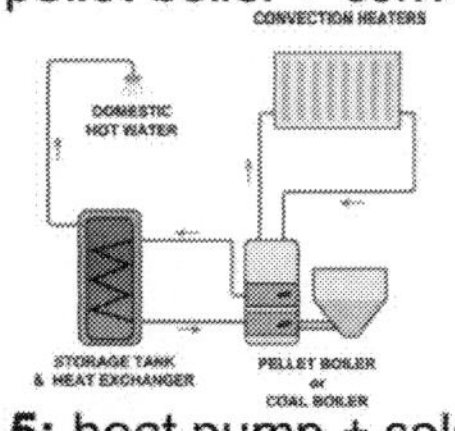

Variant 4: heat pump + electric immersion heater + floor heating

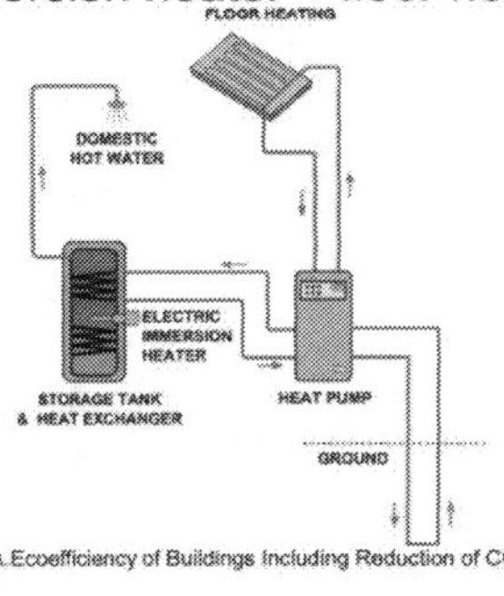

Variant 5: heat pump + solar collectors + floor heating

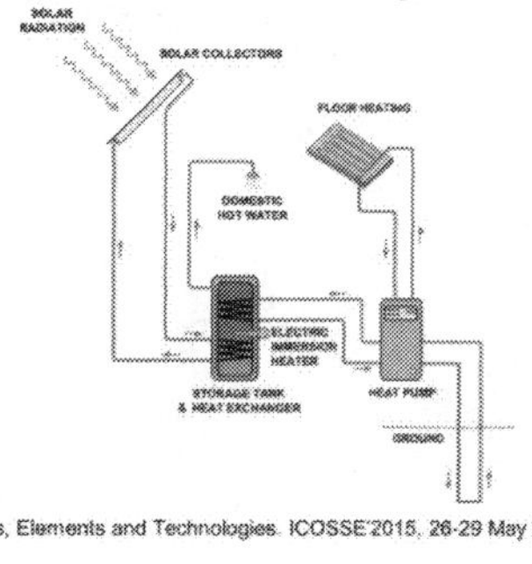

Source: TU-Gdansk [13]

3.1. Evaluation of direct and indirect CO_{2e} emissions and ecoefficiency

CO_2 emission from heating, ventilation and auxiliary electricity

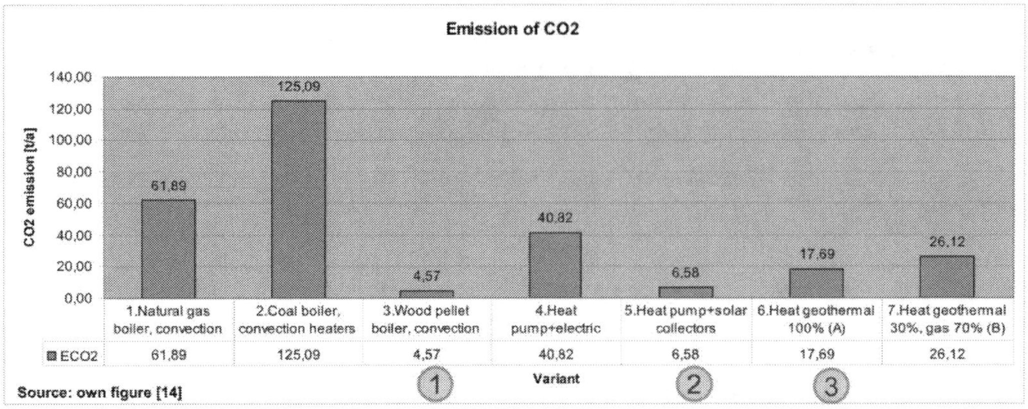

Source: own figure [14]

✓ **The best solutions** from the point of view of direct and indirect **CO_{2e} emission** is variant **3 – wood pellet boiler**

17

Sokol WA.Ecoefficiency of Buildings Including Reduction of CO2 Emission in Life Cycle of Building Materials, Elements and Technologies. ICOSSE'2015, 26-29 May 2015, Balatonfüred, HU

3.1. Evaluation of direct and indirect CO_{2e} emissions and ecoefficiency

Ecoefficiency enhancement while natural gas boiler is a reference solution

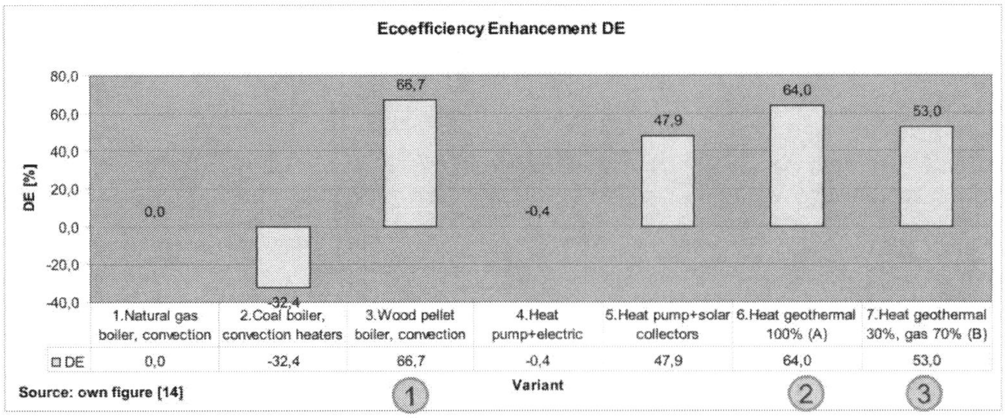

Source: own figure [14]

The best solutions from the point of view of ecoefficiency enhancement is variant **3 – wood pellet boiler**

18

Sokol WA.Ecoefficiency of Buildings Including Reduction of CO2 Emission in Life Cycle of Building Materials, Elements and Technologies. ICOSSE'2015, 26-29 May 2015, Balatonfüred, HU

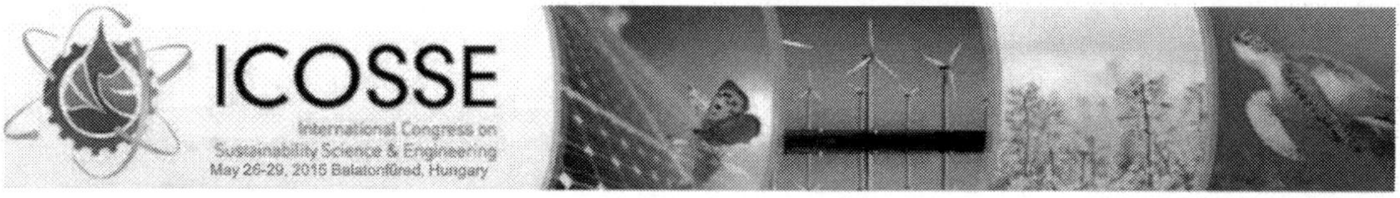

3.1. Evaluation of direct and indirect CO_{2e} emissions and ecoefficiency

Ecoefficiency enhancement while coal boiler is a reference solution

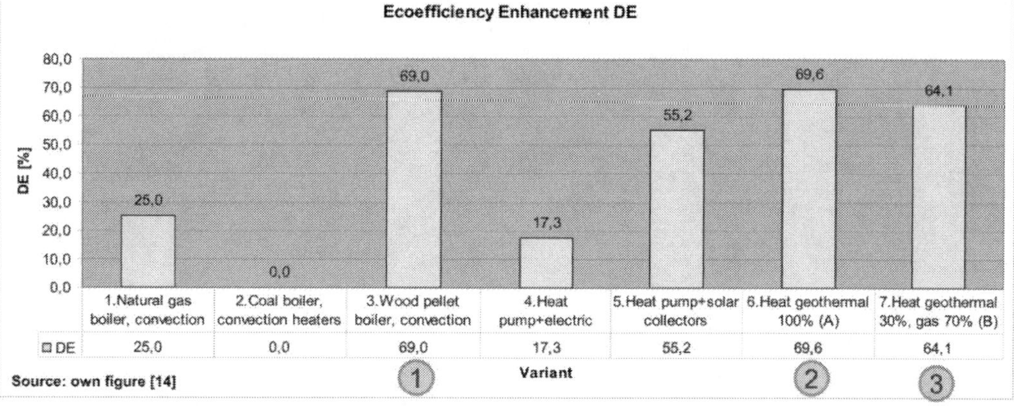

Source: own figure [14]

Wood pellet technology for heating and domestic hot water was primary recommended as ecoefficient solution with minimal demand on primary energy 19

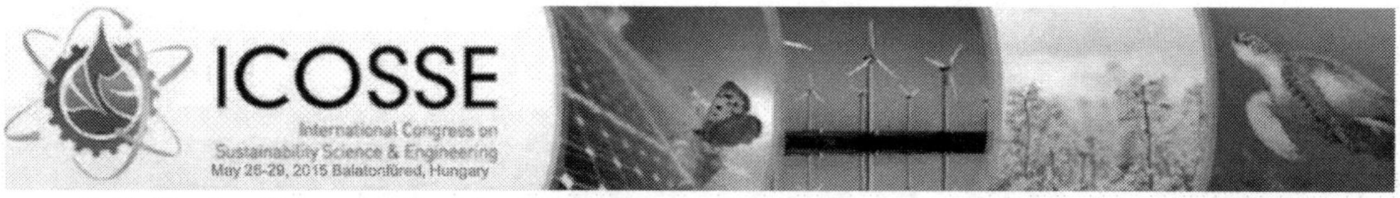

3.2. Evaluation of other indirect CO_{2e} emissions from materials, elements & technologies

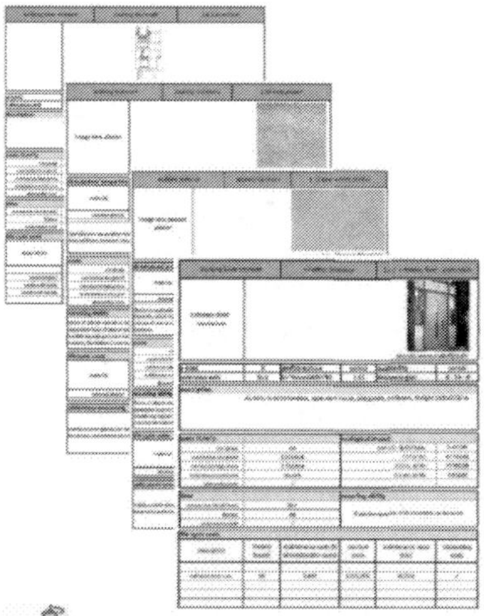

Source: Longlife Prototype Catalogue [6]

Product data and results of calculations

Products	Unit	Value	GWP kg CO2e/unit	t CO2e	Share
Basement floor	m2	201,60	43,49	8,77	1,6%
Graund floor ceiling	m2	945,65	40,33	38,14	6,8%
First floor ceiling	m2	1147,25	40,33	46,27	8,2%
Second floor ceiling	m2	1147,25	40,33	46,27	8,2%
Flat roof	m2	1147,25	54,06	62,01	11,0%
Exterior walls	m2	776,98	161,85	125,75	22,3%
Interior walls	m2	2782,06	40,86	113,67	20,2%
Exterior doors	Piece	5,00	6451,00	32,26	5,7%
Interior doors	Piece	236,00	243,00	57,35	10,2%
Windows	Piece	146,00	128,49	18,76	3,3%
Stairs	m2	88,56	40,70	3,60	0,6%
Central heating radiators	Piece	170,00	3,71	0,63	0,1%
Central heating control	Piece	1,00	7332,00	7,33	1,3%
Ventilation	Piece	1,00	2657,94	2,66	0,5%
		Total product emission	t CO2	563,47	100,0%

Source: own table

20

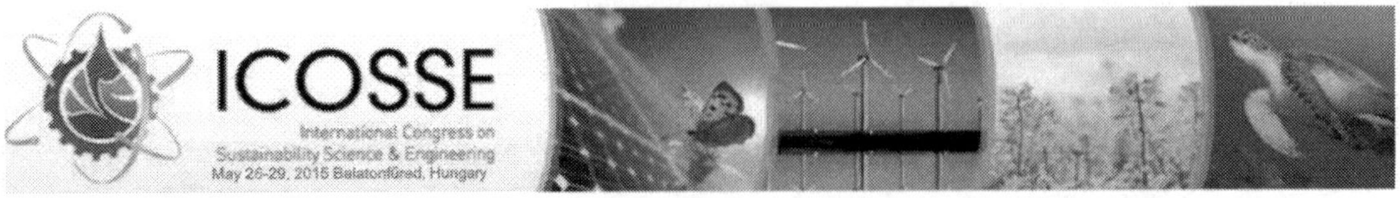

3.2. Evaluation of other_indirect CO_{2e} emissions from materials, elements & technologies

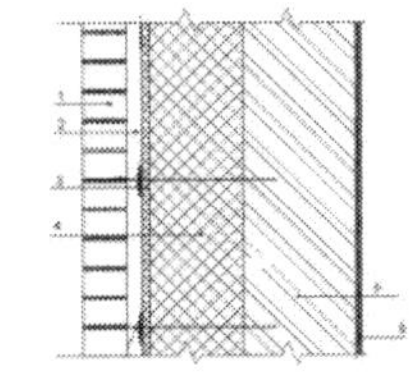

Wall U-value = 0,098 [W/m2 K]	
Layer	Width d [m]
1.Clinker brick	0,120
2.Air gap	0,035
3.Hardened mineral wool externallay	0,020
4.Mineral wool	0,250
5.Cellular concrete blocks	0,300
6.Plaster	0,015
Total	0,740

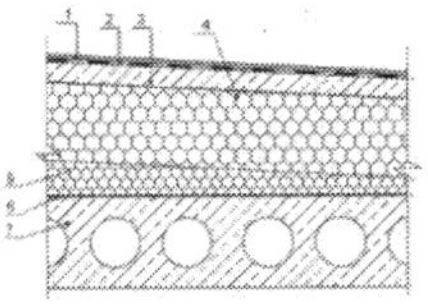

Roof (trapeozidal): U-value = 0,082 [W/m2 K]	
Layer	Width d[m]
1.Asphalt felt paper	-
3.Cement screed	0,050
4.EPS 80	0,370
6.Vapor barrier	-
7.Concrete hollow slabs	0,200
5.EPS 80 (inclined)	0,155
Total	0,620

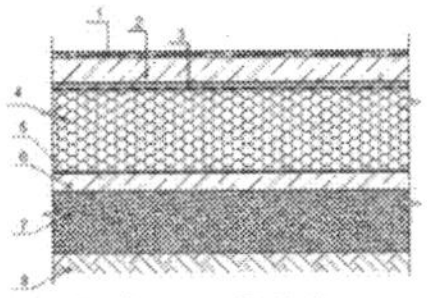

Basement floor: U-value = 0,099 [W/m2 K]	
Layer	Width d[m]
1.Flooring	0,015
2.Cement screed	0,060
3.PE foil	-
4.EPS 100	0,200
5.Waterproofing	-
6.Lean concrete	0,050
Total	0,325

Source: Building design of PST[15]

Exemplary building elements

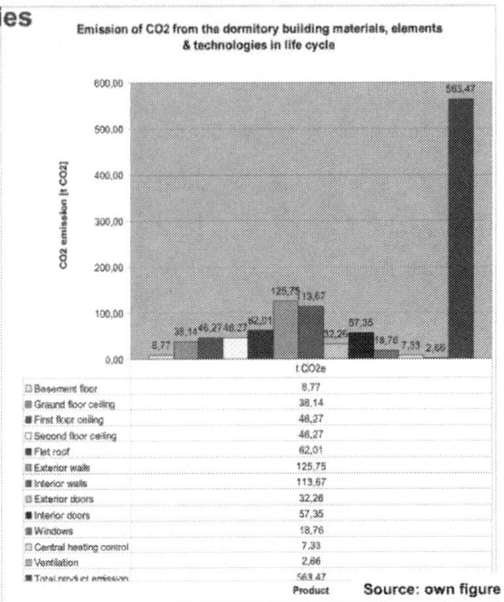

Source: own figure

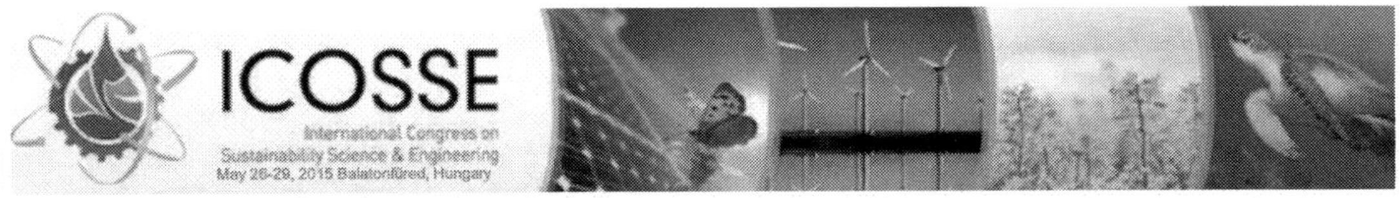

3.2. Evaluation of other_indirect CO_{2e} emissions from materials, elements & technologies

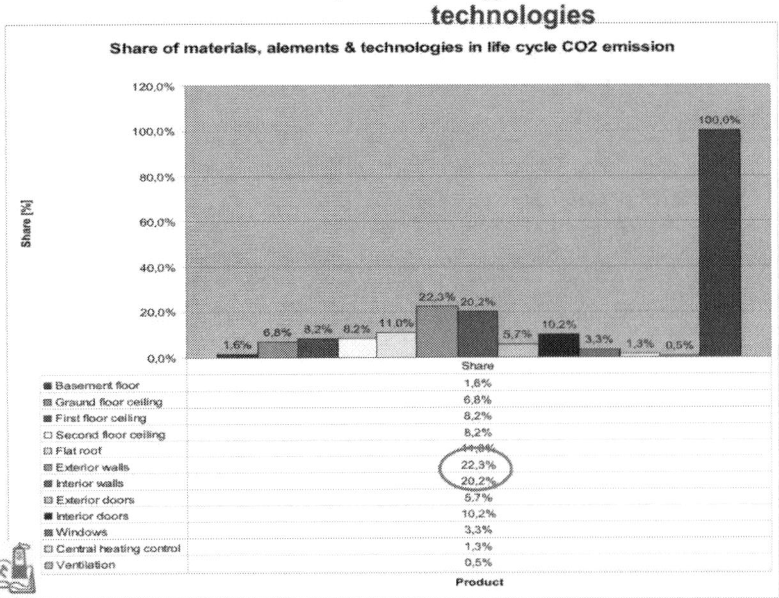

The other indirect emissions CO_{2e} from materials, elements and technologies were evaluated with medium certainty as result of risk analysis according the methodology [9]

22

Source: own figure

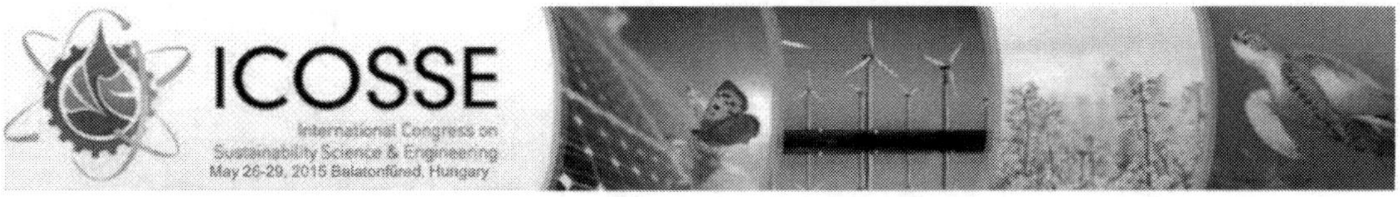

3.3. Evaluation of total CO_{2e} emissions and ecoefficiency in lifecycle

Total CO_{2e} emission/ share of product CO_{2e} emission in total emission from new dormitory building

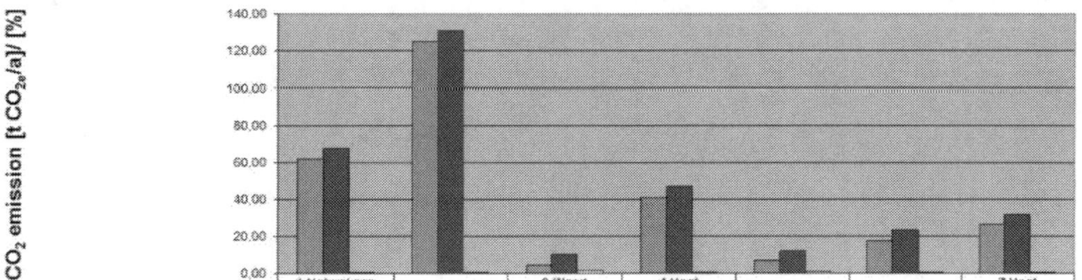

	1.Natural gas boiler, convection heaters	2.Coal boiler, convection heaters	3.Wood pellet boiler, convection heaters	4.Heat pump+electric immersion heater	5.Heat pump+solar collectors	6.Heat geothermal 100% (A)	7.Heat geothermal 30%, gas 70% (B)
Ecploitation CO2 emission	61,89	125,09	4,57	40,82	6,56	17,68	26,12
Total CO2 emision	67,52	130,72	10,20	46,45	12,21	23,32	31,75
Share of CO2 emission form products	9,1%	4,5%	123,3%	13,8%	85,6%	31,9%	21,6%

Technology variant (1) (2) (3)

Source: own figure

✓ **The best solutions** from the point of view of **total CO_{2e} emission** in lifecycle is variant **3 – wood pellet boiler** with a total emission of **10,20** [tCO_{2e}/a] because all energy for heating and domestic hot water is taken from renewable source with low auxiliary electricity than in other cases

23

Sokol WA.Ecoefficiency of Buildings Including Reduction of CO2 Emission in Life Cycle of Building Materials, Elements and Technologies. ICOSSE'2015, 26-29 May 2015, Balatonfüred, HU

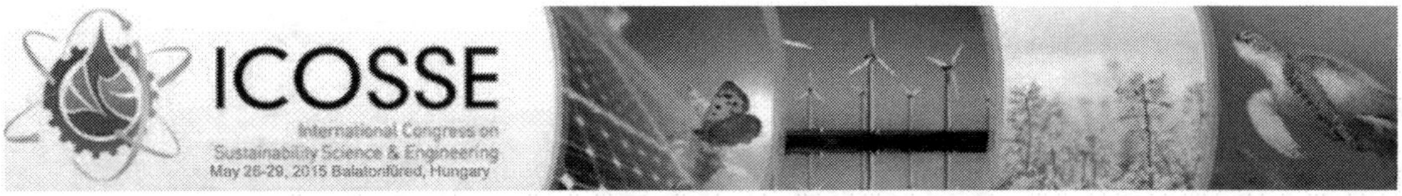

3.3. Evaluation of total CO_{2e} emission and ecoefficiency in lifecycle

Ecoefficiency enhancement while natural gas boiler is a reference solution

Ecoefficiency Enhancement DE

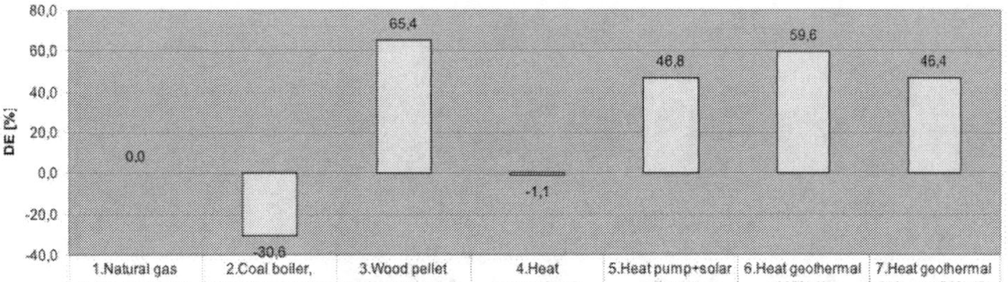

	1.Natural gas boiler, convection	2.Coal boiler, convection heaters	3.Wood pellet boiler, convection	4.Heat pump+electric	5.Heat pump+solar collectors	6.Heat geothermal 100% (A)	7.Heat geothermal 30%, gas 70% (B)
DE	0,0	-30,6	65,4	-1,1	46,8	59,6	46,4

Variant (1) (2) (3)

Source: own figure

✓ **The best solutions** from the point of view of ecoefficiency enhancement is variant **3** as well

24

Sokol WA.Ecoefficiency of Buildings Including Reduction of CO2 Emission in Life Cycle of Building Materials, Elements and Technologies. ICOSSE'2015, 26-29 May 2015, Balatonfüred, HU

3.3. Evaluation of total CO_{2e} emission and ecoefficiency in lifecycle

Ecoefficiency enhancement while coal boiler is a reference solution

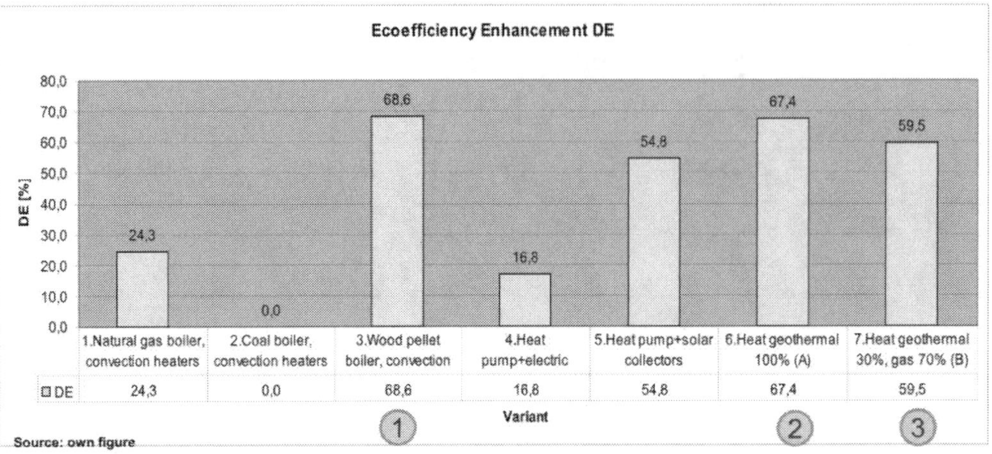

Source: own figure

However the best solutions from the point of view of ecoefficiency enhancement is variant **3** the Lithuanian legal requirements caused that finally variant **6 – geothermal heat** has been approved for implementation 25

Sokol WA.Ecoefficiency of Buildings Including Reduction of CO2 Emission in Life Cycle of Building Materials, Elements and Technologies. ICOSSE'2015, 26-29 May 2015, Balatonfüred, HU

NEW DORMITORY BUILDING

Design

Real construction process (state 22 May 2015) 26

Sokol WA.Ecoefficiency of Buildings Including Reduction of CO2 Emission in Life Cycle of Building Materials, Elements and Technologies. ICOSSE'2015, 26-29 May 2015, Balatonfüred, HU

4. REGIONAL ENVIRONMENTAL MANAGEMENT

New dormitory for TU-Klaipeda is single case but gives the possibility to disseminate the experience on implementing energy and costs efficient buildings

General remarks and specific conclusions can be formulated knowing the results of Longlife Invest and the results of other international projects.

The Europe 2020 Strategy [16] recognizes energy efficiency as the most cost effective way to reduce emissions, improve energy security and competitiveness as well as create employment. Buildings are major consumers of energy in EU [1] but not only in Europe. For example in the United States, this sector is accounted for about 41% of primary energy consumption [17].

Does mean that **buildings provide a potential for energy saving** and create new business opportunities not only for a separate organizations like TU-Klaipeda in Longlife Invest but **for sectorial and regional development.**

It worth to be strengthen by integration of efforts of significant regional actors, including administration, R&D and companies.

27

It is possible by implementing the **key non-technological innovations**: regional environmental management system (e-REMAS), business risk management, site revitalization procedure and evaluation of technology ecoefficiency. The **model of e-REMAS** is announced bellow and described more detailed in [9]

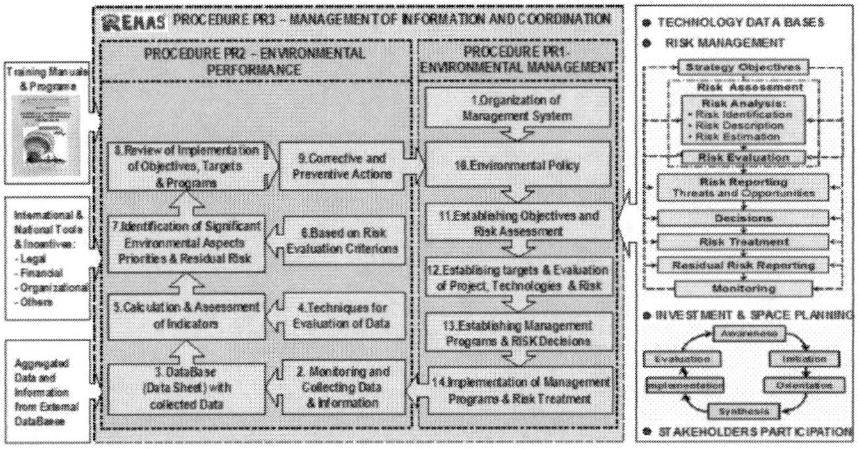

Model of Regional Environmental Management System: e-REMAS [9]

28

5. CONCLUSIONS

1. Evaluation of ecoefficiency of the building technology options is useful tool for performance the efficiency in using of resources, energy efficiency and efficiency in reduction of CO2 emission by means of one aggregated indicator.
2. The procurement procedures have to create preferences for this suppliers of materials, elements and technologies that are able to submit not only products but as well reliable data related consumption of energy and emission of CO_{2e} in lifecycle from cradle to gate.
3. To achieve durability of the results during using of the building the building owner has to assure the monitoring of consumption of all kind of energy and develop annually the report on its consumption and CO_{2e} emission.
4. Energy management procedures as part of environmental management system have to be applied for minimizing consumption of energy while using the building. Remote tools would be useful for the intelligent energy management and monitoring of the building
5. Energy efficiency and reduction of CO2 emission has to be considered as regional objective in regional environmental management system (REMAS).

ICOSSE
International Congress on
Sustainability Science & Engineering
May 26-29, 2015 Balatonfüred, Hungary

REFERENCES

1. http://www.iea.org/aboutus/faqs/energyefficiency/. Accessed on 22.05.2015
2. LONGLIFE- Sustainable, energy efficient and resource saving residential buildings in consideration of unified procedures and new and adapted technologies. http://energy-cluster.info/project_en.html. Accessed on 22.05.2015
3. Longlife Invest - The implementation of the planned Lithuanian Longlife pilot project as a dormitory for Klaipeda University. www.longlifeinvest.eu . Access on 22.05.2015
4. The Greenhouse Gas Protocol. „A Corporate Accounting and Reporting Standard" (2004)
5. ISO/TS 14072:2014(E): Environmental management — Life cycle assessment — Requirements and guidelines for organizational life cycle assessment
6. Rückert K and others. Longlife Prototype Catalogue. Gaideline. Version: 31.1.2015
7. ÖKOBAUDAT . http://www.oekobaudat.de/en.html. Accessed on 22.05.2015
8. Sokół WA. 2014. Practical Aspects of Pre-evaluation of Eco-efficiency of Environmental Technologies. Journal of Business and Economics (USA). October 2014, Volume 5, No.10, 2014, pp.1773-1784
9. Sokół W.A.(2015): Risk-Based Preevaluation Of Environmental Technologies For Ecoefficient Sustainable Revitalization Of Post-Industrial Sites. Human and Ecological Risk Assessment. An International Journal (USA), No 81/1, 2015, pp.1032-1049
10. EN 15804:2012+A1:2013 [IDT]. Sustainability of construction works – Environmental product declarations – Core rules for the product category of construction products
11. REGULATION (EC) No 1221/2009 http://ec.europa.eu/environment/emas/index_en.htm (EMAS)
12. Commission Regulation (EU) No 601/2012 of 21 June 2012 on the monitoring and reporting of greenhouse gas emissions pursuant to Directive 2003/87/EC of the European Parliament and of the Council
13. Energy demand analysis of dormitory for Klaipeda University. Gdansk University of Technology, 2013
14. Sokol WA.2015. Report on minimization of CO2 emission in the lifecycle of a new dormitory building in Klaipeda University. Central Mining Institute. Katowice, May 2015
15. Design of TU-Klaipeda dormitory building. Panevėžio statybos trestas AB, 11.09.2014
16. Europe 2020. http://europa.eu/legislation_summaries/employment_and_social_policy/eu2020/em0028_en.htm.Accessed on 22.05.2015
17. http://buildingsdatabook.eren.doe.gov/ChapterIntro1.aspx accessed on 22.05.2015

ICOSSE
International Congress on
Sustainability Science & Engineering
May 26-29, 2015 Balatonfüred, Hungary

Thanks for your attention!

All questions please address to:

GŁÓWNY INSTYTUT GÓRNICTWA

National Contact Point
for Eco-efficient Technologies and Management Systems

MINISTRY OF THE ENVIRONMENT

Central Mining Institute
40-166 Katowice Plac Gwarkow 1 Poland
www.actclean.gig.eu tel/fax: +48 32 259 22 11 w.sokol@gig.eu

Włodzimierz A.Sokół (D.Sc.Eng.)
Manager of International Projects - Director of the National Contact Point for Ecoefficient Technologies and Management Systems

31

Sokol WA.Ecoefficiency of Buildings Including Reduction of CO2 Emission in Life Cycle of Building Materials, Elements and Technologies. ICOSSE'2015, 26-29 May 2015, Balatonfüred, HU

The Role of Urbanization in Energy Sustainability Challenges

- **Conference:** International Congress on Sustainability Science Engineering ICOSSE
- **Year:** 2015
- **Proceeding:** ICOSSE 2015
- **Group:** Abstract Submissions
- **Session:**

Session 4: Energy-Water-Food Nexus: Energy

- **Time:**

Wednesday, May 27, 2015 - 3:00pm-3:20pm

Authors:

Singh, S., Purdue University
Kennedy, C.

The rapid increase of urbanization is creating stress on the world's resource requirement. Therefore, it is necessary to understand the increasing energy needs of society due to urbanization for energy sustainability. This work develops a tool for estimating energy-related CO_2 emissions due to urbanization based on a regression model. The models are developed considering climatic (heating-degree-days) and urban design (land area per person) as independent variables. The tool is applied on 3,646 urban areas for estimating impact of a) global transitioning to Electric Vehicles, b) urban density change and c) IPCC climate change scenarios on urban emissions. Results show that urban density decline can lead to significant increase in energy emissions (upto 346 % in electricity & 428 % in transportation at 2 % density decline by 2050). The tool can be improved by including more data in the model along with inclusion of other relevant emissions and climatic variables.

What Do Engineering Students Need to Know about Water Ethics?

- **Conference:** International Congress on Sustainability Science Engineering ICOSSE
- **Year:** 2015
- **Proceeding:** ICOSSE 2015
- **Group:** Abstract Submissions
- **Session:**

Session 5: Education and Social Aspects of Sustainability

- **Time:**

Wednesday, May 27, 2015 - 3:20pm-3:40pm

Authors:

Schrader, G. L., University of Arizona
Groenfeldt, D., Water-Culture Institute

"Water ethics" has been a recognized topic of water policy since Postel identified ethics as a "missing piece" in her book *Last Oasis*. A major UNESCO program on "Water and Ethics" (1997-2004) produced 11 publications/books on the topic. In 2014 UNESCO, the Water-Culture Institute, and several other organizations launched a "Water Ethics Initiative" organized around the development of a global water ethics charter process (waterethics.org). Choices about water policies, practices, and physical infrastructure apply sets of values and morals which are usually tacit. Bringing these values into policy awareness to be debated by experts and civil society is emerging as a criteria sustainability strategy with a twofold aim: (1) peer review from a science perspective, and (2) democratic vetting of the options to ensure informed public input and commitment. Learning to "see" engineering alternatives through an ethics lens can strengthen the science-policy interface, while stimulating innovative technological solutions to the challenges of sustainable water resources management.

Plotting a Course for Water Sustainability By Way of Engineering Ethics

Miriam Heller, Ph.D.

May 27, 2015
Balatonfüred, Hungary

Acknowledgements/Caveats

- AIChE support in creating an e-learning course on Engineering Ethics and Sustainable Water.
- AIChE has not yet vetted or approved this work.

Motivations for Course

Education and training in engineering ethics:

- Fosters public trust in the profession, firm, individual
- Required for licensure
- Required for accreditation, e.g., ABET
- Enhances employment opportunities
- Enables engineers to make better decisions

Focusing on sustainable water

- Addresses environmental and public safety, health, welfare and human quality of life issues
- Targets water crises erupting all over the world

Water, Water Everywhere vs. Water Sustainability

Where is Earth's Water?

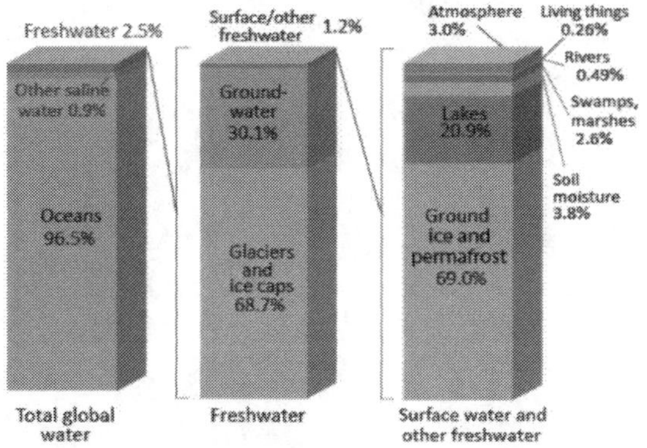

Source: Igor Shiklomanov's chapter "World fresh water resources" in Peter H. Gleick (editor), 1993, Water in Crisis: A Guide to the World's Fresh Water Resources.
NOTE: Numbers are rounded, so percent summations may not add to 100.

How Much Water on Earth?

Water

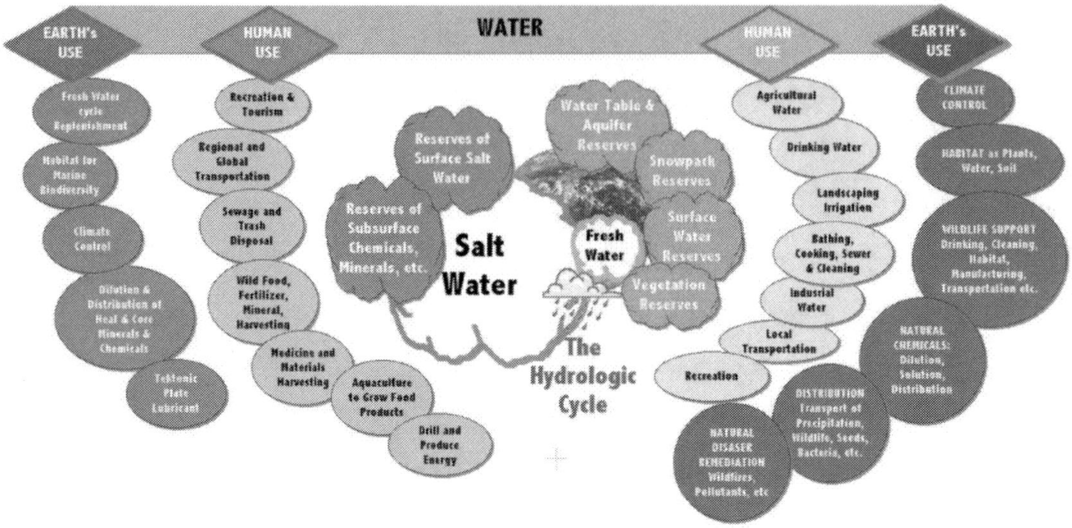

(California Green Solutions, 2015)

Engineers Influence Water Sustainability

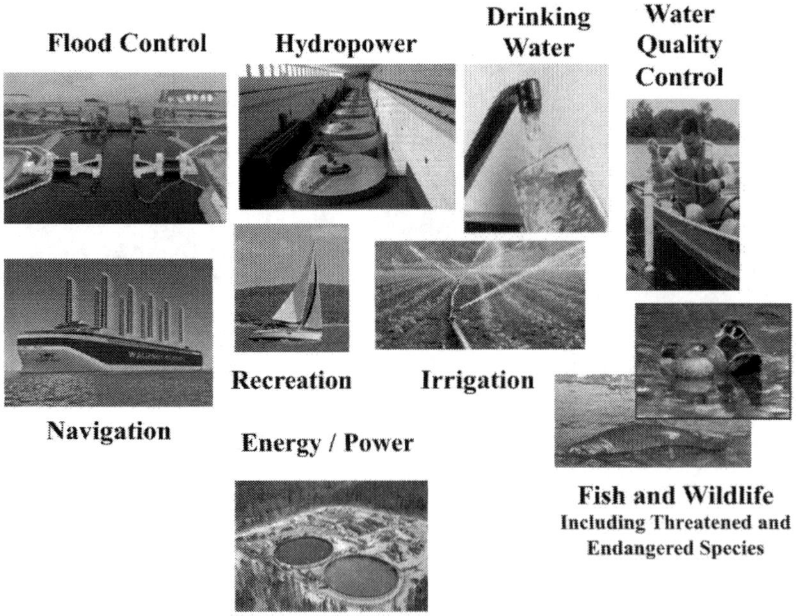

Course Goals

- Raise ethical sensitivity and ability to identify
  - Ethical dilemmas, decisions, outcomes
- Increase ethical knowledge, including
  - Theories of ethics, applied ethics, ethical reasoning
  - Professional codes of ethics
- Understand concepts of the ethics of sustainability, environment, water
- Practice making ethical judgments and decisions
- Enhance ethical discipline and commitment

Traditional Branches of Ethics

Ethics: same as moral philosophy

- Meta-ethics: *Define good/evil, right/wrong, truth/ falsity?*
- Normative ethics: *What should an engineer do? How should an engineer behave?*
- Applied ethics: *How to apply ethics to work/lives?*
- Descriptive ethics: *What morals do people follow?*
- Moral psychology (biological and psychological bases)

Traditions Governing Ethical Actions

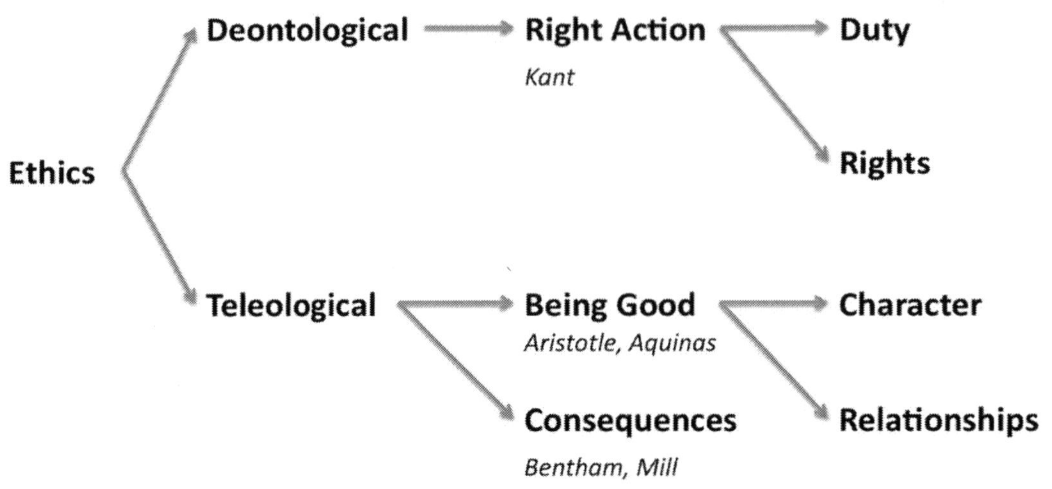

5/27/2015 Heller - Engineering Ethics for Sustainable Water

Traer, R. (2013)

National Society of Professional Engineers®

Code of Ethics for Engineers

Preamble

Engineering is an important and learned profession. As members of this profession, engineers are expected to exhibit the highest standards of honesty and integrity. Engineering has a direct and vital impact on the quality of life for all people. Accordingly, the services provided by engineers require honesty, impartiality, fairness, and equity, and must be dedicated to the protection of the public health, safety, and welfare. Engineers must perform under a standard of professional behavior that requires adherence to the highest principles of ethical conduct.

I. Fundamental Canons

Engineers, in the fulfillment of their professional duties, shall:

1. Hold paramount the safety, health, and welfare of the public.
2. Perform services only in areas of their competence.
3. Issue public statements only in an objective and truthful manner.
4. Act for each employer or client as faithful agents or trustees.
5. Avoid deceptive acts.
6. Conduct themselves honorably, responsibly, ethically, and lawfully so as to enhance the honor, reputation, and usefulness of the profession.

4. Engineers shall act for each employer or client as faithful agents or trustees.
   a. Engineers shall disclose all known or potential conflicts of interest that could influence or appear to influence their judgment or the quality of their services.
   b. Engineers shall not accept compensation, financial or otherwise, from more than one party for services on the same project, or for services pertaining to the same project, unless the circumstances are fully disclosed and agreed to by all interested parties.
   c. Engineers shall not solicit or accept financial or other valuable consideration, directly or indirectly, from outside agents in connection with the work for which they are responsible.
   d. Engineers in public service as members, advisors, or employees of a governmental or quasi-governmental body or department shall not participate in decisions with respect to services solicited or provided by them or their organizations in private or public engineering practice.
   e. Engineers shall not solicit or accept a contract from a governmental body on which a principal or officer of their organization serves as a member.

III. Professional Obligations

1. Engineers shall be guided in all their relations by the highest standards of honesty and integrity.
   a. Engineers shall acknowledge their errors and shall not distort or alter the facts.
   b. Engineers shall advise their clients or employers when they believe a project will not be successful.
   c. Engineers shall not accept outside employment to the detriment of their regular work or interest. Before accepting any outside engineering employment, they will notify their employers.
   d. Engineers shall not attempt to attract an engineer from another employer by false or misleading pretenses.
   e. Engineers shall not promote their own interest at the expense of the dignity and integrity of the profession.

2. Engineers shall at all times strive to serve the public interest.
   a. Engineers are encouraged to participate in civic affairs; career guidance for youths; and work for the advancement of the safety, health, and well-being of their community.
   b. Engineers shall not complete, sign, or seal plans and/or specifications that are not in conformity with applicable engineering standards. If the client or employer insists on such unprofessional conduct, they shall notify the proper authorities and withdraw from further service on the project.
   c. Engineers are encouraged to extend public knowledge and appreciation of engineering and its achievements.
   d. Engineers are encouraged to adhere to the principles of sustainable development[1] in order to protect the environment for future generations.

5/27/2015 Heller - Engineering Ethics for Sustainable Water

MDGs/SDGs: Human Rights to Water

The Open Working Group on Sustainable Development Goals and UN-Water proposed goal 6 "Ensure availability and sustainable management of water and sanitation for all":

6.1 by 2030, achieve universal and equitable access to safe and affordable drinking water for all.

6.2 by 2030, achieve access to adequate and equitable sanitation and hygiene for all, and end open defecation, paying special attention to the needs of women and girls and those in vulnerable situations.

6.3 by 2030, improve water quality by reducing pollution, eliminating dumping and minimizing release of hazardous chemicals and materials, halving the proportion of untreated waste water, and increasing recycling and safe reuse by x% globally.

6.4 by 2030, substantially increase water-use efficiency across all sectors and ensure sustainable withdrawals and supply of freshwater to address water scarcity, and substantially reduce the number of people suffering from water scarcity.

6.5 by 2030 implement integrated water resources management at all levels, including through transboundary cooperation as appropriate.

6.6 by 2020 protect and restore water-related ecosystems, including mountains, forests, wetlands, rivers, aquifers and lakes.

6.A by 2030, expand international cooperation and capacity-building support to developing countries in water and sanitation related activities and programmes, including water harvesting, desalination, water efficiency, wastewater treatment, recycling and reuse technologies.

6.B support and strengthen the participation of local communities for improving water and sanitation management.

Ethical Reasoning

- Reasoning about feelings
- Empathy and reason
- Critical reasoning
  - Excludes reasons solely based on belief
  - Demands a valid argument for drawing conclusions
  - Excludes rationalization
  - Reasoning by analogy
  - Deductive reasoning
  - Individual and cultural ethical relativism

Relationships in the Engineer's Ethical Community

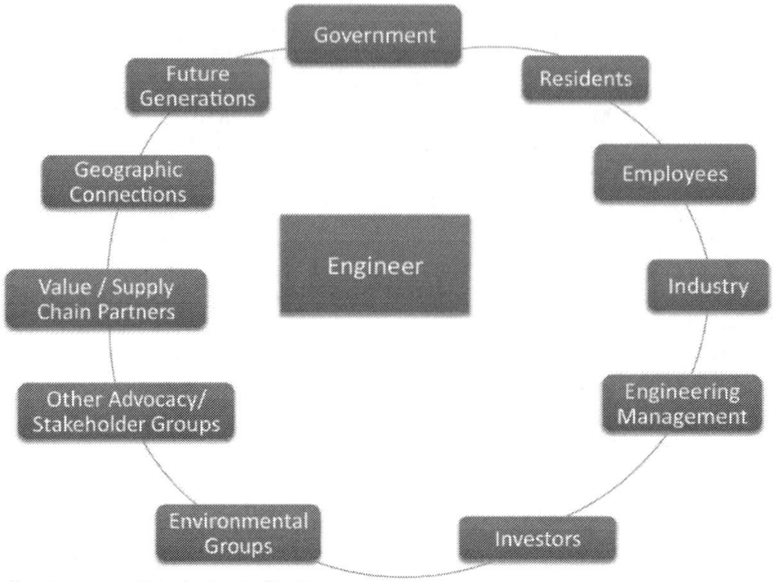

5/27/2015 Heller - Engineering Ethics for Sustainable Water

Sustainability

Sustainable development was originally defined as "development that meets the needs of the present without compromising the ability of future generations to meet their own needs"

- *The Brundtland Commission, 1987*

=> Anthropocentric Ethics

5/27/2015 Heller - Engineering Ethics for Sustainable Water

Biocentric Ethics

"If all the insects were to disappear from the earth, within 50 years all life on earth would end. If all human beings disappeared from the earth, within 50 years all forms of life would flourish."

– Jonas Salk, American medical researcher, virologist, discovered/developed first polio vaccine

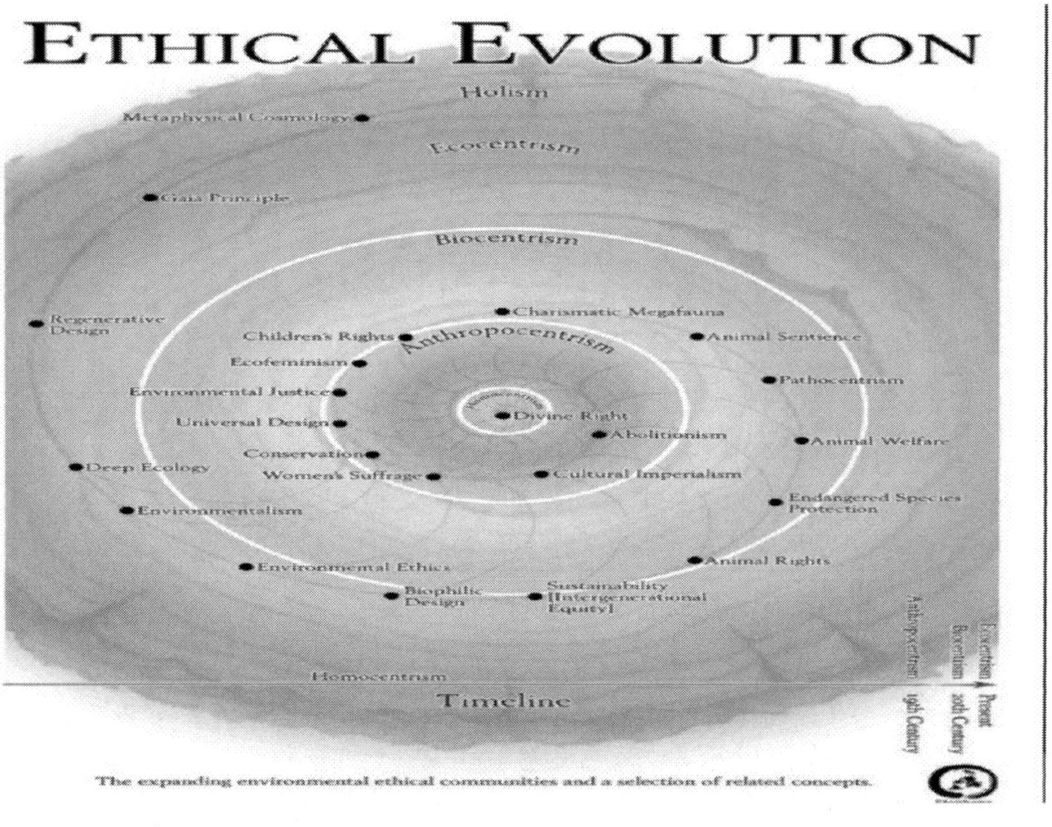

The expanding environmental ethical communities and a selection of related concepts.

Good Character, Green Virtues[1], Engineering

- Humility
  - Recognition of the elegance of natural systems and phenomena
  - Recognition of how rare it is to capture such elegance in engineered systems
  - Connected to nature needed to understand the above
- Temperance
  - Self-restraint and moderation
  - Lower consumption is better
  - Is an ancient virtue that is typically associated with weakness of will.
- Mindfulness
  - Engineering that acknowledges the scope of possible consequences of decisions even those remote in time and space.
  - Entails quantifying the consequences most likely to have effect, estimating those with potential but expected lesser effects, and then estimating the remaining uncertainty.

➡ Does this suggest the Precaution Principle?

5/27/2015 Heller - Engineering Ethics for Sustainable Water

[1]Jamieson, D.

Russian River Grapes versus Salmon: A Tale from Sonoma County, CA

- Spring and summer of 2012 – severe drought
- Late frost caused excessive water withdrawals for grape frost protection
- Reservoir and other management affords compliance with a biological opinion for minimum water flows for endangered fish
- Dead juvenile steelhead salmon found within days of the recorded withdrawal
- Grape growers, with riparian water rights, were banned from withdrawals for frost protection

5/27/2015 Heller - Engineering Ethics for Sustainable Water

Dissecting an Ethical Decision: Practice with Case Studies

- **Personal conflicts of interest** that may that affects how you view the case?
- **Factual scientific and engineering information:** What are the key facts in the case? Disputed facts are? Most plausible account? Biased presentation of information or data?
- **Ethical issues:** What is the primary ethical issue in the case? Are there other ethical issues?
- **Complicating factors**: Is there anything particularly unusual or complicated about the case? In terms of science? Or of law?
- **Relationships:** Conflicts of interest among key stakeholders?
- **Power dynamics**: will any stakeholders control general opinion?

Adapted from Santa Clara University Center, Environmental Ethics

5/27/2015 Heller - Engineering Ethics for Sustainable Water

Dissecting an Ethical Decision: Practice with Case Studies

- **Assessment of Consequences:** What *analytical tools* to use assess alternatives? What additional information and data is needed for these analyses? Will uncertainty analysis be conducted?
- **Ethical vision**: What would be a just resolution to these issues?
- **Ethical reasoning**: Which mode appears most appropriate? (commands, consequences, character)
- **Decision**: Which alternative is morally preferable?
- **Justification**: how do you justify it in terms of the ethical principles and reasoning above?
- **Communication**: How will you communicate this information to diverse audiences so it is morally reasonable?

5/27/2015 Heller - Engineering Ethics for Sustainable Water

Conclusions and Hypotheses:
Engineering ethics for sustainable water

- Promises to help engineers make better and defensible decisions through increased awareness of
  - Problem complexity
  - Relationships of decision-makers, stakeholders
  - Temporal and spatial interconnectedness
- Offers a pathway to introduce ethics education and training on sustainable water practice into curriculum
- May lead to holistic practices needed to address engineering problems today and in the future

5/27/2015 Heller - Engineering Ethics for Sustainable Water

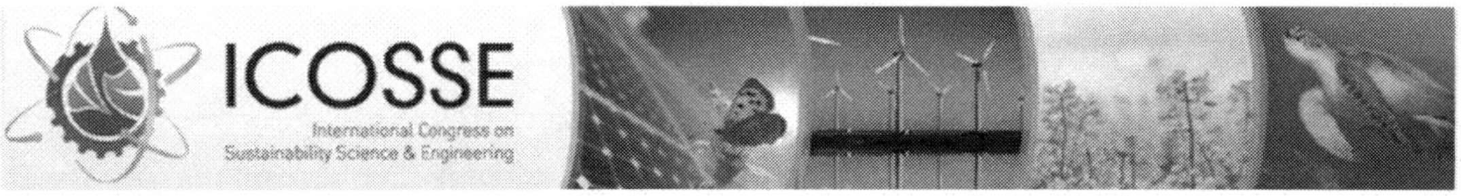

Thank you for your attention!

Miriam Heller, Ph.D.
MHITech Systems
heller@mhitech.com

References and Readings

- Californ Green Solutions. Accessed May 22, 2015 from http://www.californiagreensolutions.com/cgi-bin/gt/tpl.h,content=117.
- Our Common Future, Report of the World Commission on Environment and Development, World Commission on Environment and Development, 1987. Published as Annex to General Assembly document A/42/427, Development and International Co-operation: Environment August 2, 1987. Retrieved from http://www.un-documents.net/our-common-future.pdf on March 13, 2015.

References and Readings

- Pailla, S. (2013). Ethical Considerations of Adapting a Centralized-Water Supply Infrastructure in Low-Income Communities: Case Study of Nalgonda, India, *Viriginia Policy Review*, vol. 6(2):61-73. Retrieved from http://virginiapolicyreview.com/files/2013/08/Pailla_FINAL.pdf on 3/2/2015.
- Traer, R. (2013). doing environmental ethics (2nd ed.), Westview Press:Boulder, CO.
- Taback, H. and R. Ramanan. (2014). Environmental Ethics and Sustainability: A Casebook for Environmental Professionals, Taylor & Francis Group, LLC: Boca Raton.
- United Nations General Assembly (1987) Report of the World Commission on Environment and Development: Our Common Future. Transmitted to the General Assembly as an Annex to document A/42/427 - Development and International Co-operation: Environment. Retrieved from http://www.un-documents.net/wced-ocf.htm on 2015-03-11.
- Whitbeck, C. (2011). Ethics in Engineering Practice and Research (2nd Ed.), Cambridge University Press: NY,NY

5/27/2015 Heller - Engineering Ethics for Sustainable Water

Injecting Sustainability Across the Curriculum

John R. Schlup, Larry E. Erickson, Keith L. Hohn, and Mary Rezac

Department of Chemical Engineering
Kansas State University
Manhattan, Kansas USA

Kansas State University | College of Engineering

Biobased Products & Biofuels Graduate Certificate

- Initiated in 2001 through funding from a Department of Energy educational grant
- Result: a graduate certificate in biobased products and biofuels

Kansas State University | College of Engineering

Structure

- ➢What is a graduate certificate?
- ➢Why a graduate certificate?
- ➢Target audience?

KANSAS STATE UNIVERSITY | College of Engineering

Multi-Institutional Distance Based BioEnergy & Sustainability Education … BST Program

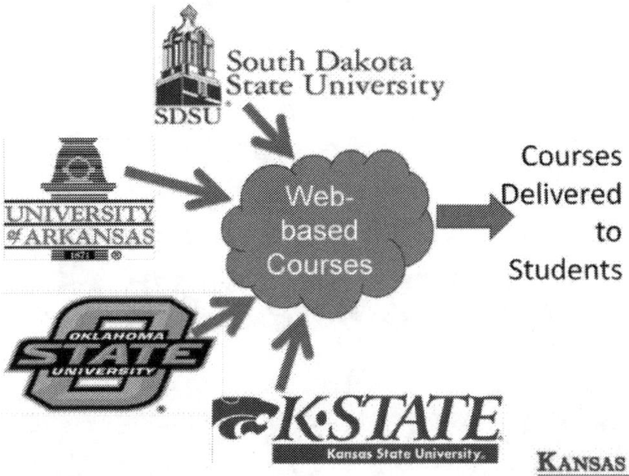

KANSAS STATE UNIVERSITY | College of Engineering

History

- August 2009
  - USDA Project Start
  - Online workspace and listserves set up
- Many options possible for web-based delivery.
  - First round of classes being taught Fall 2011
  - All new classes (8) were delivered by Spring 2013
- Leveraging AG*IDEA Structure invaluable

KANSAS STATE UNIVERSITY | College of Engineering

Structure

- 15 hrs of Graduate Courses Required
- All students take 3 core courses in
  - Feedstocks
  - Conversion
  - Sustainability Issues
- 6 hours of elective courses in any area are also required.

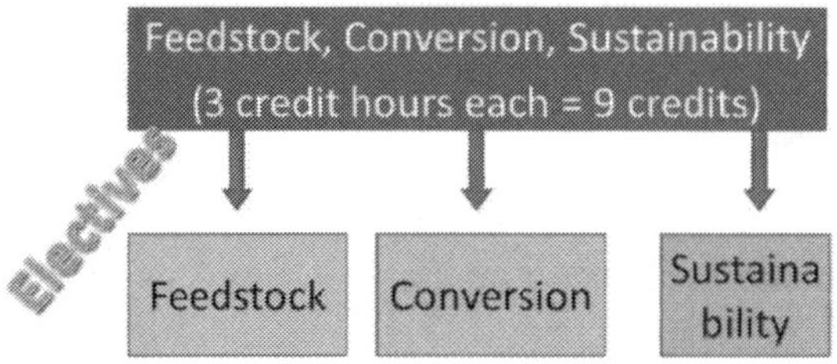

KANSAS STATE UNIVERSITY | College of Engineering

Sustainability Courses

- Bioenergy & Sustainability (Overview)
- Bioenergy & Resource Economics
- Community & Natural Resources
- Life Cycle Analysis
- Environmental & Ecological Risk Assessment
- Sustainability Seminar

Kansas State University | College of Engineering

Sustainability Courses Specific to KSU

- CHE 570/571 ... ChE Systems Design
- CHE 670 ... Sustainability Seminar
- CHE 805 ... Quantifying the Sustainability of Processes and Materials

Kansas State University | College of Engineering

CHE 570/571

- Traditional capstone design course
- Fall semester ...
  - Text book work (green chemistry & engineering, recycle, etc.)
  - Case study ... chance to use concepts
- Spring semester ...
  - Two design problems
  - Put things to work

KANSAS STATE UNIVERSITY | College of Engineering

CHE 670

Sustainability Seminar

Topics in environmental sustainability, green engineering, life cycle analysis, sustainable development, and sustainability science.

KANSAS STATE UNIVERSITY | College of Engineering

A Course in Quantifying Sustainability

Quantifying the Sustainability of Processes and Materials
(CHE 580 or CHE 805)

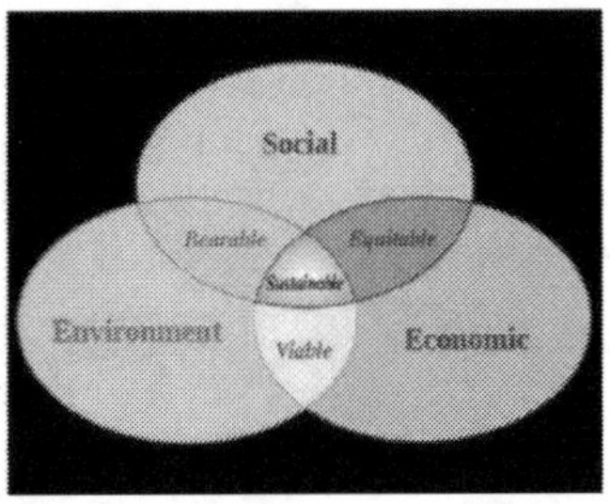

An August 2011 Intersession Course
August 1 - 19, 2011
(Both On-campus and Distance Offerings)

KANSAS STATE UNIVERSITY | College of Engineering

Topics

A simple overview of green engineering and life cycle analysis to open the course.

- Green Chemistry
- Green Engineering ... Simple Metrics
- Life Cycle Assessment, including ISO 14000
- The Argonne National Laboratory software (GREET)
- Eco-Audit Tool
- Second Law Analysis
- Sustainability Potential
- Total Cost Analysis
- Eco-efficiency & SEEBALANCE (BASF tools)

KANSAS STATE UNIVERSITY | College of Engineering

Graduate Education Programs

Kansas State University NSF I-Star IGERT Program

Motivation: Biofuels and biobased products can substantially improve environmental quality, rural economies, and national security.

Goals:

- To prepare students to become research and educational leaders who can meet the evolving needs of the bio-based energy industry through a deep appreciation of the interrelated impact of economic, technological, agricultural, and societal issues.
- To provide an integrated interdisciplinary graduate program for achieving transformative advances in the development of next-generation biorefineries.

KANSAS STATE UNIVERSITY | College of Engineering

Currently 15 students across several departments/colleges
Faculty represent technical (6), socioeconomic (6), and agricultural (4) issues

KANSAS STATE UNIVERSITY | College of Engineering

IGERT Components

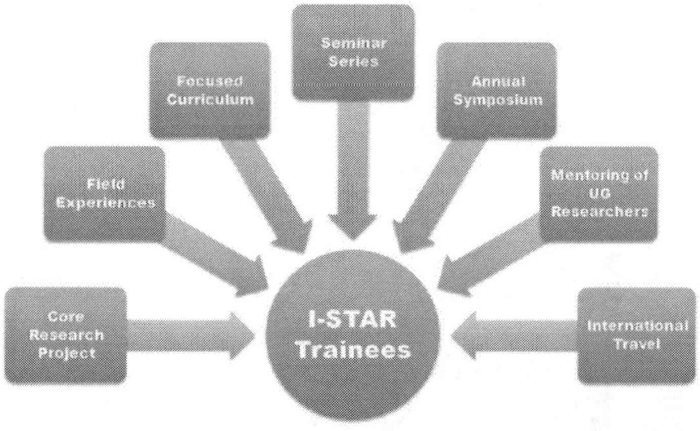

KANSAS STATE UNIVERSITY | College of Engineering

Undergraduate Research Initiatives

REU Summer Academy in Sustainable Bioenergy

- Going back to 2012
- Last year 10 undergraduate students representing 7 disciplines from 3 colleges (sociology through engineering)
- Mentored by IGERT graduate students

KANSAS STATE UNIVERSITY | College of Engineering

An Alternate REU Experience

"Earth, Wind, and Fire REU: Sustainable Energy through the 21st Century"

- Funded through the National Science Foundation

- From 2009 through 2014 … 77 undergraduate students from across the United States
- 7 disciplines in three colleges

KANSAS STATE UNIVERSITY | College of Engineering

Relevent Centers and Offices

- Center for Sustainable Energy (http://cse.ksu.edu/)
- **Consortium for Environmental Stewardship and Sustainability (CESAS) … a part of the Center for Hazardous Substance Research (http://www.engg.ksu.edu/chsr/sustainability/cesas) and (http://www.engg.ksu.edu/chsr/files/chsr/sustainability/cesas/CESAS_Brochure.pdf)**
- Urban Water Institute
- Kansas State University Office of Sustainability (http://www.k-state.edu/sustainability/)
- Kansas State University Engineering Extension (http://www.engext.ksu.edu/ecre/sustainability)

KANSAS STATE UNIVERSITY | College of Engineering

Acknowledgements

U. S. Department of Energy

U. S. Department of Transportation / Sun Grant (South Central Region)

USDA Higher Education Challenge GRANT #2009-38411-19766 (a NIFA grant)

National Science Foundation (various research grants)

CESAS

Center for Sustainable Energy (http://sustainable-energy.ksu.edu/)

Ag*IDEA

KANSAS STATE UNIVERSITY | College of Engineering

AIChE and IfS Efforts in Sustainability

- **Conference:** International Congress on Sustainability Science Engineering ICOSSE
- **Year:** 2015
- **Proceeding:** ICOSSE 2015
- **Group:** Abstract Submissions
- **Session:**

Session 5: Education and Social Aspects of Sustainability

- **Time:**

Wednesday, May 27, 2015 - 4:20pm-4:40pm

The Institute for Sustainability (IfS) is a Technological Entity of the American Institute of Chemical Engineers. IfS's mission is to serve the needs of and influence the efforts of professionals, academes, industries, and governmental bodies that contribute to the advancement of sustainability and sustainable development with the objective of promoting the societal, economic, and environmental benefits of sustainable and green engineering. IfS provides multidisciplinary scientific and technical rigor to sustainability discussions, encourages the incorporation of sustainability concepts in engineering education, advances sustainability research and development, and facilitates the development of measurement tools and frameworks to guide the design of more sustainable products and processes.

The P-graph Methodology as Tool for Studying Sustainability in the Energy-Water-Food Nexus

H. Cabezas, I. Heckl, B. Bertok & F. Friedler

University of Pannonia
Faculty of Information Technology
Department of Computer Science & Systems Technology
Veszprem, Hungary

Collaborators

COLLABORATOR	INSTITUTION
• D. Campbell	US EPA
• A. Garmestani	US EPA
• M. Hopton	US EPA
• L. Vance	US EPA
• W. Ingwersen	US EPA
• T. Eason	US EPA
• B. Bertok	University of Pannonia
• I. Heckl	University of Pannonia
• F. Friedler	University of Pannonia

Literature Sources

- Vance, L., Heckl, I., Bertok, B., Cabezas H. and F. Friedler, "A Designing Sustainable Energy Supply Chains by the P-graph Method for Minimal Cost, Environmental Burden, Energy Resources Input,"., J. Clean. Prod., in press. http://dx.doi.org/10.1016/j.jclepro.2015.02.011
- Heckl, I., Halasz, L., Szlama, A., Cabezas, H. and F. Friedler, Modeling Multi-Period Operations using the P-Graph Methodology," Comp. Aided Proc. Engr., 979-984 (2014).
- Vance, L. Cabezas, H., Heckl, I. Bertok, B. and F. Friedler, "Designing Energy Supply Chains with the P-Graph Framework under Cost Constraints and Sustainability Considerations," Comp. Aided Proc. Engr., 33, 1009-1014 (2014).
- Vance, L. Cabezas, H., Heckl, I. Bertok, B. and F. Friedler, "Synthesis of Sustainable Supply Chain by the P-Graph Framework," Ind. & Eng. Chem. Res., 52, 266-274 (2013).

Outline

- The P-Graph Framework
- Sustainability
- Sustainable Supply Chain Design
- Summary

The P-Graph Framework

P-Graph Axioms:

combinatorially feasible process networks

For given process synthesis problem, a P-graph satisfying the following five axioms is a combinatorially feasible structure:

(S1) Every final product is represented in the structure

(S2) A material represented in the structure is a raw material if and only if it is not an output of any operating unit represented in the structure

(S3) Every operating unit represented in the structure is defined in the synthesis problem

(S4) Any operating unit represented in the structure has at least one path leading to a product

(S5) If a material belongs to the structure, it must be an input to or output from at least one operating unit represented in the structure

P-Graph Process Synthesis (Friedler and Fan, 1992-)

Synthesis

Building blocks:

- –materials:
  - raw materials:
  - products:
  - intermediates:
- –operating units:

Process structure:

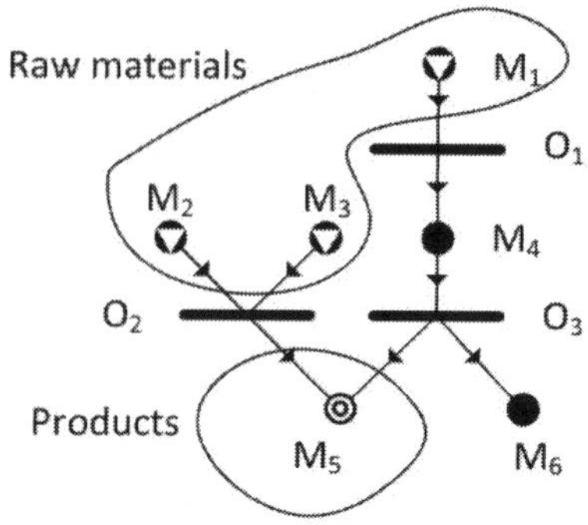

P-graph methodology: Structural Representation

Clear logical interpretation

- –Each precondition (E AND F) has to be satisfied to operate an activity (O_2)
- –Any of the activities (O_2 OR O_3) having the same effect can potentially be sufficient to initiate the consecutive step (O_1)

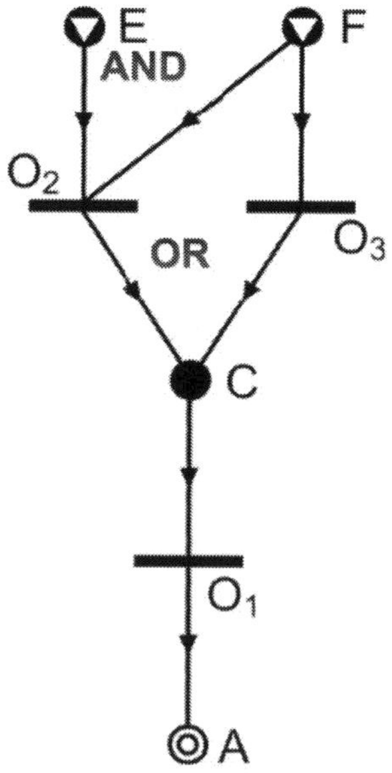

P-Graph Methodology

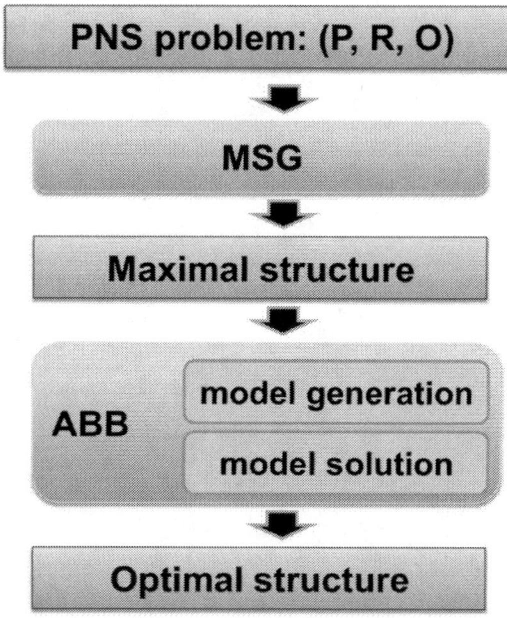

PNS Draw

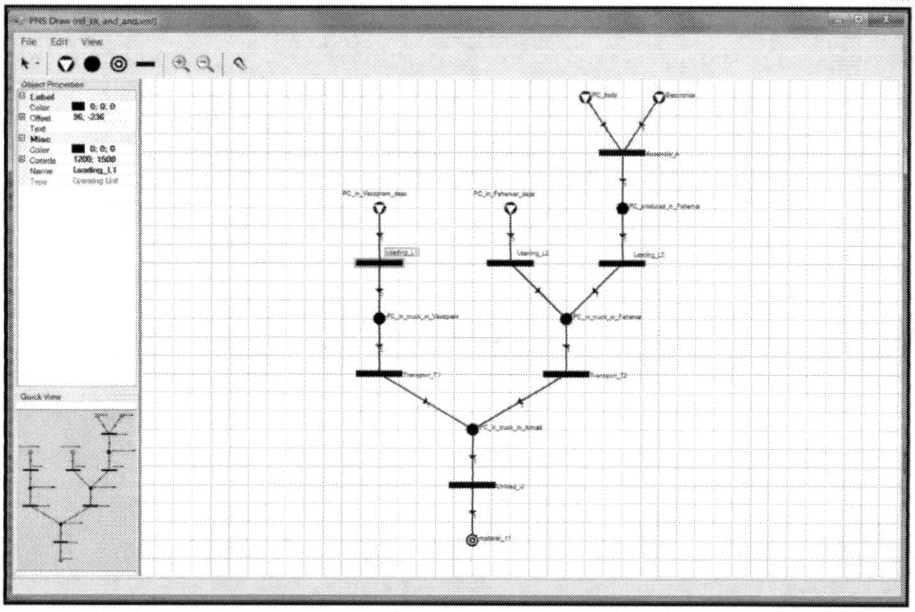

PNS Studio

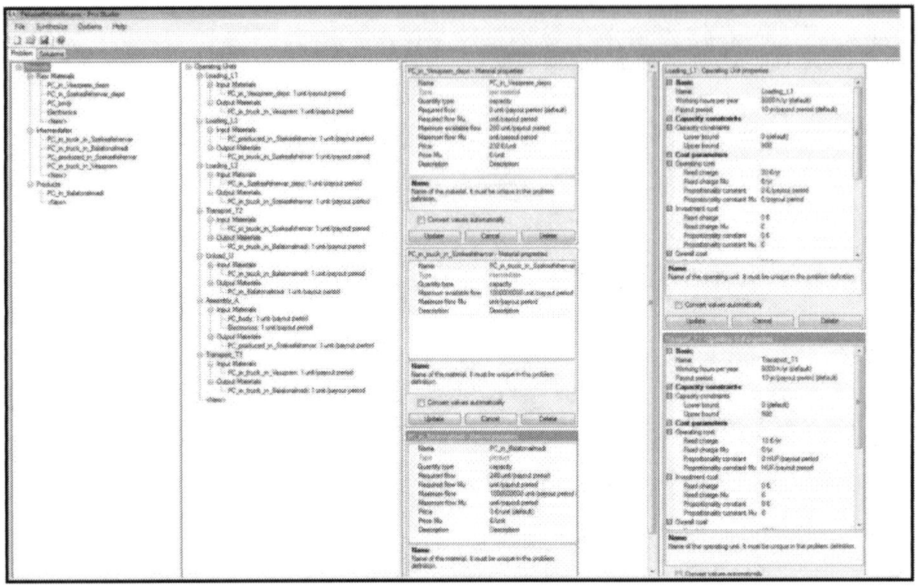

Reducing the Search Space

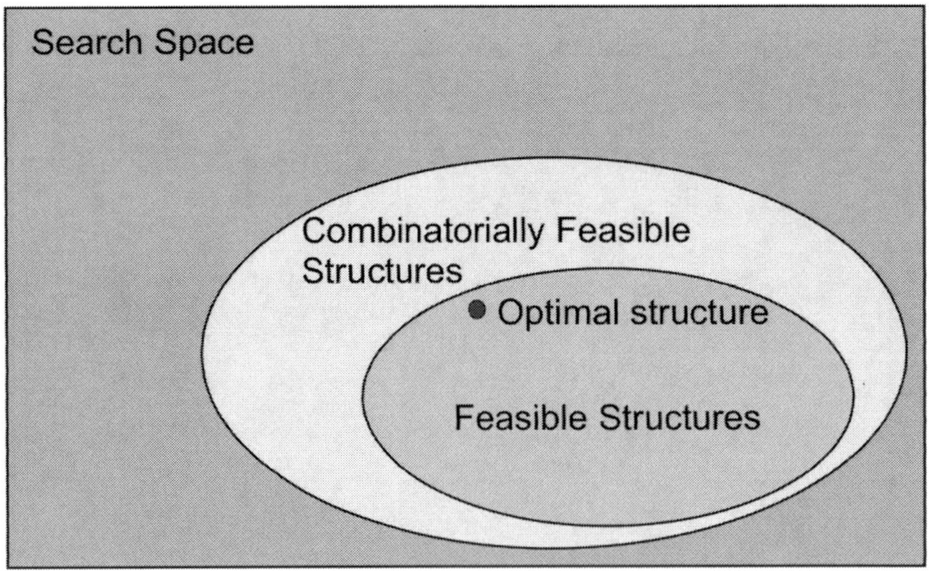

Sustainability

Ecological Footprint Basics

Land is categorized into:

- –arable land
- –forest land
- –pasture land
- –sea
- –energy land
- –built land

blog.lib.umn.edu/tupp0008/environment/2008/02/

Ecological Footprint Basics

- *Ecological Footprint* (demand) = land area required to meet level of consumption and waste generation by the human population
- *Biocapacity* (supply) = land area available to support the human population

Assumptions:

- **Can track of resources and waste generated**
- **Resource and waste flows can be converted to land area**

> *Sustainability Criteria: Ecological Footprint does not increase and Biocapacity does not decrease*
> $\Delta EF \leq 0$ *and* $\Delta B \geq 0$

Ecological Footprint (EF) and Biocapacity (BC) Calculation

For each land type i: $$EF_i = \frac{T}{Y_i} \times YF_i \times EQF$$

$$BC_i = S_i \times YF \times EQF$$

- T = Annual quantity of produced
- Y_i = local yield
- YF_i = yield factor (Y_i / Y_w)
- Y_w = world yield
- EQF_i = (Maximum Productivity i)/(Avg Productivity All *i*)

Total Footprint (EF) and Biocapacity (BC):

$$EF = \sum_i EF_i \qquad BC = \sum_i BC_i$$

Courtesy of Dr. M. Hopton, US EPA

Emergy: Basics

Emergy: the sum of all different kinds of energy previously used (directly and indirectly) to make a product

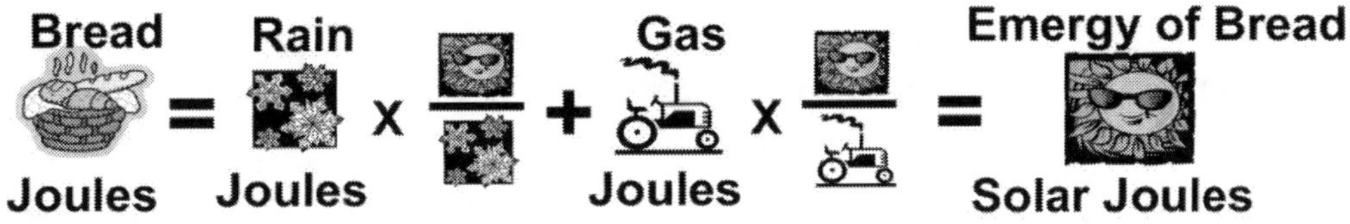

Emergy: Energy Resources

Emergy: the energy resources (in solar joules) invested by the environment in an operation or in a product

<u>Emergy</u>: the sum of all of the different kinds of energy previously used up, directly and indirectly, to make an item

> *Sustainability Criteria for Supply Chains:*
> *(Total Emergy of Inputs) → Minimum*
> *(Renewable Emergy Use) / (Total Emergy Use) → 1*

Emergy Calculation

For a material *x* produced by a process *y*:

$$Em(seJ) = b_{x,y}\left[\frac{seJ}{kg\ of\ x\ produced\ by\ process\ y}\right] \cdot [A\ kg\ of\ x\ produced\ by\ process\ y]$$

For electricity produced by process *y*:

$$Em(seJ) = b_{e,y}\left[\frac{seJ}{joule\ of\ electricity\ produced\ by\ y}\right] \cdot [A\ joules\ of\ electricity\ produced\ by\ y]$$

Sustainable Supply Chain Design

Assumptions

- The ecological footprint of the supply chain is dominated by the feedstocks and inputs
- The emergy footprint of the supply chain is also dominated by the feedstocks and inputs
- Methodology must be simple enough for wide application
- Supply chain is designed to produce 7.2 TJ/year of electricity and 18 TJ/year of heat

Raw Material & Energy Inputs

Solution structures	Raw materials					
	electricity_grid [TJ/yr]	natural_gas [m^3/yr]	area_corn [ha/yr]	area_corn_silage [ha/yr]	area_grass_silage [ha/yr]	area_wood [ha/yr]
Structure1	7.37					500.00
Structure2				117.69		367.73
Structure3					128.54	367.73
Structure4				120.03		393.66
Structure5	7.57					539.13
Structure6					131.10	393.66
Structure7		399,272.00		116.30		
Structure8			72.96	126.77		
Structure9				124.78		380.69
Structure10		399,272.00			127.02	
Structure11			72.96		138.46	
Structure12					136.29	380.69
Structure13	7.25	540,588.00				
Structure14	7.99					529.10
Structure15	8.17		102.04			
Structure16				214.45		
Structure17				125.12	98.01	
Structure18				90.05	135.88	
Structure19					234.67	
Structure20	7.95			125.00		
Structure21	8.02				136.36	

Maximal Supply Chain Structure

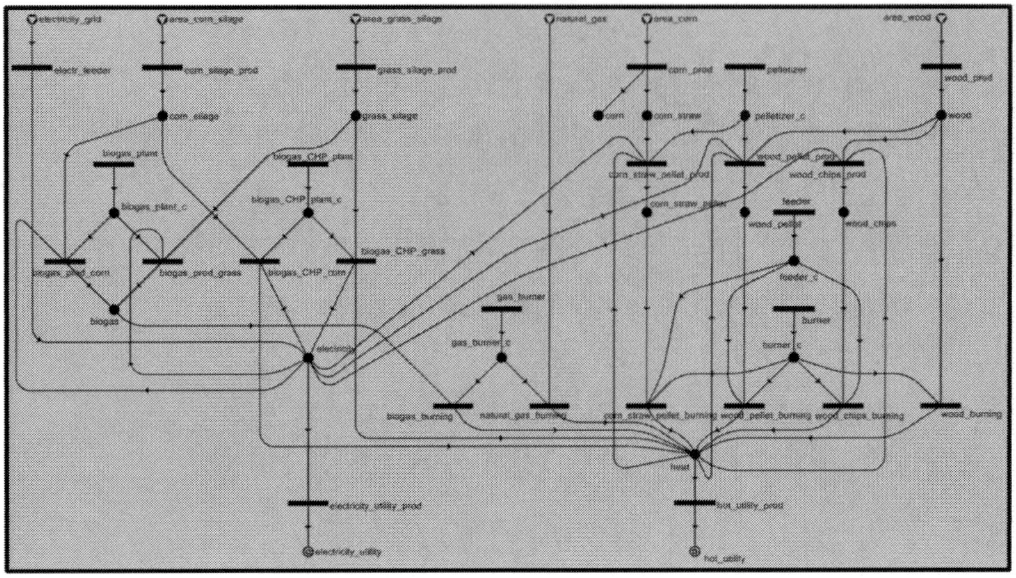

Supply Chain Structure #13: Grid Electricity & Natural Gas

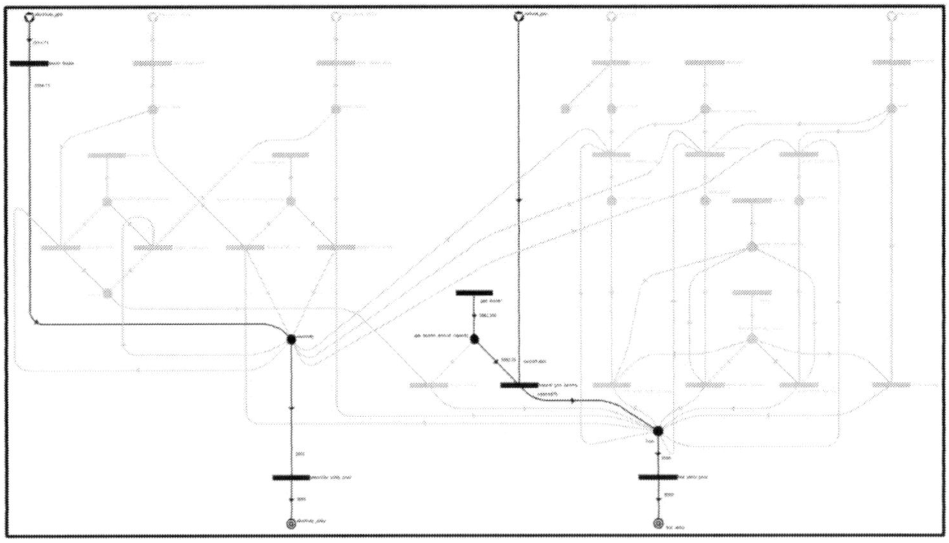

Supply Chain Structure #8: Corn Straw & Corn Silage

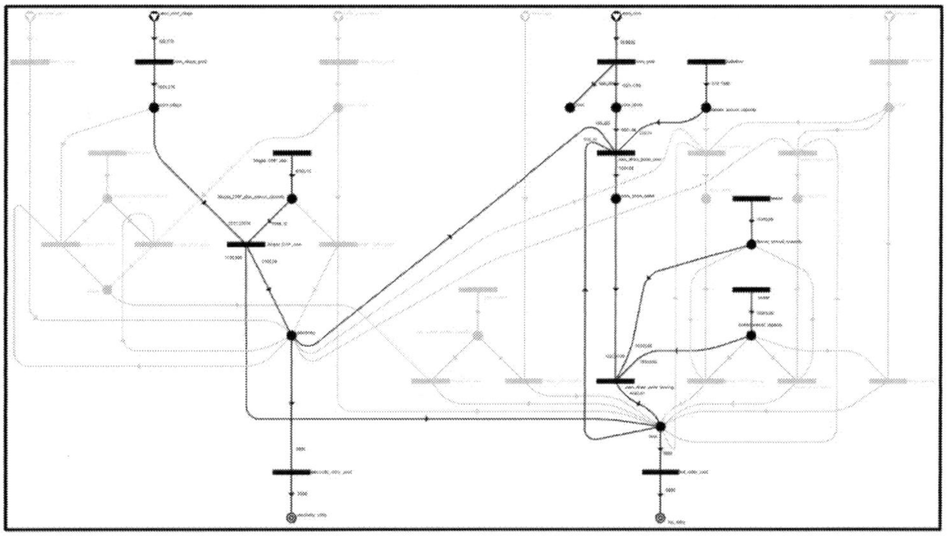

Supply Chain Structure #11: Corn Straw & Grass Silage

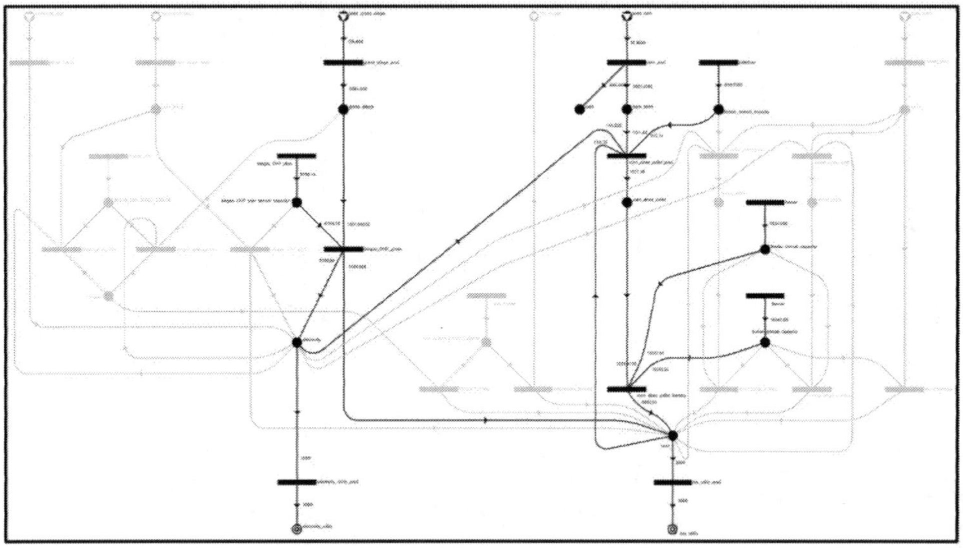

Supply Chain Structure #15: Corn Straw & Grid Electricity

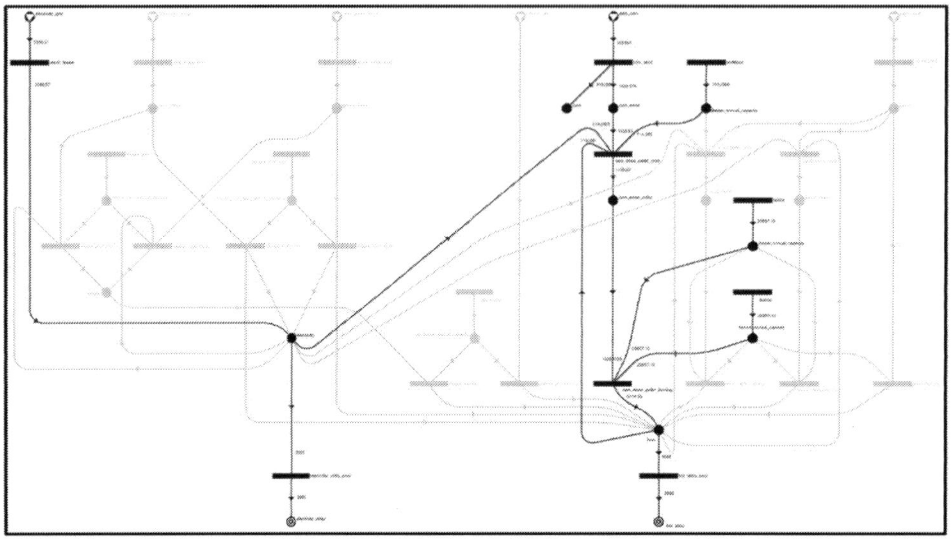

Supply Chain Structures: Heat & Electricity

Structure	Metric	% Change		Feedstock
Structure #13	Cost	0%		Natural Gas
	Footprint	0%		Grid Electricity
	Emergy	0%		
Structure #8	Cost	-3.7%		Corn Straw
	Footprint	+12%		Corn Silage
	Emergy	-89%		
Structure #11	Cost	-1.8%		Corn Straw
	Footprint	+26%		Grass Silage
	Emergy	-77%		
Structure#15	Cost	+1.3%		Corn Straw
	Footprint	+61%		Grid Electricity
	Emergy	-26%		

Supply Chain Structures

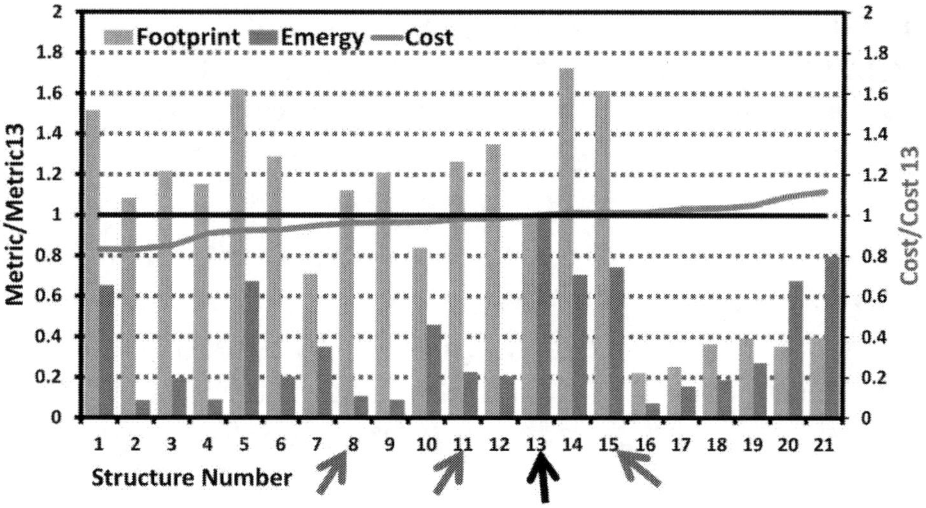

Summary

Summary

- Sustainability is about meeting human needs within the limits of the Earth so it can continue to support human existence for the indefinite future.
- Sustainability metrics based on science can provide necessary and actionable guidance.
- Application to real systems such as supply chains is both possible and essential.
- The P-Graph framework through the PNS Studio and PNS Draw provides accessible and free tools for supply chain and process design.

Methodology for a Holistic Synthesis of Sustainable Supply-Chain Networks

Zdravko Kravanja, Lidija Čuček, Žan Zore

University of Maribor,
Faculty of Chemistry and Chemical Engineering,
Laboratory for Process Systems Engineering and Sustainable Development
Smetanova 17, 2000 Maribor, Slovenia

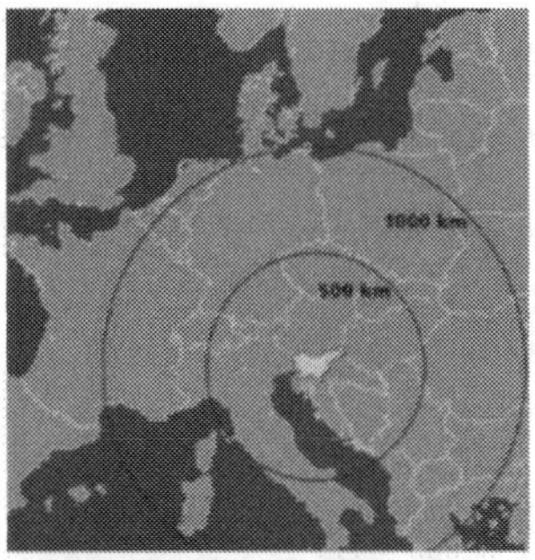

Slovenia in Pictures

Area: 20,273 km2
Population: 2.0 million
Capital city: Ljubljana
Language: Slovenian; also Italian and Hungarian in nationally mixed areas
Currency: EURO, €
Member of EU - 1 May 2004
EU Presidency for 2008

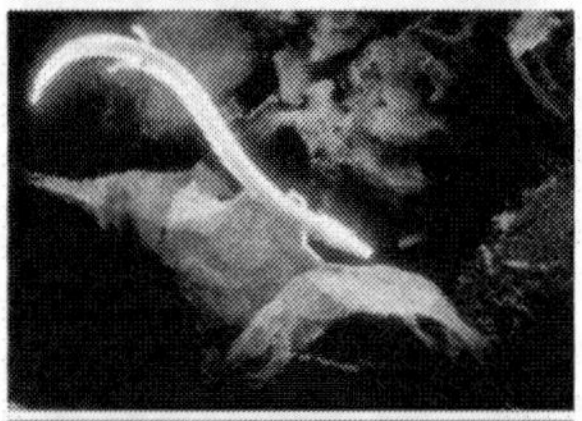

Maribor in Pictures

Univerza v Mariboru

OUTLINE

- Incentives for Sustainable Development and Applying MP
- Holistic MINLP LCA-based Sustainable System Synthesis
- Concepts of Eco- and Total Profit Systems Synthesis
  - Micro Total Profit (company's view)
  - Macro Total Profit (macro economic view)
- Case studies
  - Synthesis of a biogas production plant
  - Synthesis of a biofuel supply chain at regional level
  - EU level synthesis of a biofuel supply chain
  - Conclusion

Where are We Today?
Word Energy Consumption

Univerza v Mariboru

Quadrillion Btu=1.055 10^{18} J

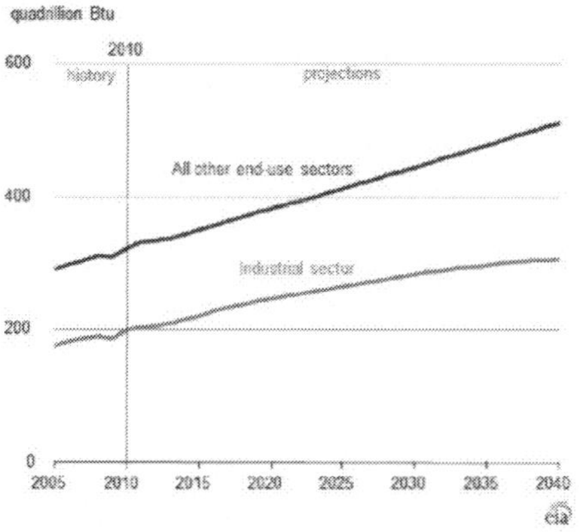

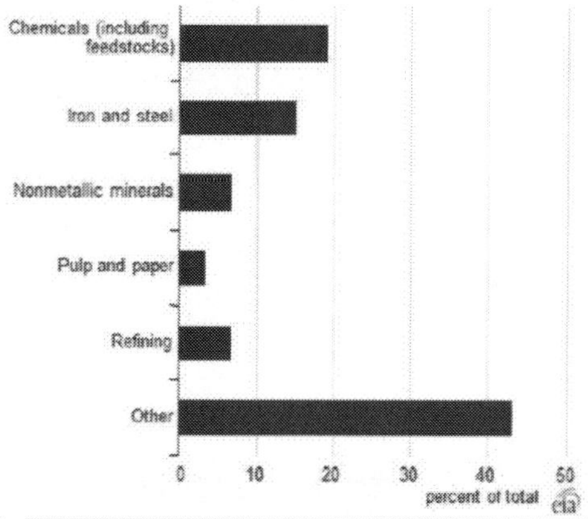

Fig. 1: World delivered end-user energy consumption

Fig. 2: Shares of world industrial sector delivered energy consumption, 2010

Source: International Energy Outlook 2013, U.S. Energy Andministration Angency, http://www.eia.gov/forecasts/ieo/industrial.cfm

(In)efficient Chemical and Process Industies Energy Consumption

Univerza v Mariboru

- Separation processes alone represent about 15%, or even 25% of total world energy consumption! Koros WJ. <energy.gatech.edu/questions/koros.php>; 2011 TUDelta. <delta.tudelft.nl/article/dow-awards-separation-by-freezing/24054>

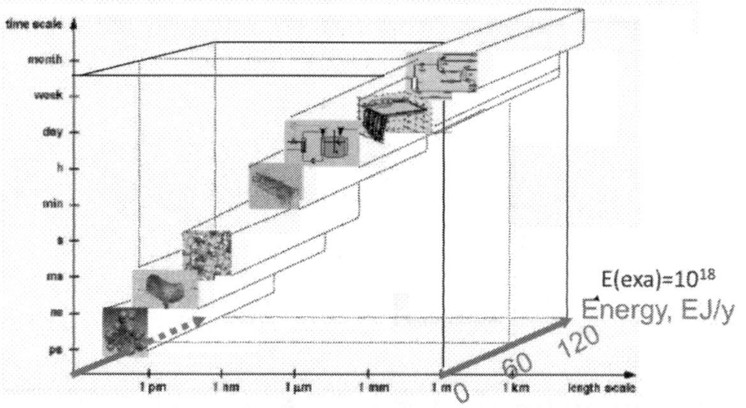

After Marquardt Wolfgang, Lars Von Wedel, and Birget Bayer. AspenWorld 2000, Orlando, FL, 2000

Figure 3: Energy involved in chemical and process industries

Unsustainable Use of Energy Resources

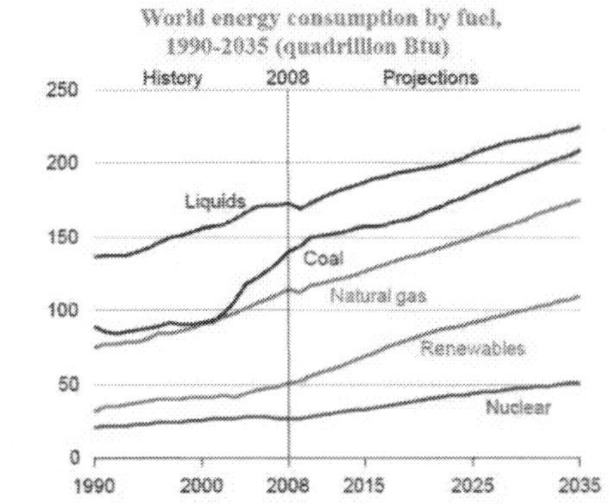

Fig. 4: World marketed energy consumption

Source: Energy Information Administration (EIA), International Energy Outlook 2011, World energy consumption by fuel 1990-2035, www.eia.doe.gov/iea/http://www.eia.gov/forecasts/ieo/highlights.cfm. Accessed 23.08.2012

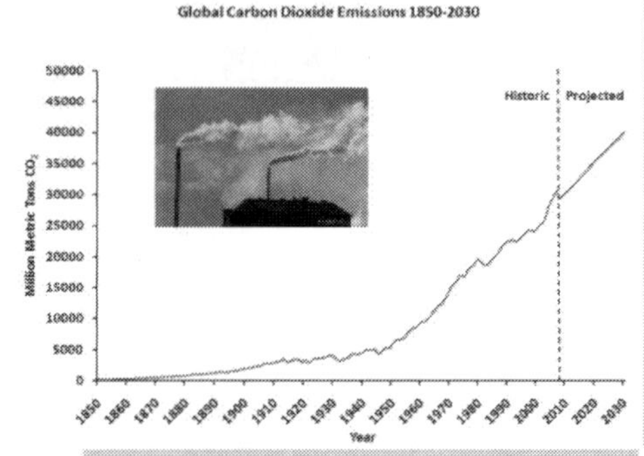

Fig. 5: Global carbon emissions from fossil fuel burning

Source: Center for climate and energy solutions, Historical global CO_2 emissions, www.c2es.org/facts-figures/international-emissions/historical. Accessed 23.08.2012

Global Temperature Rise

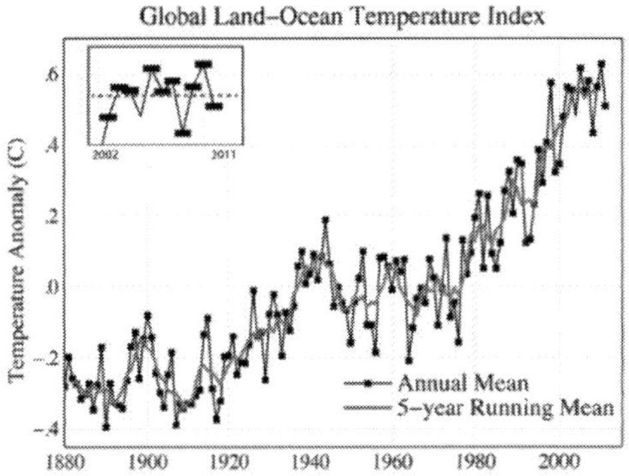

Fig. 6: Global land-ocean temperature index (data.giss.nasa.gov/gistemp/graphs_v3)

- The imbalance of the last decade heat has gone into deep oceans
- If climate forcing in 2050 reaches 4.5 W/m^2, what consequences might be expected?

Univerza v Mariboru

Global Damage Due to CO_2 Emissions

Global damage:

Global Social Cost of Carbon (GSCC) **158 \$/tCO_2**

x

Global emission **33 Gt CO_2/y**

=

5-6 trillion \$/y

Global damage due to CO_2 at least 5-6 trillion \$/y!!

1/10 of the global GDP (69 trillion \$/y) Trillion = 10^{12}

Univerza v Mariboru

Global Damage Due to Human Unsustainable Practice

- CO_2 emission – **global warming with unknown consequences** 1/10 GDP
- NO_x emission – **eutrophication, smog formation, ozone depletion, also global warming and biodiversity loss** (damage even higher than by CO_2) >1/10 GDP
- Biodiversity loss – **irreversible due to the extinction of species** (extinction rate is up to 140,000 species per year) ?/10 GDP

Net GDP (€/y) = GDP – Eco-loss = GDP – GDP/2 = GDP/2

Conclusion: Global BDP is significantly overestimated!

Stagnation when (ΔGDP – ΔEco-loss)<0

Sustainable development considerably improves global economics!

Sustainable Development: Blue Map – New Scenario for CO_2 Emissions

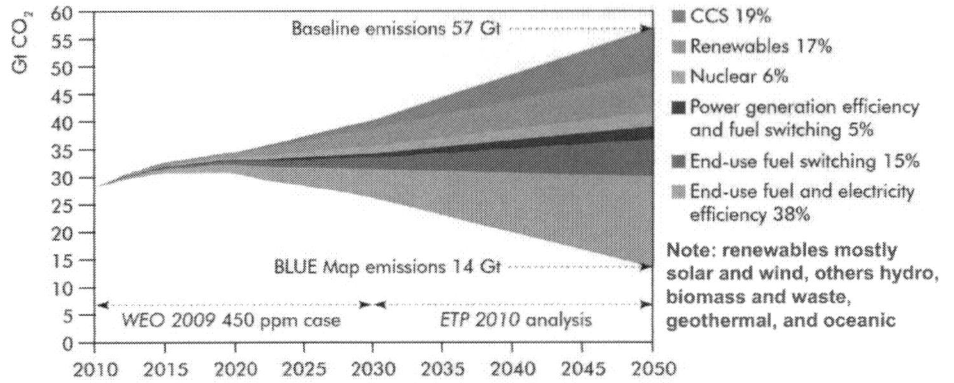

Fig. 7: Blue Map scenario and key technologies for reducing CO2 emissions
OECD/IEA, Energy Technology Perspectives 2010, Scenarios & Strategies to 2050, http://www.iea.org/techno/etp/etp10/English.pdf

Is this sufficient and economicaly viable?
New innovations are necessary to satisfy future raw-materials and energy consumption!

Incentives

Key idea for today and tomorrow

Holistic systems approach with its mathematical programming can provide efficient solution framework for the sustainable chemical and other supply chains management

Holistic Systems Approach

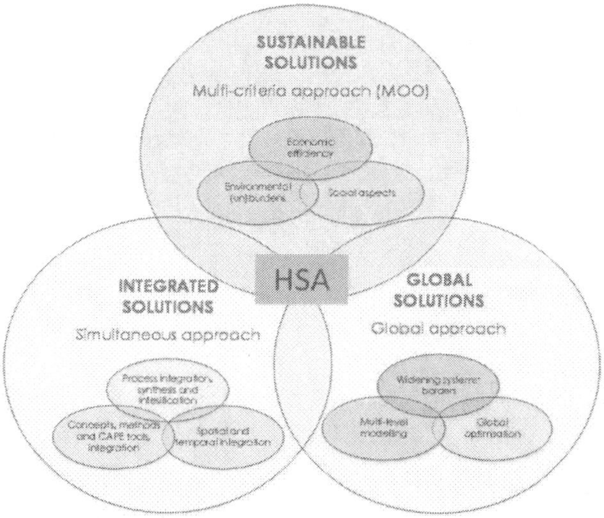

Fig. 8: Elements of the holistic systems approach

LCA-Synthesis

ICOSSE 2015, Balatonfűred, Hungary, 26-29 May 2015

13

LCA-based System Synthesis

Incomplete measurements for sustainability is one of the major limitations of LCA methodology

Consequences: poor or even wrong solutions and decisions!

More advanced concept and measurements are needed

Besides the direct burdening, also indirect unburdening effects have to be considered

MO-Synthesis

ICOSSE 2015, Balatonfűred, Hungary, 26-29 May 2015

14

LCA-based System Synthesis
New Concept: Direct and Indirect Effects

Univerza v Mariboru

***Raw materials**, which* ***burden** the environment if they are processed*
DIRECT impacts (BURDEN)

***Products**, which* ***burden** the environment related to processing, disposal, and transportation*
DIRECT impacts (BURDEN)

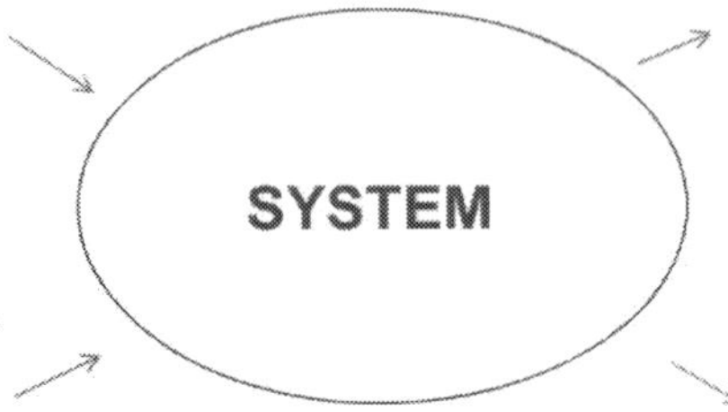

***Raw materials**, which mainly* ***unburden** or benefit the environment , e.g. utilization of waste rather than deposit*
INDIRECT impacts (UNBURDEN)

***Products**, which also unburden or benefit the environment due to products' substitution*
INDIRECT impacts (UNBURDEN)

TOTAL effects = DIRECT (burdening) + INDIRECT effects (unburdenning)

LCA-based Synthesis Approach
Direct vs. Total Environmental Effects

Univerza v Mariboru

Direct Effects:

1. Footprints
2. Sustainability Index
3. Eco-cost (Vogtländer et al., 2010)

+ Indirect effects →

Total Effects: (Kravanja, COCHE, 2012)

1. Total Footprints (Čuček, Varbanov, Klemeš, Kravanja, Energy, 2012)
2. Total Sustainability Index (Kravanja, Čuček, APEN, 2013)
3. Eco-profit and Total Profit (Čuček, R. Drobež, B. Pahor, Z. Kravanja, CCE, 2012)

Eco-profit and Total Profit

- Composite criterion in monetary unit
- Evaluate investment needed to prevent burdenning of the environment rather than environmental damage and loss
- Zero waste assumption

Total Effects in Composite-Criterion: Total Profit

R_B – raw materials, which only burden the environment if they are processed (direct effects)

R_{UNB} – raw materials, which mainly unburden or benefit the environment when they are used , e.g. utilization of waste *(direct +indirect effects)*

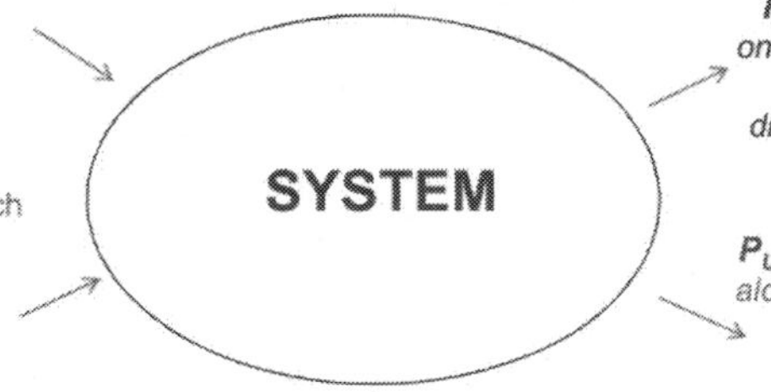

P_B – set of products, which only burden the environment related to processing, disposal, and transportation (direct effects)

P_{UNB} – set of products which alco unburden or benefit the environment (direct +indirect effects)

Eco-cost (€/yr) : $$EC = \sum_{i \in R_B} q_{m_i}^{R_B} \cdot c_i^{d,R_B} + \sum_{j \in P_B} q_{m_j}^{P_B} \cdot c_j^{d,P_B}$$

Eco-benefit (€/yr): $$EB = \sum_{i \in R_{UNB}} q_{m_i}^{R_{UNB}} \cdot c_i^{R_{UNB},t} + \sum_{j \in P_{UNB}} q_{m_j}^{P_{UNB}} \cdot f_j^{S/P_{UNB}} \cdot c_j^{S,t}$$

Eco-profit(€/yr) = Eco-benefit - Eco-cost

Total profit (€/yr) = Economic profit + Eco-profit

$$TP = (R - E - D) + (EB - EC)$$

Eco-cost coefficients: Delft University of Technology, <www.ecocostsvalue.com> Čuček, Drobež, Pahor, Kravanja, 2012

Apply Total Profit MINLP Synthesis Approach to Supply Chains

$$
\begin{aligned}
&a)\quad \max \quad TP = P^{Economical}(y,x) + P^{Eco}(y,x)\\
&b)\quad s.t \quad h_{ls}(x,y) = 0\\
&c)\quad \quad g_{ls}(x,y) \le 0\\
&\quad x \in X = \{x \in R^n : x^{LO} \le x \le x^U\}\\
&\quad y \in Y = \{0,1\}^m
\end{aligned}
\quad \left.\right\} \begin{matrix}\forall l \in Levels\\ \forall S \in Supply\,chains\end{matrix}
$$

MILP or MINLP!

a) Objective function as Total Profit

b) Equality constraints: mass and energy balances, design equations

c) Inequality constraints: product specifications, operational, environmental and feasibility constraints

d) Logical disjunctive constraints for selection of sustainable alternatives

Simultaneous

Total Profit-based Synthesis

ADVANTAGES

- Direct solution procedure with composite objective
- No iterative solution of Patero curve needed
- Very large-sized problems can be solved

SOME PROBLEMS

- Definition of Eco-cost coefficients for a given problem as they are geographically dependent
- Is compromizing company's economic profit acceptable for increasing other company's eco-profit?

Company's view of Total Profit – Micro Total Profit

Macro economic view of Total Profit – Macro Total Profit

Macro vs. Micro Total Profit

Univerza v Mariboru

$$TP = P^{Economic} + P^{Eco}$$

Makro and Mikro

$$P^{\text{Economic Makro}} = R^{market\,price} - E - D$$

$$P^{\text{Economic Mikro}} = R^{market\,price} - E - D + R^{subsidy} - c^{tax}$$

$$P^{Eco\,Makro} = \sum_{i \in R_B} q_{m_i}^{R_B} \cdot c_i^{d,R_B} + \sum_{j \in P_B} q_{m_j}^{P_B} \cdot c_j^{d,P_B} - \Big(\sum_{i \in R_{UNB}} q_{m_i}^{R_{UNB}} \cdot c_i^{R_{UNB},t} + \sum_{j \in P_{UNB}} q_{m_j}^{P_{UNB}} \cdot f_j^{S/P_{UNB}} \cdot c_j^{S,t} \Big)$$

$$P^{Eco\,Mikro} = \sum_{p \in R_B} (q_{m_{p,tot}}^{R_B} - q_{m_p}^{R_B}) \cdot c_p^{R_B} + \sum_{j \in P_B} q_{m_j}^{P_B} \cdot c_j^{d,P_B} - \Big(\sum_{i \in R_{UNB}} q_{m_i}^{\text{Mikro } R_{UNB}} \cdot c_i^{R_{UNB},t} + \sum_{j \in P_{UNB}} q_{m_j}^{\text{Mikro } P_{UNB}} \cdot f_j^{S/P_{UNB}} \cdot c_j^{S,t} \Big)$$

Unspent company's waste — Selling products — Converted company's waste within company — Company's products spent within company

Macro vs. Micro Total Profit

Univerza v Mariboru

$$TP = P^{Economic} + P^{Eco}$$

Makro and Mikro

$$TP^{Makro} = (R^{market\,price} - E - D) +$$
$$+ \Big(\sum_{i \in R_B} q_{m_i}^{R_B} \cdot c_i^{d,R_B} + \sum_{j \in P_B} q_{m_j}^{P_B} \cdot c_j^{d,P_B} - \Big(\sum_{i \in R_{UNB}} q_{m_i}^{R_{UNB}} \cdot c_i^{R_{UNB},t} + \sum_{j \in P_{UNB}} q_{m_j}^{P_{UNB}} \cdot f_j^{S/P_{UNB}} \cdot c_j^{S,t} \Big)\Big)$$

$$TP^{Mikro} = (R^{market\,price} - E - D + R^{subsidy} \cancel{- c^{tax}}) +$$
$$+ \Big(\sum_{p \in R_B} (q_{m_{p,tot}}^{R_B} - q_{m_p}^{R_B}) \cdot c_p^{R_B} + \sum_{j \in P_B} q_{m_j}^{P_B} \cdot c_j^{d,P_B} -$$
$$- \Big(\sum_{i \in R_{UNB}} q_{m_i}^{Mikro\ R_{UNB}} \cdot c_i^{R_{UNB},t} + \sum_{j \in P_{UNB}} q_{m_j}^{Mikro\ P_{UNB}} \cdot f_j^{S/P_{UNB}} \cdot c_j^{S,t} \Big)\Big)$$

Univerza v Mariboru

Demonstration Case Study

Synthesis of biogas processes with auxiliary facilities

Univerza v Mariboru

Scheme of demonstration case study

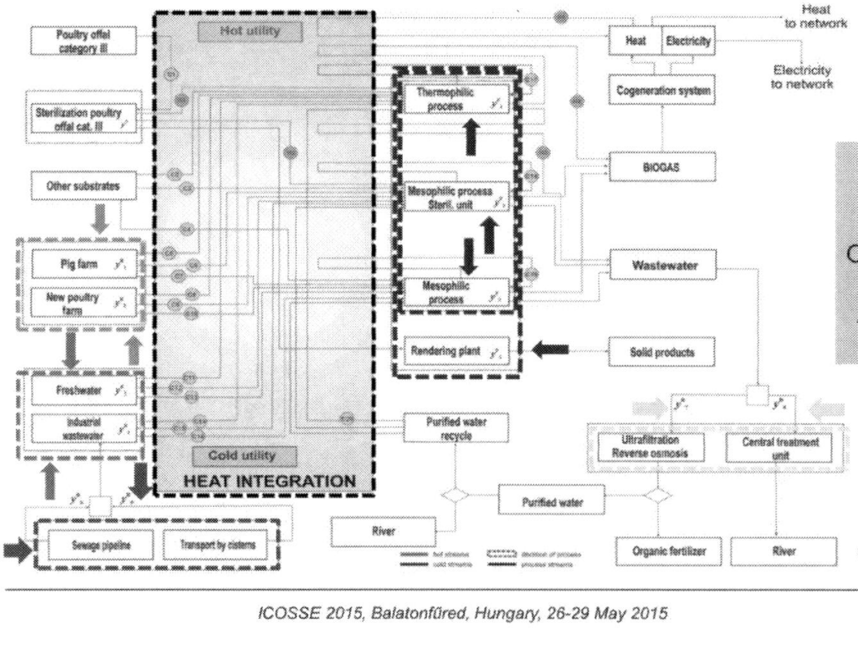

Fig. 9: Biogas from Organic and Animal Waste

study

Univerza v Mariboru

Total Effects: Eco-profit and Total Profit Optimisation

Table 3: Different optimisation schemes with Eco-profit for Biogas problem

Maximization	Company's Economic Profit	Company's Total Profit	Macro Total Profit
Company's Economic Profit (M€/y)	**3.122**	3.053	1.080
Company's Eco-profit (M€/y)	1.469	1.764	1.709
Company's Total Profit (M€/y)	4.591	**4.817**	2.789
Macro Economic Profit (M€/y)	- 1.594	- 1.591	- 1.464
Macro Eco-profit (M€/y)	2.668	2.924	3.509
Macro Total Profit (M€/y)	1.074	1.333	**2.045**
Income (M€/y)	3.287	3.243	3.677
Depreciation (M€/y)	1.937	1.910	2.720
Investment (M€)	20.727	20.599	17.050
Operating cost (M€/y)	4.881	4.835	5.141
Subsidy (M€/y)	4.716	4.644	2.738
Biogas capacity (MW)	4.215	4.150	2.447
Rate of manure in inlet substrates (%)	71.0	72.0	93.0
Cogeneration of heat (%)	45.0	45.0	45.0
The amount of used wastes (t/y)	122,861	121,180	114.667

Univerza v Mariboru

Direct Effects: Economical and Net Profit Optimization

Čuček, Drobež, Pahor, Kravanja, CCE, 2012
Kravanja, Čuček, APEN, 2013

Maximisation of the company's economic profit

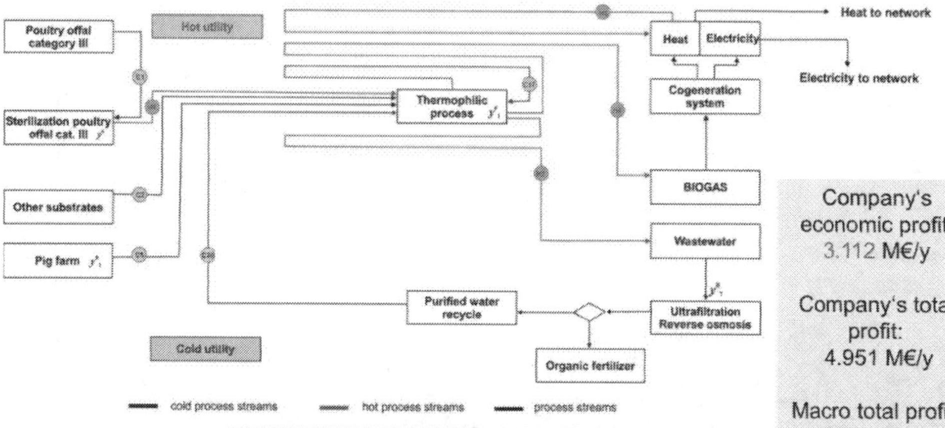

Fig. 10: Optimal Biogas production flowsheet

Company's economic profit: 3.112 M€/y

Company's total profit: 4.951 M€/y

Macro total profit: 1.074 M€/y

Total Effects: Eco-profit and Total Profit Optimization

Univerza v Mariboru

Maximization of the company's total profit

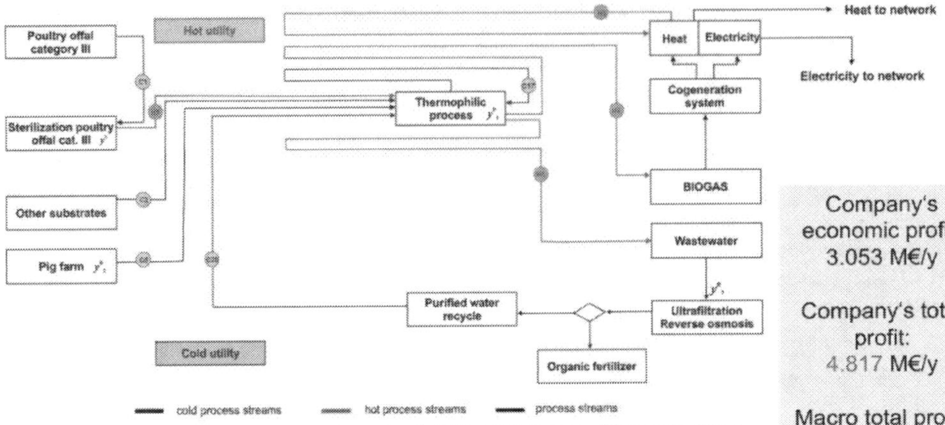

Fig. 11: Optimal Biogas production flowsheet

Company's economic profit: 3.053 M€/y

Company's total profit: 4.817 M€/y

Macro total profit: 1.333 M€/y

Total Effects: Eco-profit and Total Profit Optimization

Univerza v Mariboru

Maximization of the makro total-profit

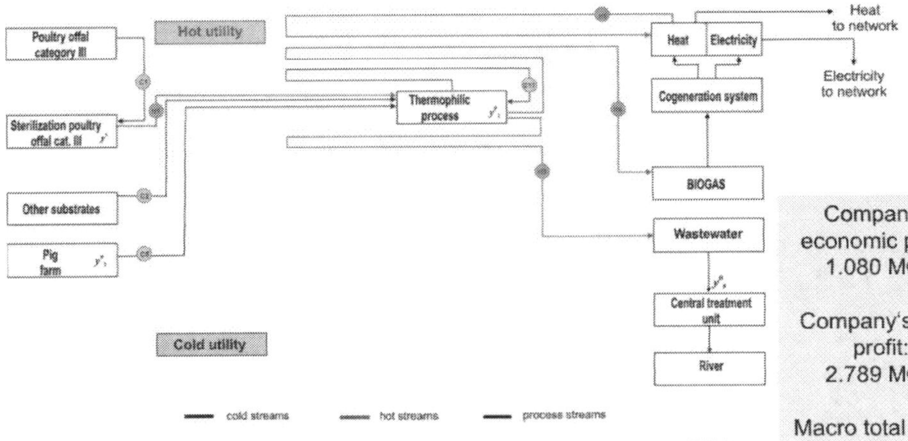

Fig. 12: Optimal Biogas production flowsheet

Company's economic profit: 1.080 M€/y

Company's total profit: 2.789 M€/y

Macro total profit: 2,045 M€/y

Multi-objective Optimization: Company's Economic vs. Macro Eco-profit

Univerza v Mariboru

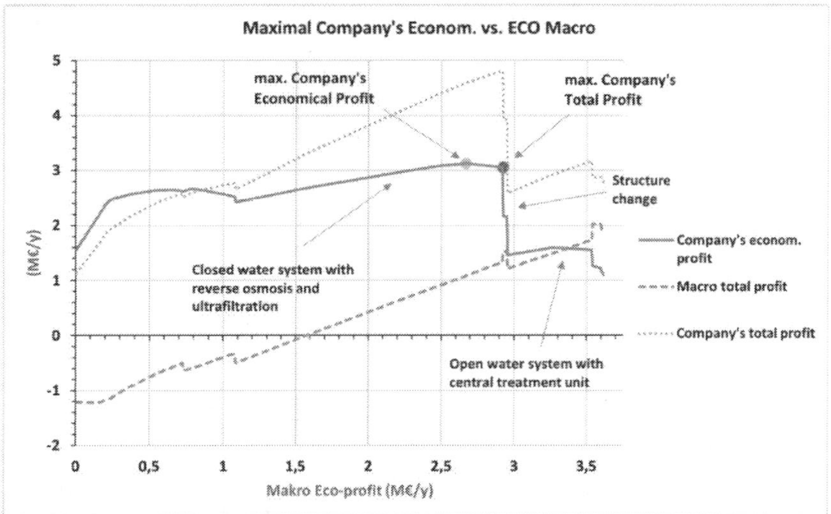

Fig. 13: Micro Economic Profit vs. Macro Eco-Profit

Multi-objective Optimization: Company's Economical vs. Company's Eco-profit

Univerza v Mariboru

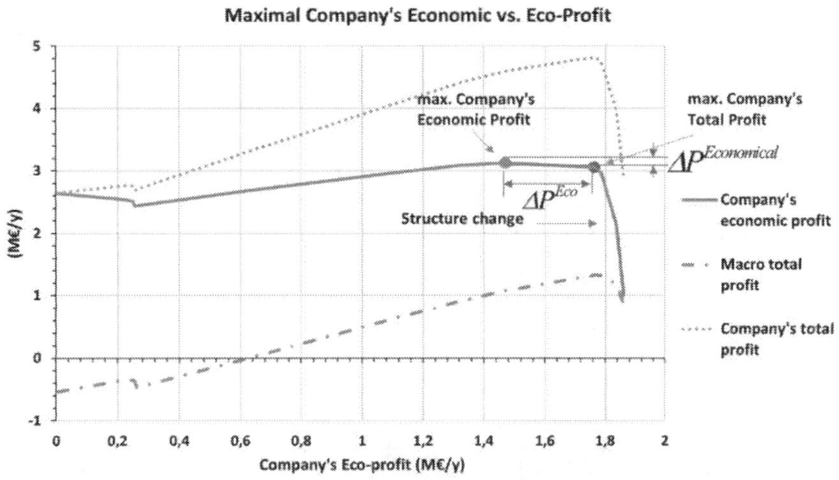

Fig. 14: Micro Economic Profit vs. Micro Eco-Profit

Multi-objective Optimization: Macro Economic vs. Macro Eco-profit

Univerza v Mariboru

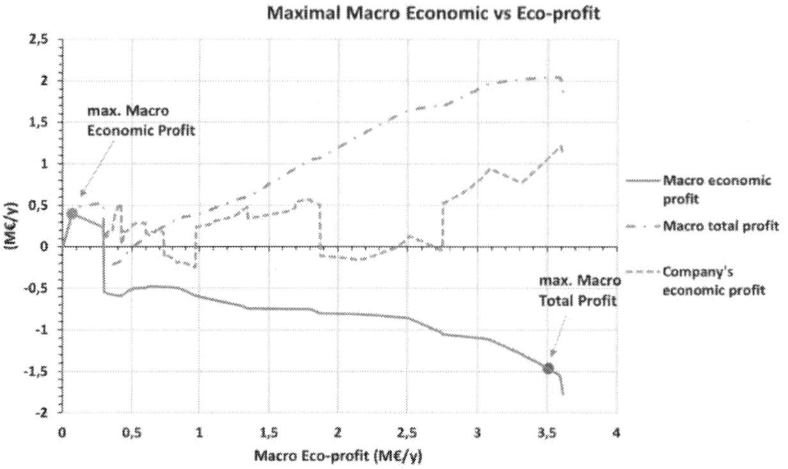

Fig. 15: Macro economic profit is positive when choosing rendering plant. As soon biogas plant is chosen the macro profit (without subsidies) become negative.

Multi-objective Optimization: Company's Total vs. Macro Total Profit

Univerza v Mariboru

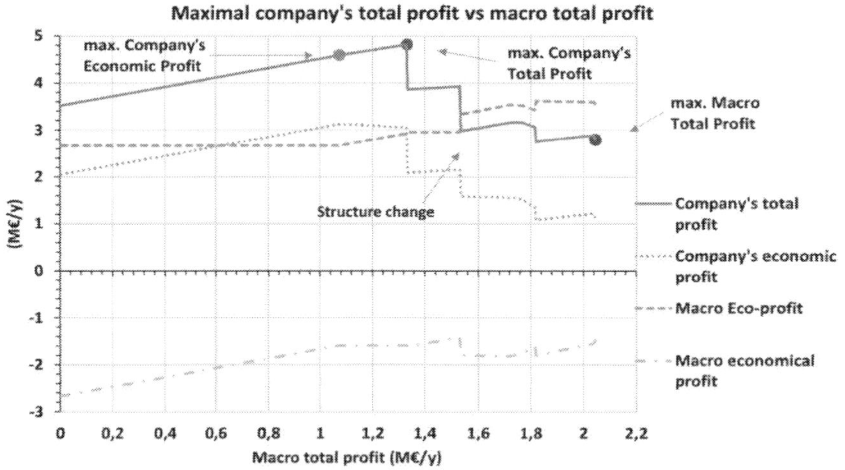

Fig. 16: Graph shows us that subsidies and tax shemes work well to the certain point. After the maximum company's total profit there is not any 'good' solutions from company's angle.

Univerza v Mariboru

Biorefinery Supply Network

Univerza v Mariboru

Supply Chain Network Design

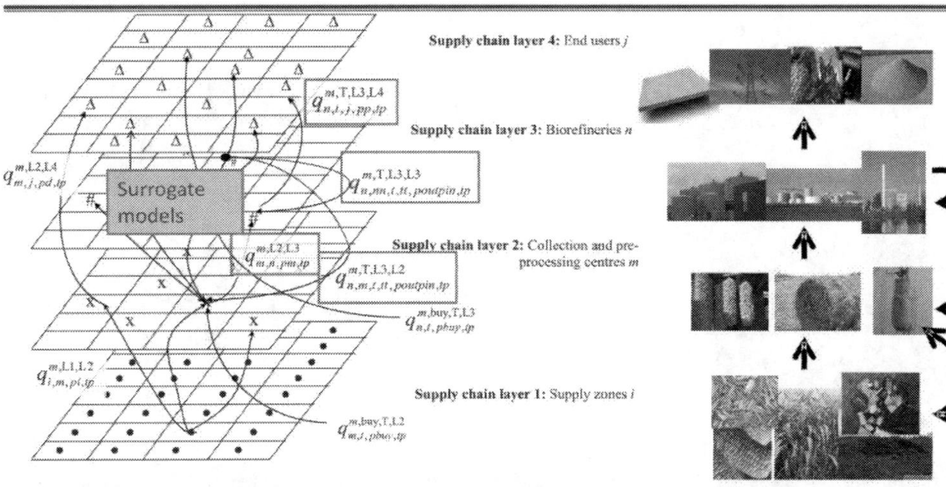

Fig. 16: Four-layer supply chain network design (Čuček et al., 2014)

Čuček, L., Martín, M., Grossmann, I.E., Kravanja, Z., 2014. Multi-period Synthesis of Optimally-Integrated Biomass and Bioenergy Supply Network. Computers and Chemical Engineering 66, 57-70.

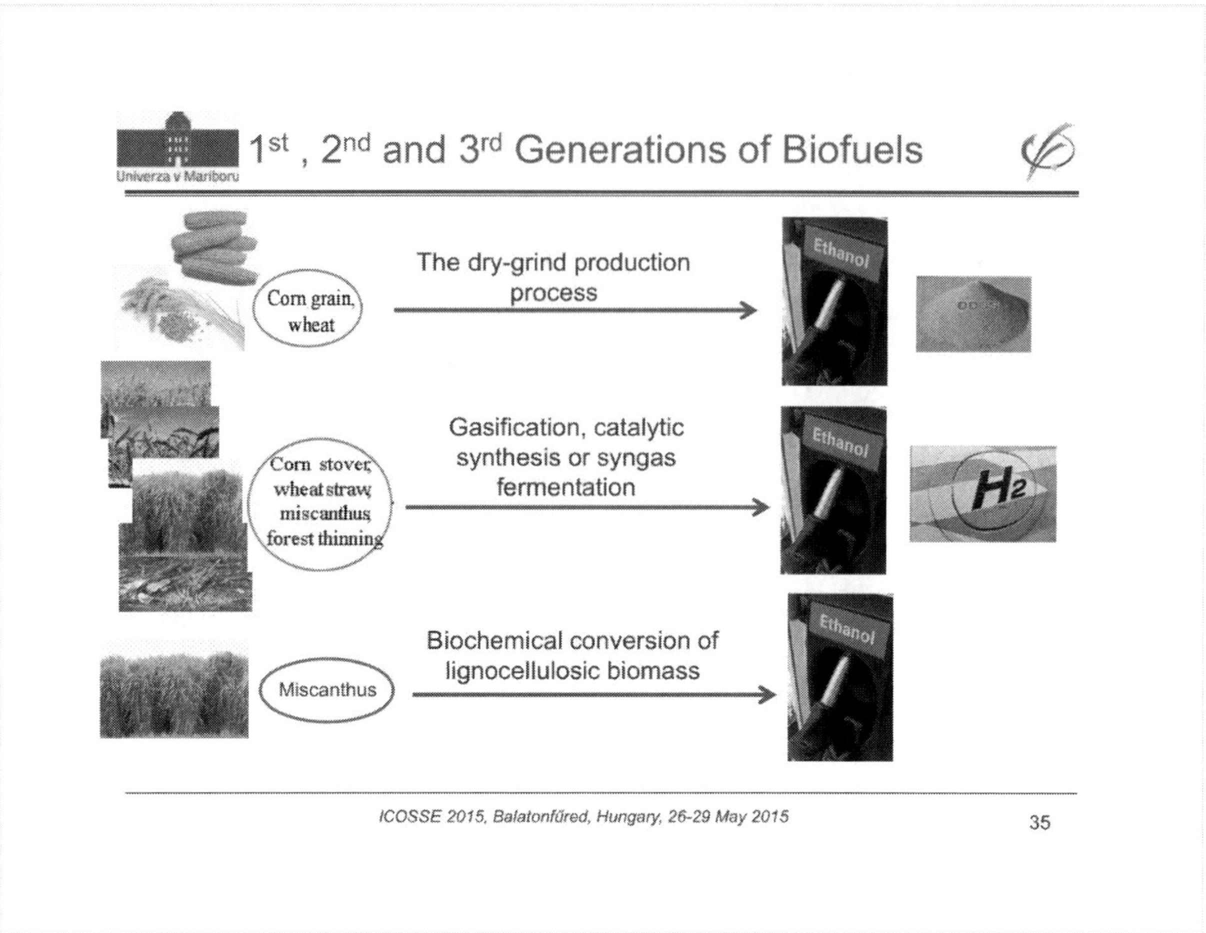

1st , 2nd and 3rd Generations of Biofuels
Univerza v Mariboru
Corn grain, wheat
The dry-grind production process
Ethanol
Corn stover, wheat straw, miscanthus, forest thinning
Gasification, catalytic synthesis or syngas fermentation
Ethanol
H2
Miscanthus
Biochemical conversion of lignocellulosic biomass
Ethanol
ICOSSE 2015, Balatonfüred, Hungary, 26-29 May 2015
35

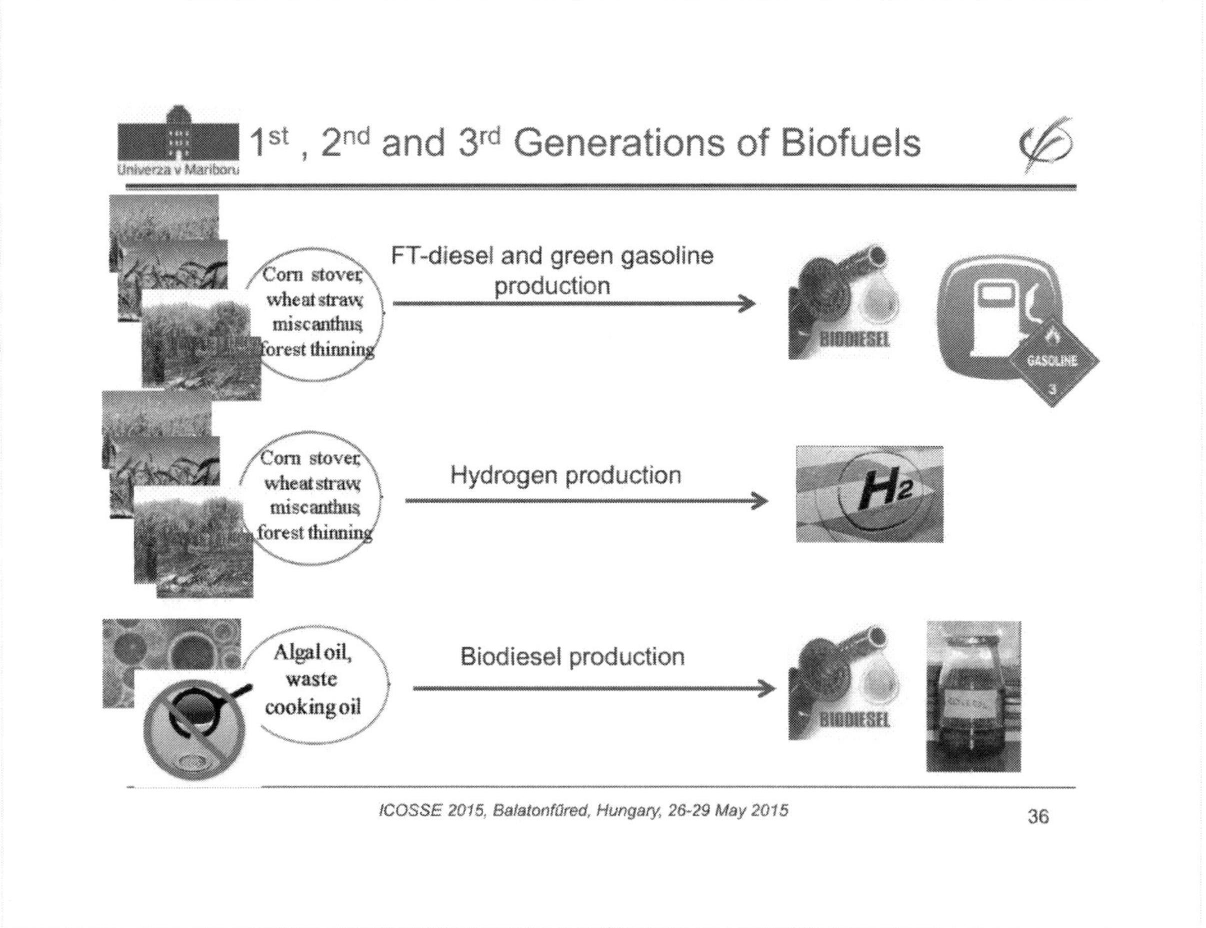

1st , 2nd and 3rd Generations of Biofuels
Univerza v Mariboru
Corn stover, wheat straw, miscanthus, forest thinning
FT-diesel and green gasoline production
BIODIESEL
GASOLINE
3
Corn stover, wheat straw, miscanthus, forest thinning
Hydrogen production
H2
Algal oil, waste cooking oil
Biodiesel production
BIODIESEL
ICOSSE 2015, Balatonfüred, Hungary, 26-29 May 2015
36

Integrated Biorefineries

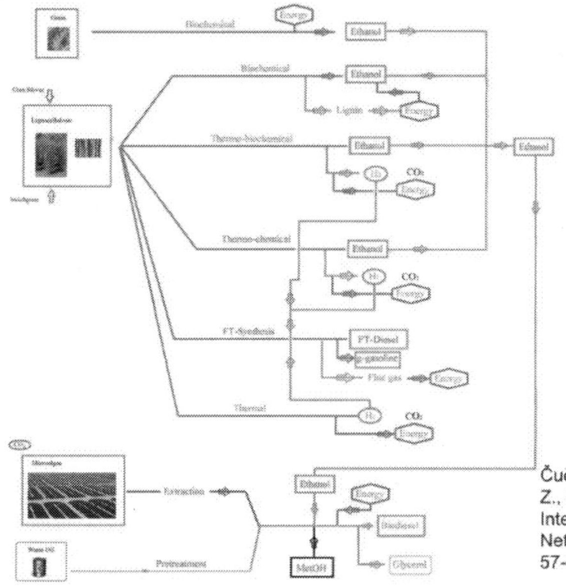

Fig. 17: Integration possibilities

Čuček, L., Martín, M., Grossmann, I.E., Kravanja, Z., 2014. Multi-period Synthesis of Optimally-Integrated Biomass and Bioenergy Supply Network. Computers and Chemical Engineering 66, 57-70.

Mathematical Model

- Defined for four layers
  - Layer 1: Supply and harvesting of biomass and waste
  - Layer 2: Collection, pretreatment (e.g., drying), preprocessing (e.g., oil extraction) and storage
  - Layer 3: Processing (different technologies at possible plant location) and storage
  - Layer 4: Usage (at possible demand locations) and includes all the transportation links between and within the layers
- Includes
  - Mass balances
  - Production and conversion constraints
  - Cost functions (transportation, operating and investment cost)
- The objective is maximizing the macro economic profit, company's economic and totalprofit

Univerza v Mariboru

Objective function I

Economic profit with subsidies and taxes (company's view):

$$P = \sum_{n \in N} \sum_{(pp,t) \in PPT} \sum_{j \in J} \sum_{tp \in TP} q_{n,t,j,pp,tp}^{m,net,T,L3,L4} \cdot c_{pp,tp}^{market\ price} +$$ Income from selling products

$$\sum_{m \in M} \sum_{pd \in PD} \sum_{j \in J} \sum_{tp \in TP} q_{m,j,pd,tp}^{m,L2,L4} \cdot c_{pd,tp}^{price} +$$ Income from selling food

$$\sum_{i \in I} \sum_{pi \in PI} \sum_{tp \in TP} q_{i,pi,tp}^{m,L1} \cdot c_{pi}^{cost} -$$ Raw material cost (biomass and waste)

$$\sum_{m \in M} \sum_{t \in T} \sum_{pbuy \in PBUY} \sum_{tp \in TP} q_{m,t,pbuy,tp}^{m,buy,L2} \cdot c_{pbuy}^{cost} -$$ Raw material cost (purchased)

$$\sum_{n \in N} \sum_{t \in T} \sum_{pbuy \in PBUY} \sum_{tp \in TP} q_{n,t,pbuy,tp}^{m,buy,L3} \cdot c_{pbuy}^{cost} -$$

$$c^{tr} - c^{storage} - c^{operating} - c^{depriciation} +$$ Transportation, storage, operating and investment cost

$$\sum_{n \in N} \sum_{(pp,t) \in PPT} \sum_{j \in J} \sum_{tp \in TP} q_{n,t,j,pp,tp}^{m,net,T,L3,L4} \cdot c_{pp,tp}^{subsidy} - \sum_{n \in N} \sum_{(pb,t) \in PBT} \sum_{j \in J} \sum_{tp \in TP} q_{n,t,j,pb,tp}^{m,T,L3} \cdot c_{pb,tp}^{tax}$$ Subsidies and taxes

Univerza v Mariboru

Objective function II

Economic profit without subsidies and taxes (macro view):

$$P = \sum_{n \in N} \sum_{(pp,t) \in PPT} \sum_{j \in J} \sum_{tp \in TP} q_{n,t,j,pp,tp}^{m,\text{net},\text{T},\text{L3},\text{L4}} \cdot c_{pp,tp}^{\text{price}} +$$ Income from selling products

$$\sum_{m \in M} \sum_{j \in J} \sum_{pd \in PD} \sum_{tp \in TP} q_{m,j,pd,tp}^{m,\text{L2},\text{L4}} \cdot c_{pd,tp}^{\text{price}} -$$ Income from selling food

$$\sum_{i \in I} \sum_{pi \in PI} \sum_{tp \in TP} q_{i,pi,tp}^{m,\text{L1}} \cdot c_{pi}^{\text{cost}} -$$ Raw material cost (biomass and waste)

$$\sum_{m \in M} \sum_{t \in T} \sum_{pbuy \in PBUY} \sum_{tp \in TP} q_{m,t,pbuy,tp}^{m,\text{buy},\text{L2}} \cdot c_{pbuy}^{\text{cost}} -$$ Raw material cost (purchased)

$$\sum_{n \in N} \sum_{t \in T} \sum_{pbuy \in PBUY} \sum_{tp \in TP} q_{n,t,pbuy,tp}^{m,\text{buy},\text{L3}} \cdot c_{pbuy}^{\text{cost}} -$$

$$c^{\text{tr}} - c^{\text{storage}} - c^{\text{operating}} - c^{\text{depreciation}}$$ Transportation, storage, operating and investment cost

Objective function II

Eco-profit and total profit without subsidies and taxes (macro view):

$$R^{\mathrm{Eco}} = \sum_{p \in R_{\mathrm{UNB}}} q_p^m \cdot c_p^{\mathrm{s}} + \sum_{p \in P_{\mathrm{UNB}}} q_p^m \cdot f_p^{S/P_{\mathrm{UNB}}} \cdot c_p^{\mathrm{s}}$$

Eco-benefit

$$c^{\mathrm{Eco}} = \sum_{p \in R_{\mathrm{B}} \cup R_{\mathrm{UNB}} \cup P_{\mathrm{B}} \cup P_{\mathrm{UNB}}} q_p^m \cdot c_p^{\mathrm{s}} + \sum_{p \in R_{\mathrm{B}} \cup R_{\mathrm{UNB}} \cup P_{\mathrm{B}} \cup P_{\mathrm{UNB}}} q_p^m \cdot l_p \cdot D_p \cdot c_p^{\mathrm{s,tr}}$$

Eco-cost

$$P^{\mathrm{Eco}} = R^{\mathrm{Eco}} - c^{\mathrm{Eco}}$$

Eco-profit

$$P^{\mathrm{T}} = P^{(\mathrm{Econ})} + P^{\mathrm{Eco}}$$

Total Profit

Regional-Size Example

Regional Testing Example

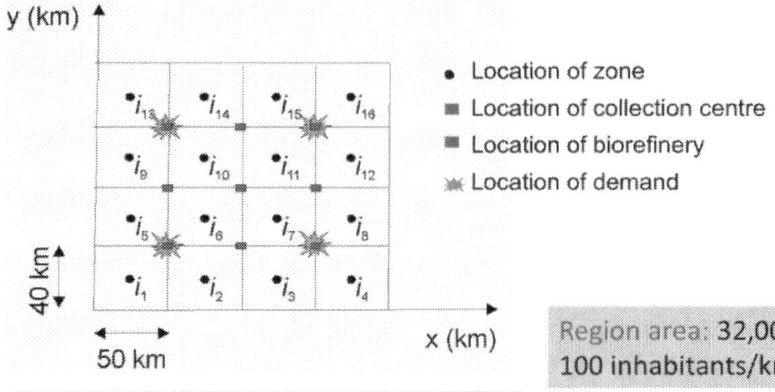

Region area: 32,000 km^2
100 inhabitants/km^2

Regional plan with 16 zones

Results

Up to 10 % of the total area of each zone could be dedicated to biofuels production can be greater than or equal to demand for food – corn and wheat (≥ 100 %)

Table 1: Main results from optimisation

Maximization	Micro economic profit	Micro total profit	Macro total
Micro economic profit (M$/y)	**7,329**	7,308	7,033
Micro total profit (M$/y)	7,329	**7,308**	7,033
Macro economic profit (M$/y)	4,610	4,588	4,416
Macro ECO profit (M$/y)	-0.173	-0.173	876
Macro total profit (M$/y)	4,610	4,588	**5,292**
Raw materials (kt/y)	Algae: 8,792 Switchgrass: 1,142	Algae: 8808 Switchgrass: 1,168	Algae: 8,808 Cooking oil: 11 Forest residue: 613
Technologies	- GSF - Oil extraction - BioDP from algal oil with methanol	- GSF - Oil extraction - BioDP from algal oil with methanol	- GSF - Oil extraction - BioDP from algal and cooking oil with methanol
Investment cost (M$)	2,280	2,718	2,314
Production cost (M$/y)	1,104	1,125	1,809
Transportation cost (M$/y)	91	85	77
Water consumption (kt/y)	3,597	3,597	3,263
Biofuels (kt/y)	Bioethanol: 364 Biodiesel: 4.228 Hydrogen: 39	Bioethanol: 364 Biodiesel: 4,228 Hydrogen: 39	Bioethanol: 192 Biodiesel: 4.238 Hydrogen: 20
Solution time (CPU-s)	307.899	306.945	133.010
Optimality gap (%)	0.52	0.82	0.98

Univerza v Mariboru

Multi-objective Optimization: Mikro Economic vs. Macro Eco-profit

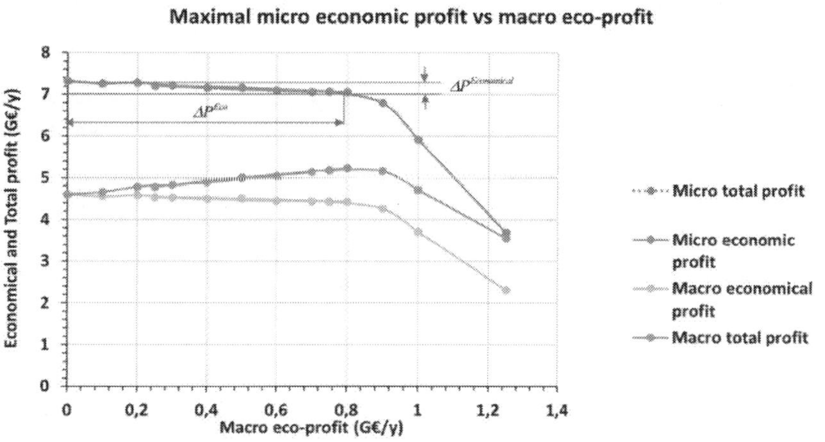

Univerza v Mariboru

European Continental-Scale Case Study

Continental Example – EU Supply Network for the Production of Biofuels

The model consists of 1,163,000 single equations, 35,450,000 single variables, and 28,300 discrete variables

Modelling system GAMS 24.2, mathematical programming solver GUROBI 4.0, and computer with 141 GB of RAM and processor Intel(R) Xeon(R) E5-2640 v2 were used.

Up to 8 % of existing and 2 % of additional area of each zone could be dedicated to biofuels and food

EU-28 with 140 zones

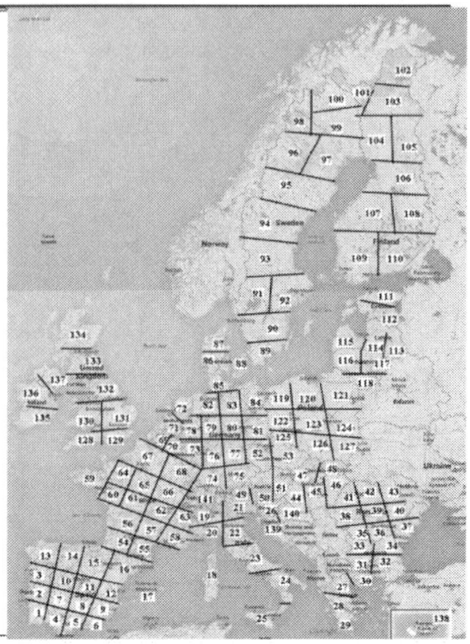

Results – EU and Country Level

Table 2: Main results from optimisation - production can be greater than or equal to demand for food – corn and wheat (≥ 100 %) and biofuels (≥ 10 %)

Maximization	Micro total profit		Macro total	
Micro total profit (M$/y)	**216,841**		141,274	
Macro total profit (M$/y)	123,769		**145,413**	
Raw materials (kt/y)	Corn grain: 85,041 Corn stover: 51,025 Wheat: 127,808 Wheat straw: 130,365	Miscanthus: 249,779 Algae: 69,129 Cooking oil: 750	Corn grain: 85,041 Corn stover: 51,025 Wheat: 127,808 Wheat straw: 129,116	Miscanthus: 59,042 Forest residue: 1,140 Algae: 67,736 Cooking oil: 750
Technologies	Gasification and syngas fermentation Gas Fischer-Tropsch Oil extraction BioDP from algal and cooking oil with methanol		Gasification and syngas fermentation Gas Fischer-Tropsch Oil extraction BioDP from algal and cooking oil with methanol	
Investment cost (M$)	122,941		83,420	
Production cost (M$/y)	133,249		113.626	
Transportation cost (M$/y)	40,000		24.628	
Water consumption (kt/y)	209,552		133,773	
Biofuels (kt/y)	Bioethanol: 92,309 Green gasoline: 5,682 Biodiesel: 33.901	FT-diesel: 21,374 Hydrogen: 7,721	Bioethanol: 55,760 Green gasoline: 2,759 Biodiesel: 33.233	FT-diesel: 10,379 Hydrogen: 5,855
Solution time (CPU-s)	7,307		9,457	
Optimality gap (%)	3.48		4.40	

Univerza v Mariboru

Redistribution of Gasoline production: Profit vs. CO_2-based Total Profit

0.135 EUR/kg CO_2 eq, www.ecocostvalue.com

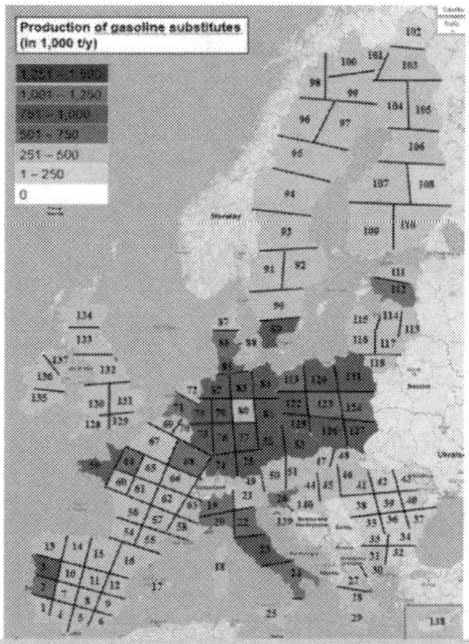

Micro total profit optimization

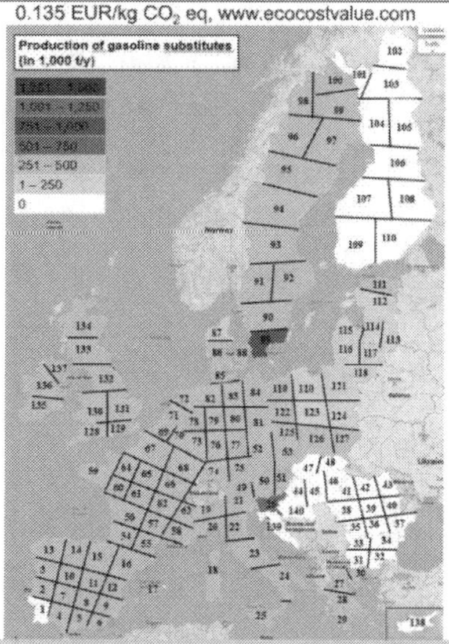

Macro total profit optimization

Univerza v Mariboru

Redistribution of Diesel production: Profit vs. CO_2-based Total Profit

0.135 EUR/kg CO_2 eq, www.ecocostvalue.com

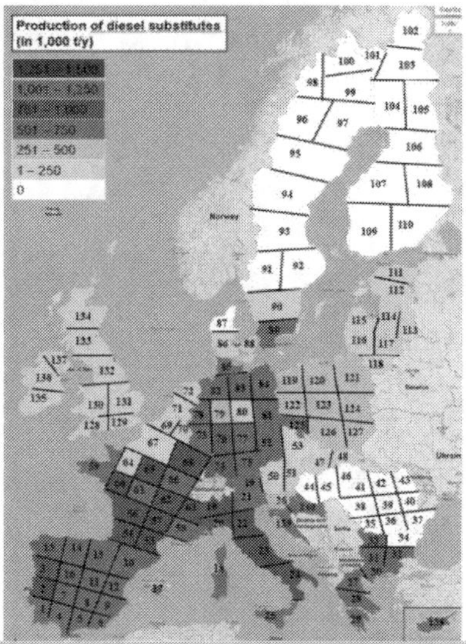

Micro total profit optimization

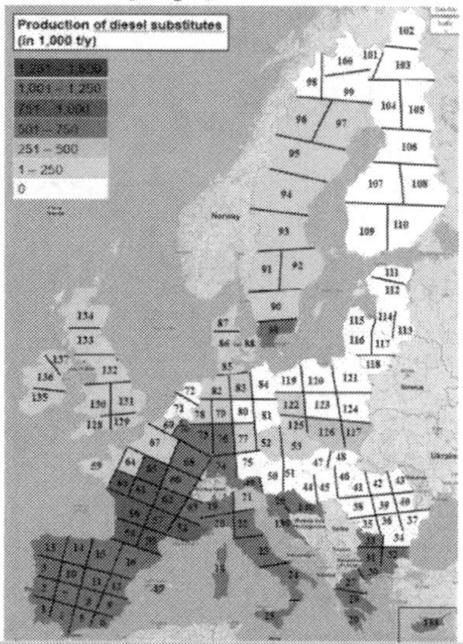

Macro total profit optimization

Univerza v Mariboru

Conclusion

- The role of holistic approach was highlighted for sustainable systems synthesis.
- Micro Total Profit to optimize systems from company's view
- Macro Total Profit to optimize systems from macroeconomic view
- Different solutions are obtained when maximizing economical and total profit
- Total Profit trade-off is established at large increase of Eco Profit and only small decrease of Economic profit
- Micro Total Profit can be used as a uniform Economic-Environmental criterion

Univerza v Mariboru

Acknowledgements

Prof. Ignacio E. Grossmann, Prof. Mariano Martin

Prof. Jiří Klemeš, Prof. Petar Varbanov,

Prof. Lam Hon Loong

Prof. Zorka Novak Pintarič, Dr. Bojan Pahor, Dr. Rozi Drobež

Univerza v Mariboru

ESCAPE 2016, Portorož, Slovenia

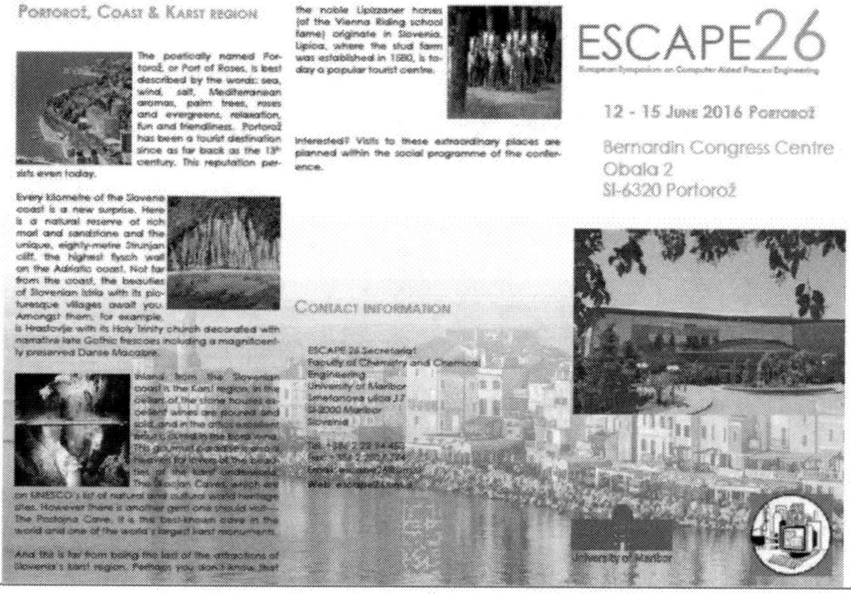

PORTOROŽ, COAST & KARST REGION

The poetically named Portorož, or Port of Roses, is best described by the words: sea, wind, salt, Mediterranean aromas, palm trees, roses and evergreens, relaxation, fun and friendliness. Portorož has been a tourist destination since as far back as the 13th century. This reputation persists even today.

Every kilometre of the Slovene coast is a new surprise. Here is a natural reserve of rich marl and sandstone and the unique, eighty-metre Strunjan cliff, the highest flysch wall on the Adriatic coast. Not far from the coast, the beauties of Slovenian Istria with its picturesque villages await you. Amongst them, for example, is Hrastovlje with its Holy Trinity church decorated with narrative late Gothic frescoes including a magnificently preserved Danse Macabre.

the noble Lipizzaner horses (of the Vienna Riding school fame) originate in Slovenia. Lipica, where the stud farm was established in 1580, is today a popular tourist centre.

Interested? Visits to these extraordinary places are planned within the social programme of the conference.

ESCAPE26

12 - 15 JUNE 2016 PORTOROŽ

Bernardin Congress Centre
Obala 2
SI-6320 Portorož

CONTACT INFORMATION

ESCAPE 26 Secretariat
Faculty of Chemistry and Chemical Engineering
University of Maribor

University of Maribor

Univerza v Mariboru

2nd SEE SDEWES 2016, Piran, Slovenia

2nd South East Europe conference on Sustainable Development of Energy Water and Environment Systems (SDEWES)

Univerza v Mariboru

ECOS 2016

29th international conference on
Efficiency, Cost, Optimization, Simulation and Environmental Impact of Energy Systems

June 19-22, 2016
Portorož, Slovenia

Univerza v Mariboru

Thank you

Maximizing Sustainability of Ecosystem Model through Socio-Economic Policies Derived from Multivariable Optimal Control Theory

Doshi, R.[1,3] Diwekar, U.[2,3] Benavides, P.[2,3] Yenkie, K.[2,3] Cabezas, H.[4]

[1]Northview High School, Duluth, GA - USA
[2]Department of Bio Engineering, University of Illinois, Chicago, IL 60607 - USA
[3]Center for Uncertain Systems: Tools for Optimization & Management (CUSTOM), Vishwamitra Research Institute, Clarendon Hills, IL 60514 - USA
[4] National Risk Management Research Laboratory, Office of Research and Development, U.S. Environmental Protection Agency. Cincinnati, Ohio 45255- USA

1

Outline

2

- Introduction
- Objective
- Methodology
- Sensitivity Analysis
- Optimal Control
- Questions?

TWO GLOBAL TRENDS

3

Population Explosion | **Consumption Increase**

How do we…

feed the human world

without

devastating the environment

and

ruining the climate

in the process?

4

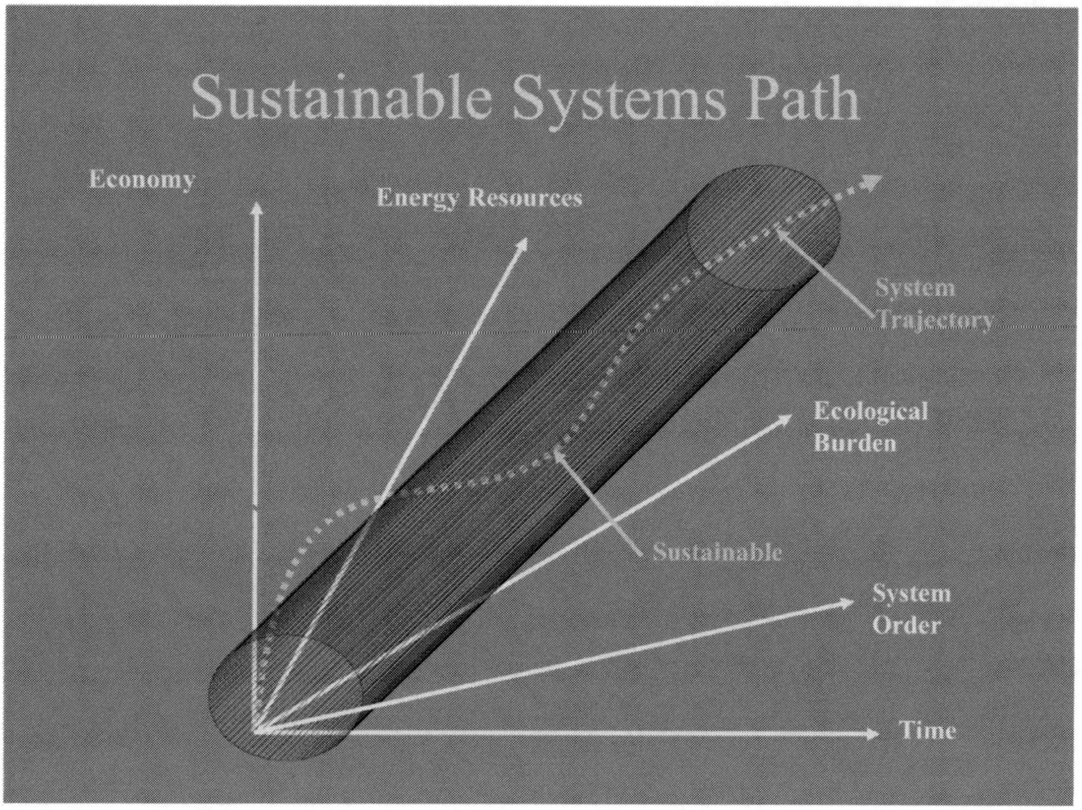

Sustainability

6

- The development that meets the needs of the present without compromising the ability of future generations to meet their own needs.
- PATH THROUGH TIME - MANY INFLUENCING DIMENSIONS

Cabezas, Heriberto, Christopher W. Pawlowski, Audrey L. Mayer, and N. Theresa Hoagland. "Sustainability: Ecological, Social, Economic, T echnological, and Systems Perspectives." Clean Techn. Environ. Policy 5 (2003): 167-180. Print.

Approach

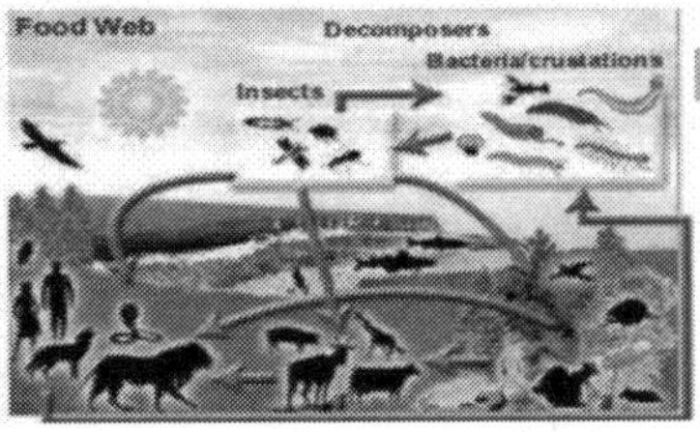

System modeling: Integrated ecological-economic-social model

Scenario studies:

- Modeling of possible future developments
- Identification of undesirable trends

Policy development:

- Identification of the appropriate objectives
- Use of optimization and control theory to develop policy guidelines

Integrated ecological economical food web energy model: A Global Supply Chain

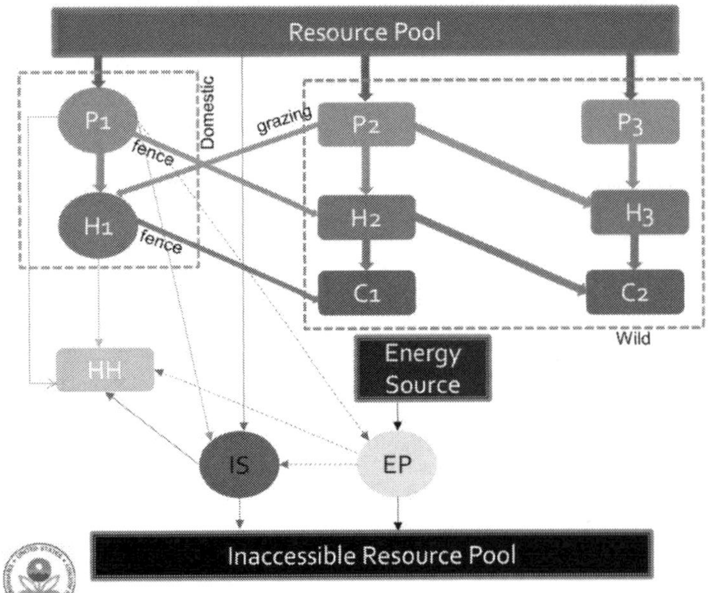

Natural Resource Pool

Primary Producers

Herbivorous Animals

Carnivorous Animals

Human Households

Industrial Sector

Energy Producers

Inaccessible resource

8

Why this model?

9

ECONOMIC$

+

ECOLOGY

Mathematics

10

- Three general types of equations in the model:
  - Food web's mass flows → Adapted Lotka-Volterra equations
  - Human demand for resources → Macroeconomic equations
  - All other algebraic equations
    - ✓ 81 constant parameters
    - ✓ 19 time dependent state variables described by differential equations
    - ✓ 2000 Lines of Matlab code

Sustainability

"Development that meets the needs of the present without compromising the ability of the future generations to meet their own need"

-- The world commission on environment and development

Path optimization to attain sustainability for

Fisher Information (FI)

12

- Mathematical measure of sustainability
- Taken from information theory
- Measure of order/organization of a system

$$FI_t = \frac{1}{T_c}\int_0^{T_c} \frac{a(t)^2}{v(t)^4}\,dt \qquad (1)$$

$$v(t) = \sqrt{\sum_{i=1}^{n}\left(\frac{dx_i}{dt}\right)^2} \qquad (2)$$

$$a(t) = \frac{1}{v(t)}\left[\sum_{i=1}^{n}\frac{dx_i}{dt}\,\frac{d^2x_i}{dt^2}\right] \quad (i=1,2,\ldots,n) \qquad (3)$$

Model Simulation

- Project focuses on consumption increase scenario
- Consumption increase (CI) leads to extinction while base case is sustainable
- Not seen: CI leads to extinction of various mass compartments

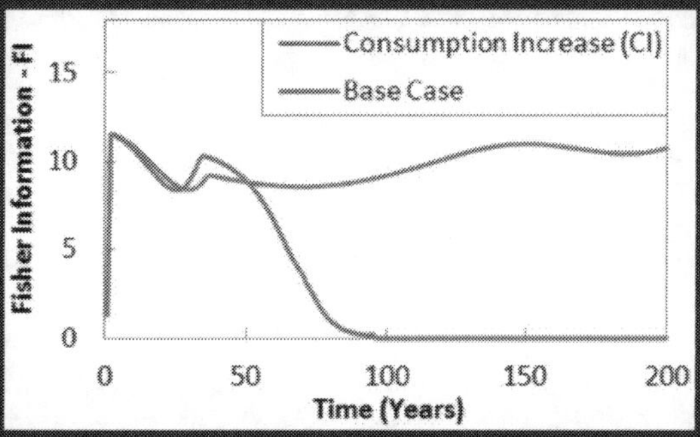

13

Sensitivity Analysis

14

Category	Name	Parameter	Mass-Flow	Meaning	Selected?
Pollution-Fee Related	Discharge Fee	d_{fee}	IS→IRP	Discharge fee imposed by government to the Industry sector for waste disposal.	Yes
	Energy Production Fee	d_{EE}^{fee}	EP→IRP	Discharge fee imposed by government to the energy producers.	No
Economic-Price Related	Plant Consumption	k_{P1}	P1→HH	Parameter relating the price of P1 to the demand of P1 by human households	Yes
	Animal Consumption	l_{H1}	H1→HH	Parameter relating the price of H1 to the demand of H1 by human households	Yes
	Agricultural Fee	F_{P1H1}	P1→H1	Parameter relating the price of P1 and the demand of P1 by H1	No
	Plant Material Fee	S_{P1IS}	P1→IS	Parameter relating the price of P1 to the demand of IS	No
	Industrial Energy Consumption	S_{EEIS}	EP→IS	Parameter relating the price of energy to the demand of IS	No
	Household Energy Consumption	O_{EE}	EP→HH	Parameter relating the price of energy to the demand of energy by Human household	No
	Household Biofuel Consumption	K_{EE}	P1→EP	Parameter relating the price of P1 to the demand of energy by Human household	No
Technologic–Mass Related	Theta	θ_{P1}	P1→IS	Amount of P1 required to produce a unit of industrial product IS	Yes
	Khat	$\hat{k}$	P2→H1	Constant value specified by the government. Represents the grazing of P2 by H1.	Yes
	Bioenergy Production Efficiency	λ_{biom}	P1→EP	Amount of biomass (P1) needed to produce a unit of bioenergy	Yes

Plant Consumption Sensitivity Analysis

15

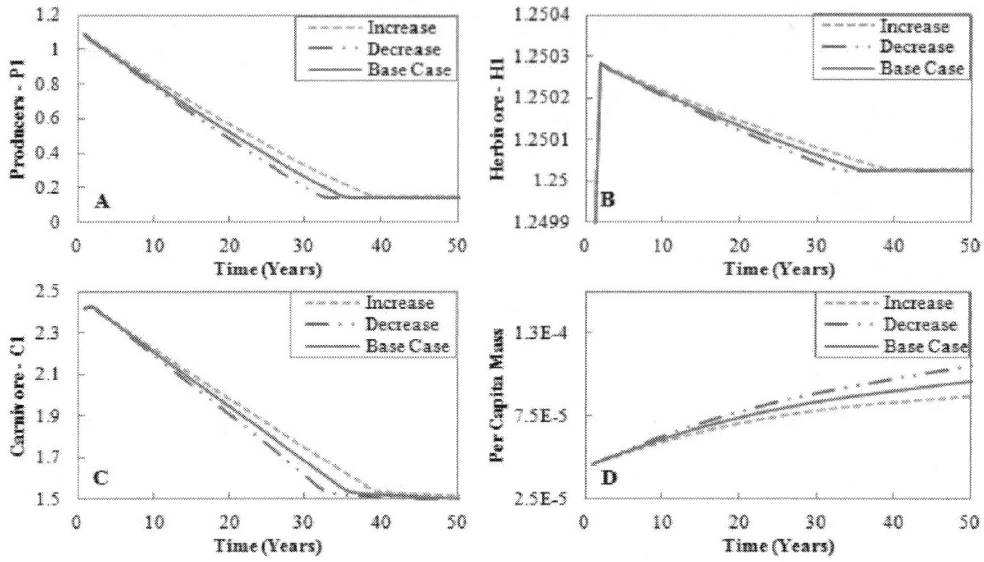

Optimal Control

16

- We want to extend the life span of mass compartments that are heading for extinction.
- By dynamically controlling certain parameters, sustainability is increased.

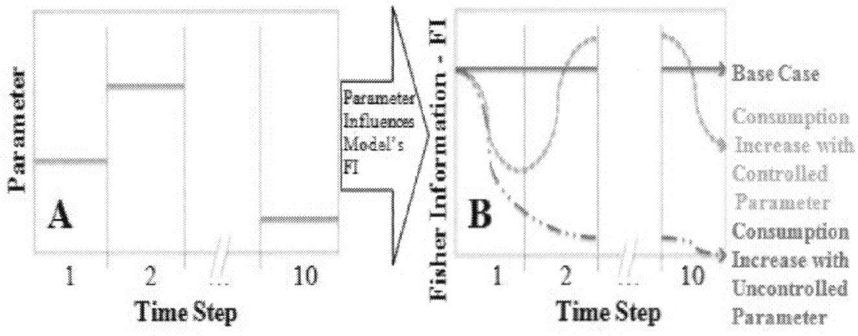

Objective Function

17

- Here is objection function:

$$J = Min \sum_{i=1}^{N} (I(i) - I_c(i))^2$$

- Defined as follows:
  - J = objective Function
  - N = number of time steps
  - I(i) = current FI at time step i
  - I_c (i) = ideal FI at time step i

Non-linear Programming (NLP) Optimizer

18

- The optimal control problem is discretized into a series of decision variables solved using NLP at each time step.
- Initial values of the decision variable (parameters) are known
- The model calculates the objective function and the optimizer tries to satisfy optimality conditions (Karush-Kuhn-Tucker conditions, KKT)
- Optimizer calculates a new value for the decision variable
- Iterative sequence continues until the optimization criteria (KKT) are met

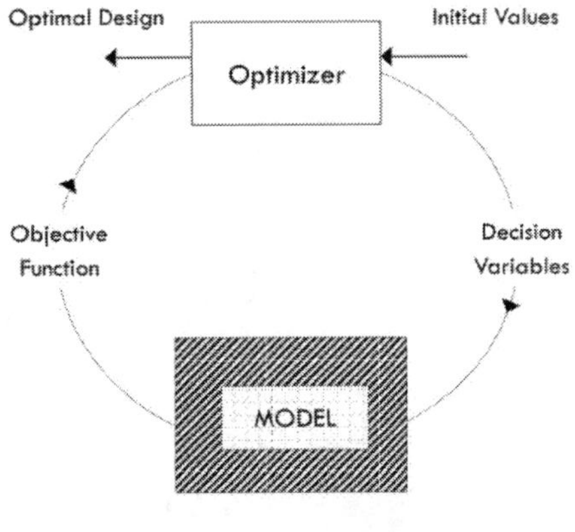

Model Adapted from: Diwekar, Urmila M. *Introduction to Applied Optimization*. Norwell, MA: Kluwer Academic, 2003. Print.

Plant Consumption Single Variable Optimal Control

19

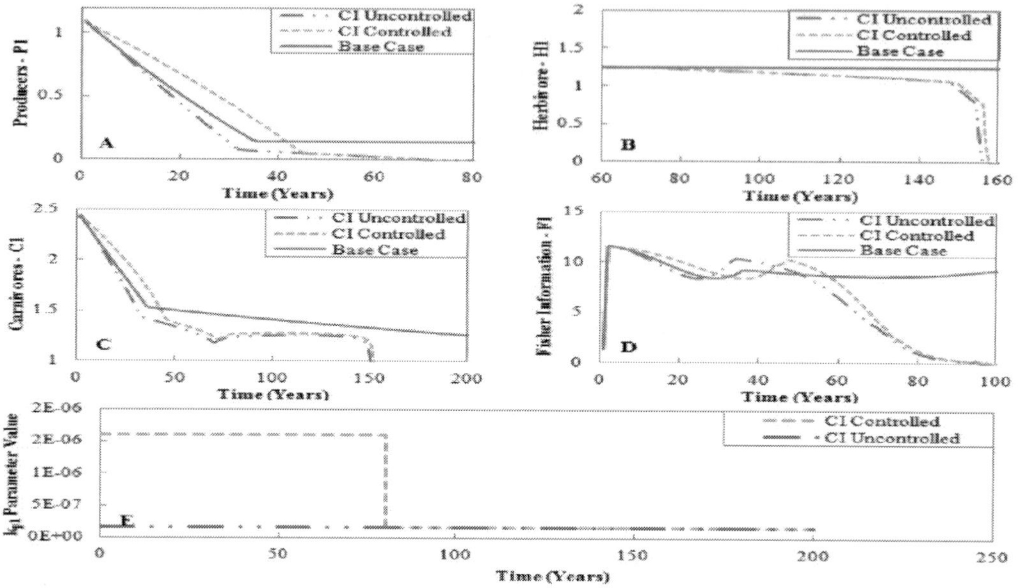

Multivariable Optimal Control for the Consumption Increase Scenario (CI)

20

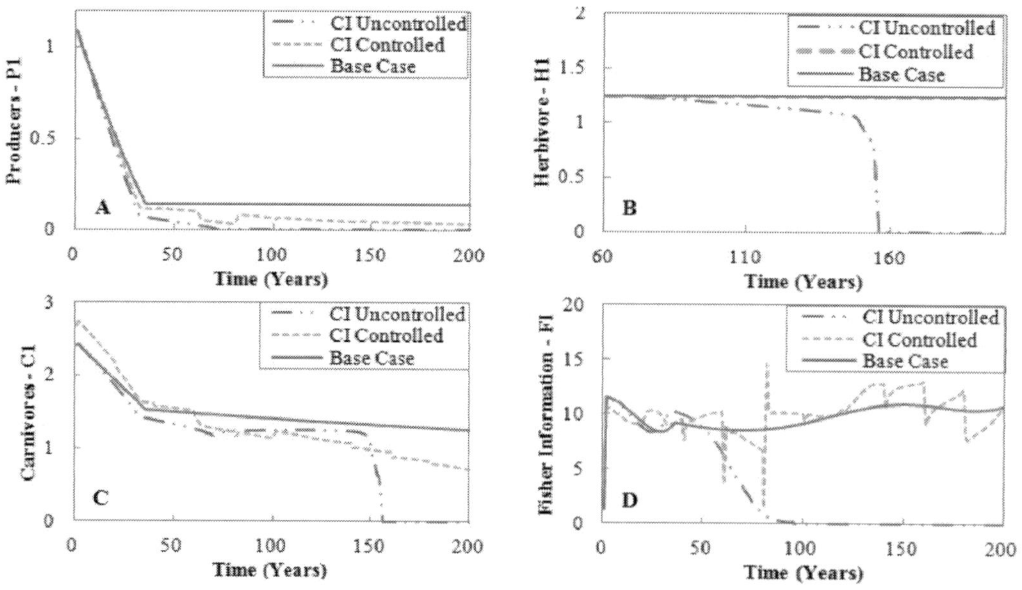

Control Profiles for 6 Parameters

21

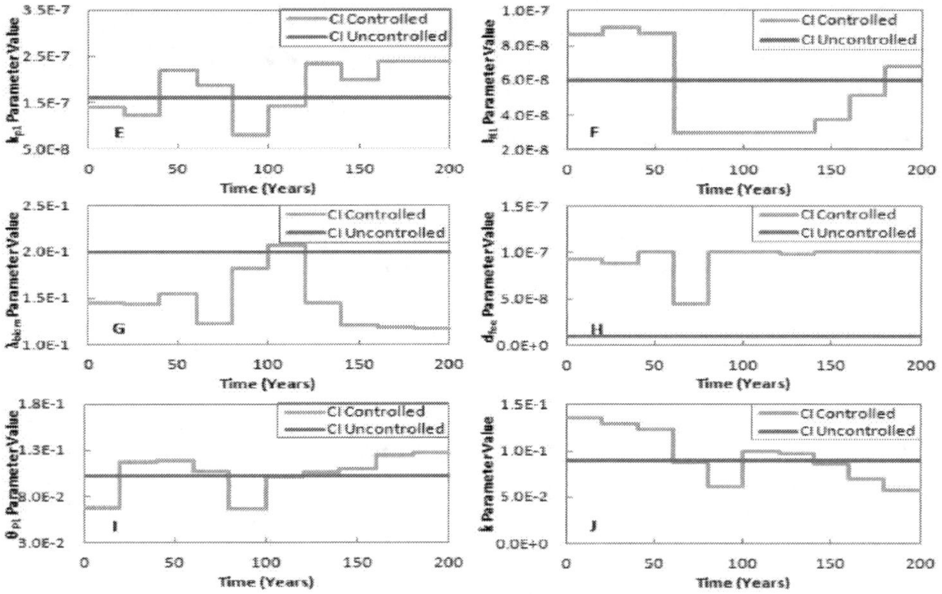

Conclusion

22

- Even while simulating consumption increase, all mass compartments manage to avoid extinction by sustaining past 200 years.
- Sustainability of even a simple ecosystem may not be intuitive
  - Use of biomass as a source of energy accelerates the extinction of species.
- Applying optimal control techniques to ecosystem management simulations can help with determining qualitative conclusions about various regulatory approaches to sustainability.

Life Cycle Assessment of Food Loss Associated with Current U.S. Consumption Compared to the Recommended USDA Food Patterns

- **Conference:** International Congress on Sustainability Science Engineering ICOSSE
- **Year:** 2015
- **Proceeding:** ICOSSE 2015
- **Group:** Abstract Submissions
- **Session:**

Session 6: Energy-Water-Food Nexus: Supply Chains

- **Time:**

Wednesday, May 27, 2015 - 4:30pm-4:50pm

Authors:

Kim, D., University of Arkansas
Thoma, G., University of Arkansas
McNeill, S.

Life cycle assessment was conducted to quantify the potential environmental impacts of food loss associated with current U.S. consumption patterns compared to the recommended USDA Food Patterns (FP). The life cycle estimation of GHG emissions associated with this lost food from production to disposal is, in total, 369 million tons CO_2e per year (3.31 kg CO_2e capita^{-1} day^{-1}) based on the current consumption patterns, and it increases to 427 million tons CO_2e per year (3.83 kg CO_2e capita^{-1} day^{-1}) with adoption of the USDA FP recommendations. Overall environmental impacts from food losses would be greater if U.S. citizens were to adhere to the USDA dietary guidelines with the largest increases observed among the total vegetables, fruit and juices, and milk and dairy food groups. As a conclusion, we call attention to the fact that accumulated reductions of food waste across the entire life cycle will have greater environmental benefits.

unicat

Unifying Concepts in Catalysis

http://www.unicat.tu-berlin.de/

Case Study:

How to Increase the Performance of the Methane Oxidative Coupling Process?

Berlin University of Technology (TU Berlin), Chair of Process Dynamic and Operation, www.dbta.tu-berlin.de

4th International Congress on Sustainability Science & Engineering, May 2015, Balatonfüred, Hungary

Contents [contributions to sustainability]

- **OCM reactions and characteristics [Local solution]**
- **Need for a miniplant study [Process scale analysis]**
- **Study of catalyst and reactor [Less CO_2 production]**
- **Hybrid separation process [Less energy]**
- **Alternative downstream units [Process Intensification]**
- **Integrated reactors [Less energy and less emission]**
- **Integrated processes [Less energy and less emission]**
- **Conclusions**

2 unicat

Ethylene Production and Local Solution

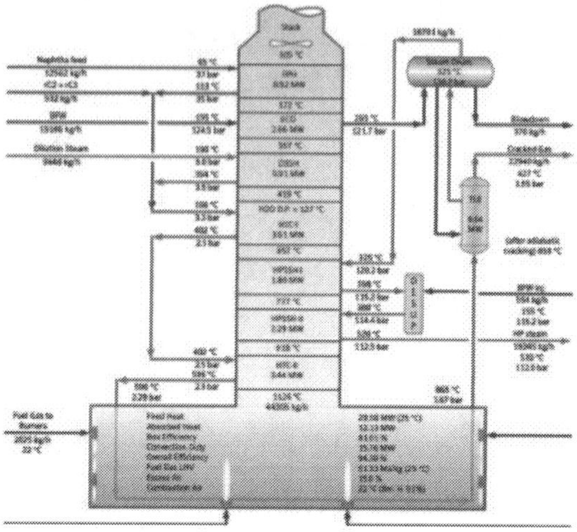

3 unicat

Oxidative Coupling of Methane (OCM) to Ethylene

- **Worldwide research efforts to convert CH_4 into chemicals (C_2H_4, ...)**
- **Oxidative Coupling of Methane (OCM) is a promising direct conversion route of methane to ethylene but has not reached industrial practice yet**

Desired: $2\ CH_4 + O_2 \longrightarrow$ 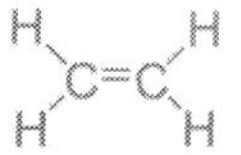 $+\ 2\ H_2O \quad \Delta G\ (800°C) = -\ 306\ kJ\ mol^{-1}$

Undesired: $2\ CH_4 + 3\ O_2 \longrightarrow 2\ CO + 4\ H_2O \quad \Delta G\ (800°C) = -\ 1222\ kJ\ mol^{-1}$

$2\ CH_4 + 4\ O_2 \longrightarrow 2\ CO_2 + 4\ H_2O \quad \Delta G\ (800°C) = -\ 1602\ kJ\ mol^{-1}$

⇒ **kinetic control by means of a catalytic process needed**

4 unicat

Mechanism of Oxidative Coupling of Methane

Highly exothermal reactions 700-900 C

Unstable catalyst and complicated reactor design

Trade-off methane conversion and ethylene yield

Low yield towards ethylene <30%

Complicated separation

OCM: Surface and gas phase reactions

5 unicat

OCM Miniplant and Process Scale Analysis

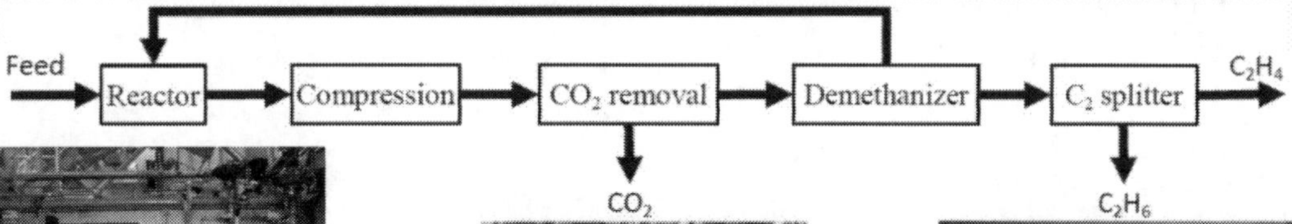

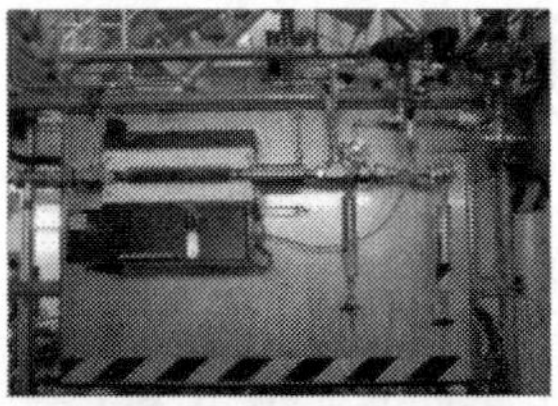

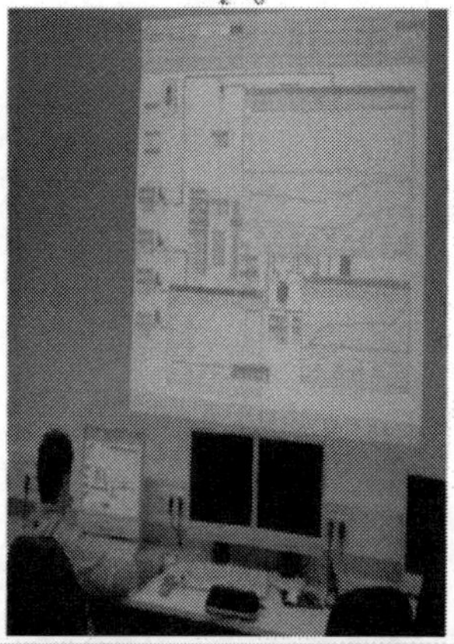

6 unicat

Multi Scales Approach

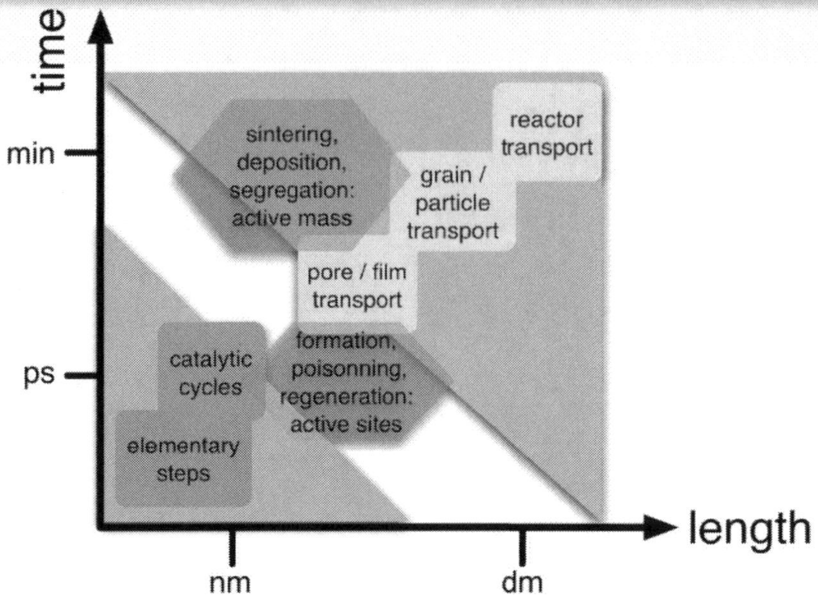

Catalyst Preparation: Two Different Preparations of Mn-Na2WO4/SiO2 Catalyst

Incipient wetness impregnation catalyst [1]

Amorphous SiO_2 (Silical Gel Davisil grade 636)

Aqueous solution $Mn(NO_3)_2 \times 4H_2O$

Aqueous solution $Na_2WO_4 \times 2H_2O$

Drying and calcination

Sol-gel [2] **catalyst**

Pre-hydrolysis of Tetraethyl Orthosilicate (TEOS)

Sodium acetate anhydrous, Manganese (II) acetate tetrahydrate, Ammonium metatungstate hydrate

Mixing the solutions and PH control

Drying and calcination

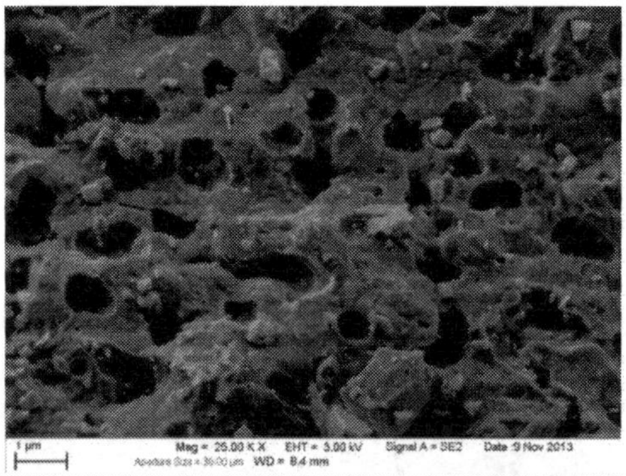

[1] Wang et al., Journal of Catalysis 155 (1995) 390
[2] Godini et al., Journal of Catalysis Today 236 (2014) 12-22

Characterization of Mn-Na2WO4/SiO2 Catalysts

Preparation method	α-cristobalite (SiO_2)	Quartz (SiO_2)	Tungsten oxide (Na_2WO_4)	Braunite ($MnMn_6SiO_{12}$)
Incipient Wetness Impregnation	x	x	x	x
Sol-gel	x	x	x	x

	Na	Mn	W
Sol-gel catalyst [%]	0.73	0.71	0.24
Theoretical bulk concentration [%]	0.71	0.72	0.35
Incipient wetness impregnated catalyst [%]	12.5	3.1	0.8

Catalyst prepared by	Sol-gel method	Incipient wetness impregnation method
BET surface [m^2/g]	1.1	2.5
Bulk density [g/cm^3]	1.2	0.6

Performance of the Catalysts in Membrane Reactor

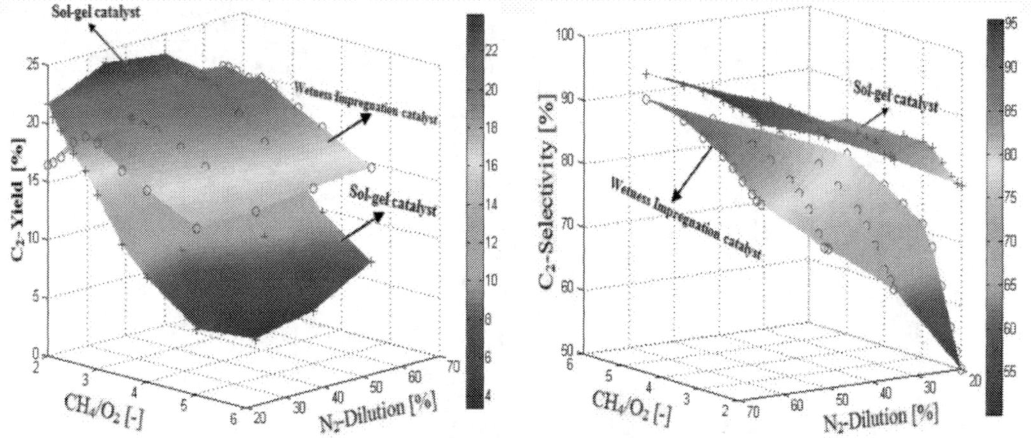

20% ethylene-yield, 78% C_2-selectivity
under very low (20%) diluted reaction atmosphere

Support Variation for Mn-Na_2WO_4/SiO_2

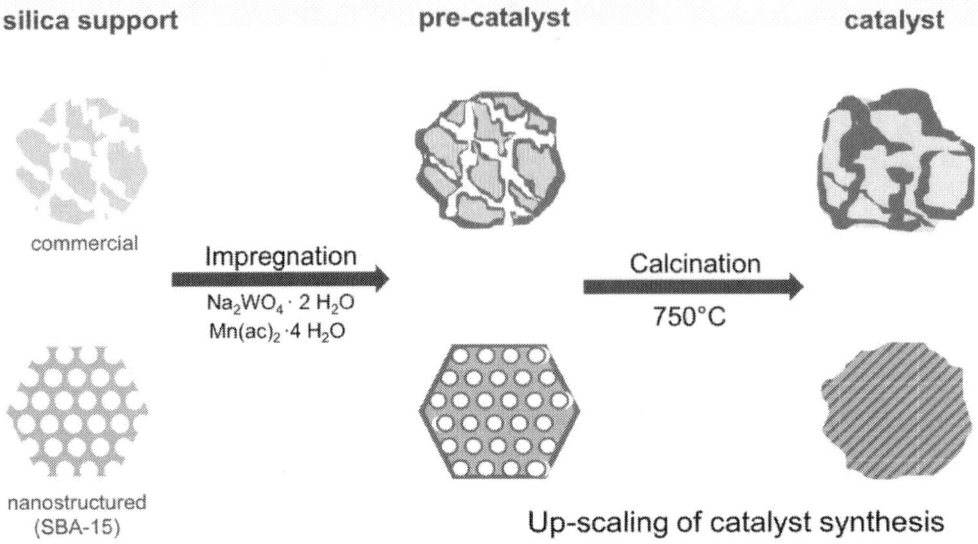

Up-scaling of catalyst synthesis

Large Scale OCM catalyst preparation via UniCat granulator

Alternative OCM Reactors

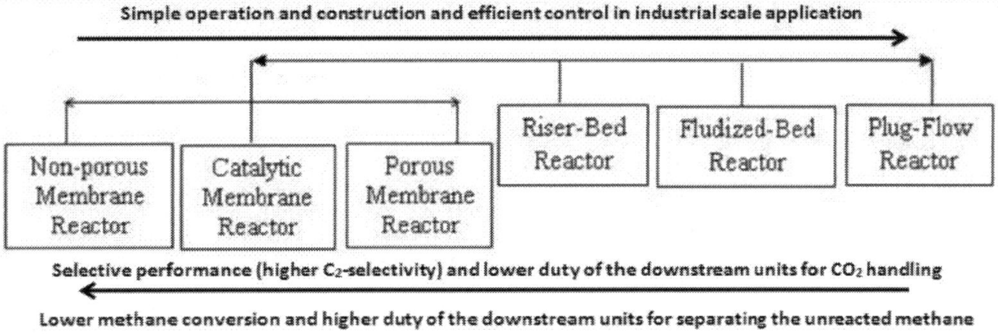

Selected spectrum of the alternative OCM reactor concepts

13 unicat

UniCat Miniplant in TU-Berlin

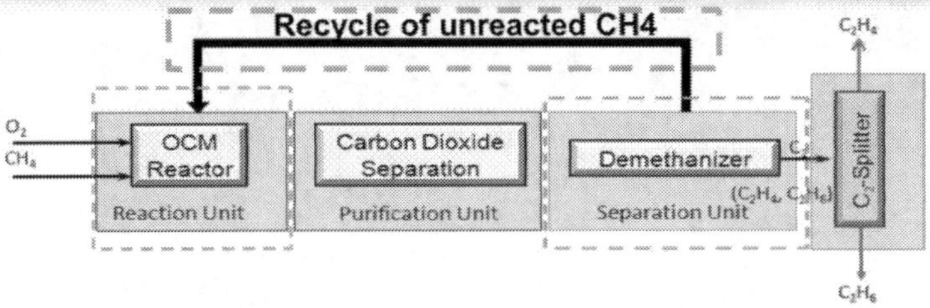

Catalyst development

Reactor Concepts: Fixed Bed Reactor, Fluidized Bed Reactor, Membrane Reactor and Fluidized Bed Membrane Reactor

Carbon Dioxide Separation Technologies: Polymer membrane, Chemical and physical absorbers, Adsorption

Separating the Unreacted Methane: Reactor and process integration

14 unicat

Dimensions & Operating Conditions in UniCat Mini-Plant

Equipment	Duty	Dimension	Operating condition
Fluidized bed reactor	Achieving an isothermal performance for OCM reactions	40 mm ID 40 cm height 45 gram catalyst	700-880 °C, 1-6 bar Typical (820 °C, 1 bar)
Membrane and fixed bed reactor	Achieving a selective performance for OCM reactions	7 mm ID 20 cm length 4 gram catalyst	700-880 °C, 1-6 bar Typical (800 °C, 1 bar)
Absorption column	Using different absorbent for selective CO_2 removal	40 mm ID 500 cm height	20-40 °C, 1-40 bar Typical (30 °C, 10 bar)
Desorption column	Regenerating the absorbent	100 mm ID 400 cm height	110-130 °C, 1-3 bar Typical (120 °C, 1 bar)
Polymeric membrane	Using different polymers and operating conditions for energy efficient CO_2 removal	44 Sheets of 8cm×7cm in 7 compartments	20-50 °C, 1-40 bar Typical (25 °C, 10 bar)
Adsorption	Using different adsorber (e.g. Zeolite 4A) for ethylene removal (some parts of the CO_2 and most of ethane are removed)	60 mm ID 100 cm height	Adsorption 40 °C, 1 bar Desorption 200 °C, 1 bar

15 unicat

View of the OCM Miniplant

Performance Indicators of the Fluidized-bed Reactor

2%Mn-2.2%Na_2WO_4/SiO_2 catalyst, average particle size 250-450 μm

Bed height: 10 cm, $CH_4:O_2$ = 1.7 to 10, Feed Flow rate: 0.1 l/min – 40 l/min, 45 g catalyst

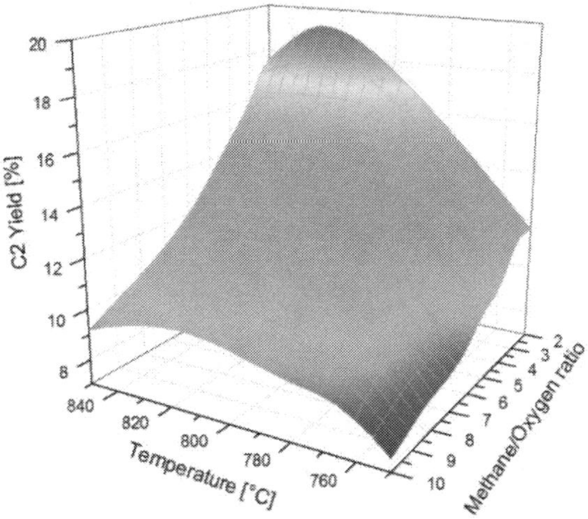

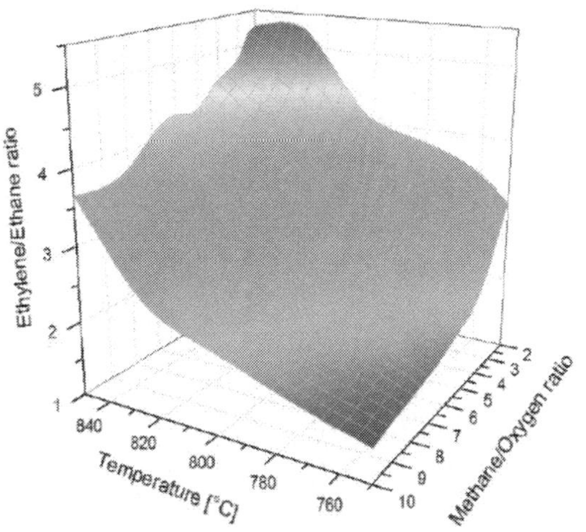

Contours were made by using 48 experimental data

19.6% maximum yield

OCM Membrane Reactor Set-up

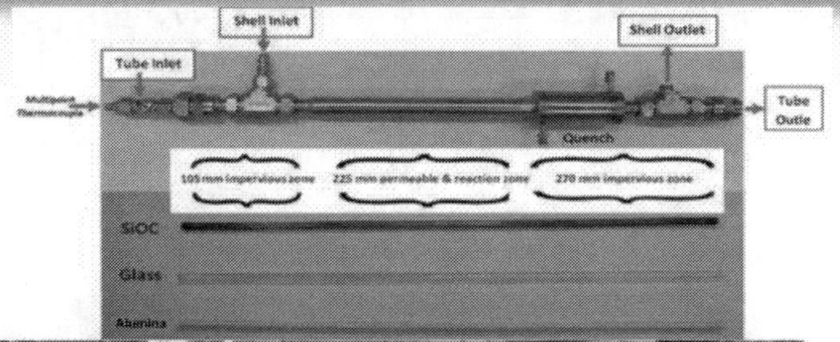

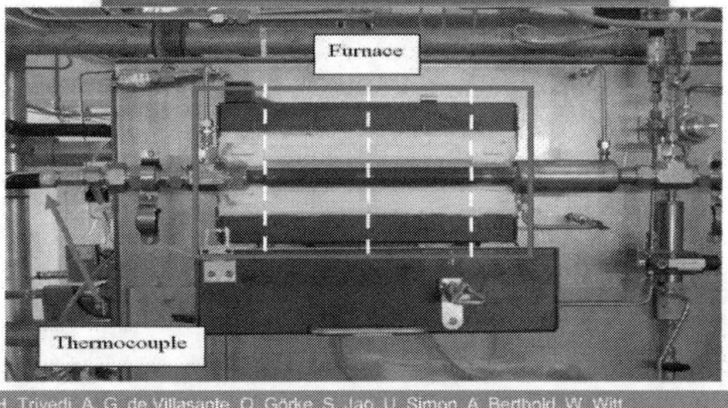

Selective Performance of the Membrane Reactor: Less CO2

2%Mn-5%Na_2WO_4/SiO_2 Catalyst, Average Particle Size: 250-450 µm

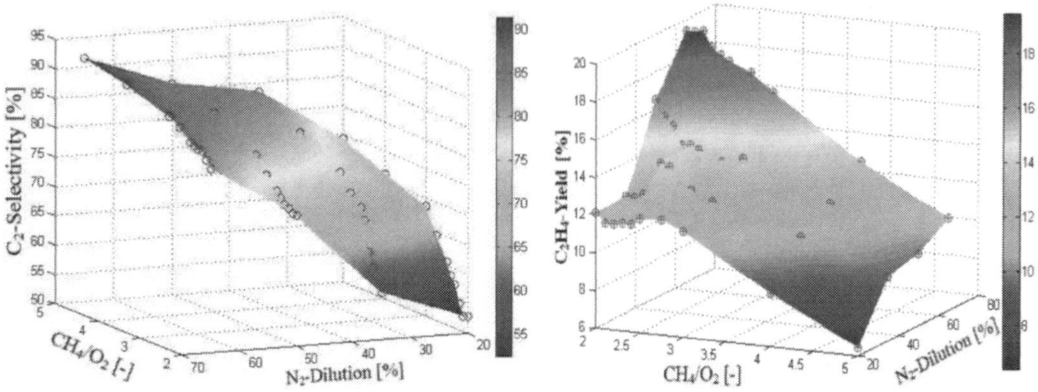

25.5% C_2-Yield (20% C_2H_4-Yield) was achieved under 20% nitrogen dilution

20% C_2H_4-Yield was achieved at 64% C_2H_4-Selectivity

H. R. Godini, A. Gili, O. Görke, U. Simon, K. Hou, G. Wozny, *Energy & Fuels 28* (**2014**) 877 - 890

Absorption-Desorption (Energy Efficiency)

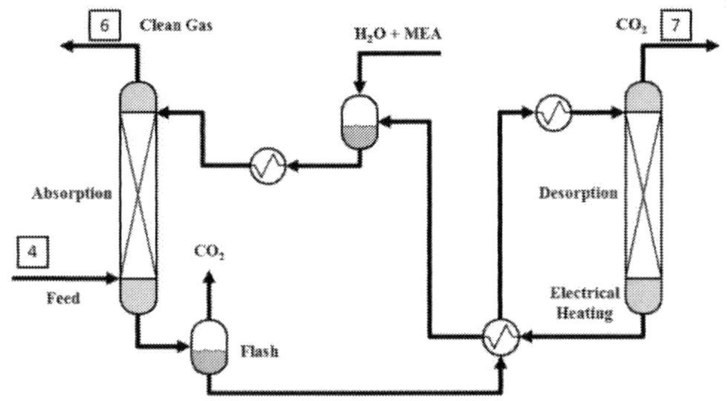

- Online gas analysis
- Titration for liquid analysis
- 22 control loops
- More than 150 actuators and sensors
- Hybrid with membrane separation

Process Integration with Dry Methane Reforming

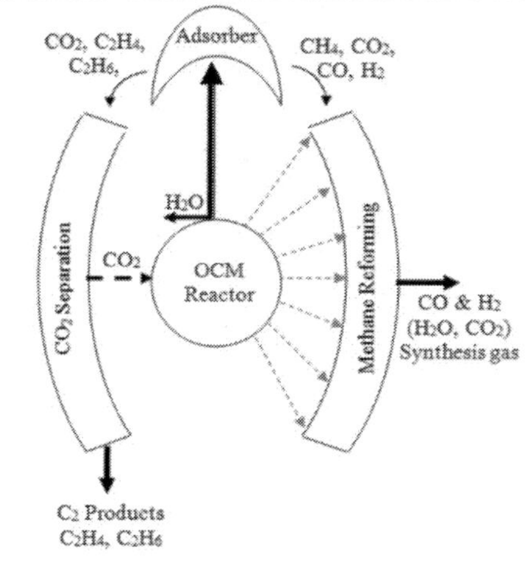

21 unicat

Reactor Integration: Dual-Membrane Reactor

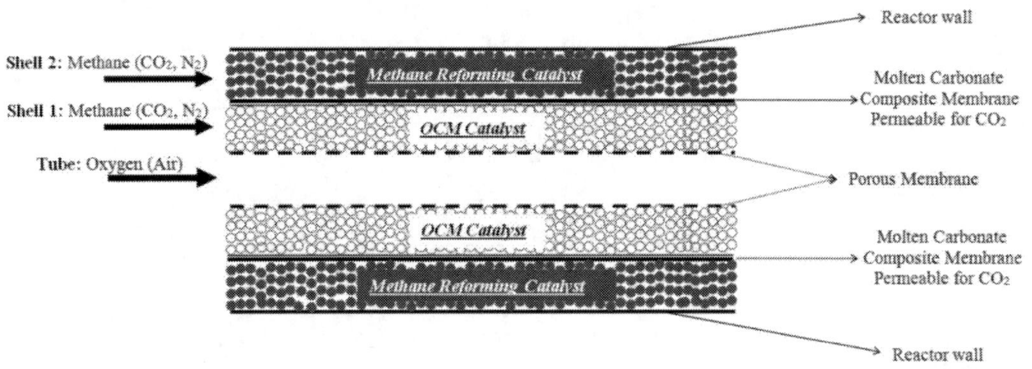

H.R. Godini, S. Xiao, M. Kim, O. Görke, S. Song, G. Wozny, Chemical Engineering and Processing 74 (2013) 153–164

22 unicat

Conclusions and Remarks

- OCM might be a local solution for ethylene production
- Efficient concurrent design and miniplant process analysis
- Catalyst-reactor design for better thermal-selective performance
- Hybrid downstream units for improving the energy efficiency
- Analysing alternative downstream units and integration potentials
- Integrated reactors for Energy efficiency and less emission
- Integrated processes for CO_2 free emission & energy efficiency

OCM case study: Looking for sustainable solution from different perspectives

27 unicat

Acknowledgment

Thank you for the attention!

The support from the Cluster of Excellence „Unifying Concepts in Catalysis" coordinated by the Berlin Institute of Technology - Technische Universität Berlin and funded by the German research foundation - Deutsche Forschungsgesellschaft – DFG is sincerely acknowledged

BASF The Chemical Company

EVONIK INDUSTRIES

SIEMENS

ThyssenKrupp Uhde

Helmholtz-Zentrum Geesthacht
Zentrum für Material- und Küstenforschung

SCHMIDT + HAENSCH

Endress+Hauser

unicat

International Congress on Sustainability Science and Engineering (ICOSSE '15)
Balatonfured, Hungary
May 26-29, 2015

Reduction of Excess Activated Sludge by a High Pressure Jet Device (HPJD)

Masaaki Hosomi [1], Toshikazu Suenaga [1], Hiroyuki Yoshino [1], Xie Li [1], Tadahiro Fujii [3], Hiroshi Satoh [2], Shohei Riya [1] and Akihiko Terada [1]

1) TUAT; Tokyo Univ. of Agr. & Technol. Department of Chemical Engineering
2) TEPSCO; Tokyo Electric Power Services co., ltd
3) DPK; Dojo-kankyo Process Laboratory co., ltd

Activated Sludge – Excess Sludge treatment accounts for 18-57%

2

76 million ton/yr - disposal costs US $1.2 billion /yr
(US $ 160 /ton)
(2009, Japan Ministry of the Environment)
Lifetime of a final disposal site ··· 13 years

Reduction of excess sludge is an engineering challenge

Solubilization of Activated Sludge (biological, physicochemical)

Ozone treatment
Hydrothermal treatment
Bead mill etc...

Problems

- High cost and low efficiency
- Complexity of a device
- Limited treatment capacity

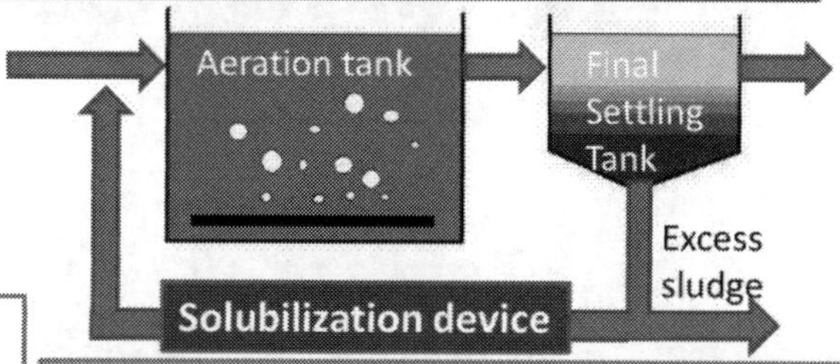

Development of a new solubilizaton device ; HPJD

4

HPJD: High-Pressure Jet Device (simple, cost-effective, maintenance-free)

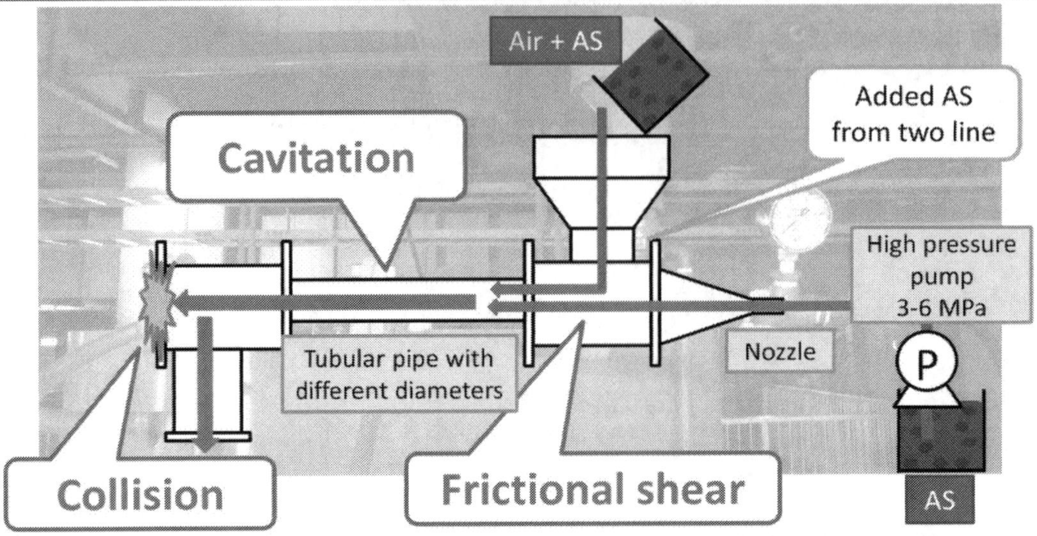

Mechanism of solubilizaton: Dual supply of excess activated sludge from these two points causes turbulent mixing of the sludge with air and generates microbubbles in the pipe, resulting in friction, cavitation and collision of bacterial cells in the sludge. Such multiple effects damage bacterial cells.

7

Batch treatment of AS by HPJD

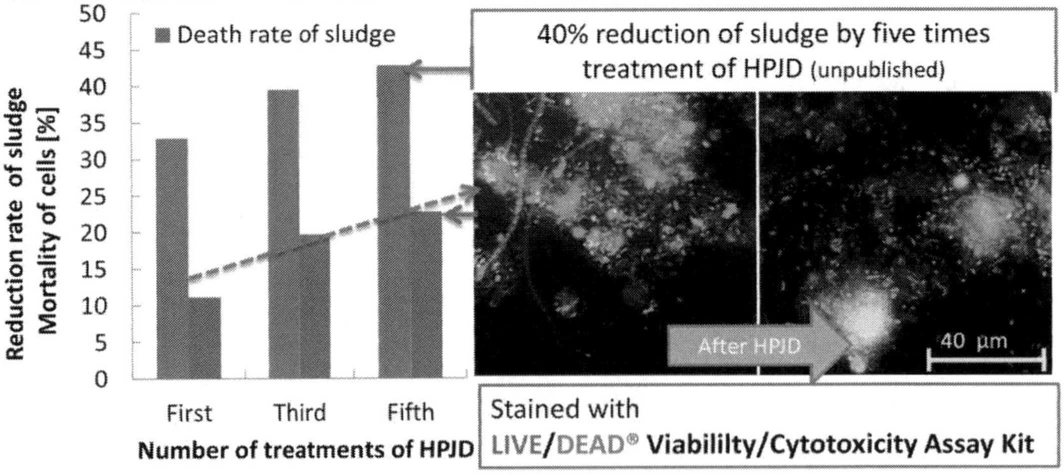

HPJD is an effective technique to reduce AS

Application of HPJD to a pilot-scale AS system

5

Questions

1. How much is the excess sludge reduced by HPJD in AS processes?
2. Are there any adverse effects of HPJD on treatment performance of AS processes?
3. Can HPJD change the microbial community of AS?
4. How much can HPJD reduce O&M cost?

Pilot-scale AS system

Set up in "Kitatama Wastewater Treatment Plant (Tokyo, Japan) "

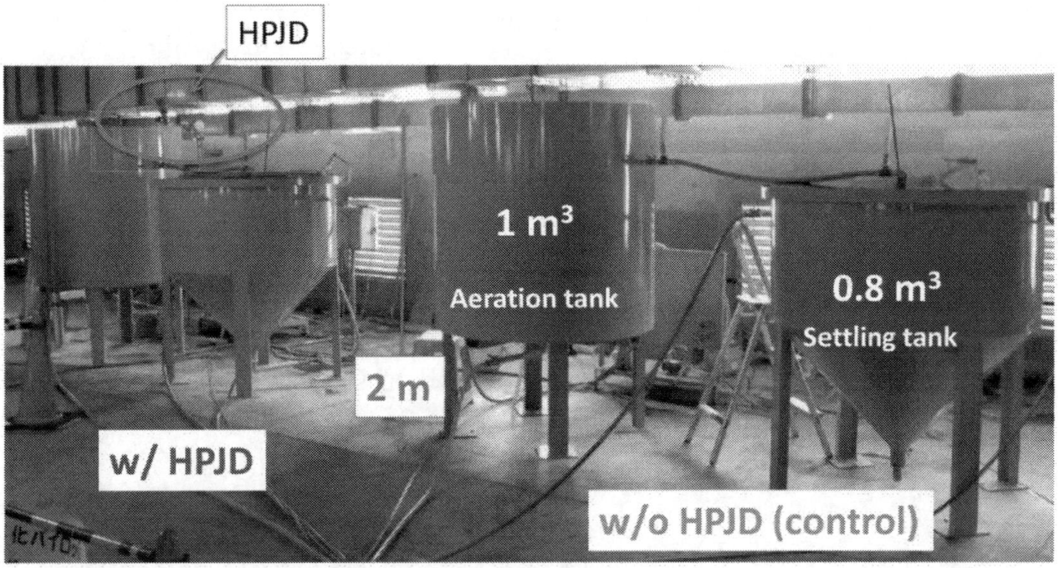

MLSS, sludge volume index (SVI), total organic carbon (TOC), inorganic carbon (IC), total nitrogen (T-N), ammonium (NH_4^+), nitrite (NO_2^-) and nitrate (NO_3^-).

Comparison of cumulative excess sludge

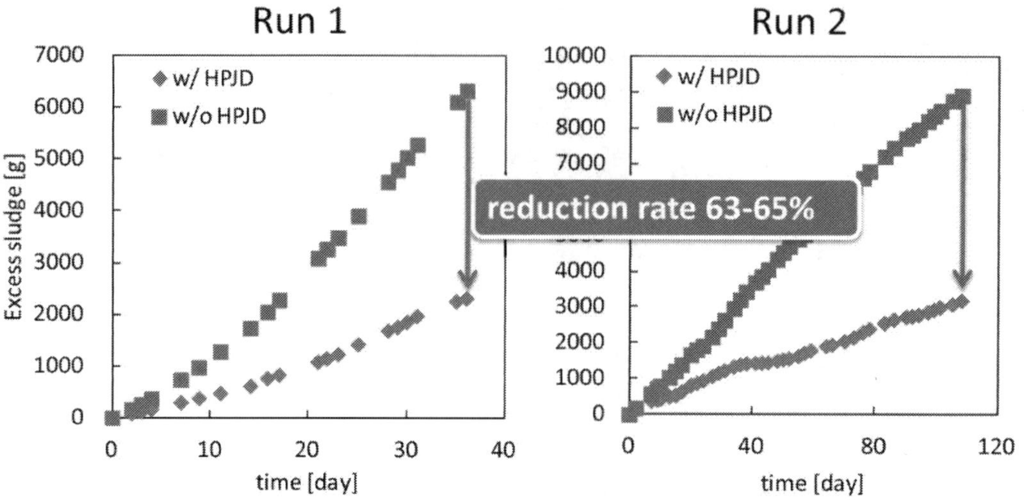

Excess sludge volume[g]	w/ HPJD	w/o HPJD
Run 1 for 37 days	1769	5059
Run 2 for 108 days	1278	3463

Efficiency of Sludge Reduction and Energy Consumption

Technology	Reduction [%]	**Energy [Wh/g-MLSS]**	References	Scale
Ozonation	100	**3**	Yan et al. (2009)	4.6 L, CSTR
Beadmill	89	**3.78**	Rai et al. (2008)	3L, CSTR
HPH	42	**1.40**	Lan et al. (2013)	1260 L, SBR
HPJD	65	**0.14**	In this study	1000 L, CSTR

*HPH = High Pressure Homogenizer

Compared to typical sludge reduction technologies, HPJD has sufficient sludge reduction potential and significantly low energy consumption.

Comparison of MLSS in aeration tank

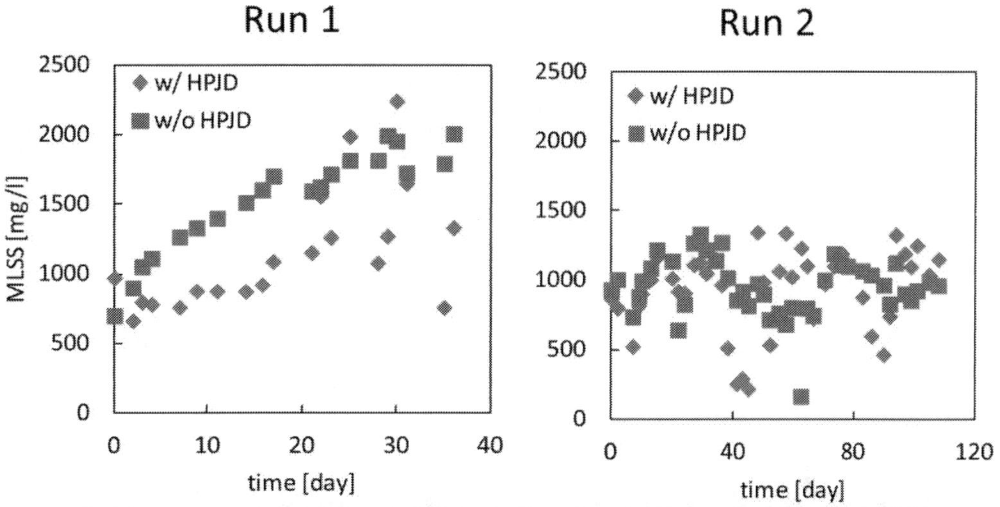

Mean MLSS [mg/l]	w/ HPJD	w/o HPJD
Run 1	1146	1532
Run 2	932	952

Effect of HPJD on treatment performance

TOC removal rate

TOC or NH_4^+ removal rate [%]		w/ HPJD	w/o HPJD
TOC	Run 1	69.4	68.0
	Run 2	58.9	52.3
NH_4^+	Run 1	99.8	99.5
	Run 2	86.9	88.5

NH_4^+ removal rate

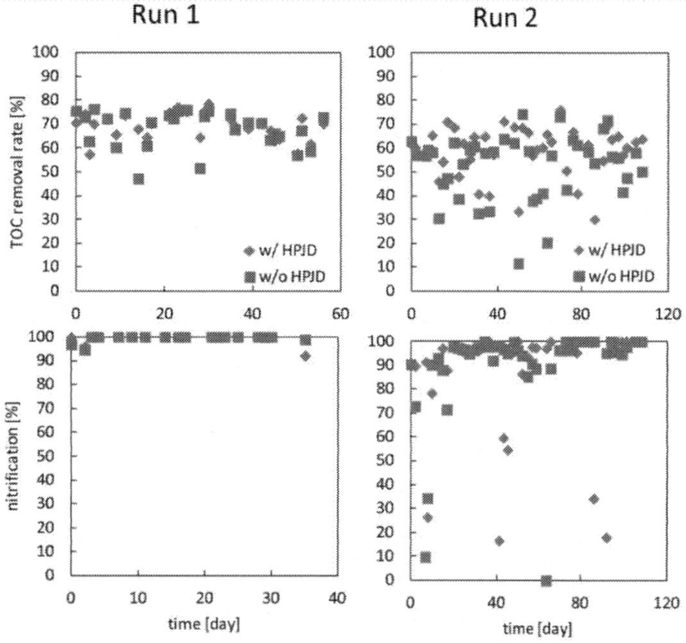

Effect of HPJD on SVI

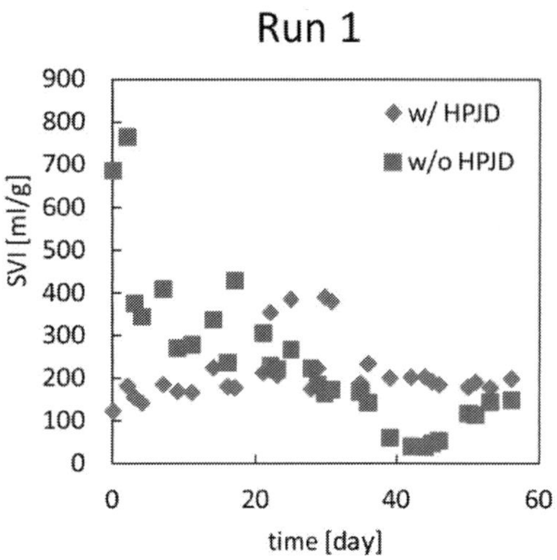

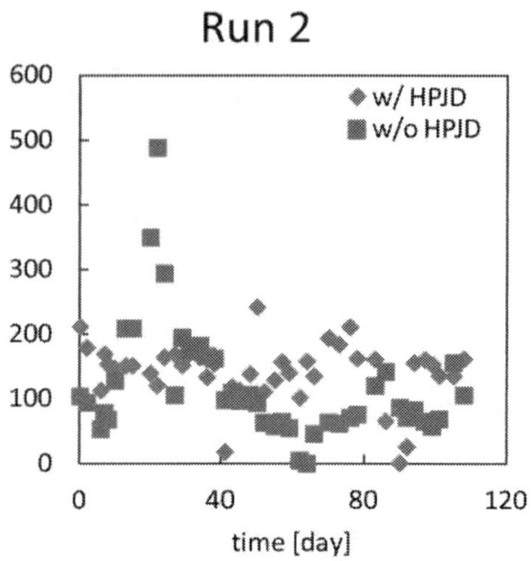

Mean SVI [ml/g]	w/ HPJD	w/o HPJD
Run 1	215	242
Run 2	143	119

No adverse effect on SVI

Effects of HPJD on effluent SS

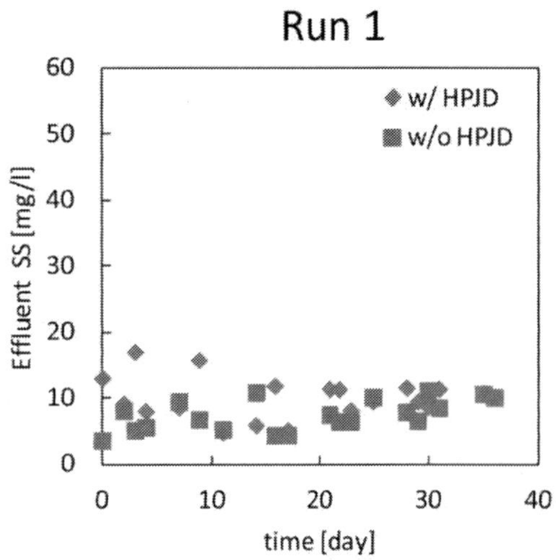

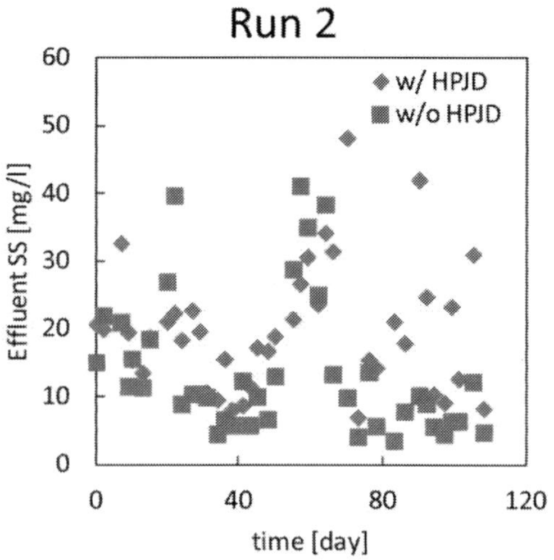

Mean Effluent SS [mg/l]	w/ HPJD	w/o HPDJ
Run 1	10.1	7.5
Run 2	19.9	14.0

Particle size of AS (A: HPJD; B: w/o HPJD)

HPJD may destruct the floc of sludge and make it small-size particle of sludge

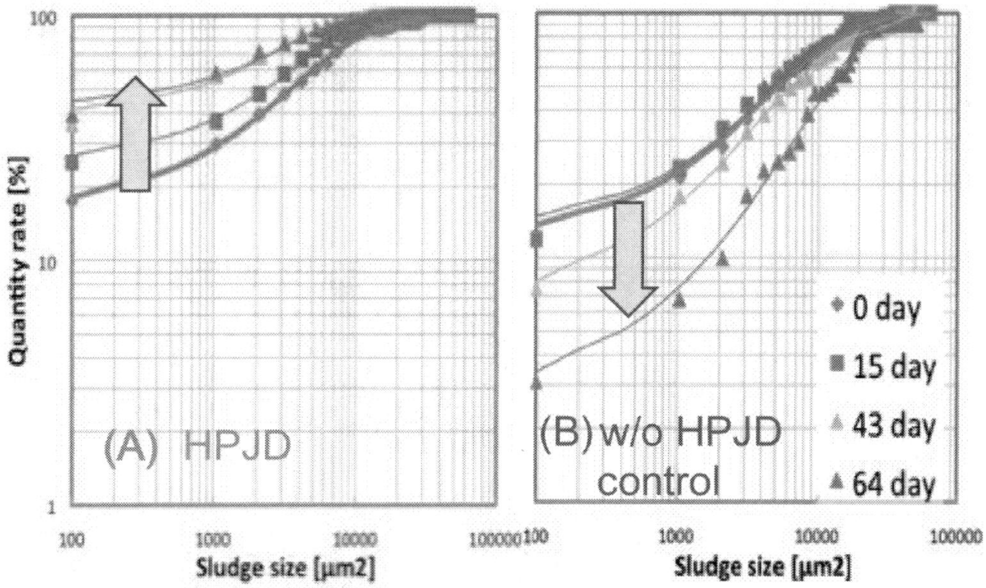

Prokaryote community

Next generation sequencer analyis

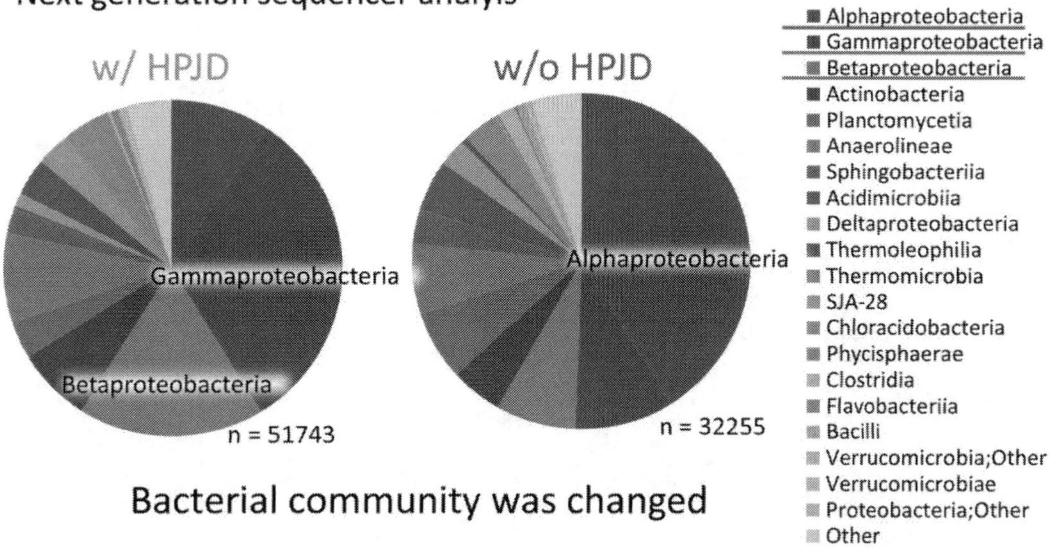

Bacterial community was changed

However water treatment efficient was not changed

14

Eukaryote community

Microscopy analysis

w/ HPJD

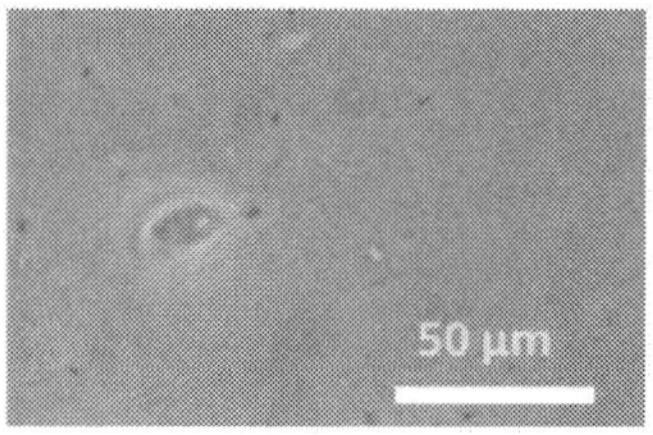

Bodo sp.
(Protozoa)
~30 µm

Prey on suspended bacteria

w/o HPJD (control)

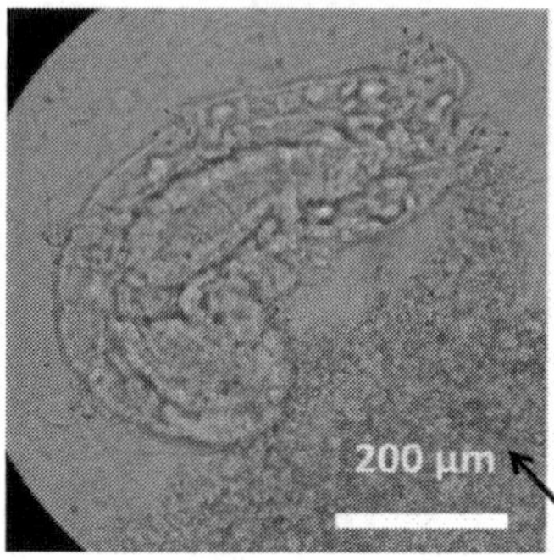

Macrobiotus
(Metazoa)
~800 µm

Prey on flocculated bacteria

Floc disaggregation of sludge by HPJD treatment may cause microbial community changes.

15

Expenses of O & M cost in publicly owned wastewater treatment plant

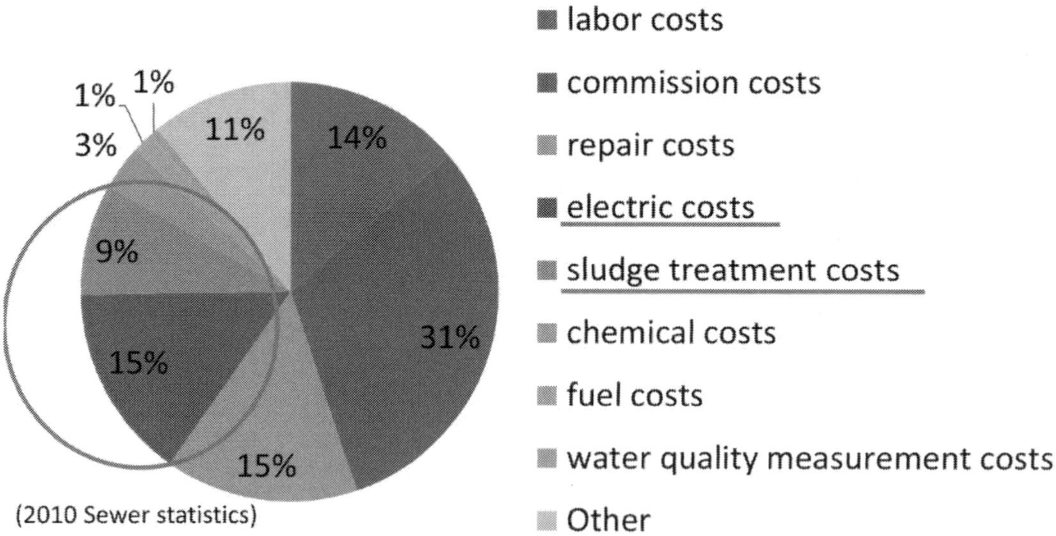

(2010 Sewer statistics)

16

Aeration Rate and DO concentration

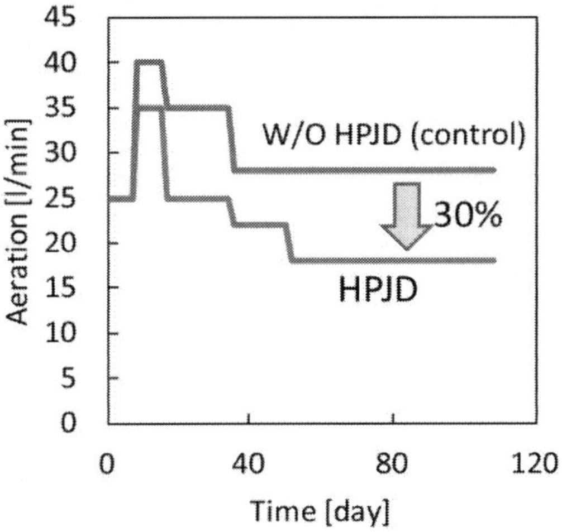

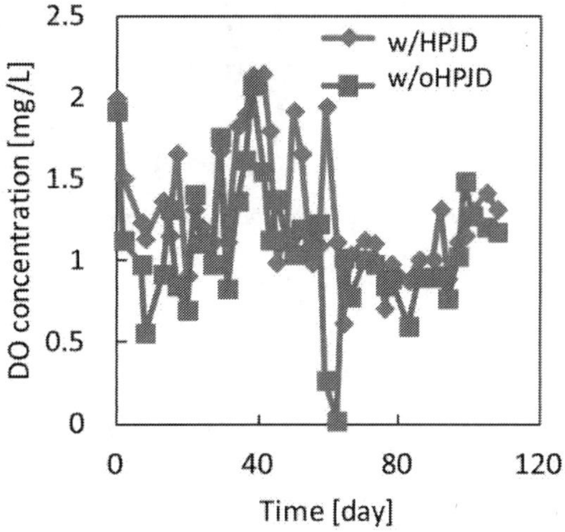

*Aeration rate: Rate of air to be blown to aeration tank

HPJD can decrease about 30% of aeration cost .

17

Cost Estimation

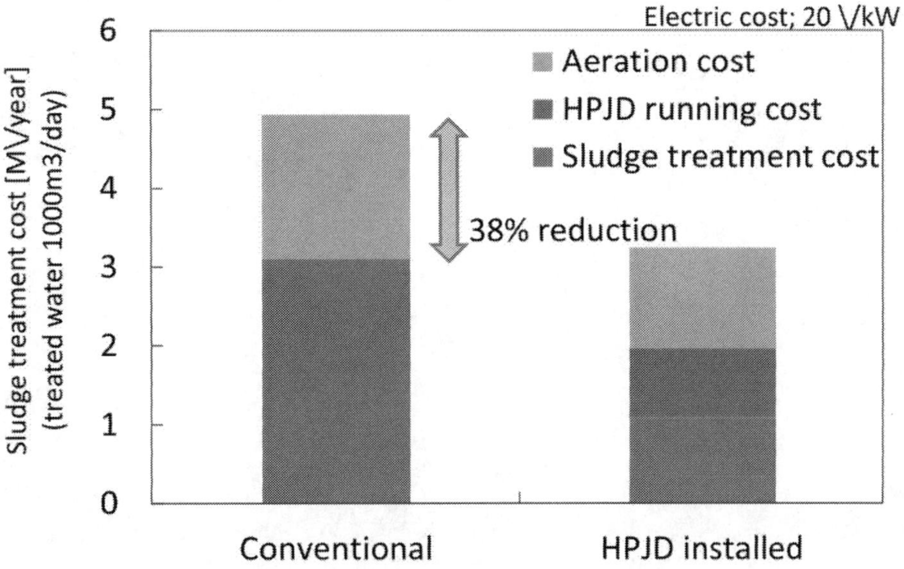

Introduction of HPJD results in 38% cost reduction

18

Conclusions

1. **63% to 65% of excess activated sludge was reduced by HPJD for 37 and 108 day operation of pilot-scale activated sludge processes.**

2. **Application of HPJD did not adversely affect settling of activated sludge and treatment performances regarding organic carbon removal and nitrification**

3. **Application of HPJD resulted in emergence of different microbial communities compared to those in the control system without HPJD.**

4. **Introduction of HPJD into activated sludge treatment plant results in 38% O & M cost reduction.**

Acknowledgements: We have had the convenience of experiment from Tokyo Kitatama Water Reclamation center. We show gratitude here greatly.

Acknowledgements: This work was supported by the Adaptable and Seamless Technology Transfer Program through the target-driven R&D of Japan Science and Technology Agency (AS2311489E), a Grant-in-Aid for Challenging Exploratory Research (24656311) and a Grant-in-Aid for Scientific Research (A) (26249076) from the Ministry of Education, Culture, Sports, Science, and Technology of Japan.

Biochemical Engineering Journal 100 (2015) 1–8

Contents lists available at ScienceDirect

Biochemical Engineering Journal

journal homepage: www.elsevier.com/locate/bej

High-pressure jet device for activated sludge reduction: Feasibility of sludge solubilization

Toshikazu Suenaga[a,1], Mio Nishimura[a,1], Hiroyuki Yoshino[a], Hiroaki Kato[a], Minoru Nonokuchi[b], Tadahiro Fujii[c], Hiroshi Satoh[d], Akihiko Terada[a,*], Masaaki Hosomi[a]

[a] Department of Chemical Engineering, Tokyo University of Agriculture and Technology, Naka-cho 2-24-16, Koganei, Tokyo 184-8588, Japan
[b] Tokyo Electric Power Service Company, 1-7-12 Shinonome, Koto, Tokyo 135-0062, Japan
[c] DPK, 5780-8 Shinyoshida, Tsuzuki, Yokohama, Kanagawa 223-0056, Japan
[d] R&D Center, Tokyo Electric Power Company, 4-1 Egasaki-cho, Tsurumi, Yokohama, Kanagawa 230-8510, Japan

ARTICLE INFO

Article history:

ABSTRACT

The objective of this study was to investigate the feasibility of a high-pressure jet device (HPJD) as an

P-Graph Approach to Optimal Crisis Operations in Multi-Product Systems

R. R. Tan, M. F. D. Benjamin, C. D. Cayamanda, K. B. Aviso
Chemical Engineering Department
De La Salle University, Manila, Philippines

ICOSSE, May 26-29, 2015, Balatonfured, Hungary

Outline of Presentation

- ❑ Climate change issues
- ❑ Food, water and energy nexus
- ❑ P-Graph methodology
- ❑ Aluminum complex case study
- ❑ Implications, conclusions and further research opportunities

Exceeding the Limits

(Rockstrom et al., 2009)

- ❑ Atmospheric CO_2 levels now exceed 400 ppm
- ❑ Global GHG emissions continue to grow, fuelled by economic and demographic trends
- ❑ Climate change has complex links with other issues – e.g., biodiversity loss, water stress, land use

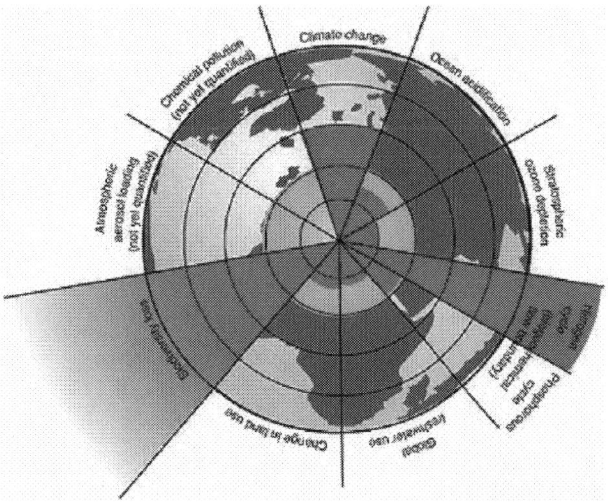

3

CO_2 Abatement Technology "Wedges"

(Source: International Energy Agency)

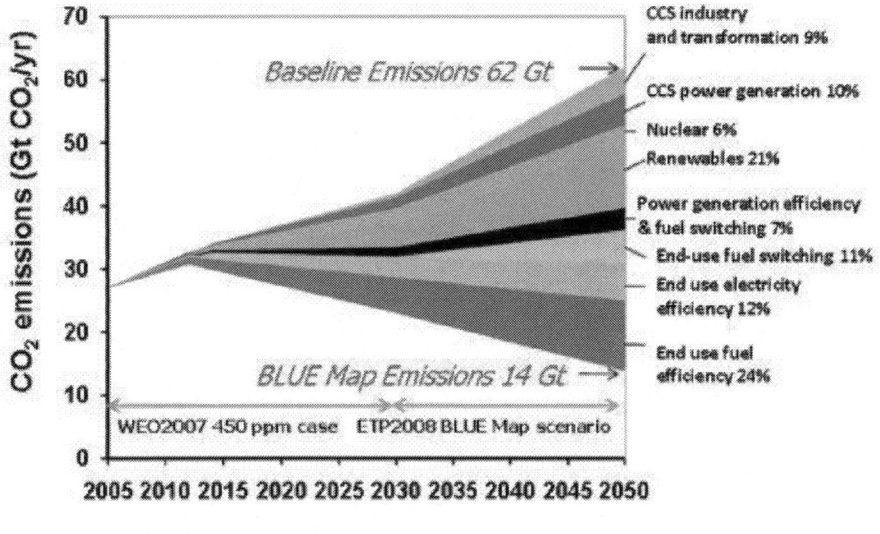

4

Energy Security Issues

Energy security and supply resilience are now essential aspects of climate change adaptation strategies.

Unfortunately, many low-carbon technologies may themselves be vulnerable to climatic effects.

5

Risk Analysis
(Kaplan and Garrick, 1981)

- ❑ What can go wrong?
- ❑ What is the likelihood?
- ❑ What are the consequences?

System connectivity and interdependencies have made rigorous modeling essential for the third question.

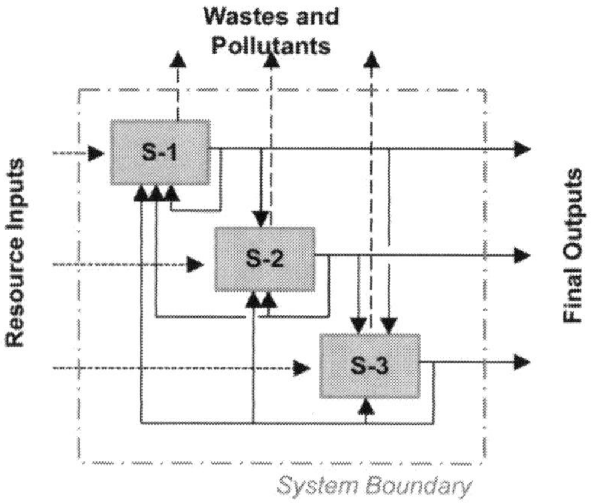

6

Why Process Systems Engineering?

Focus of PSE on the development of systematic, rigorous techniques to aid decision-making for specific classes of problems.

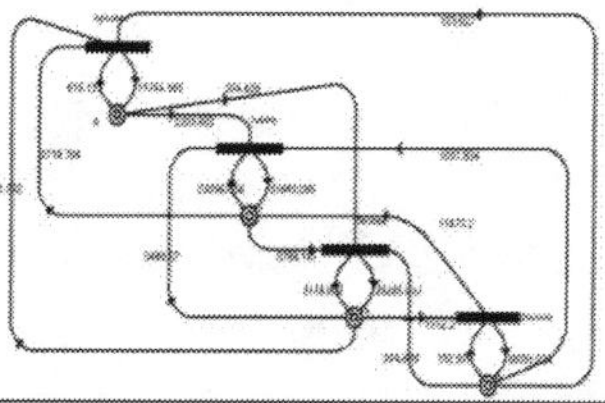

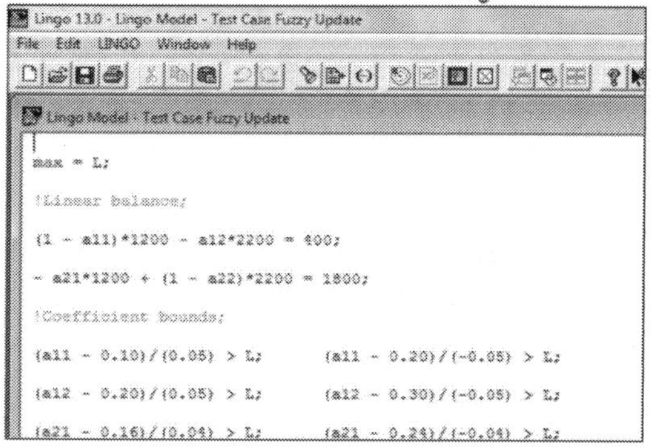

7

P-Graph Methodology

- Graph theoretic framework for process network synthesis (PNS)
- Mathematical foundations were outlined in a series of papers by Friedler et al. (1992a, b; 1993)
- Numerous PNS applications have emerged as surveyed by Lam (2013)
- Software and resources available at www.p-graph.com

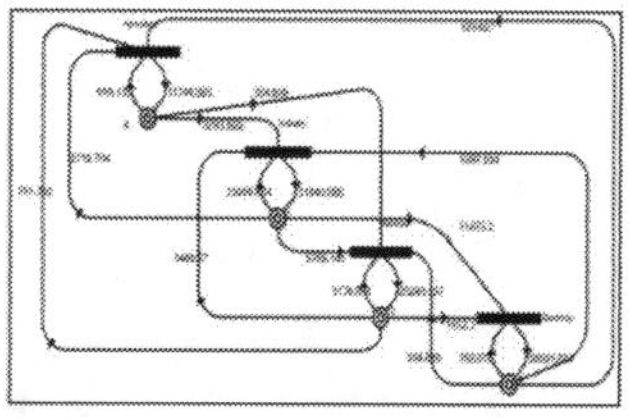

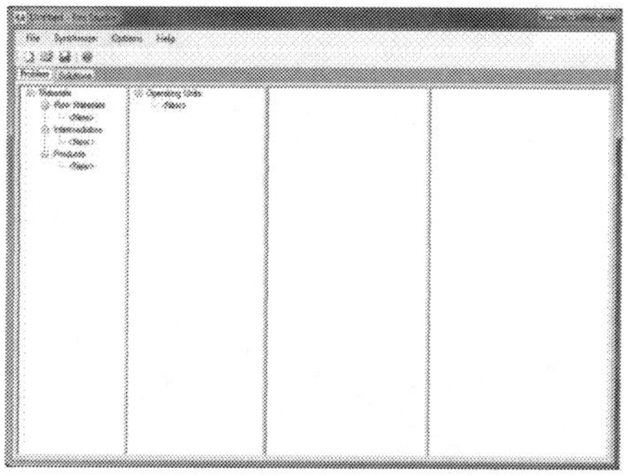

8

P-graph Methodology

(Friedler et al., 1992a, b; 1993)

Algorithm	Description
Maximal Structure Generation (MSG)	Generation of union of all combinatorially feasible networks
Solution Structure Generation (SSG)	Generation of all feasible networks from MSG
Accelerated Branch-and-Bound (ABB)	Determination of optimal solution with efficient search

9

Crisis Operation of Polygeneration Plant

(Tan et al., 2014)

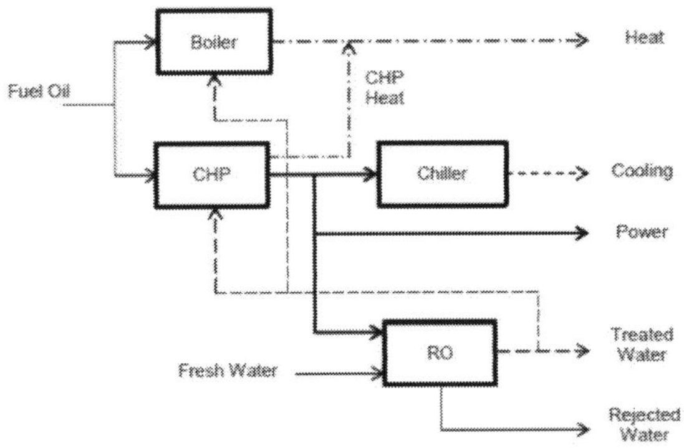

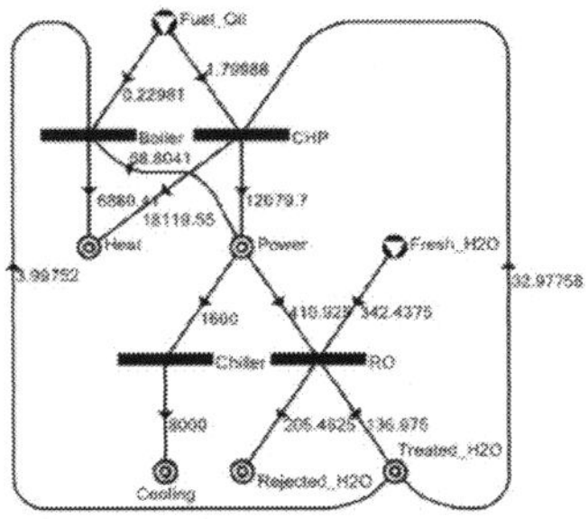

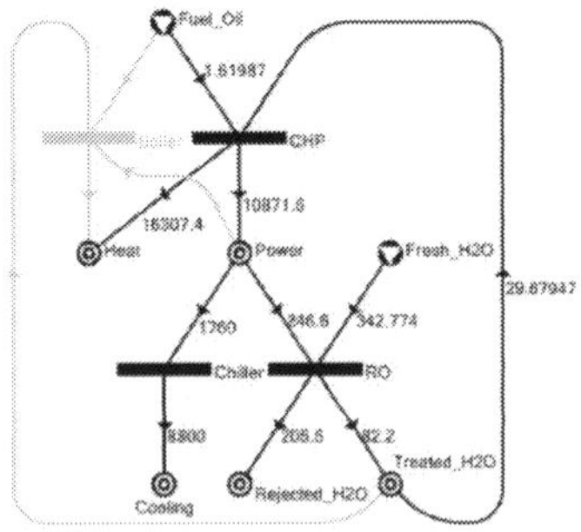

Crisis Operation of Economic System
(Aviso et al., 2015)

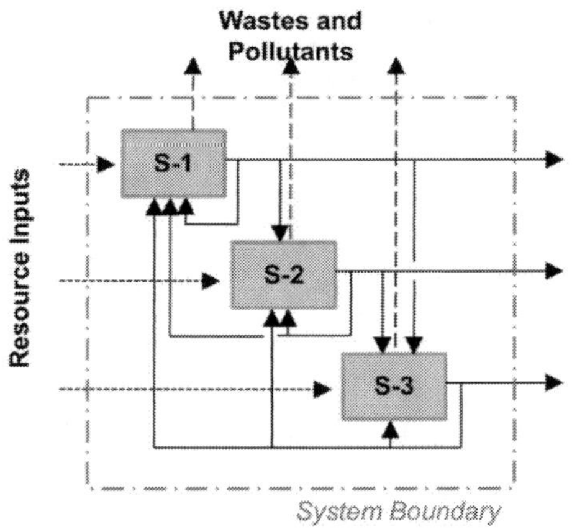

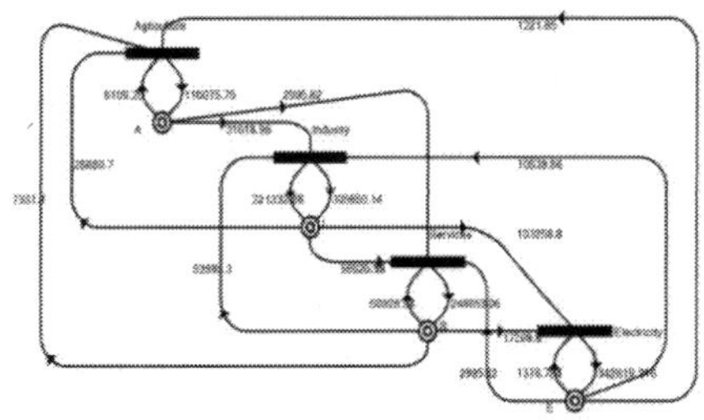

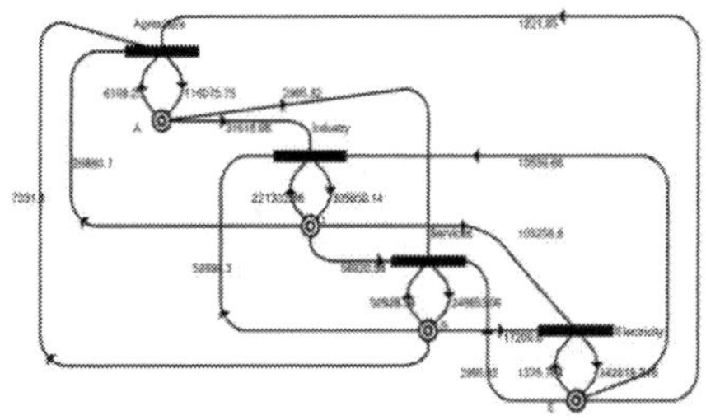

Problem Statement

- Given an industrial complex comprised of *n* plants and *n* products
- Given a crisis event that results in the reduction in availability of the *k*th product
- Given that each plant has a minimum level of feasible partial load operation
- The problem is to determine the optimal allocation of production capacity during a crisis, so as to maximize total value of final products

12

Aluminum Case Study

- Aluminum production is highly electricity intensive due to power requirements for electrolytic primary production.
- Data in this case study is based on Hendrickson et al. (2006) and Evans et al. (2009).
- Baseline production levels of 0.738, 0.228 and 0.034 t/h of shape-cast, extruded and rolled aluminum, respectively.
- Minimum partial load limits are 0.6, 0.5 and 0.8 for shape casting, extrusion and cold/hot rolling, respectively.
- It is assumed that changes in water availability reduces electricity supply by 2%, thus requiring adjustments in production levels.
- Relative values of shape cast, extruded and rolled aluminum are 1.25, 1.1 and 1, respectively.

13

Baseline State of Aluminum Complex

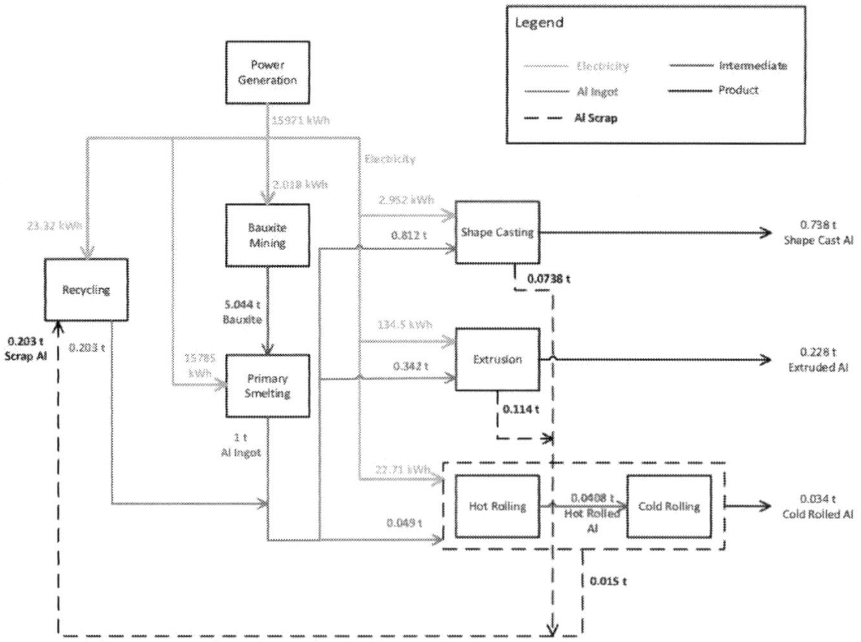

14

P-graph Representation of Aluminum Complex at Baseline State

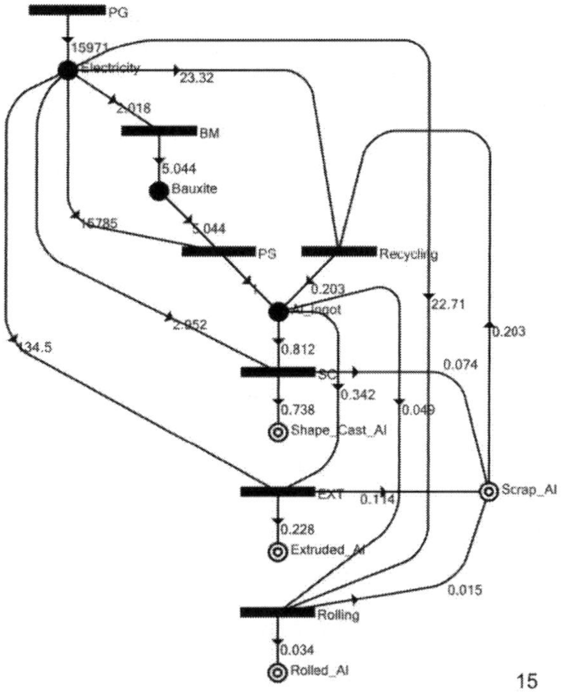

15

Optimal Allocation During Crisis

Process	Operating Capacity
Power Generation	0.980
Mining	0.981
Primary Smelting	0.981
Recycling	0.954
Shape Casting	1.000
Extrusion	0.945
Cold Rolling	0.800

Product Output	Reduction from Baseline
Shape Cast Aluminum	0.00%
Extruded Aluminum	5.53%
Rolled Aluminum	20.00%

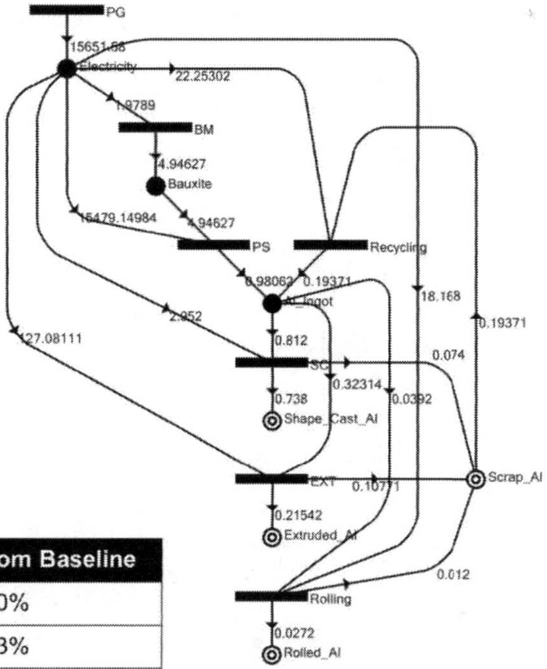

16

Sensitivity Analysis

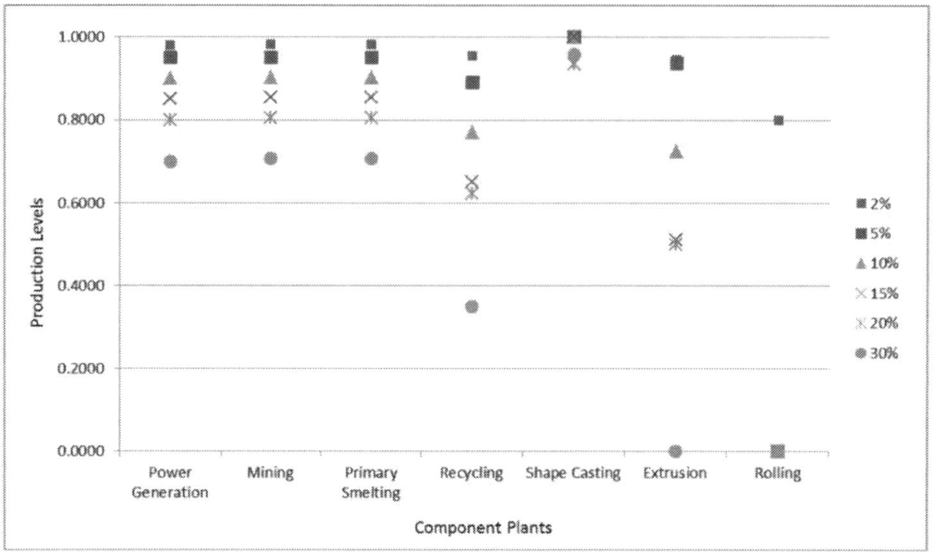

17

Conclusions and Further Work

- Resilient industrial systems will be needed to cope with future climatic effects
- We have developed a P-graph based approach for determining optimal operating strategies for industrial complexes during climate-induced crises
- Future work will focus on:
  - Dynamic or multi-period extensions
  - Multi-scenario extensions
  - Nested models

18

Acknowledgement

- ❑ We are grateful for the financial support of the Philippine Commission on Higher Education via the PHERNet Sustainability Studies Program.
- ❑ Special thanks to Dr. Hon Loong Lam of the University of Nottingham Malaysia Campus for introducing us to P-graphs in 2012.

19

References

1) Aviso K. B. et al. 2015. *Journal of Cleaner Production* 92: 308–317.
2) Evans, A., et al. 2009. *Renewable and Sustainable Energy Reviews* 13: 1082–1088.
3) Friedler, F., et al. 1992. *Chemical Engineering Science* 47: 1973–1988.
4) Friedler, F., et al. 1992. *Computers & Chemical Engineering* 16: 313–320.
5) Friedler, F., et al. 1993. *Computers & Chemical Engineering* 17: 929–942.
6) Hendrickson et al. 2006. *Environmental Life Cycle Assessment of Goods and Services. An Input-Output Approach*. RFF Press, Washington DC.
7) Kaplan, S. and Garrick, B. 1981. *Risk Analysis* 1: 11–27.
8) Lam, H. 2013. *Current Opinion in Engineering* 2: 475–486.
9) Rockstrom et al., 2009. *Nature* 461: 472–475.
10) Tan, R. R., et al. 2014. *Applied Energy* 135, 402–406.
11) Varbanov. P. S. 2014. *Current Opinion in Chemical Engineering* 5: 15–21.

20

The Authors

Raymond R. Tan
Professor

C. D. Cayamanda
Research Assistant

M. F. D. Benjamin
Ph.D. Student

Kathleen B. Aviso
Associate Professor

21

Thanks for your attention

Comments and questions are welcome

Or contact me via e-mail:

Raymond R. Tan
raymond.tan@dlsu.edu.ph

22

Green Propellant Development through the SBIR Process

Gregory E. Ogden
Ogden Engineering & Associates, LLC

May 28, 2015

Outline

- OE&A
- Rocket Engine Primer
- Hypergols
- Green Propellant Development
- Summary

ICOSSE May 28, 2015

Ogden Engineering & Associates

Goal: To provide Chemical and Environmental Engineering Services to Government and Industrial Clients

- Transition faculty research to industry
- Teaming for success
- Target SBIR/STTR programs
  - All branches of government
  - Chemical/Process Engineering related projects

ICOSSE May 28, 2015

OE&A's Propulsion Roots

- 1st Phase I on HTP2003
  - Presented at 2004 JANNAF meeting
  - Met future HGF Teaming Partners
- 2004 Call for Green Propellants
  - Formed Diverse Team
    - University Research Institution
    - Aerospace Prime Contractors

ICOSSE May 28, 2015

Rocket Engine Primer

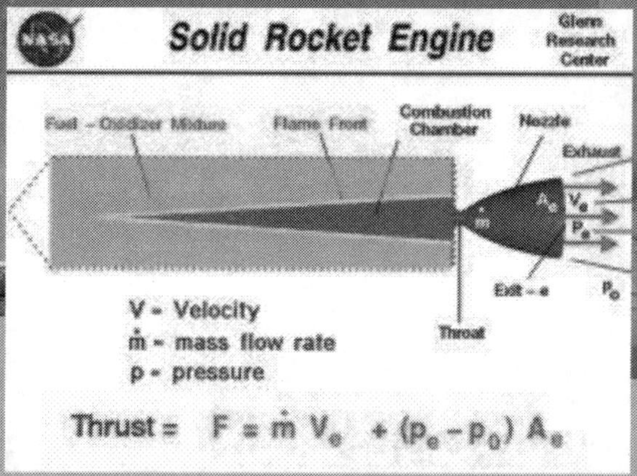

http://www.grc.nasa.gov/WWW/K-12/airplane/srockth.html

http://chapelle.me/photos/2007/family_webpage_2007_dec/thumbs_d/rnd_y_model_rockets_jpg.html

ICOSSE May 28, 2015

Liquid Propellants

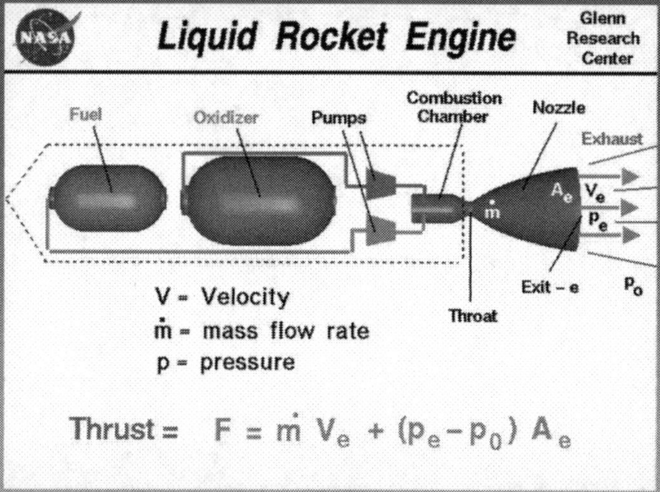

http://blog.spacetec.org/wp-content/uploads/2010/12/NASA_bipropellant-rocket-

ICOSSE May 28, 2015

Hypergols

- Fuel and Oxidizer combination
  - Ignites upon contact
- Common fuels include Hydrazine and its derivatives (MMH and UDMH)
- Common oxidizers
  - Inhibited red fuming nitric acid (IRFNA)
  - Red fuming nitric acid (RFNA)
  - Mixed oxides of nitrogen (i.e. MON-3)

ICOSSE May 28, 2015

Hypergol Issues

- Toxicity
  - Exposure during manufacture
  - Maintenance
- Leaks
  - Spontaneous ignition
- Corrosion
  - Oxidizers degrade infrastructure

ICOSSE May 28, 2015

Drawbacks of Hydrazine

- Toxic Vapor
  - LC_{50} Rat: 747 mg/m3
  - PEL: 1.3 mg/m3
  - TLV: 0.013 mg/m3
  - LD_{50} Rat: 60 mg/kg (ingested)
- Probable Human Carcinogen
- Skin contact may produce chemical burns/severe dermatitis
- High Melting Point (1°C)

ICOSSE May 28, 2015

Working with Hydrazine based Fuels

Handling Hydrazine

Fueling satellites with MMH &AH

ICOSSE May 28, 2015

http://www.arianespace.com/images/missionup-dates/2005/missionupdate

Aircraft Maintenance

Source:
http://www.belgianwings.be/webpages/navigator/news/Special%20Features/rocourt%202006/f16a_fa048_hydrasine_verw_e.jpg

ICOSSE May 28, 2015

Ongoing Need for Green Propellants

- Antares Rocket Explosion, October 2014
  - Space Station Cargo Rocket
- 1 Year Cleanup Effort
  - Requires soil and water remediation

ICOSSE May 28, 2015

Source: http://www.baltimoresun.com/business/bs-bz-wallops-update-20141205-story.html

Green Alternative

- Ionic Liquids
  - "Rediscovered" in late 1990's
- Physical Characteristics
  - High Density
  - Non Toxic (skin irritant)
  - Negligible Vapor Pressure

  - Viscous
  - Not-really hypergolic

ICOSSE May 28, 2015

Hypergol Analysis

- Definitions
  - IDT
    - Ignition delay time
    - Time from injection to flame
  - CDT
    - Chemical Delay Time
    - Time between gas evolution & flame evolution
    - Rapid screening tool

ICOSSE May 28, 2015

Drop on Drop Measurements (Oxidizer Lead)

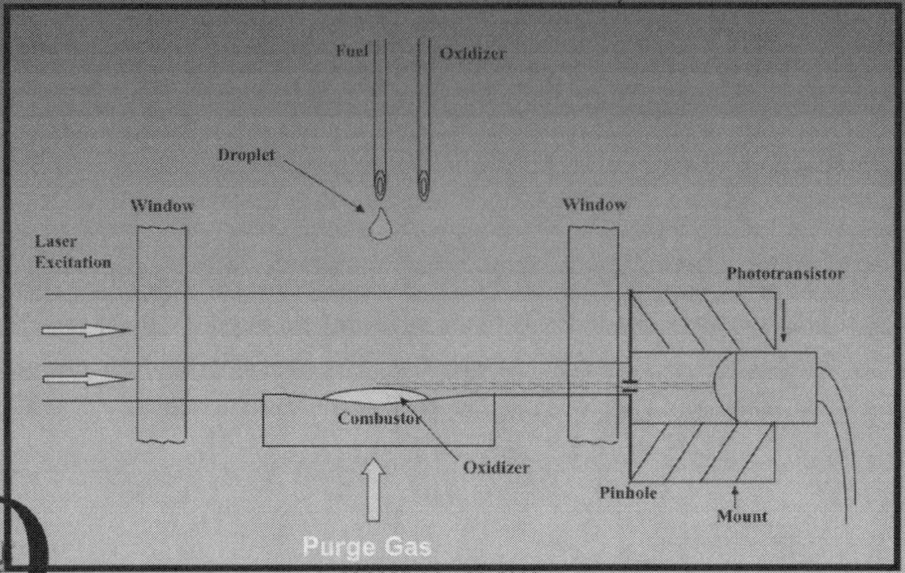

ICOSSE May 28, 2015

HGF Optimization

- CDT Screening Study
- Conducted >1,000 tests in 8 months
  - in open lab environment
    - Evaluated Additives
    - Preheat Temperature
    - Fuel to Oxidizer ratio

→Minimized both CDT & IDT

ICOSSE May 28, 2015

Green Propellant Features

- Reactivity only part
- Toxicity
  - Non-toxic ionic liquids
- Sensitivity
  - Meet U.S. DOT, UN Shipping Requirements

ICOSSE May 28, 2015

UN Series 3 Testing

- Evaluates response to initiating stimuli
  - TB7002 Initial UN Hazardous Classification Procedures for articles and substances
  - A measure of sensitivity
  - Required to ship fueled components

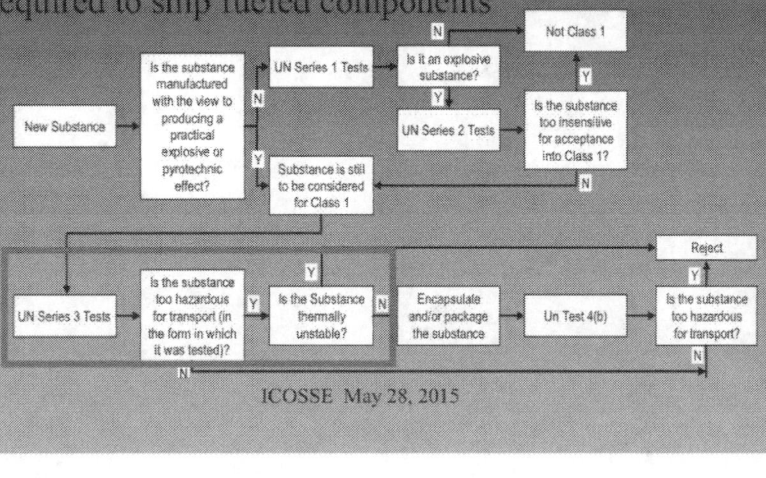

ICOSSE May 28, 2015

Thermal Stability 3(c)

- Materials exposed to elevated temperatures
  - Heat to 75°C
  - Measure ΔT
  - Ignition, explosion or >3°C temp rise (self heating) → failure
- Required for Class 1 admission

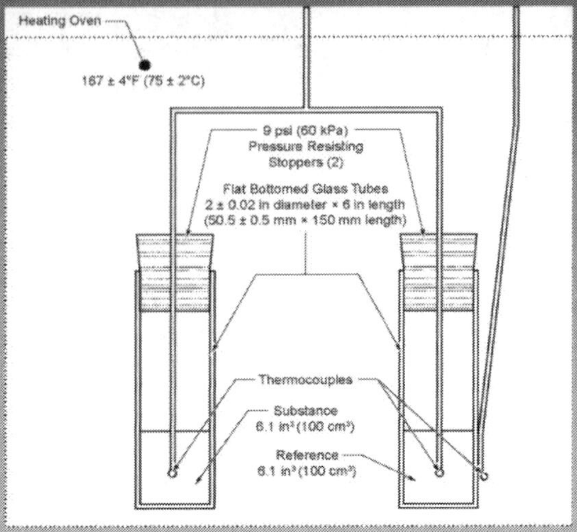

ICOSSE May 28, 2015

Impact Test 3(a)(i)

- Drop weight onto sample
- Measures impact sensitivity
  - Determine height needed to cause explosion

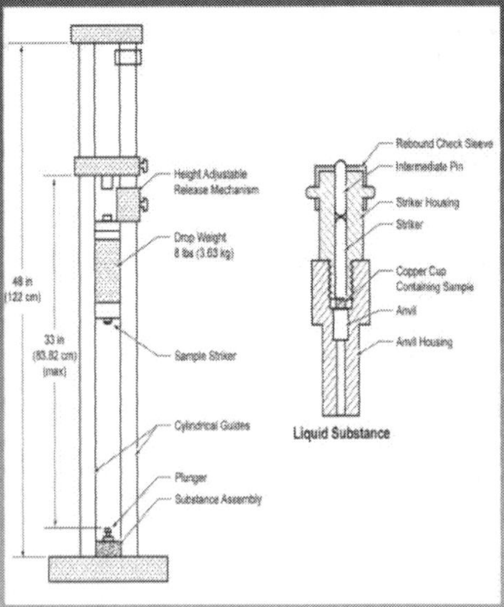

ICOSSE May 28, 2015

Friction Test 3(b)(i)

- Scrape ceramic piece across material
  - Evaluates Friction Stability
  - Max Load 360N
  - 80N Limit to pass test

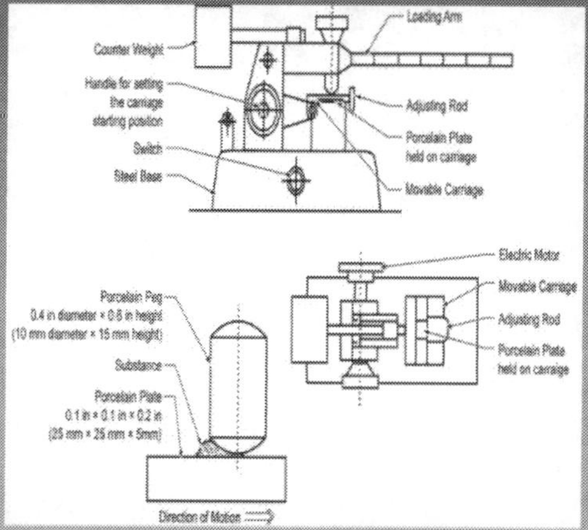

ICOSSE May 28, 2015

Small Scale Burn Test 3(d)(i)

- Determines response to unconfined fire
  - Fuel placed in container set on a burning base
  - Looking for detonation

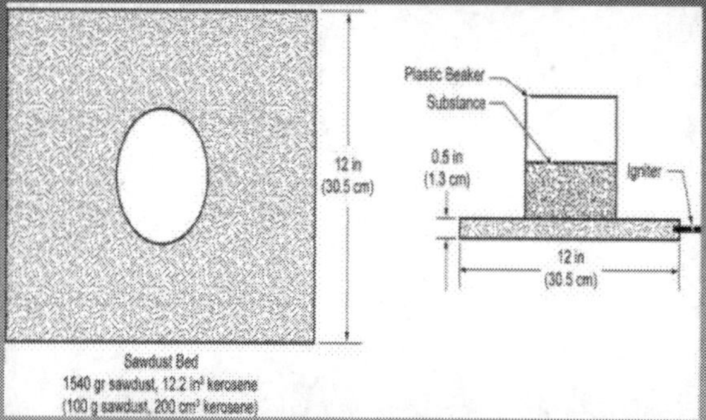

ICOSSE May 28, 2015

Spray Test

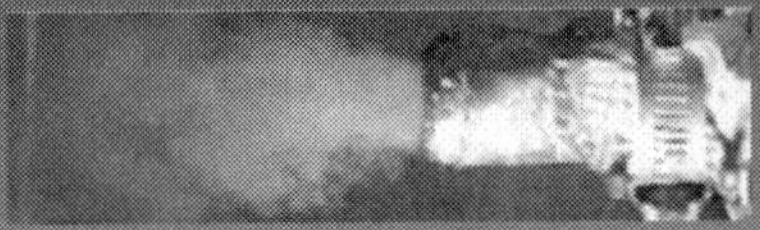

Oxidizer Lead

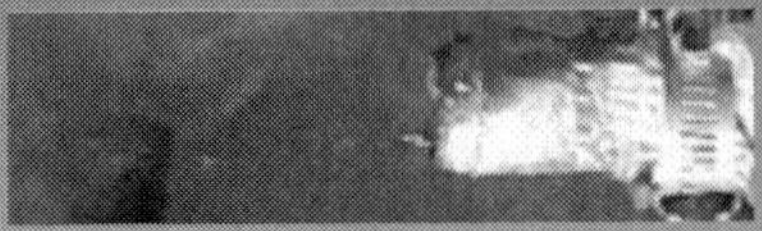

0.2 s burn

Fuel Rich spray (no oxidizer)

ICOSSE May 28, 2015

Project Outcomes

- Optimized Fuel in shirt sleeve environment
- Successful Scale-up
  - Injector Test
  - 5lb Engine Test
- Industrial support from prime contractor
- Strong SB⇔University collaboration
- Initial testing indicates insensitive fuel

ICOSSE May 28, 2015

References

- McGeehon, Dale, “Green Propellants”, MDA TechUpdate, Summer 2010
- http://www.grc.nasa.gov/WWW/K-12/airplane/srockth.html
- http://www.baltimoresun.com/business/bs-bz-wallops-update-20141205-story.html
- http://www.belgianwings.be/webpages/navigator/news/Special%20Features/rocourt%20 2006/f16a_fa048_hydrasine_verw_e.jpg
- http://kookbang.dema.mil.kr/newspaper/tmplat/upload/20130827/thumb1/BBS_201308 270628340650.jpg
- http://chapelle.me/photos/2007/family_webpage_2007_dec/thumbs_d/md_y_model_roc kets_jpg.html
- http://blog.spacetec.org/wp-content/uploads/2010/12/NASA_bipropellant-rocket-
- http://www.arianespace.com/images/missionup-dates/2005/missionupdate

ICOSSE May 28, 2015

PAUL SCHERRER INSTITUT

Paul Scherrer Institute, Switzerland

Global Energy Scenarios 2050

Orchestrating or Improvising the Energy Transition

28 May 2015

Martin Densing, Stefan Hirschberg, Evangelos Panos, Kathrin Volkart

PAUL SCHERRER INSTITUT

Development of energy scenarios of PSI & WEC

2011: Global Transport Scenarios 2050
- only transport sector

2013/14: World Energy Scenarios 2050
- whole global energy system

2015+: Extensions, e.g.
- Multi-objective optimization
- Open-source energy-system model

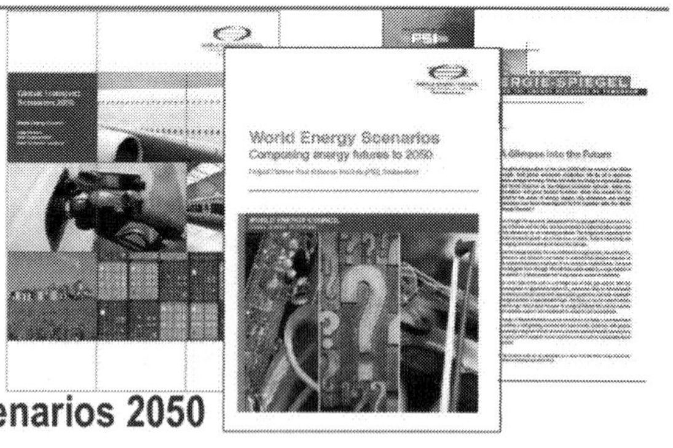

Methodology for World Energy Scenarios 2050

- **2 scenarios** with coherent storylines of social, political and economical drivers
  - Developed in workshops 2012-13 (international practitioners' & academics' view)
  - e.g., markets expansion / innovation → GDP → energy intensity → demand
- **Analyse-Tool: Energy-system model GMM** (Global Multi-regional MARKAL):
  - Supply/demand-model with a detailed representation of resources, technologies, energy flows, and technological change [Barreto 2001; Gül 2007; Densing & Turton 2012]
  - Cost-optimization of the energy system (market allocation = competitive equilibrium with inelastic useful-energy demands; total final consumption is elastic)
  - Non-cost and behavioural factors are modelled as side constraints

Energy system model: Flow of energy carriers

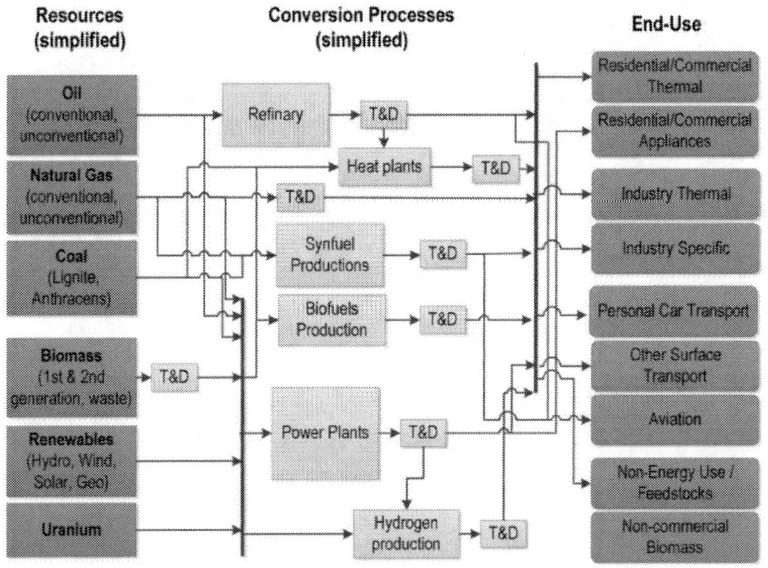

GMM has approx. 400 technologies per region

PSI (2013): Energy-Mirror

page 3

15 world regions (aggregated to 8 regions in WEC report)

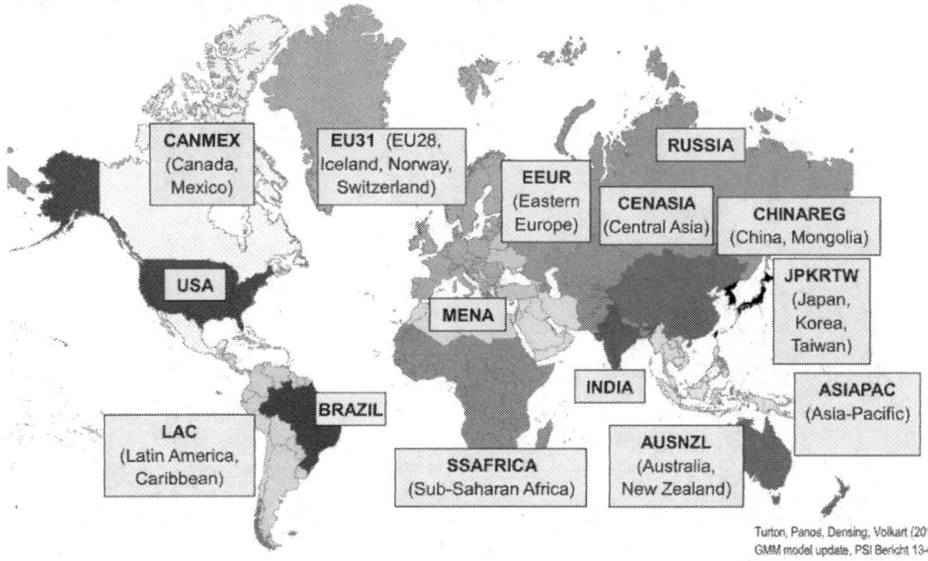

Turton, Panos, Densing, Volkart (2013): GMM model update, PSI Bericht 13-03

Separation criteria, e.g.:

- Economic size (e.g. single countries USA, China, India, Brazil)
- Fossil resources (→ see next slide)

page 4

Regional partition: Primary energy in year 2010

Domestic production of primary energy supply

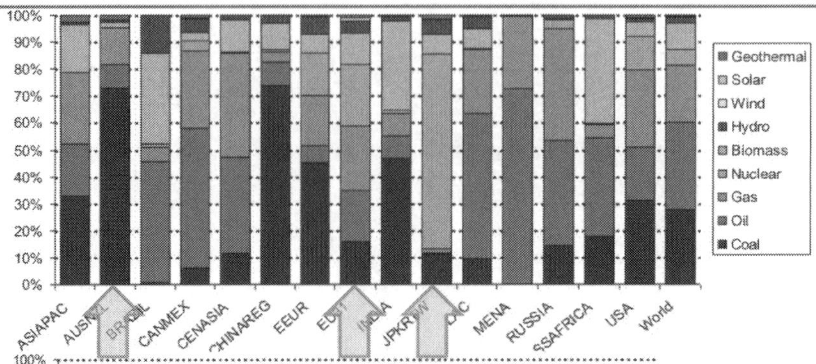

Net import in % of total primary energy supply

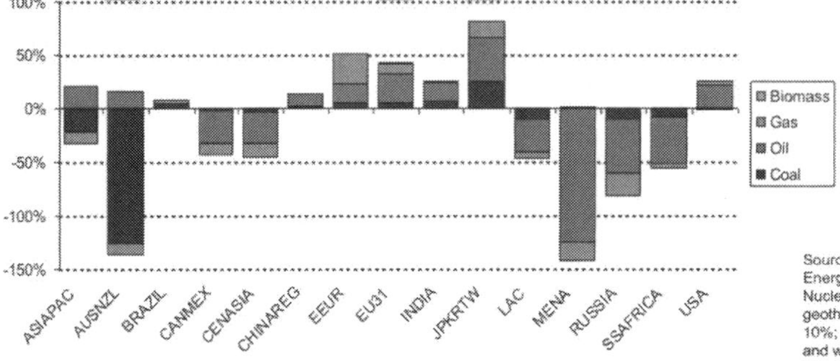

Source: IEA (2012), Energy Balances. Nuclear eff. = 33%; geothermal eff. = 10%; solar, hydro and wind: produced energy.

BRAZIL: Bio -0.4%, Oil: 0.1%; CHINAREG: Gas 0.5%; INDIA: Gas 1.5%; LAC: Bio -0.4%; EU31: Bio: 0.3%

Assumption in the 2 scenarios

	Jazz: market-oriented future	Symphony: regulation-oriented future
Goals	• Affordable energy access through free markets • High income ➔ Mainly adaption to environmental damages	• Secure energy access • Targeted regulation through states and international organizations ➔ Mainly avoidance of environmental damages
Economic Growth	**GDP growth has priority** (3.5% annual average to 2050, PPP)	**Less GDP growth** (3.1%)
Population	**Increase** (8.7 billion in 2050)	**Stronger Increase** (less income) (9.3 billion)
Climate Policy	**CO_2-markets develop slowly** (CO_2 price in 2050: 23-45 \$/tCO_2)	**Stronger coordinated state action** (70-80 \$/tCO_2)
Energy Efficiency / Intensity	**Efficiency increases based on market criteria**	**State promotion of efficiency measures and energy savings**
Unconventional Resources (shale gas/oil etc.)	**Expanded opening of markets.** High incentive because high energy demands.	**Regulation** (for water use, market access). Fewer incentives because less demand.
Renewable Energy	**Limited promotion.** „The market" selects the technologies.	**Selected state promotion.**
Non-renewable Energy	**Limited Support:** • CCS market-driven (pilot in 2030) • Nuclear under construction not all operational	**State support:** • CCS available from 2020 • Nuclear, large hydro power

PSI (2013): Energy-Mirror

Primary energy supply: 2010-2050

Jazz
(consumer-driven and market-oriented)

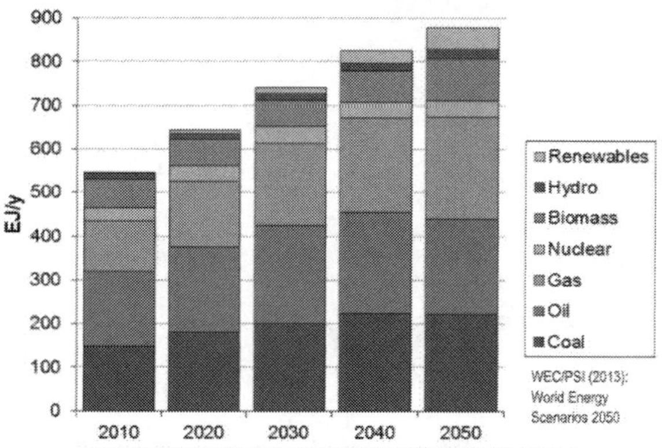

Renewables: output of electricity and heat; Biomass: primary supply incl. waste; Nuclear: 33% efficiency

Symphony
(voter driven and more regulation)

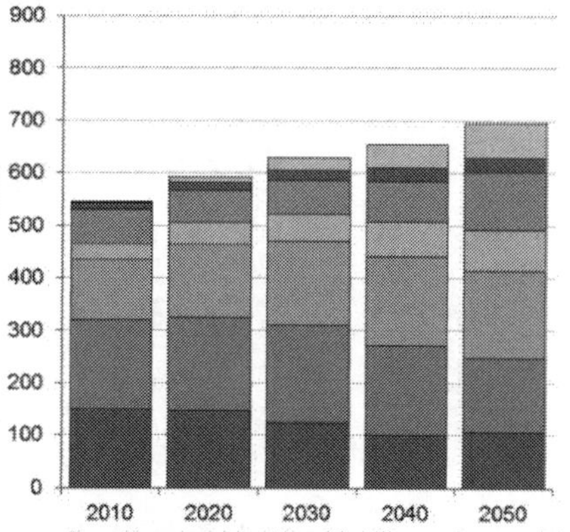

Renewables: output of electricity and heat; Biomass: primary supply in
Nuclear: 33% efficiency

- Shale gas: 32% of global gas resources [EIA 2013]
- Coal: E.g. also Coal→Fuel

- More renewables (e.g. solar)
- Hydropower limited; nuclear power
- Coal remains in the energy mix (CCS)

Primary energy supply by world region

Jazz
(consumer driven and market-oriented)

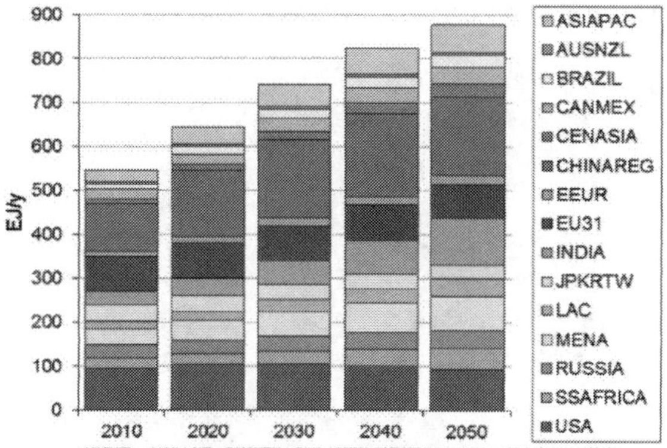

AUSNZL = AUS + NZL; CANMEX = CAN + MEX; CHINAREG = CHN + MNG; EU31 = EU28 + ISL + NOR + CHE, JPKRTW = JAP + KOR + TWN

Symphony
(voter driven and more regulation)

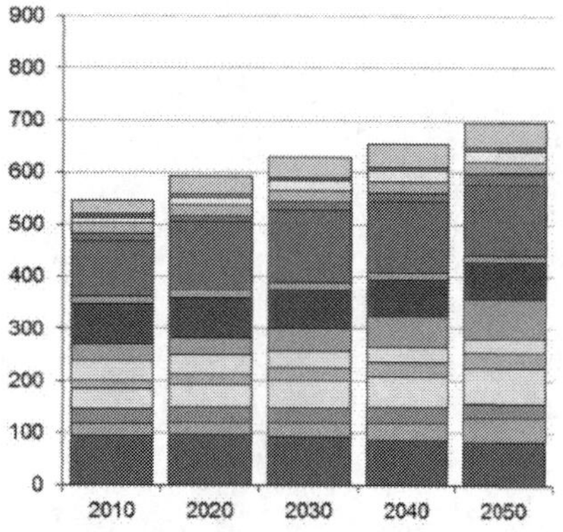

- Increase of primary energy in China, India, MENA and Southeast-Asia

WEC/PSI (2013): World Energy Scenarios 2050

PAUL SCHERRER INSTITUT PSI

CO_2 emissions (from energy sector)

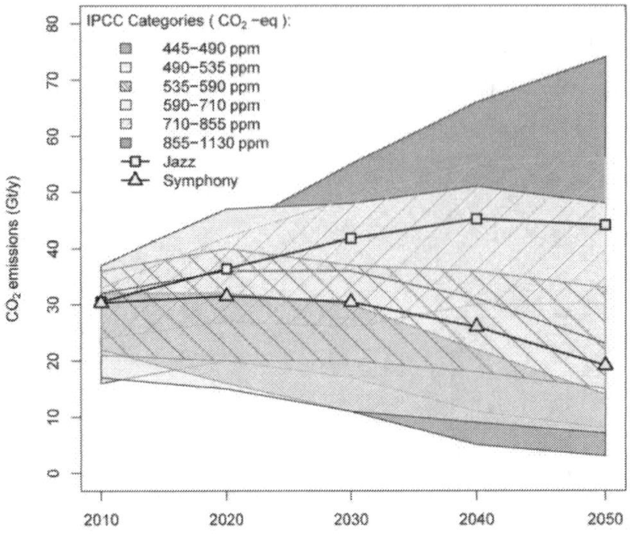

WEC/PSI (2013): World Energy Scenarios 2050

Focus of climate policy:

- Jazz: **Adaptation** (mitigation comes later, 2030+)
- Symphony: **Mitigation** (also after 2050); +2° degree may be achieved [AR5, IPCC 2013]

PAUL SCHERRER INSTITUT PSI

Electricity production in the world region

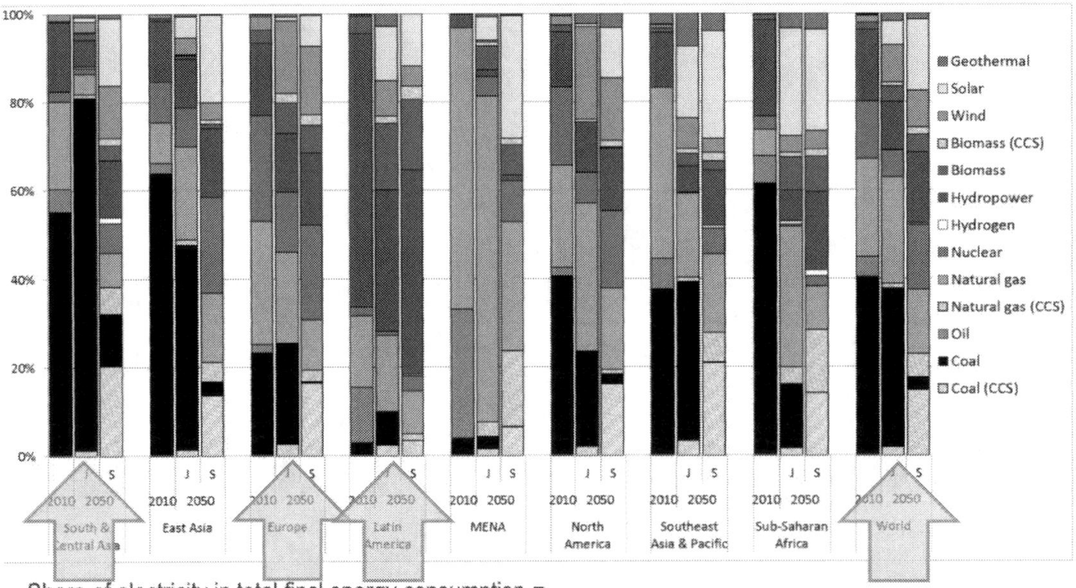

Share of electricity in total final energy consumption =

2010: 17% , **Symphony 2050:** 32%, **Jazz 2050:** 27%, **Switzerland 2010:** 24%

PSI (2013): Energy-Mirror
TFC: includes non-energy use

PAUL SCHERRER INSTITUT PSI

Investment in electricity generation (cumulative)

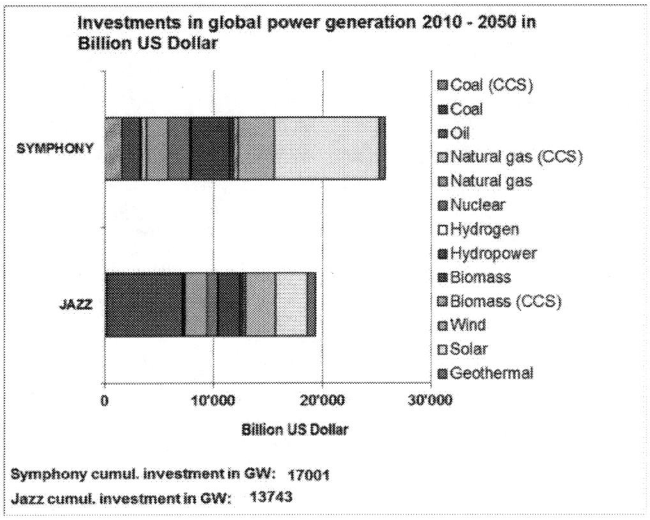

PSI (2013): Energiespiegel

- Electricity production 2050: 54 PWh **in Jazz**, only 48 PWh **in Symphony**
- Solar energy 2050: 2'000 GW new **in Jazz**, 5'900 GW **in Symphony** (2008→2012: 15→97 GW)

IEA, PVPS 2012

PAUL SCHERRER INSTITUT PSI

Population without access to electricity

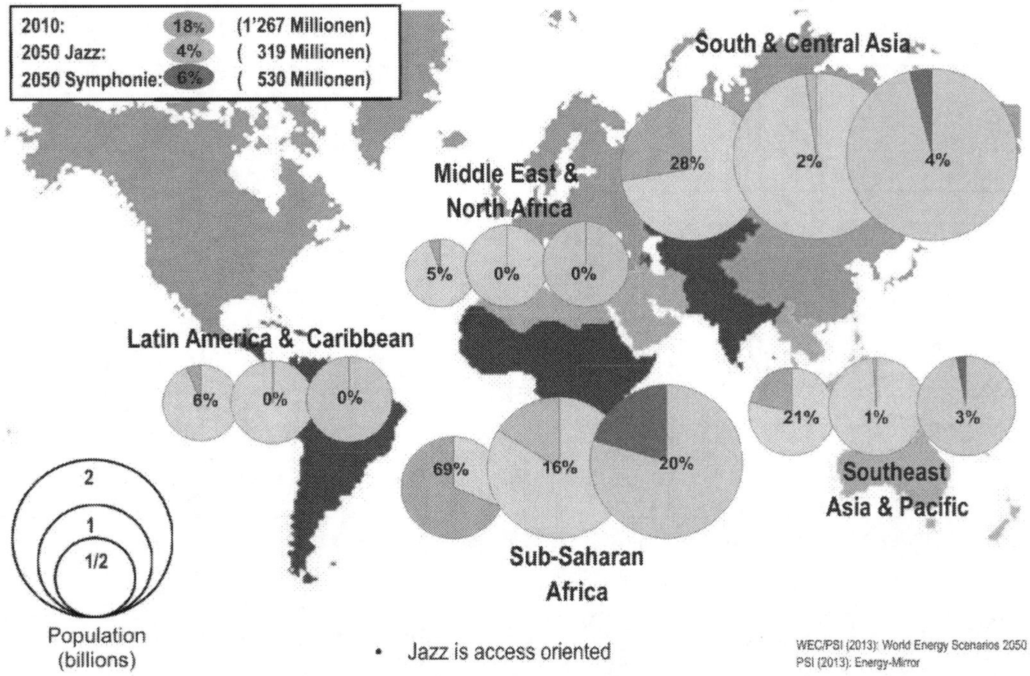

- Jazz is access oriented

WEC/PSI (2013): World Energy Scenarios 2050
PSI (2013): Energy-Mirror

Income per person

GDP per person (1'000 $\$_{2010}$ MER):

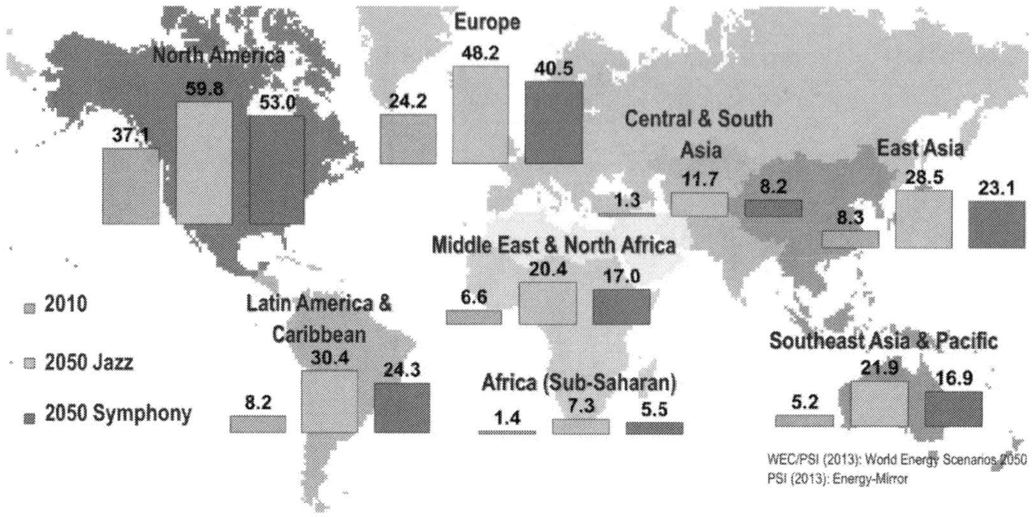

GDP growth is partially offset by population growth in some developing countries. E.g.:
Sub-Saharan Africa: **2010:** 860 mio. people; **2050: Jazz** 1.6 bio. people, **Symphony** 2 bio.

Comparison with other global energy scenarios

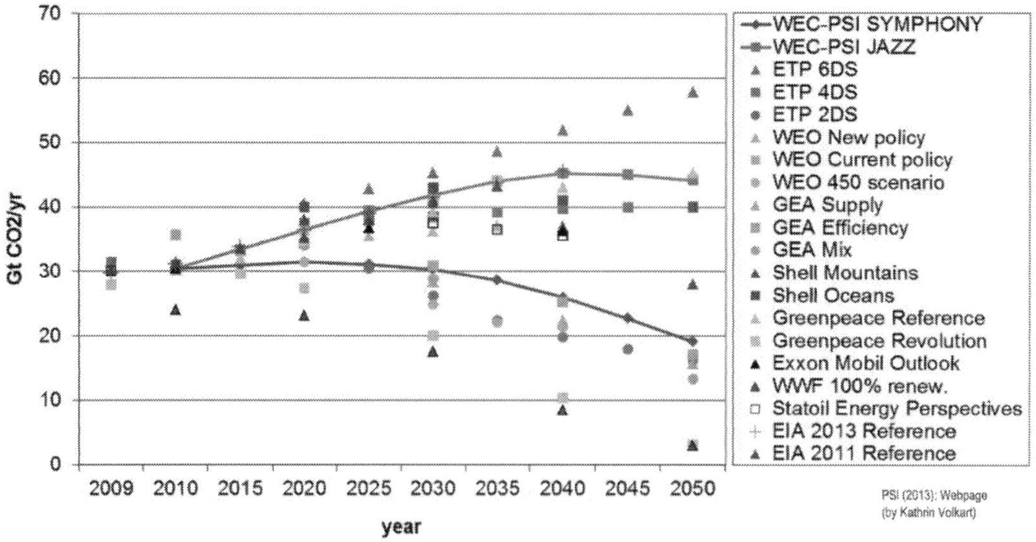

- PSI/WEC scenarios are **explorative** (and not **normative** with a CO_2 upper bound)

PAUL SCHERRER INSTITUT PSI

World Energy Scenarios: Conclusions for 2050

1. **Complexity of energy system** increases (e.g. by stochastic renewables)
2. **Energy efficiency** is crucial (in case demand is outstripping supply)
3. **Coal/Gas/Oil** are not vanishing (e.g. used in road travel / ships / aviation)
4. **Regional priorities/potential/current state:** No 'one-size-fits-all' solution to the energy trilemma
   - Affordable energy access
   - Environmental sustainability of energy system
   - Security of energy supply
5. **450ppm-CO_2eq target** difficult (with for all acceptable CO_2 prices)
6. **CCUS** (CO_2-capture, utilisation & storage) can challenge renewables
7. **Key technological uncertainties:** CCUS, solar energy and electric storage
8. **Balancing of energy trilemma** implies difficult trade-offs
9. **Energy markets** require investments and integration of all consumers, whereas **energy policy** should ensure that energy and CO_2-markets deliver
   [9. is opinion of WEC]

PSI, 01.06.2015

page 15

PAUL SCHERRER INSTITUT PSI

Thank you!

PSI, 1. Juni 2015

PAUL SCHERRER INSTITUT PSI

Selected key scenario indicators

		1990	2010	2050 Jazz	2050 Symphony
CO_2	Gigatons / year	21.0	30.5	44.1	19.1
CO_2 China & India	% world	14%	31%	40%	35%
Primary energy	Exajoule	367	546	879	696
Primary energy China & India	% world	14%	25%	33%	30%
Final consumption	Exajoule	252	373	629	491
Final consumption China & India	% world	15%	24%	33%	31%
GDP	Trillion $\$_{2010}$	34	63	201	172
GDP China & India	% world	3%	12%	26%	24%
Population	Billion	5.3	6.9	8.7	9.4
Population China & India	% world	38%	37%	33%	32%
GDP / capita (income)	$\$_{2010}$	6'814	9'162	23'139	18'317
CO_2 / capita	ton / year	4.0	4.4	5.1	2.0
CO_2 / GDP (CO_2-intensity)	kg / $\$_{2010}$	0.62	0.48	0.22	0.11
Final consumption / capita	Gigajoule / year	47	54	72	52
Renewables / Primary energy	%	13%	15%	19%	29%
Renewables / Electr. production	%	19%	20%	31%	48%
Electricity / Final consumption	%	14%	17%	27%	32%

PSI, 01.06.2015

PSI (2013): Energy-Mirror page 17

PAUL SCHERRER INSTITUT PSI

Primary Energy: Comparison with other studies

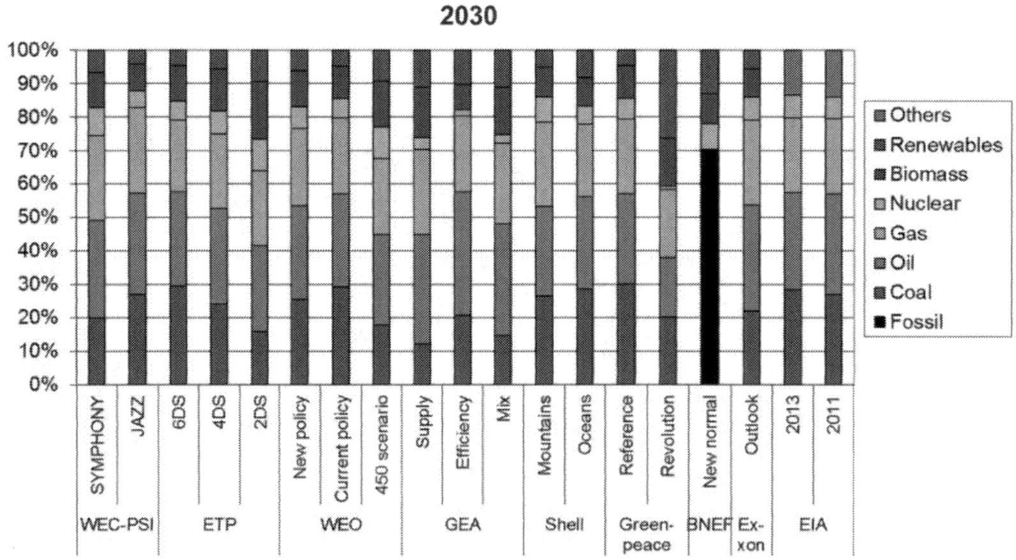

PAUL SCHERRER INSTITUT PSI

Summary of selected results

Energiespiegel No. 22, Titlepage

Scenario outcomes (and implied decisions) lead only partially to sustainability:

•Symphony (stronger government role; voters' viewpoint):
- reduced burden on global climate (less CO_2) and less depletion of fossil resources
- more energy security (more renewables, more large hydro)
- more CCS and more nuclear (esp. non-Europe)
- higher investments, especially in electricity sector (e.g., more capacity and more backup for intermittent renewables)

•Jazz (opening of markets; consumers' viewpoint):
- improved energy access and increases average income
- greater climate burden (more CO_2, but less CCS); more unconventional gas/oil

•Moreover:
- very stringent global climate change mitigation targets (e.g., <2 degrees) are challenging without additional ambitious policy action
- CCS may be a competitor and complement to deployment of renewables
- disruptive technology breakthroughs could change the picture

PSI, 01.06.2015

page 19

PAUL SCHERRER INSTITUT PSI

Scenario quantification framework

- Evolutions of key scenario drivers are expressed in a **coherent storyline** of future economic and social developments
- Scenario storylines quantified with GMM

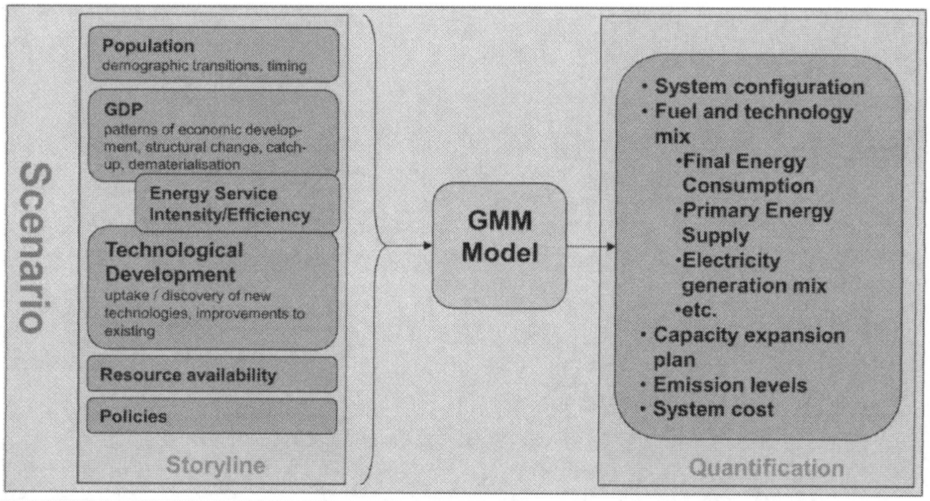

page 20

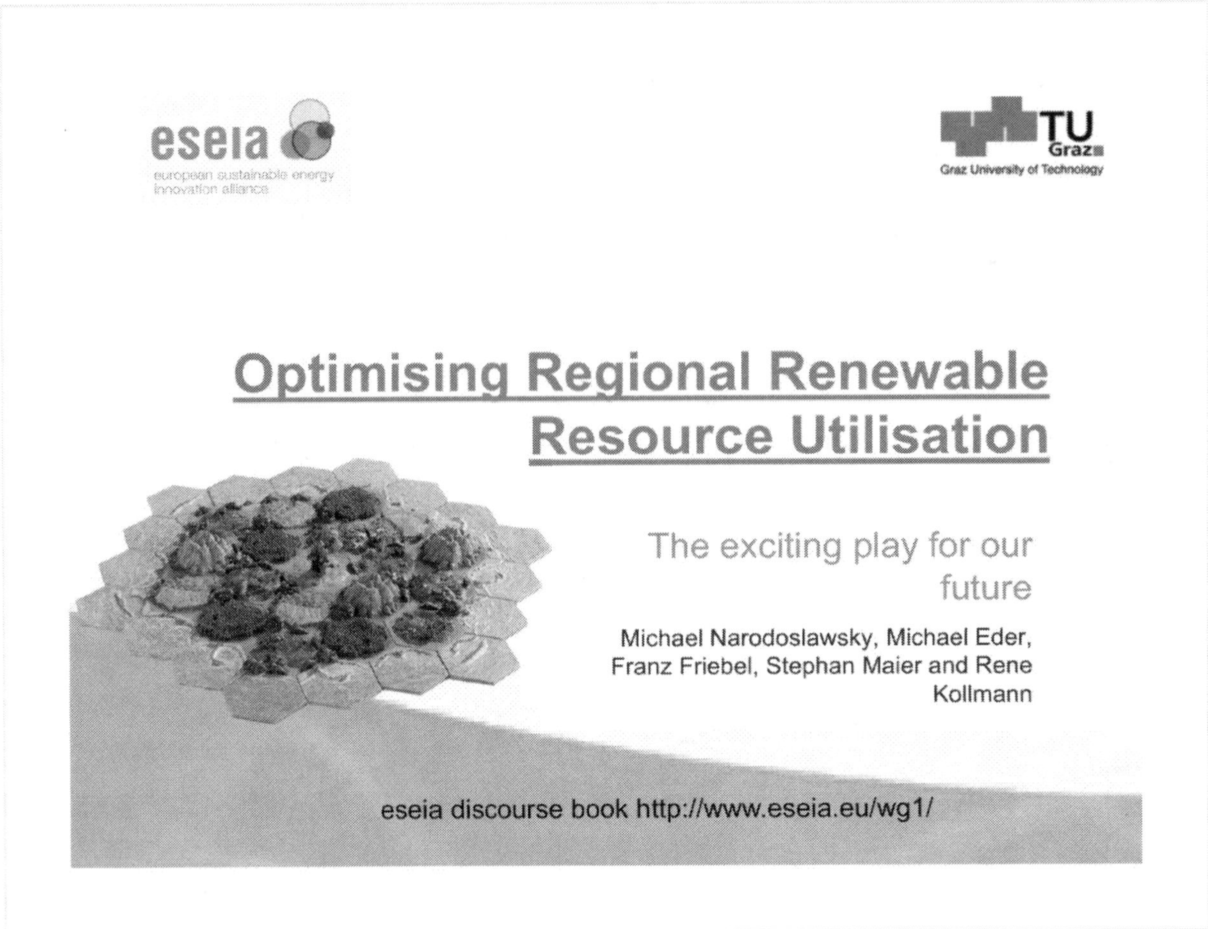
eseia
european sustainable energy innovation alliance
TU Graz
Graz University of Technology
Optimising Regional Renewable Resource Utilisation
The exciting play for our future
Michael Narodoslawsky, Michael Eder, Franz Friebel, Stephan Maier and Rene Kollmann
eseia discourse book http://www.eseia.eu/wg1/

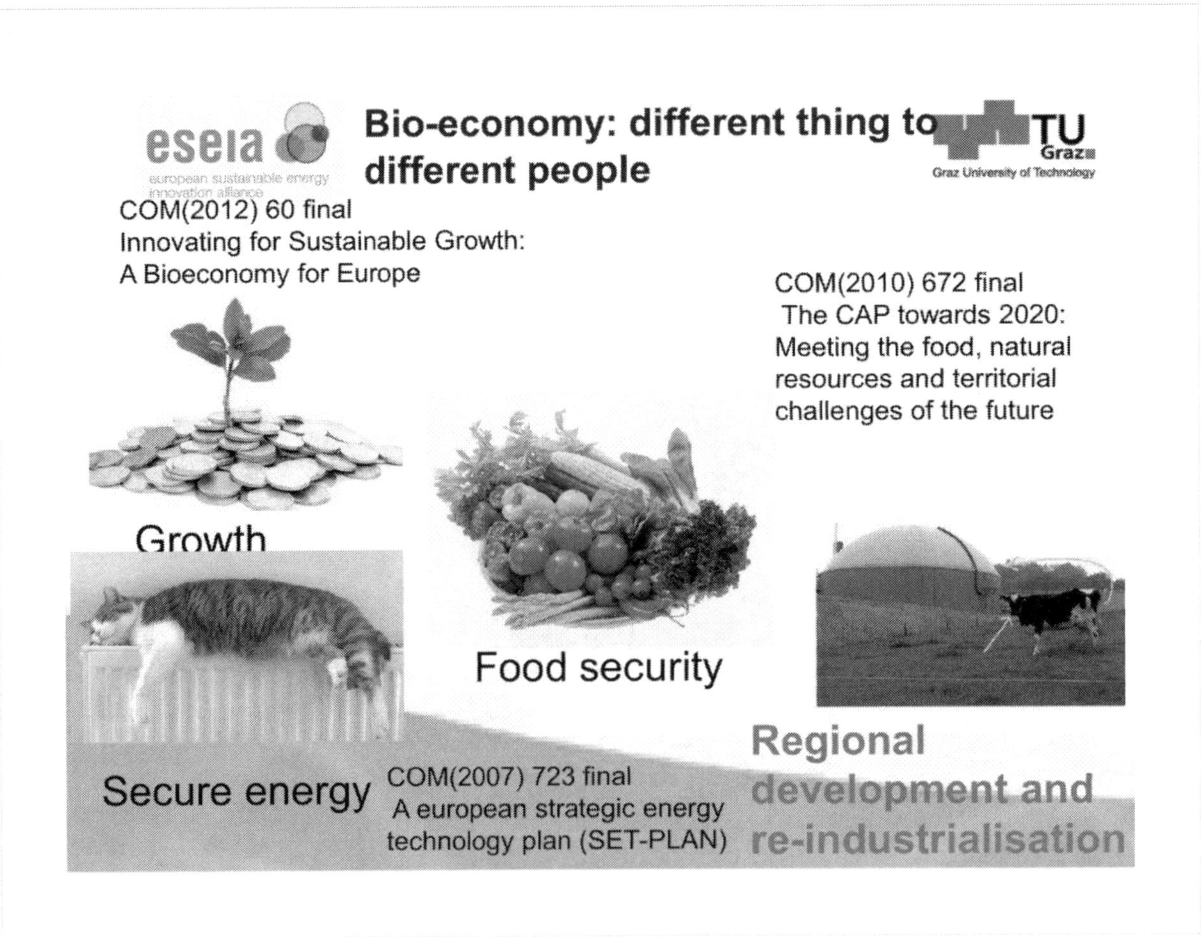
eseia
european sustainable energy innovation alliance
Bio-economy: different thing to different people
TU Graz
Graz University of Technology
COM(2012) 60 final
Innovating for Sustainable Growth:
A Bioeconomy for Europe
COM(2010) 672 final
The CAP towards 2020:
Meeting the food, natural resources and territorial challenges of the future
Growth
Food security
Secure energy
COM(2007) 723 final
A european strategic energy technology plan (SET-PLAN)
Regional development and re-industrialisation

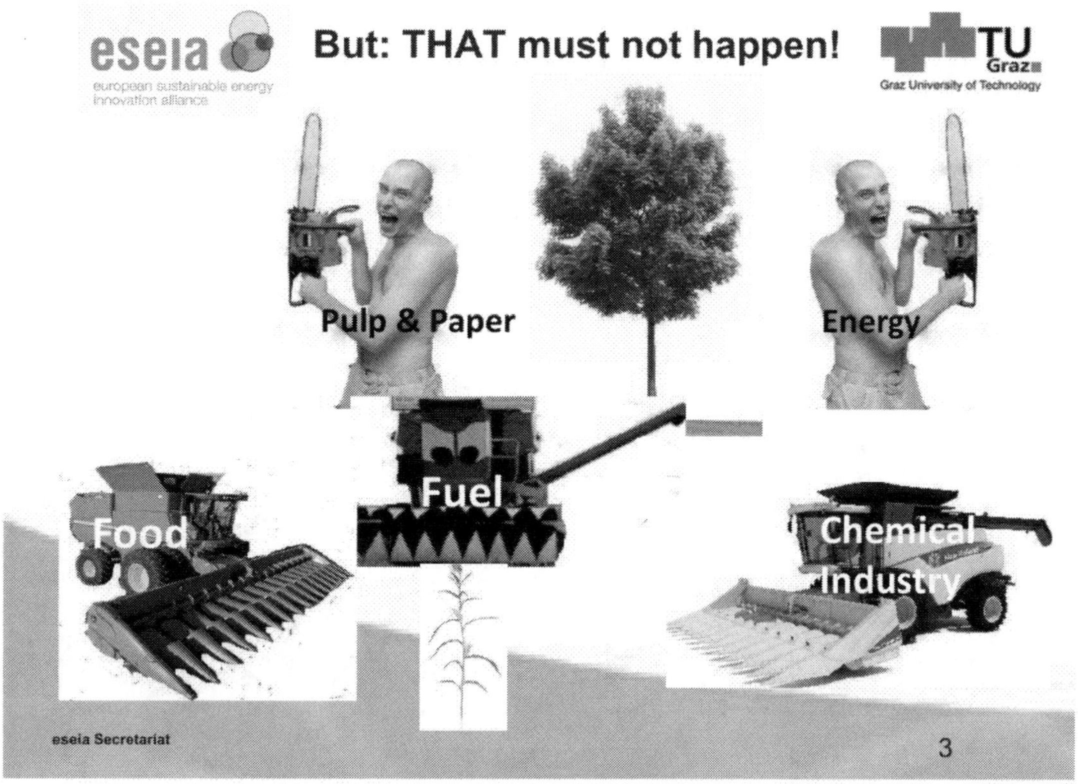
eseia
european sustainable energy innovation alliance
But: THAT must not happen!
TU Graz
Graz University of Technology
Pulp & Paper
Energy
Fuel
Food
Chemical Industry
eseia Secretariat
3

eseia
We need a rational use of key resources
TU Graz
Graz University of Technology
Pig
Pearls
Do not throw „Pearls to Pigs“
Energy and plastic bags can be made from different sources:
Manure
Grass
Straw
(Waste-) Wood

eseia — european sustainable energy innovation alliance

The "Logistical Trap"

TU Graz — Graz University of Technology

Conversion	Material	Humidity [%w/w]	Energy content [MJ/kg]	Density [kg/m³]	Energy density [MJ/m³]
Incineration	Straw (grey)	15	15	100-135	1.500-2.025
	Wheat	15	15	670-750	10.050-11.250
	Rape seed	9	24.6	700	17.220
	Wood chips	40	10.4	235	2.440
	Wood pellets	6	14.4	660	9.500
Biogas production	Grass silage	60-70	3.7	600-700	2.220-2.590
	Manure	95	0.7	1000	700
	Light fuel oil	0	42.7	840	36.000

eseia — european sustainable energy innovation alliance

Understanding "new" resources

TU Graz — Graz University of Technology

Taking transport density into account…

- …5,7 km transport of manure and
- …12 km transport of straw with tractor…
- …or 40 km transport of wood chips with lorry…
- …or 475 km transport of pellets with train
- …or 7.800 km transport of crude oil with ship or pipeline consume 1 % of the transported energy

Raw material provision must become closer!

eseia
european sustainable energy innovation alliance

The „Centrality trap“

TU Graz
Graz University of Technology

Plant size	Upper and lower land requirement in km²	Transport distance in km with land use	
		30 %	10 %
Very small 50 t/a	0.125	0.4	0.6
	.03	0.2	0.3
Small 5.000 t/a	12.5	3.6	6.3
	3.3	1.9	3.2
Medium 50.000 t/a	125	11.5	20
	33	5.9	10.2
Industrial 500.000 t/a	1,250	36	63
	333	18.8	32.6

Upper land requirement calculated with 4 t/ha.a (e.g. wood chips), lower limit with 15 t/ha.a (e.g. corn or miscanthus)

eseia
european sustainable energy innovation alliance

The „Intermittency trap“

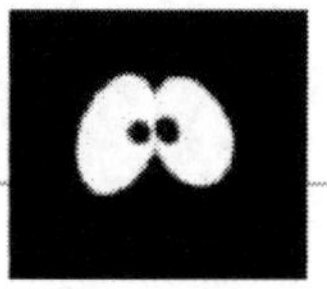

TU Graz
Graz University of Technology

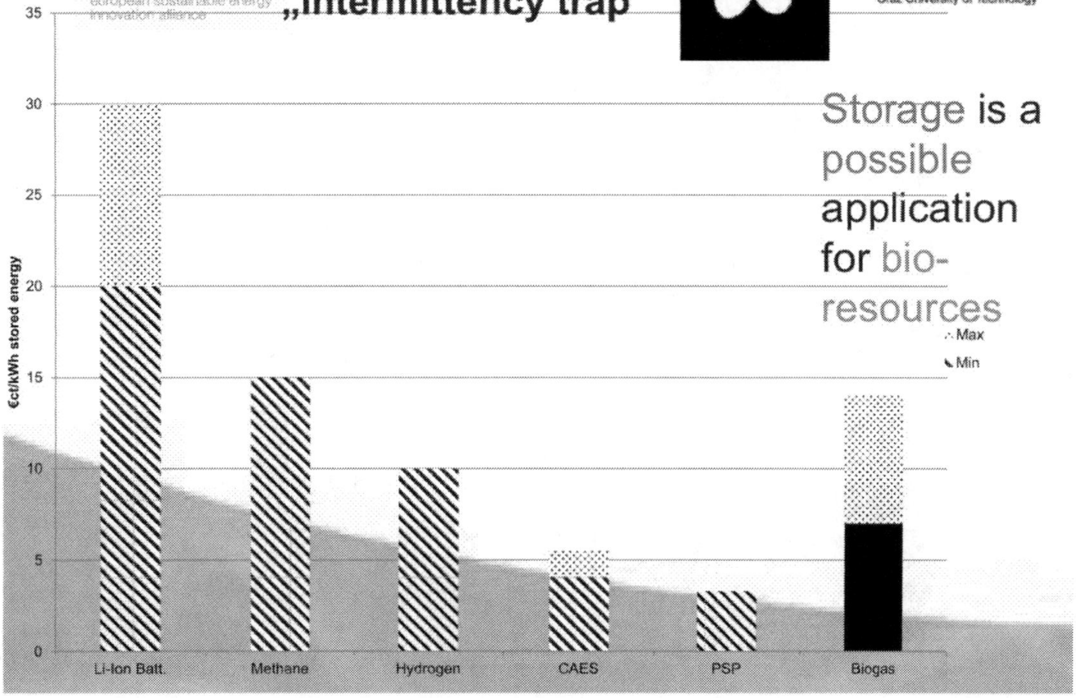

The bio-resource service challenge

Type of service	Service	Possible other resources
Social	Nutrition	None
	Jobs and development for rural regions	None
	Social stability for rural regions	None
Economic	Stability for energy distribution grids	Smart grids, hydro power, pumped hydro power, hydrogen, compressed air energy storage, (fossil resources)
	Transport fuel	Electricity (using battery storage), hydrogen, synthetic fuels, (fossil resources)
	High temperature industrial heat	(fossil fuels), H_2 from excess electricity
	Feedstock for synthetic materials and plastics	(fossil resources), sequestered CO_2 plus H_2 from excess electricity
	Feedstock for conventional bio-based products	None
Environmental	Reduction of greenhouse gas emissions	Wind and hydro power, solar thermal systems, photovoltaic, oceanic power, geothermal energy
	Preserving soil fertility	None
	Preserving water and nutrient cycles	None
	Preserving bio-diversity	None

Technical challenges

Learn your „old dogs“ new tricks!

- Increase raw material flexibility of existing technologies
- Increase tolerance for lower grade resources
- Increase range of useful products from current technologies

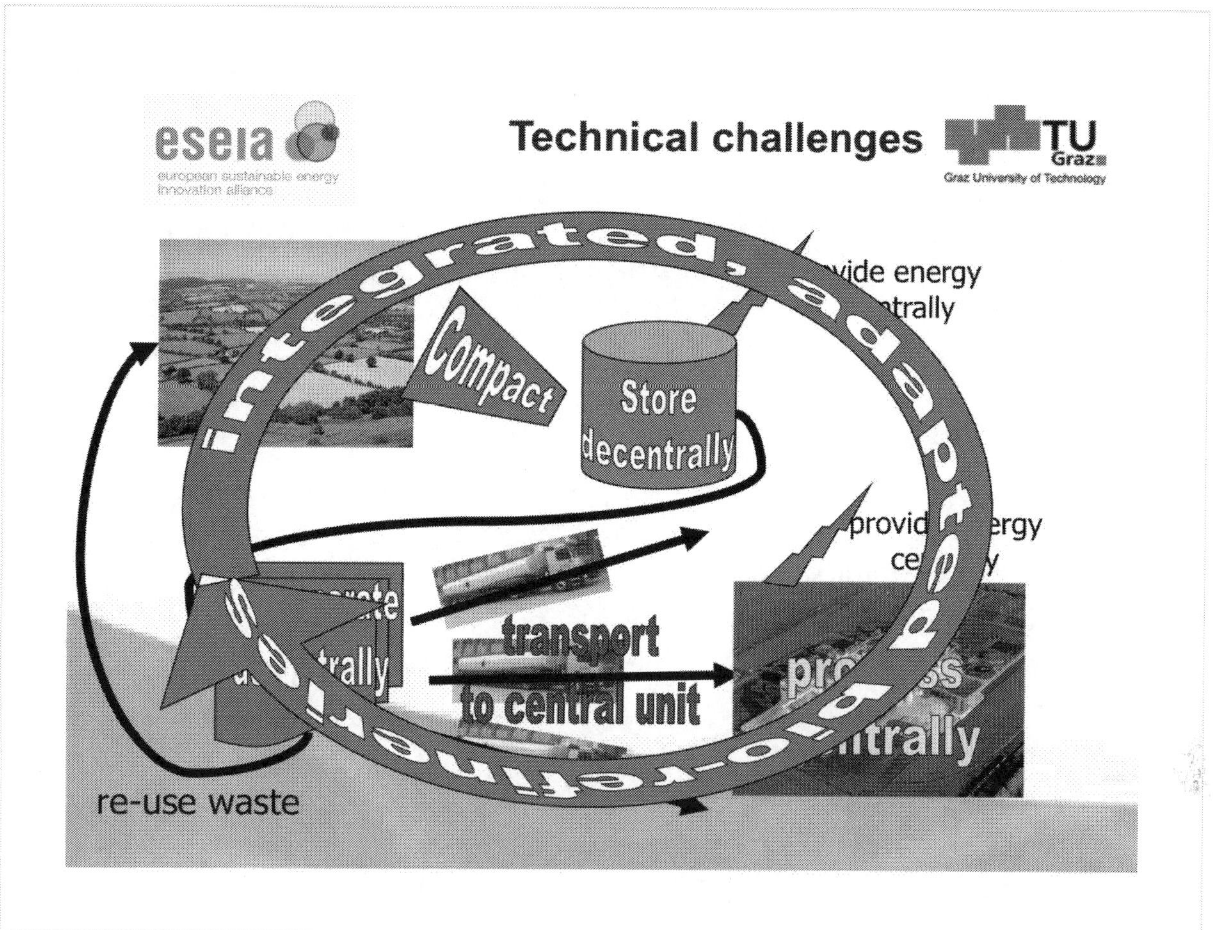
eseia
european sustainable energy
innovation alliance
Technical challenges
TU Graz
Graz University of Technology
integrated, adapted bio-refineries
Compact
Store
decentrally
transport
to central unit
re-use waste

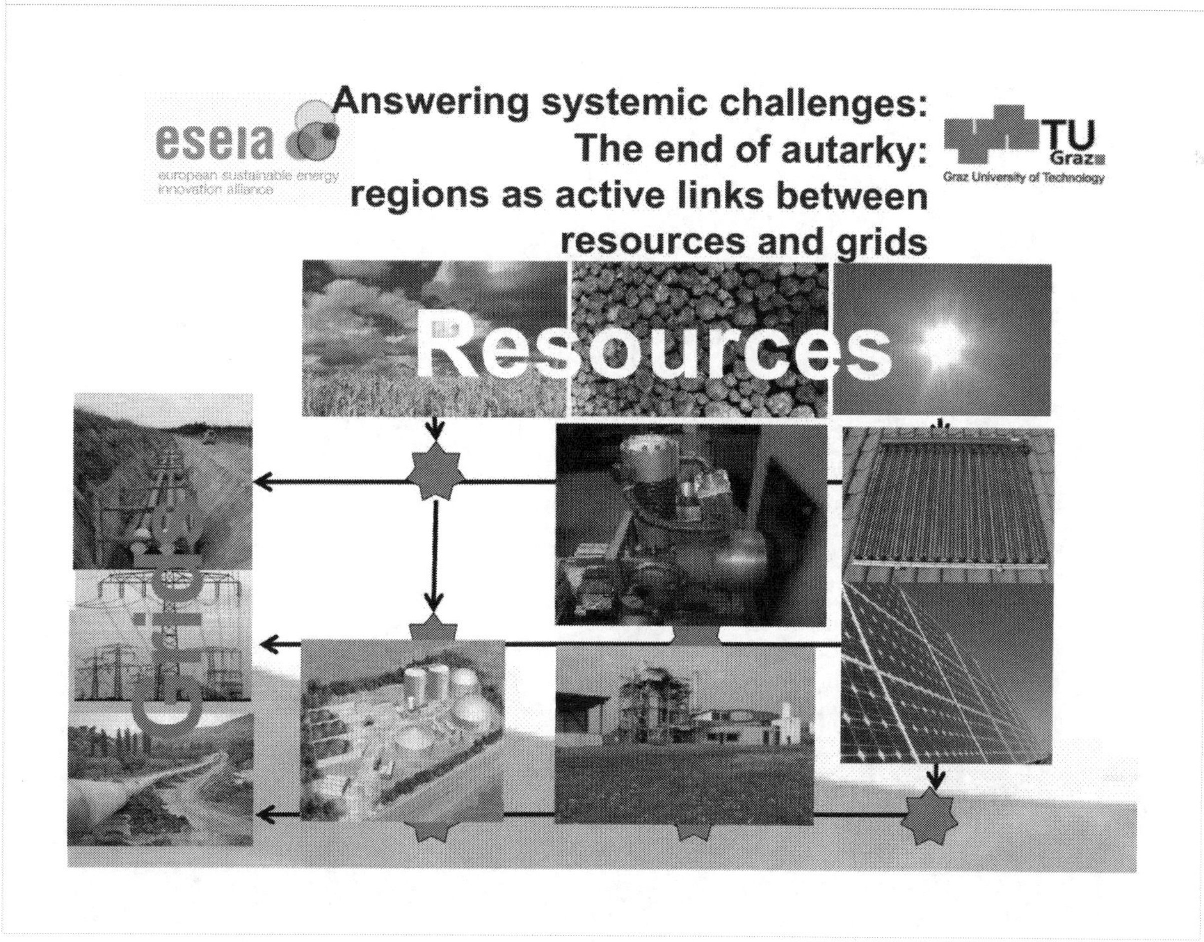
eseia
european sustainable energy
innovation alliance
Answering systemic challenges:
The end of autarky:
regions as active links between
resources and grids
TU Graz
Graz University of Technology
Resources
Grid

Frame regional resource plans

- Modernise existing bio-resource utilisation sectors (e.g. food industry, pulp & paper industry, timber industry) to become versatile bio-refineries
- Favour pathways leading to material products, establishing longer value chains and generating more social benefit from bio-resources.
- Keep as much material close to the place where bio-resources are grown thus closing material cycles with least transport effort and retaining valuable nutrients for preserving eco-system functions.
- Use intersections of energy distribution grids for bio-refineries based on secondary bio-resources (residues from agriculture) and tertiary bio-resources (wastes from industry and society) in order to contribute to stabilising grids

energsustainsoc.com

Thank you for your attention!

And if you want to evaluate yourself or your process ecologically or optimise your regional energy system, visit www.fussabdrucksrechner.at

Optimization of Sustainable Biodiesel Washing Process from an Appropriate Technology Based Approach for Developing Regions

4th International Congress on Sustainability Science & Engineering
Balatonfüred, Hungary

Maxwell Croft
Undergraduate Researcher
University of Kentucky College of Engineering

Jeffrey Seay, Ph.D., P.E.
Associate Professor of Chemical Engineering
University of Kentucky College of Engineering

Project Motivation

What is Appropriate Technology?

- Role Appropriate Technology (AT) plays in renewable energies.

- Importance of and application in developing regions.
  - While many countries need alternative energy solutions, most developing countries are in short supply of reliable energy.

4th International Congress on Sustainability Science & Engineering Balatonfüred, Hungary May 28, 2015

Background

UK
UNIVERSITY OF
KENTUCKY
College of Engineering
Appropriate Technology and Sustainability Research Team

- The global demand for alternative energies has become steadfast throughout the past few decades.
  - Due to issues ranging from diminishing fossil fuels to environmental pollution.
- Underdeveloped countries
  - Access to reliable energy is key for economic development and improving standards of living.
- Sub-Saharan Africa
  - Rural areas have little access to reliable energy.
  - Many areas are poverty stricken.
  - Abundance of viable agricultural resources.
  - Resources can be refined into viable biofuel such as biodiesel.

• 4th International Congress on Sustainability Science & Engineering Balatonfüred, Hungary May 28, 2015 •

Biodiesel

UK
UNIVERSITY OF
KENTUCKY
College of Engineering
Appropriate Technology and Sustainability Research Team

- Sustainable solution for providing a liquid fuel source.
  - Generate electric power.
  - Provide mobility through use as a transportation fuel.
- Composed from locally available feedstocks.
- Low cost and simple process.

• 4th International Congress on Sustainability Science & Engineering Balatonfüred, Hungary May 28, 2015 •

Case Study

UK
University of Kentucky
College of Engineering
Appropriate Technology and Sustainability Research Team

- The African Center for Renewable Energy and Sustainable Technology (ACREST).
  - UKATS partnered with this NGO located in Bangang Village, Cameroon.
  - Field research study was conducted in May of 2012 & 2014.
- Main focus was to implement an AT based process to convert locally sourced vegetable oil into biodiesel.

ACREST Washing Approach

UK
University of Kentucky
College of Engineering
Appropriate Technology and Sustainability Research Team

Production of biodiesel from locally available vegetable oils. → Removing soap produced as a side product of the transesterification reaction. → Production must be sustainable and cost effective for underdeveloped regions.

ACREST
African center for Renewable Energy & Sustainable Technology

● 4th International Congress on Sustainability Science & Engineering Balatonfüred, Hungary May 28, 2015 ●

Biodiesel Production

- Three Major Steps to Produce Biodiesel:
  1. Filtration
     - The raw oil is filtered to improve the purification level of the biodiesel.
  2. Transesterification
     - The oil is mixed with an alcohol and base catalyst, then reacted.
  3. Washing
     - Catalyst and the free fatty acid react to form soap.
     - This is where the emulsion is formed, leaving a final product of biodiesel, unreacted oil, methanol, water, and a soapy emulsion.

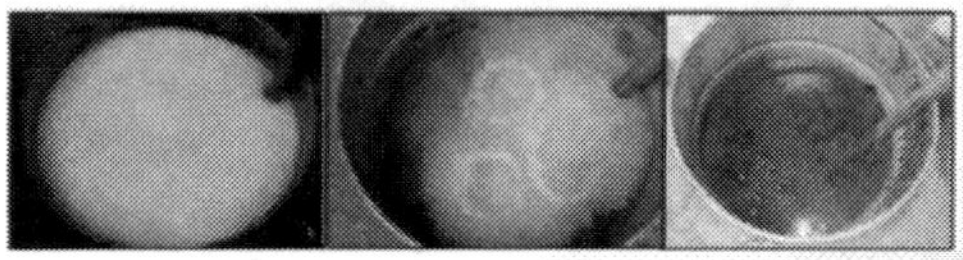

Triglyceride + 3 [R-OH] <-> Glycerol + R_1-$COOCH_3$ + R_2-$COOCH_3$ + R_3-$COOCH_3$

Industrial Production & Purification Techniques

- Industrial Production vs. AT Production:
  - Industry has access to modern manufacturing processes, electricity, and industrial resources.
  - AT production is prime for tropical regions, which are rich in biological resources, such as oil seeds.
- There are many conventional techniques that industry relies on to purify biodiesel, such as:
  - Pressurized Water Washing.
  - Decantation.
  - Separative Membranes.

Industrial Production & Purification Techniques

- A primary technical challenge that is often overlooked for AT based biodiesel is removing the soap produced as a side product in the transesterification reaction.
  - Diesel engines could encounter problems ranging from clogged filters to engine knocking if the soap is not properly removed.
- Physical Phenomena Focus:
  - Soap and Biodiesel Interfacial Properties.
    - Liquid-Liquid extraction.
  - The Chemical Reaction.
    - Between solvent and soap molecules.
  - The Surface Chemistry.
    - Eliminating emulsions formed.

• 4th International Congress on Sustainability Science & Engineering Balatonfüred, Hungary May 28, 2015 •

Preliminary Experiments

- Initial wash experiments were conducted to determine the effectiveness of these additives:
  - Vinegar
  - Salt
  - Glycerin
- Experimental parameters used to wash the crude biodiesel.
  - Different wash additives and concentrations.
  - Feed to solvent ratio.
  - Wash solution temperature.
- Experiments examined:
  - Biodiesel yield.
  - Emulsion thickness.
  - Soap content (ppm).

• 4th International Congress on Sustainability Science & Engineering Balatonfüred, Hungary May 28, 2015 •

Preliminary Experimental Results

UK UNIVERSITY OF KENTUCKY College of Engineering Appropriate Technology and Sustainability Research Team

- Experimental Results Determined:
  - Acetic Acid yielded the best results in eliminating the emulsion and lowering the soap content.
  - The wash solution should be heated to 55°C before adding to biodiesel for best results.
  - AT based washing method requires one week of settling time.
- Further wash experiments were conducted to determine the effectiveness for these locally available wash additives:
  - Calcium Carbonate.
  - Fermented Lemon Vinegar.
  - Calcium Carbonate & Acetic Acid.
  - Fermented Pineapple Vinegar.

• 4th International Congress on Sustainability Science & Engineering Balatonfüred, Hungary May 28, 2015 •

Preliminary Experimental Results

UK UNIVERSITY OF KENTUCKY College of Engineering Appropriate Technology and Sustainability Research Team

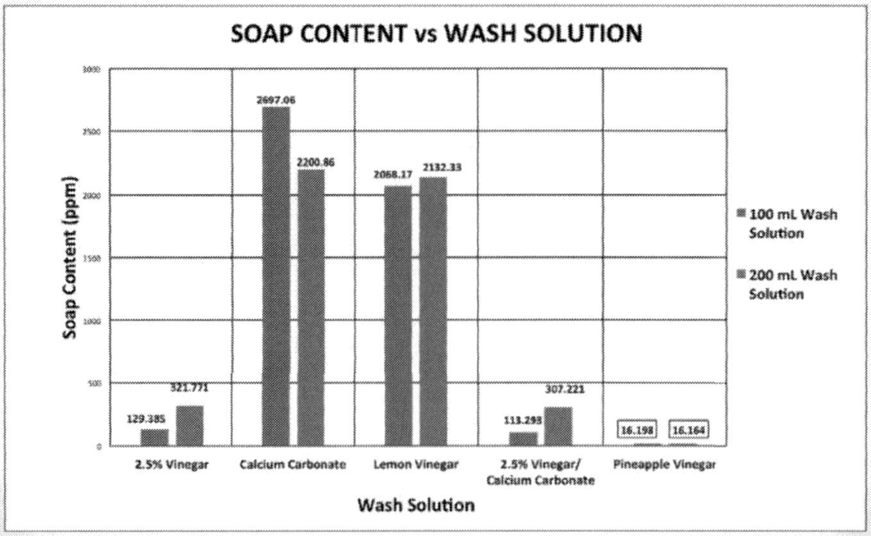

• 4th International Congress on Sustainability Science & Engineering Balatonfüred, Hungary May 28, 2015 •

Experimental Method

- Statistical design of experiments was conducted by response surface methodology.
- The purpose was to minimize the soap content by varying:
  - Feed to Solvent Ratio of Biodiesel to Wash Solution.
  - Wash Solution Concentration of Acetic Acid.
- After reaching the appropriate settling time, the soap content was analytically measured by titrations.

● 4th International Congress on Sustainability Science & Engineering Balatonfüred, Hungary May 28, 2015 ●

Conclusions & Discussion

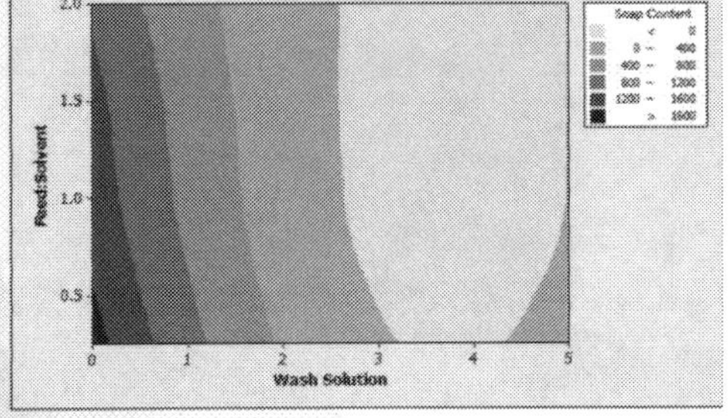

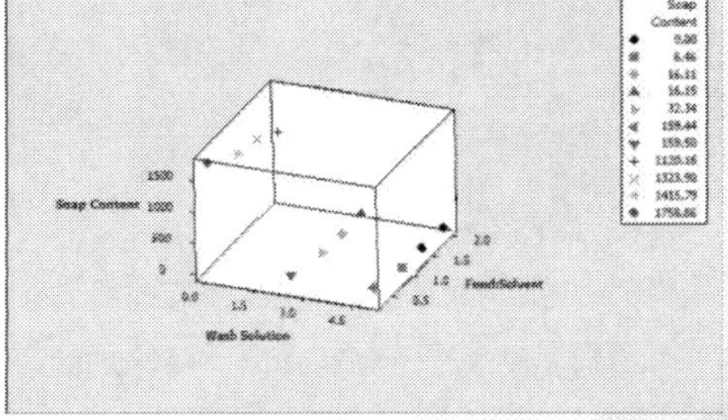

● 4th International Congress on Sustainability Science & Engineering Balatonfüred, Hungary May 28, 2015 ●

Conclusions & Discussion

UK UNIVERSITY OF KENTUCKY College of Engineering Appropriate Technology and Sustainability Research Team

The response surface methodology confirmed the AT based biodiesel washing process in that:

- Wash Concentration:
  - Washing with acetic acid (vinegar) at a concentration of 2.5 to 5,0% yielded a soap content under the purity requirements.
  - Washing with water at any Feed:Solvent ratio will form a thick emulsion and yield a high soap content.
- Feed to Solvent Ratio:
  - Washing with acetic acid at a Feed:Solvent ratio of 1:1 to 1:2 will yield a soap content under the purity requirements.
  - Washing with acetic acid at a Feed:Solvent ratio of 1:0.25 will yield a soap content of 160ppm, which is above the purity requirements.

• 4th International Congress on Sustainability Science & Engineering Balatonfüred, Hungary May 28, 2015 •

Conclusions & Discussion

UK UNIVERSITY OF KENTUCKY College of Engineering Appropriate Technology and Sustainability Research Team

Factors contributing to a sustainable washing process:

- AT wash solution is required to yield operational biodiesel.
  - How to determine the appropriate concentration?
- Determining biodiesel purity levels:
  - How will this fuel be used?
  - What type of engine will the biodiesel be used for?
- The Feed:Solvent ratio provides the appropriate surface area required for each purification level.

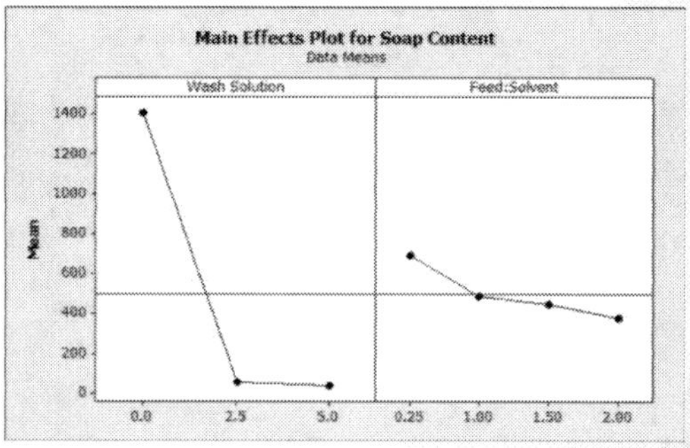

• 4th International Congress on Sustainability Science & Engineering Balatonfüred, Hungary May 28, 2015 •

Appropriate Technology Purity Testing

UK UNIVERSITY OF KENTUCKY College of Engineering Appropriate Technology and Sustainability Research Team

- Industry test biodiesel purity by means of:
  - Titrations.
  - Fourier Transform Infrared Spectrometer (FTIR).
  - Gas Chromatograph (GC).
- Developing regions test biodiesel purity by means of:
  - Phenylalanine red test.
  - Measuring the pH of the initial and final washed product.
- The AT purity test can determine:
  - Soap Content
    - By degree of haze formed in the biodiesel layer.
  - Soapy Emulsion
    - If present, emulsion will form between the liquid interface.
  - pH
    - By degree of color change, as compared to the purity standard.

Future Work

UK UNIVERSITY OF KENTUCKY College of Engineering Appropriate Technology and Sustainability Research Team

ODAM

- Start next case study in Tamil Nadu, India this summer.
  - Partnering with the Organization of Development Action and Maintenance (ODAM).
  - Investigate locally available feedstocks for biodiesel oil and washing solutions.

Acknowledgments

UK UNIVERSITY OF KENTUCKY College of Engineering Appropriate Technology and Sustainability Research Team

The authors would like to thank the African Centre for Renewable Energy and Sustainable Technology (ACREST).

- Partnership with the University of Kentucky Appropriate Technology and Sustainability Research Team.
  - Photos from William Croft.

ACREST
African center for Renewable Energy & Sustainable Technology

• 4th International Congress on Sustainability Science & Engineering Balatonfüred, Hungary May 28, 2015 •

Questions?

UK UNIVERSITY OF KENTUCKY College of Engineering Appropriate Technology and Sustainability Research Team

• 4th International Congress on Sustainability Science & Engineering Balatonfüred, Hungary May 28, 2015 •

ICOSSE 2015

Climate Change Mitigation by Municipal Solid Waste Syngas Substitution for Chemical Feedstock

Byung Chul Shin[1], M.H. Jang[1], J. W. Shin[1], J.H. Gu[2], I.S. Lee[3]

[1] R&D Center, GreenConTech Co., Ltd, Gyeonggi, South Korea
[2] Plant Engineering Center, Institute for Advanced Engineering, Gyeonggi, South Korea
[3] R&D Center, Samsung-BP Chemicals Co., Ltd, Ulsan, South Korea

Acetic Acid Process

○ Carbon Monoxide

$$\text{B-C Oil} + O_2 \longrightarrow CO + H_2$$

○ Acetic Acid Reaction

$$CH_3OH + CO \xrightarrow{\text{Catalyst}} CH_3COOH$$

○ Vinyl Acetate Monomer

$$C_2H_4 + CH_3COOH \quad ½\ O_2 \xrightarrow{\text{Catalyst}} CH_3COOC_2H_3$$

2

Background of Waste Gasification Project

✓ **To make Cost Strength in Current Acetic Acid Value Chain**

- The most important key in acetic acid value chain is competitive CO production cost
- CO supply from fossil fuel has weakened the competitiveness due to high variable cost
- Persistent effort to secure non-petroleum energy and diversity of CO source

✓ **Future Growth Engine for Business Diversity & Value Creation Enterprise**

- New business such as gas and derivatives utilizing competitive CO economics
- Explore overseas market based on domestic experience

✓ **Play a Leading Role of Green Growth in Petrochemical Industry**

- Greenhouse gas reduction by saving conventional fossil fuel energy
- Minimization the waste production through waste-to-energy

3

GHG Emission Reduction by Waste Gasification

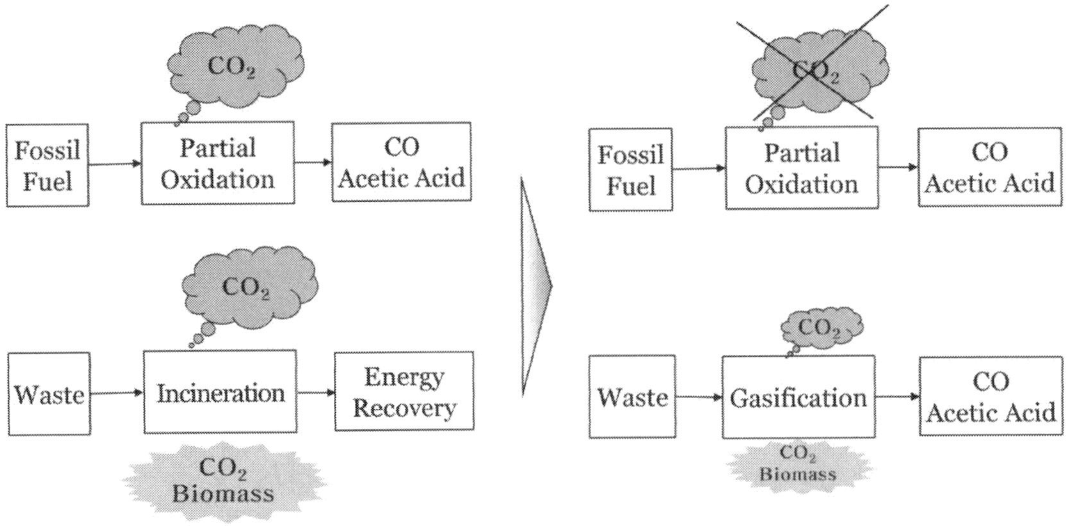

4

Syngas process from MSW Gasification

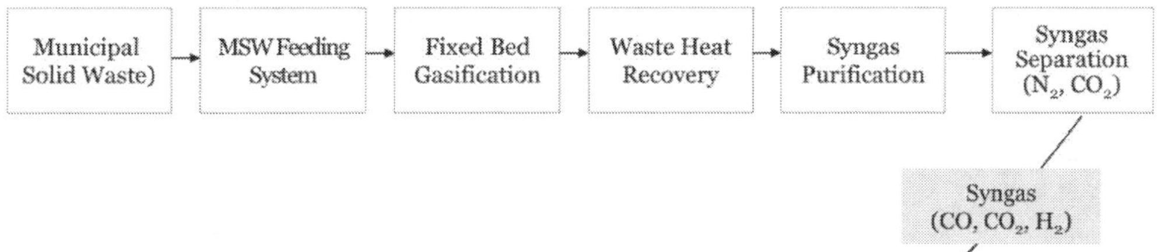

Syngas process from Fossil Fuel (Existing)

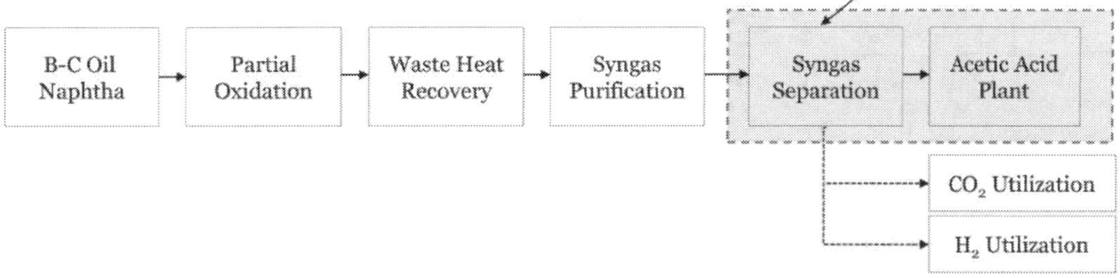

MSW Gasification by Pilot Test

○ Physiochemical Properties of Waste

	Combustible. wt %						Moisture wt %	Ash wt %	LHV kcal/kg
	C	H	O	N	S	Cl			
Industrial MSW	39.3	5.5	17.7	0.4	0.1	0.2	2.6	34.2	4,000
Waste Sludge	6.7	1.1	4.0	0.1	0.2	0.0	80	7.9	205
ASR	45.0	4.0	17.4	1.1	0.4	1.0	9.6	21.5	4,007

○ Composition of Industrial Waste

	Paper	Vinyl	Plastics	Wood	Rubber	Styrofoam	Textile	Leather	Others	Inert
wt %	32.0	11.56	10.22	7.78	6.71	2.45	1.51	0.31	17.23	10.23

Syngas Characteristics from MSW Gasification by Pilot Test

	Input Material				Product Syngas	
	Waste Input ton/day		Auxiliary Fuel Nm^3/h	Oxygen Nm^3/h	Cold Gas Efficiency %	$CO + H_2$ %
Case 1	MSW	2.3	6.5~8	60.5~75.8	65.5~72.9	64.6~71.8
Case 2	MSW	2.2	7.7~12.2	64.1~75.6	70.2~76.5	70.1~75.2
Case 3	ASR	1.9	5.9~6.4	70.8~72.3	71.7~93.4	72.8~76.9
Case 4	MSW +ASR	2.9	7.6~11.3	77.3~87.2	57.9~76.1	65.7~75.2

$$\text{Cold Gas Efficiency, \%} = \frac{\text{Potential energy of gas made x 100}}{\text{Total heat input}} = \frac{LHVgas \times Vgas}{LHV_{fuel} \times Mfuel}$$

ASR : Automotive shredder residue

Commercial Design Basis for Feasibility Study

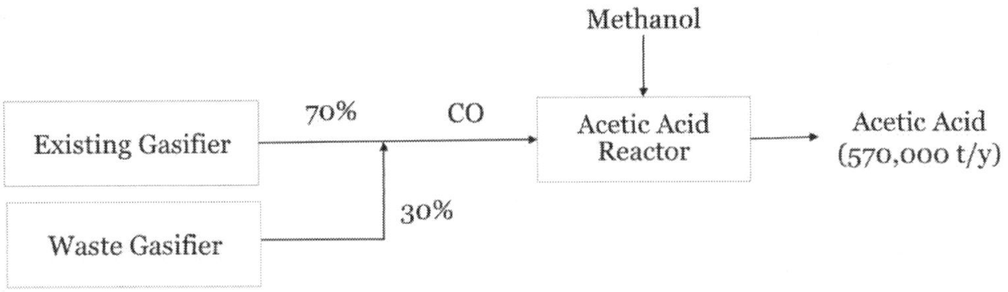

- **UNIT #1**
  - Mixed Waste : 240 ton/day (= MSW 190+Sludge 50)
  - LHV, Waste : 3,200 kcal/kg
  - CO Yield : 0.52 kg/kg-waste

- **UNIT #2**
  - ASR : 170 ton/day
  - LHV : 4,007 kcal/kg
  - CO Yield : 0.83 kg/kg-waste

Mass Balance of Waste Gasification by Process Simulation

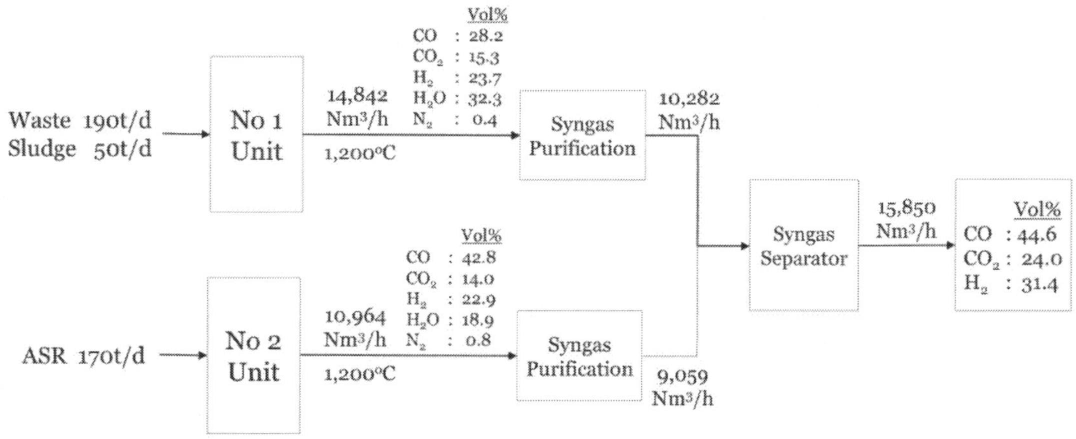

Greenhouse Gas Emission Reduction

○ Project Boundary

The project boundary contains all anthropogenic GHG emissions by sources under the control of the project participants that are significant and reasonably attributable to the project activity.

○ GHG Emission reduction

$$ER_y = BE_y - PE_y - LE_y$$

ER_y = GHG emission reduction at y year, tCO_2/y
BE_y = Baseline emissions at y year, tCO_2/y
PE_y = Project emission at y year, tCO_2/y
LE_y = Project leakage at y year, tCO_2/y

○ Baseline Emission

$BE_y = BE\_FC_y + BE\_EC_y + BE\_SEP_y + BE\_INC_y$

BE_y	=	Baseline emission in existing gasification facility, tCO_2/y
BE_FC_y	=	GHG emission by fossil fuel consumption, tCO_2/y
BE_EC_y	=	GHG emission by electrical consumption, tCO_2/y
BE_SEP_y	=	CO_2 emission from existing fossil fuel gasifier, tCO_2/y
BE_INC_y	=	GHG emission from waste incineration, tCO_2/y

- Emission from waste incineration

$$BE\_INC_y = [\ \sum M_{i,y} \times (SW_j \times dm_j \times CF_j \times FCF_j \times Of_j) \times 10^{-10} \times 44/12 + \sum M_{i,y} \times (\ EF_{N2O,i} \times GWP_{N2O} + EF_{CH4,i} \times GWP_{CH4}\ )\ ] \times 10^{-6}$$

BE_INC	=	GHG emission from waste incineration, $t\text{-}CO_2/y$
$M_{i,j}$	=	Quantity of waste, t/y (i = MSW, sludge, ASR, j = waste material component)
SWj	=	each waste material component, %
dm_j	=	dry matter content, %
CF_j	=	Total carbon content in waste material component j, %
FCF_j	=	Carbon content from fossil fuel in waste material component j, %
Of_j	=	Combustion efficiency, %
EF_{N2O}, EF_{CH4}	=	Emission coefficient of incineration type, $g\text{-}N_2O$ or $g\text{-}CH_4/t$-waste
GWP_{N2O}, GWP_{CH4}	=	Global worming potential, 210 $t\text{-}CO_2/t\text{-}N_2O$ or 23 $t\text{-}CO_2/t\text{-}CH_4$

11

- Acetic acid production capacity : 570,000 t-AA/y
- CO substitution from waste gasification : 30% of total CO feedstock

Baseline Emission Source			CO_{2eq} Emission
Acetic acid plant	Feed	B-C Oil	· CO_2 Separation & Recovery BE_SEP = 6,800 $t\text{-}CO_2/y$ · Emission by electricity consumption BE_EC = 4,200 $t\text{-}CO_2/y$
	CO process	Gasification Partial oxidation	
	Sub-total		11,000 t-CO2/y
Waste Plant	Feed	Ind. MSW Sludge, ASR	· CO_2 emission from fossil fuel component BE_INC_{FF} = 206,200 $t\text{-}CO_2/y$ · CO_2 emission from biomass component BE_INC_{BF} = 81,800 $t\text{-}CO_2/y$
	Treatment process	Incineration & Heat recovery	
	Sub-total		288,000 $t\text{-}CO_2/y$
Baseline Emission for 30% CO production			BE = 299,000 $t\text{-}CO_2/y$

12

○ Project Emission

$$PE_y = SEP_y + EC_y + FC_y - ISO_y$$

PE	=	GHG emission by project execution, tCO_2/y
SEP	=	CO_2 emission from waste gasification process, tCO_2/y
EC	=	Emission by electrical consumption in waste gasification, tCO_2/y
FC	=	Emission by fossil fuel consumption in waste gasification, tCO_2/y
ISO	=	Quantity of CO_2 equivalent to CO embedded in acetic acid, tCO_2/y

○ Project Leakage

$$LE_y = EF_{CO2,MC} \times \sum (M_{i,y} \times NCV_{i,y})$$

LE_y	=	GHG emission by leakage, tCO_2/y
$EF_{CO2,MC}$	=	Emission coefficient of most carbon intensive fossil fuel, tCO_2/GJ
$M_{i,y}$	=	Quantity of waste type i, t/y (i = MSW, sludge, ASR)
NCV_i	=	Net calorific value of waste type i, GJ/t

○ Excluding carbon neutral

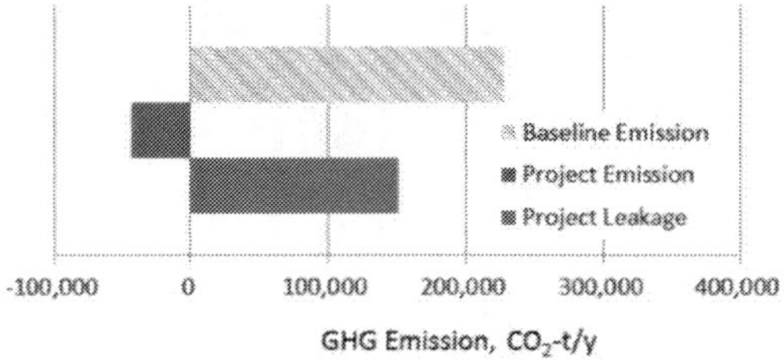

Ext. incinerator heat recovery rate	Emission Reduction, %	
	Carbon intensive fuel (coal)	Carbon least fuel (LNG)
85%	46	76
80%	51	79

○ Including carbon neutral

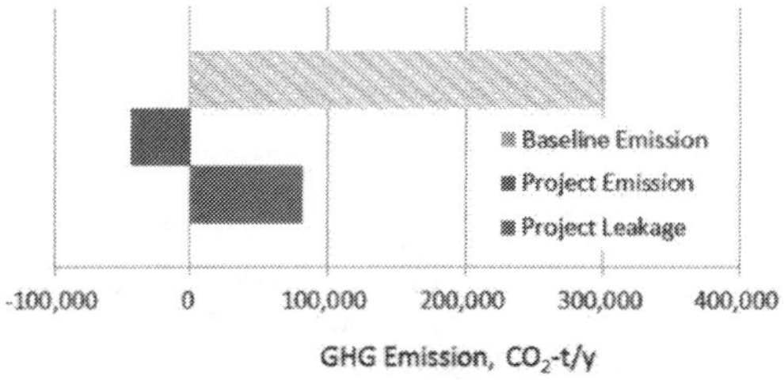

Ext. incinerator heat recovery rate	Emission Reduction, %	
	Carbon intensive fuel (coal)	Carbon least fuel (LNG)
85%	61	83
80%	64	85

Thank you

Advances in Process Integration Research for the Hybrid Power System Supply Planning and Demand Management

- **Conference:** International Congress on Sustainability Science Engineering ICOSSE
- **Year:** 2015
- **Proceeding:** ICOSSE 2015
- **Group:** Abstract Submissions
- **Session:**

Session 10: Energy Generation Life-Cycles and Ecological Impact

- **Time:**

Thursday, May 28, 2015 - 1:10pm-1:40pm

Authors:

Wan Alwi, S. R., Universiti Teknologi Malaysia
Mohammad Rozali, N. E., Universiti Teknologi Malaysia
Abd Manan, Z., Universiti Teknologi Malaysia
Klemeš, J. J., University of Pannonia

The International Energy Agency has estimated a renewable power escalation of 40 %, and expected the Renewable Energy (RE) to make up almost a quarter of global power mix. A Hybrid Power System (HPS) combines the use of intermittent electrical energy from various sources including the RE with "purchased" or outsourced power. Modeling tools for the optimal HPS supply planning and demand management have been relatively established. However, complementary tools that can provide planners, decision-makers, energy managers and electrical as well as power engineers with graphical and visualisation insights that are vital for better conceptual understanding of the problems have just been recently developed. This paper presents the recent developments on the insight-based graphical and algebraic Process Integration (PI) tool for the optimal HPS supply planning and demand management. Known as the Power Pinch Analysis (PoPA), the tool has been among the next-generation PI technique for resource conservation.

Investigation of CO2 Capture on Aminopolymer Impregnated MCM-36

- **Conference:** International Congress on Sustainability Science Engineering ICOSSE
- **Year:** 2015
- **Proceeding:** ICOSSE 2015
- **Group:** Abstract Submissions
- **Session:**

Session 10: Energy Generation Life-Cycles and Ecological Impact

- **Time:**

Thursday, May 28, 2015 - 1:40pm-2:00pm

The bulk of the research into solid adsorbents has currently been focused on increasing the amine loading of the sorbents used in order to directly increase CO_2 capture capacity and selectivity, without investigating the changes that occur to the pore environment upon impregnation. However, increasing the amine loading in a porous solid can have drastic effects on the diffusion of CO_2, making a material that seems promising in the lab infeasible in practice. We have recently reported on the effects of impregnation on the kinetics and diffusion of CO_2 in MCM-36. These results suggest that maximum capacity is in a region where the diffusion limitations incurred through impregnation are not yet severe enough to decrease capture capacity. Our results suggest that a more detailed study of the changes to diffusion with amine loading may allow for the detection of more optimum conditions for maximum capture performance at lower amine loadings.

Marrying Life Cycle Inventory Data and Green Chemistry Measures to Advance Sustainable Chemicals Production

- **Conference:** International Congress on Sustainability Science Engineering ICOSSE
- **Year:** 2015
- **Proceeding:** ICOSSE 2015
- **Group:** Abstract Submissions
- **Session:**

Session 10: Energy Generation Life-Cycles and Ecological Impact

- **Time:**

Thursday, May 28, 2015 - 2:00pm-2:20pm

There are several well-known examples in green chemistry where process changes to improve a target synthesis have induced the use of highly toxic chemicals two or more steps upstream, yet no automated approach has been developed to measure the inherent toxicity of these intermediate chemical inputs. Here we present a new modeling algorithm adapted from LCA that measures inherent risk (toxicity) of inputs rather than impacts from emissions that occur over the life cycle and applied this algorithm to all organic chemicals in the ecoinvent LCI database. For each target chemical we have calculated a new metric of 'life cycle chemical risk' and compared this to the toxicity of the target chemical itself. Results show that there is very low correlation ($r<0.01$) between these sets of values, suggesting that current measures of inherent chemical risk are missing large upstream effects and thus do not provide chemists with adequate design guidance.

Synthesis and Design of a Sustainable CO_2 Utilization Network

Rebecca Frauzem, Kasper Fjellerup, Rafiqul Gani

Department of Chemical and Biochemical Engineering,
Technical University of Denmark, Søltofts Plads Building 229,
2800 Kgs. Lyngby, Denmark

Thursday, May 28, 2015
ICOSSE 2015

DTU Chemical Engineering
Department of Chemical and Biochemical Engineering

Overview

- Introduction
- Project Motivation
- Methodology
- Combined Carbon Capture and Utilization
- Sustainability Criteria
- Case Study
- Generation of Alternatives
- Conclusions and Future Work

Environmental Scenario

Global population growth → Increased energy demands and manufacturing capacities → Increased emissions and greenhouse gas emissions

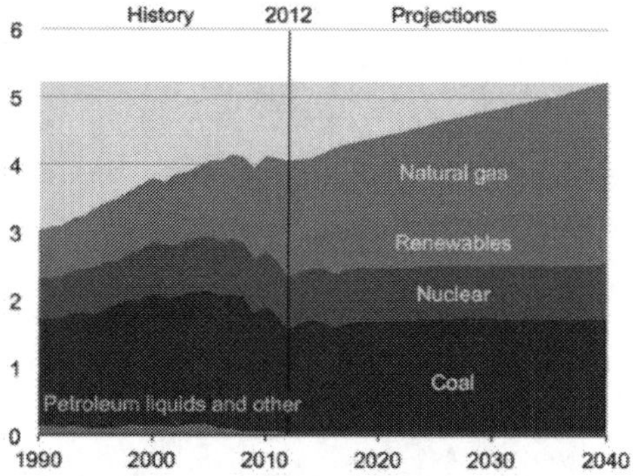

Electricity generation by fuel from 1990-2040 (trillion kWh)

Global greenhouse gas emissions (Gt CO_2-eq/year)

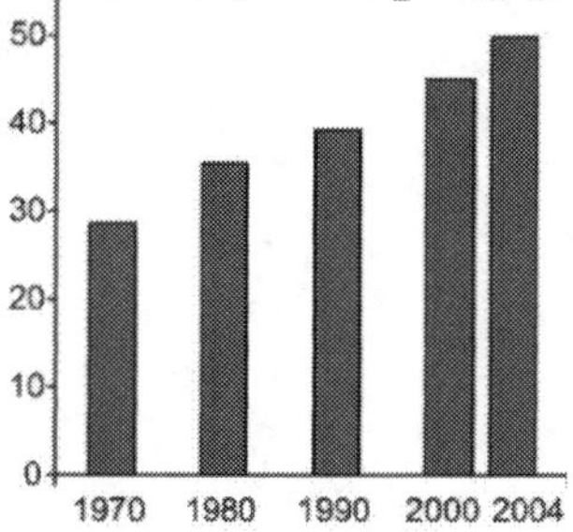

(left) EIA, 2013. International energy outlook 2013, Washington, DC: EIA. Available at: http://www.eia.gov/ieo

(right) IPCC, 2007. Climate Change 2007: Mitigation of Climate Change B. Metz et al., eds., New York: Cambridge University Press.

3 **DTU Chemical Engineering, Technical University of Denmark** Rebecca Frauzem ICOSSE 2015 Oral Presentation 28/05/2015

Greenhouse Gases & Global Warming

- Global warming is attributed to the increased amount of greenhouse gases in the atmosphere
- Of these, carbon dioxide (CO_2) is in the largest quantities
- In order to mitigate environmental impact, methods of reducing emissions (especially CO_2) need to be developed

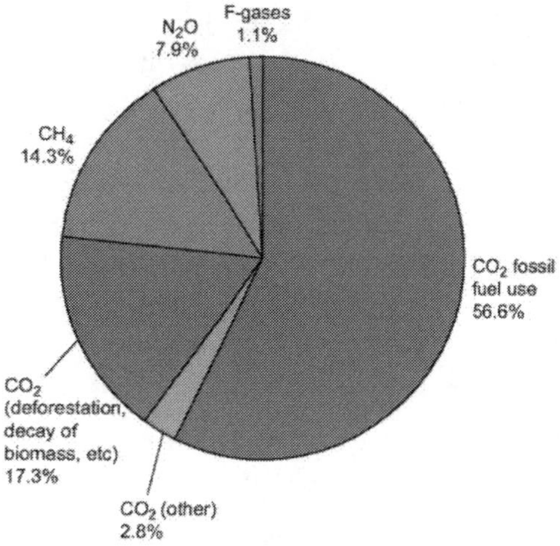

IPCC, 2007. Climate Change 2007: Mitigation of Climate Change B. Metz et al., eds., New York: Cambridge University Press.

4 **DTU Chemical Engineering, Technical University of Denmark** Rebecca Frauzem ICOSSE 2015 Oral Presentation 28/05/2015

CO_2 Mitigation

- 2 Methods of long-term reduction: Carbon Capture and Storage (CCS) and **Utilization**
- CCS is prominently discussed and being implemented
- 2 targets for improvement of CCS
  - Capture technology- reduction of additional energy, solvents or raw materials
  - Storage methods- early stages of development
- Conversion (one element of utilization) involves use of chemical reactions to convert CO_2 to valuable product

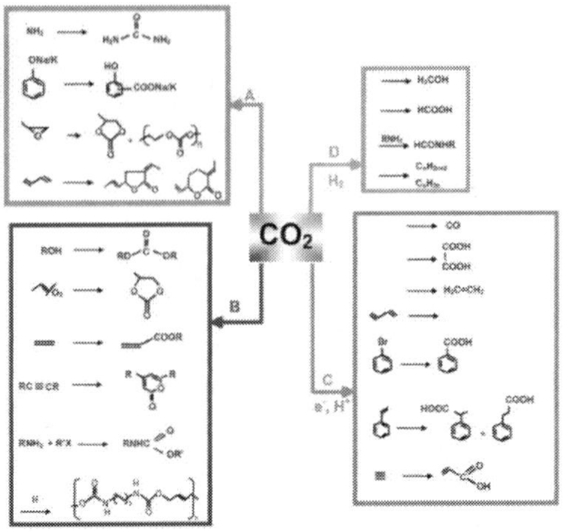

Aresta, M., 2010. Carbon Dioxide as Chemical Feedstock, Carbon Dioxide as Chemical Feedstock. doi:10.1002/9783527629916

Motivation and Objective

- Environmental concerns demand attention
- Need methods to address greenhouse gas emissions (primarily CO_2)
- This work focuses on a methodology for formulating a sustainable CO_2 capture and **utilization** network
  - Focuses on the sustainable chemical conversion of CO_2 to valuable products
  - Incorporating carbon capture and conversion within the network
  - Goal: Non-trade-off solutions with a Net CO_2 reduction ($Net_{CO2} = \Sigma(Outflow_{CO2}) - \Sigma(Inflow_{CO2})$ should be negative)

DTU

Methodology

Stage 1- Superstructure-based Process Synthesis
- Generation of Feasible Reactions using Reaction Path Synthesis
- Development of Superstructure to link carbon capture, CO_2 (raw material) and Product Portfolio
- Mathematical Description of Superstructure
- Optimization

Stage 2 – Process Design
- Base Case Design from Superstructure
- Detailed Design Using Simulation Software
- Analysis of the Process

Stage 3 Innovative & More Sustainable Design
- Using Process Intensification Methods Target Hotspots
- Redesign
- Analyze New Design

DTU

Current State-of-the-Art

- Carbon Capture and Storage (CCS)
  - Currently the primary method of addressing CO_2 emissions
  - Capture primarily (though not exclusively) through amine (MEA) absorption
  - Storage through
    - Geological formations
    - Tanks
  - Capture is done to a high purity (99.99%)

Combined Carbon Capture and Utilization

- There are high quality and low quality emissions
- Currently, all sources are purified to the same, high purity
- However, for conversion, this high purity is not necessary
- By linking the carbon capture and conversion processes, it is possible to optimize for the overall sustainability and not the individual sustainability

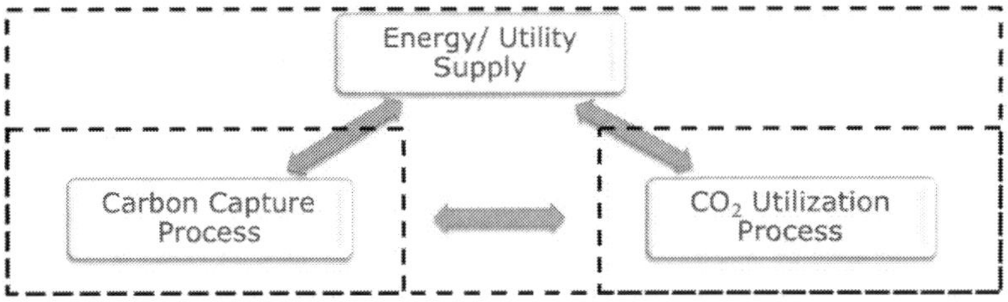

Sustainability Criteria

- Non-trade-off solution
  - Economic benefits (looking at capital costs, investments, operating costs)
  - Process benefits (reduce losses of both heat and material)
  - Environmental impacts
    - Net CO_2 Emissions

$$Net_{CO2} = \Sigma(Outflow_{CO2} + Indirect_{CO2}) - \Sigma(Inflow_{CO2})$$

- This takes into account the CO_2 flow in the streams as well as the amount indirectly produced through energy requirements (heating)

Case Study

- Methanol Synthesis
- 3 Steps:
  - Carbon Capture
  - Methanol Synthesis
  - Combined Carbon Capture and Methanol Synthesis
- This shows the utility of combined capture and utilization and how to apply this

Case Study – Carbon Capture

- Industrial standard is amine absorption using MEA (monoethyl amine)
- Solvent recovery is energy intensive
  - Reboiler duties are extremely high due to the high recovery and purity requirements that are set
- Base Case Design for a CO_2 Capture facility in Canada
  - 150 kWh
  - 1.14 Mt/yr CO_2
- First, the carbon capture process was optimized using the purity requirements that are industrial standard
- Second, the purity and recovery conditions are manipulated and analyzed

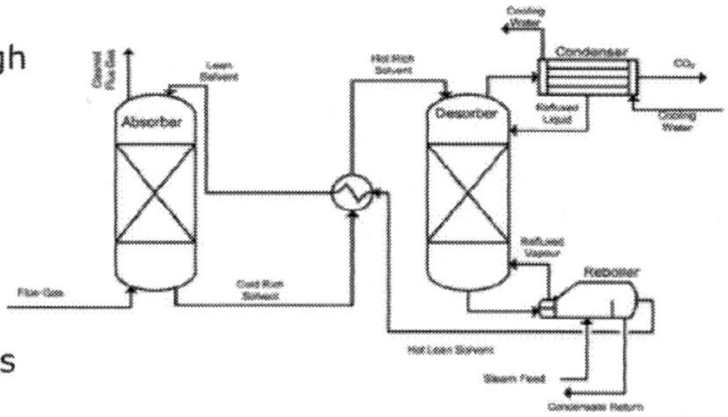

MacDowell, N. et al., 2010. An overview of CO2 capture technologies. Energy & Environmental Science, 3(11), pp.1645–1669.

DTU

Case Study – Carbon Capture

	Exit CO2 conc. [molar fraction]	Overall CO2 capture [%]	Utility cost [$/yr]	Capital investment [$]	Indirect carbon emission [kg CO2 equivalent/kWh]	Carbon footprint [kg CO2 equivalent/kWh]
High Purity	0.9988	99.98	1.91E+08	2.29E+06	5.73	4.9
Low Purity	0.9722	99.78	2.41E+07	1.48E+06	0.608	-0.26

- From the two scenarios for Carbon Capture that are simulated and analyzed:
  - Reducing the purity constraint results in a CO_2 reducing process
  - Tenfold decrease in utility cost
  - Almost tenfold decrease in indirect CO_2 emissions

DTU

Case Study – Methanol Synthesis

- Methanol currently synthesized via synthesis gas
  - Two step process: reforming (4 types of reforming possible: bi-reforming, tri-reforming, dry reforming, and oxy-reforming) and synthesis
  - Reforming unit is energy intensive (results in large indirect emissions)
- Direct hydrogenation over a Cu/ZnO-multicomponent catalyst is an alternative (this is simulated and analyzed)
  - Takes into account kinetics
  - Large recycle to maintain high selectivity (therefore, lower conversions)
  - Manipulation of feed, variation of operating conditions and integration of heat flow performed

Case Study – Methanol Synthesis

- Feasible and sustainable conversion process
- Heat integration shows a **-1.125 $kgCO_{2,net}/kg_{MeOH}$** net CO_2

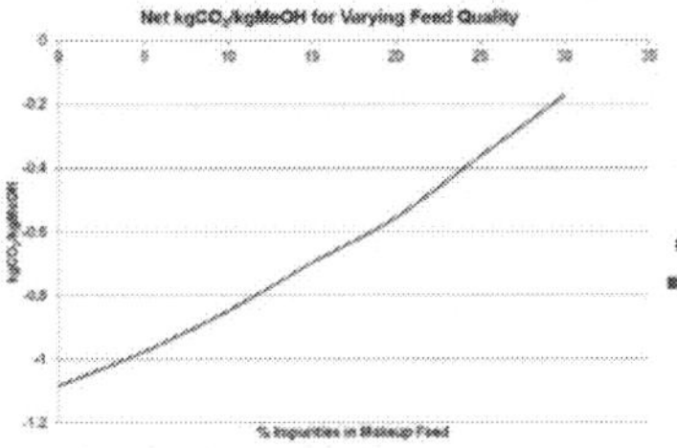

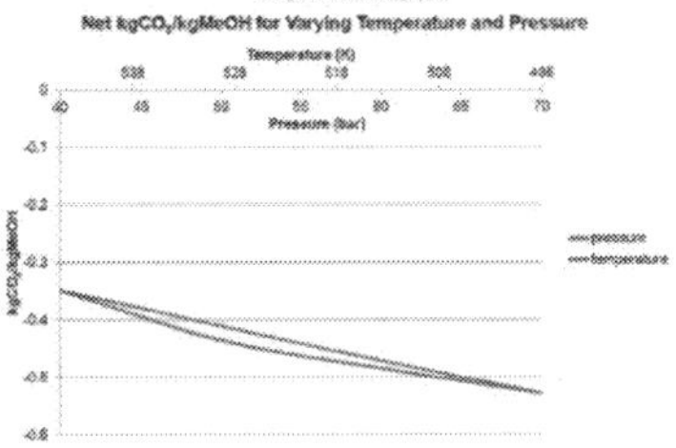

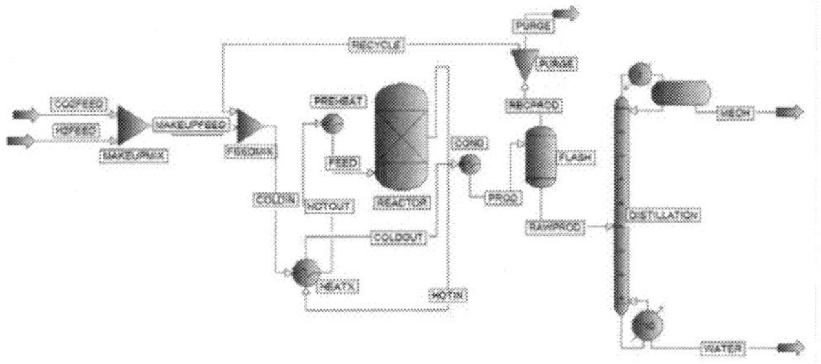

(top left) Frauzem, R., 2014. Techno-economic evaluation of a CO2 -conversion process. Master's Thesis

(bottom left) Frauzem, R., 2014. Techno-economic evaluation of a CO2 -conversion process. Master's Thesis

(right) Ushikoshi, K., Mori, K., Kubota, T., Watanabe, T., Saito, M., 2000. Methanol synthesis from CO_2 and H_2 in a bench-scale test plant. Appl. Organomet. Chem. 14, 819–825. doi:10.1002/1099-0739(200012)14:12<819::AID-AOC90>3.0.CO;2-#

Case Study – Combined Carbon Capture and Methanol Synthesis

- It has been shown individually that the carbon capture process and the methanol synthesis processes are sustainable
- Now, the link between them is formed to determine the overall sustainability of a capture and conversion process
- Produce 0.414 Mt/year of Methanol
- Payback within a year of operation

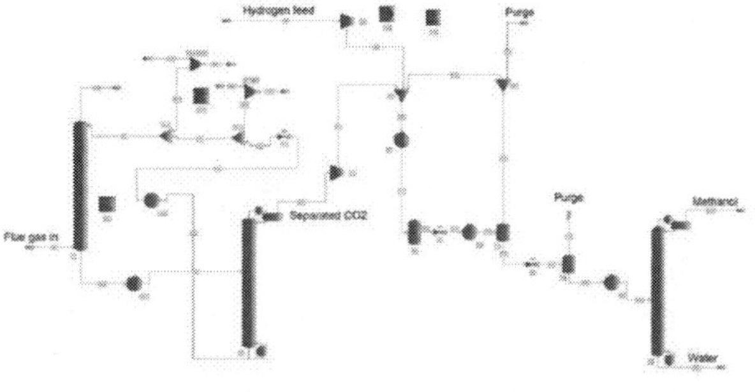

Generating Utilization Alternatives

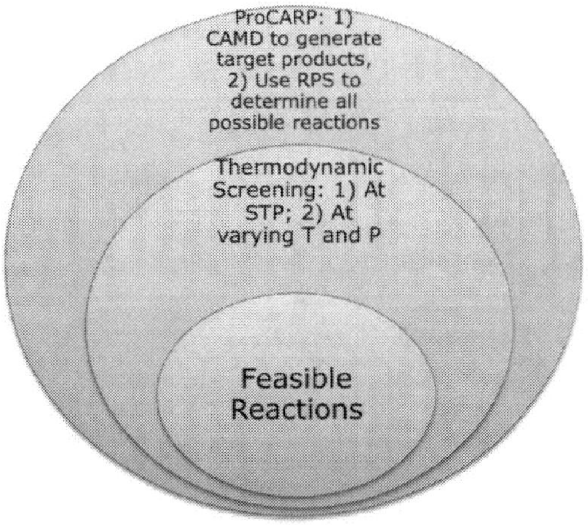

- Use reaction path synthesis (RPS)
  - From the generation to the screening of CO_2 reactions
  - Create a reaction tree for CO_2 to describe all possible conversion and how the connect to one another

CO_2 Reaction Tree

- From the generated reactions, those that are thermodynamically feasible are collected
- These are being analyzed to create a reaction tree
- Using this, a network of conversion reactions is formed
- Using this, a superstructure can be generated

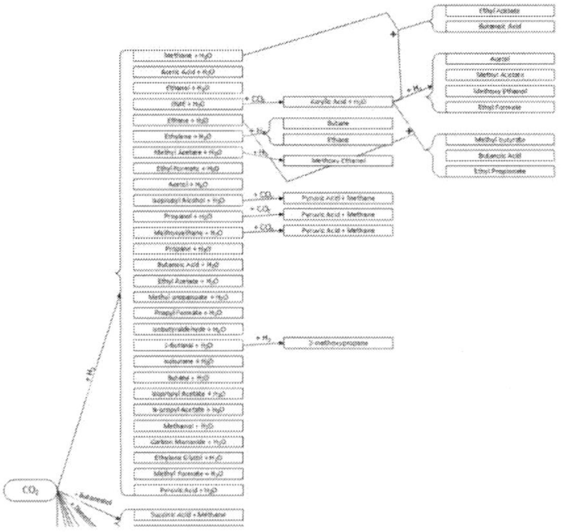

Conclusions

- A methodology for the sustainable utilization of CO_2 is being developed
  - Involves a superstructure-based approach linking CO_2 (raw material) and valuable chemical products
  - Creates a network of conversion processes
- Using reaction path synthesis, a reaction tree is formed for the feasible CO_2 reactions
  - These can be used as the starting point for superstructure definition
- Case studies have been performed to show the feasibility of combined capture and utilization

Future Work

- Further processes need to be studied as case studies
  - Have done methanol and DMC
  - Other alternatives from the reaction tree should be investigated
- For further opportunities, the reaction tree can be expanded
  - The tree can be constantly expanded upon with more information
- Data is being collected to determine the demand for the chemical products
  - The amount of CO_2 that is needed to produce these via conversion will be calculated
  - The power plants that would generate this amount of CO_2 (the amount of emissions that would thereby be eliminated)

Thanks!

Works Cited

Aresta, M., 2010. Carbon Dioxide as Chemical Feedstock, Carbon Dioxide as Chemical Feedstock. doi:10.1002/9783527629916

Babi, D.K., Woodley, J.M. & Gani, R., 2014. Achieving More Sustainable Designs through a Process Synthesis-Intensification Framework. Computer Aided Chemical Engineering, 33, pp.31–36.

EIA, 2013. International energy outlook 2013, Washington, DC: EIA. Available at: http://www.eia.gov/ieo

Fjellerup, Kasper. Master's Thesis (in progress)

Frauzem, R., 2014. Techno-economic evaluation of a CO2 -conversion process. Master's Thesis

Harper, P.M. & Gani, R., 2000. A multi-step and multi-level approach for computer aided molecular design. *Computers and Chemical Engineering*, 24, pp.677–683.

IPCC, 2007. Climate Change 2007: Mitigation of Climate Change B. Metz et al., eds., New York: Cambridge University Press.

Kongpanna, P., Pavarajarn, V., Gani, R., Assabumrungrat, S., 2014. Techno-economic evaluation of different CO_2-based processes for dimethyl carbonate production. Chem. Eng. Res. Des. doi:10.1016/j.cherd.2014.07.013

Kubota, T. et al., 2001. Kinetic study of methanol synthesis from carbon dioxide and hydrogen. Applied Organometallic Chemistry, 15(2), pp.121–126. Available at: http://doi.wiley.com/10.1002/1099-0739(200102)15:2<121::AID-AOC106>3.0.CO;2-3.

MacDowell, N. et al., 2010. An overview of CO2 capture technologies. Energy & Environmental Science, 3(11), pp.1645–1669.

Nguyen, Bui Huu Tuan. Master's Thesis (in progress)

Quaglia, A. et al., 2014. Systematic network synthesis and design: Problem formulation, superstructure generation, data management and solution. Computers & Chemical Engineering, 72, pp.68–86. Available at: http://linkinghub.elsevier.com/retrieve/pii/S0098135414000787.

Ushikoshi, K., Mori, K., Kubota, T., Watanabe, T., Saito, M., 2000. Methanol synthesis from CO_2 and H_2 in a bench-scale test plant. Appl. Organomet. Chem. 14, 819–825. doi:10.1002/1099-0739(200012)14:12<819::AID-AOC90>3.0.CO;2-#

Wilcox, J., 2012. Carbon Capture, New York: Springer.

Hydrogen Production and Integration with Methanol Synthesis

- Hydrogen (H_2) is the second reactant in various conversion reactions
- However, current production is not sustainable (according to our criteria) and also expensive
- Therefore, alternative sources and production methods have been investigated
- Possible alternative sources:
  - Off gas: using membranes or other separation technology, H2 can be removed from off-gas
  - If a process can replace an existing process, that hydrogen can be implement

Hydrogen Production and Integration with Methanol Synthesis

- Additionally, alternatives to reforming can be implemented for the production of H2
  - Wind/electric
  - Nuclear/steam electrolysis
  - Solar thermal
  - Biomass
  - Hydroelectric
- From work done one the combination of hydrogen production and methanol synthesis, the following was discovered:
  - Net CO2 for these renewable scenarios is negative
  - Capital cost and production cost for hydroelectric, solar thermal and biomass is less that steam reforming; for wind/electric and nuclear/steam electrolysis these are higher

DTU

DMC (Dimethyl Carbonate) Production Process

- DMC is a secondary processing step in the network
  - Involves inputs from previous steps along with CO_2
- Base case from thermodynamic analysis – DMC via urea synthesis

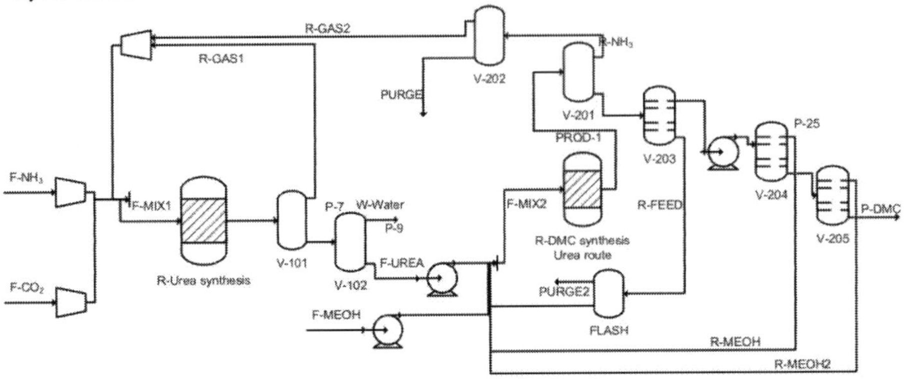

Kongpanna, P., Pavarajarn, V., Gani, R., Assabumrungrat, S., 2014. Techno-economic evaluation of different CO_2-based processes for dimethyl carbonate production. Chem. Eng. Res. Des. doi:10.1016/j.cherd.2014.07.013

DTU

DMC (Dimethyl Carbonate) Production Process

- Analysis of simulation reveals a hot spot in the downstream separation
  - Process intensification performed
  - Targeted energy of downstream separation
  - Alternatives: Pervaporation, Membrane reactor, Reactive distillation

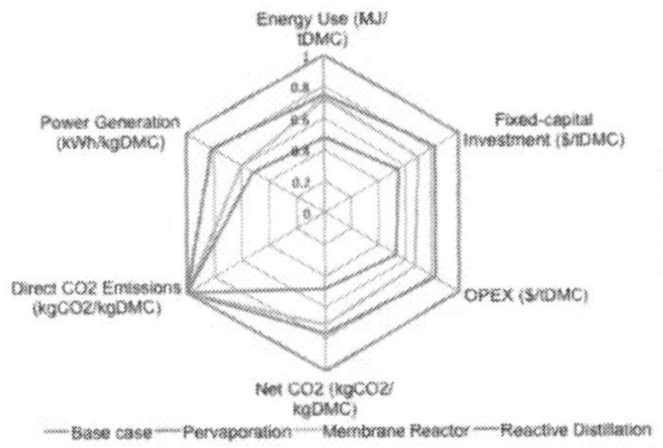

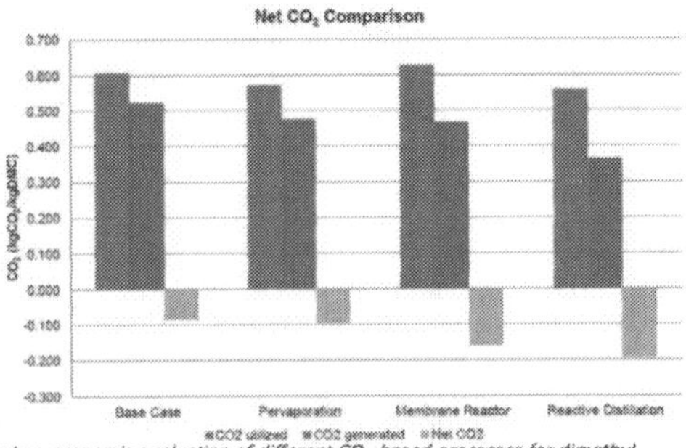

Kongpanna, P., Pavarajarn, V., Gani, R., Assabumrungrat, S., 2014. Techno-economic evaluation of different CO_2-based processes for dimethyl carbonate production. Chem. Eng. Res. Des. doi:10.1016/j.cherd.2014.07.013

Comparison of Alternative Supply Chains for Energy Production from Marginal Biomasses

- **Conference:** International Congress on Sustainability Science Engineering ICOSSE
- **Year:** 2015
- **Proceeding:** ICOSSE 2015
- **Group:** Abstract Submissions
- **Session:**

Session 10: Energy Generation Life-Cycles and Ecological Impact

- **Time:**

Thursday, May 28, 2015 - 2:40pm-3:00pm

Authors:

Tugnoli, A., University of Bologna
Paolucci, N.
Porcelli, R.

In the present contribution a methodological approach to the analysis and optimization of the industrial-scale supply chain of biomasses for energy purposed is applied to the case of cultivations in marginal lands. The optimal supply chain configuration for a given territory is identified by a two-tier sustainability analysis, which considers both the environmental and the economic points of view. The first tier performs a quick evaluation of the supply chain, based on simplified assumptions on the parameters characterizing the geographical territory. The second tier allows for the inclusion of spatially explicit parameters of the territory and realizes a more detailed optimization my multi-objective optimization methods. The results obtained from the two tiers of the model provide quantitative information, to support quick and effective decision making on the optimal configuration of the supply chain in terms of plant size, location, transport logistics and cultivation.

Sustainable Design and Manufacturing: A Multiscale Complex Systems Approach

Yinlun Huang
Dept. of Chem. Eng. & Mat. Sci.
Wayne State University
Detroit, MI 48202

Int'l Congress on Sustainability Science and Engineering
Balatonfured, Hungary
May 26-29, 2015

Outline

- Introduction
- Multiscale Complex Systems Approach
- Sustainable Nanomaterial Design and Manufacturing (from micro to macroscale)
- Technology Assessment for Sustainable Biodiesel Manufacturing (from meso to macroscale)
- Industrial Zonal Sustainability (macroscale)
- SMART CN – Multidisciplenary, Multinational Collaboration
- Concluding Remarks

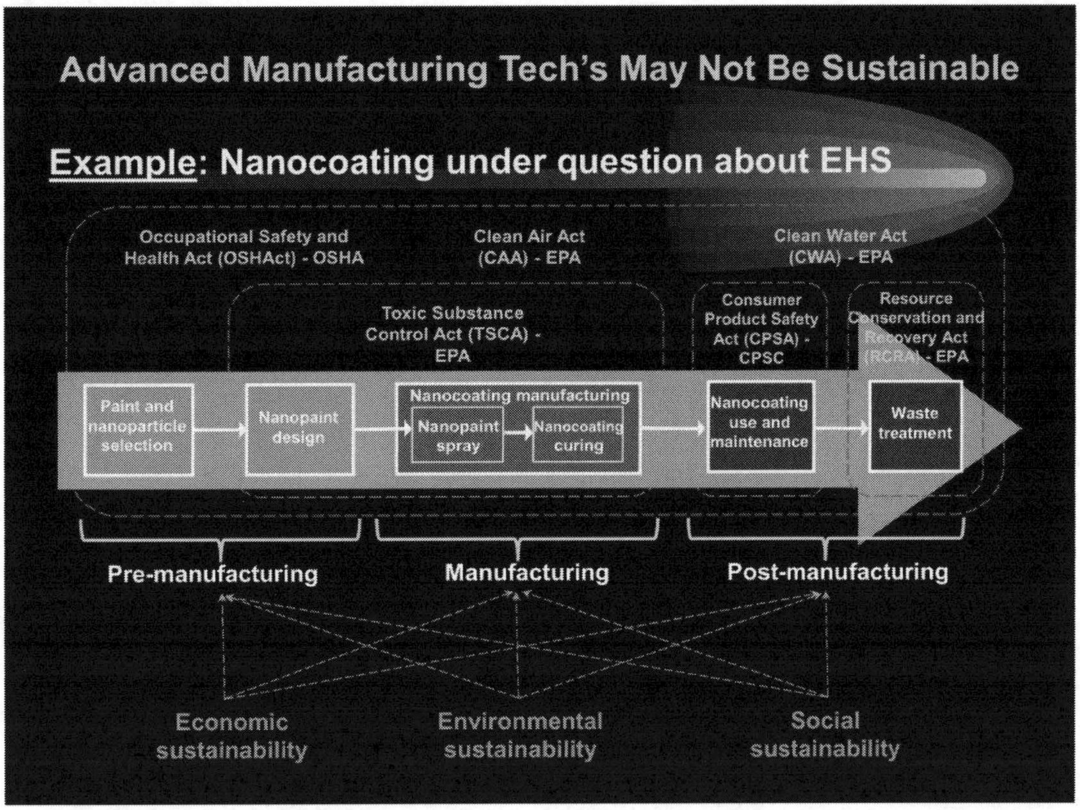
Advanced Manufacturing Tech's May Not Be Sustainable
Example: Nanocoating under question about EHS
Occupational Safety and Health Act (OSHAct) - OSHA
Clean Air Act (CAA) - EPA
Clean Water Act (CWA) - EPA
Toxic Substance Control Act (TSCA) - EPA
Consumer Product Safety Act (CPSA) - CPSC
Resource Conservation and Recovery Act (RCRA) - EPA
Paint and nanoparticle selection
Nanopaint design
Nanocoating manufacturing
Nanopaint spray
Nanocoating curing
Nanocoating use and maintenance
Waste treatment
Pre-manufacturing
Manufacturing
Post-manufacturing
Economic sustainability
Environmental sustainability
Social sustainability

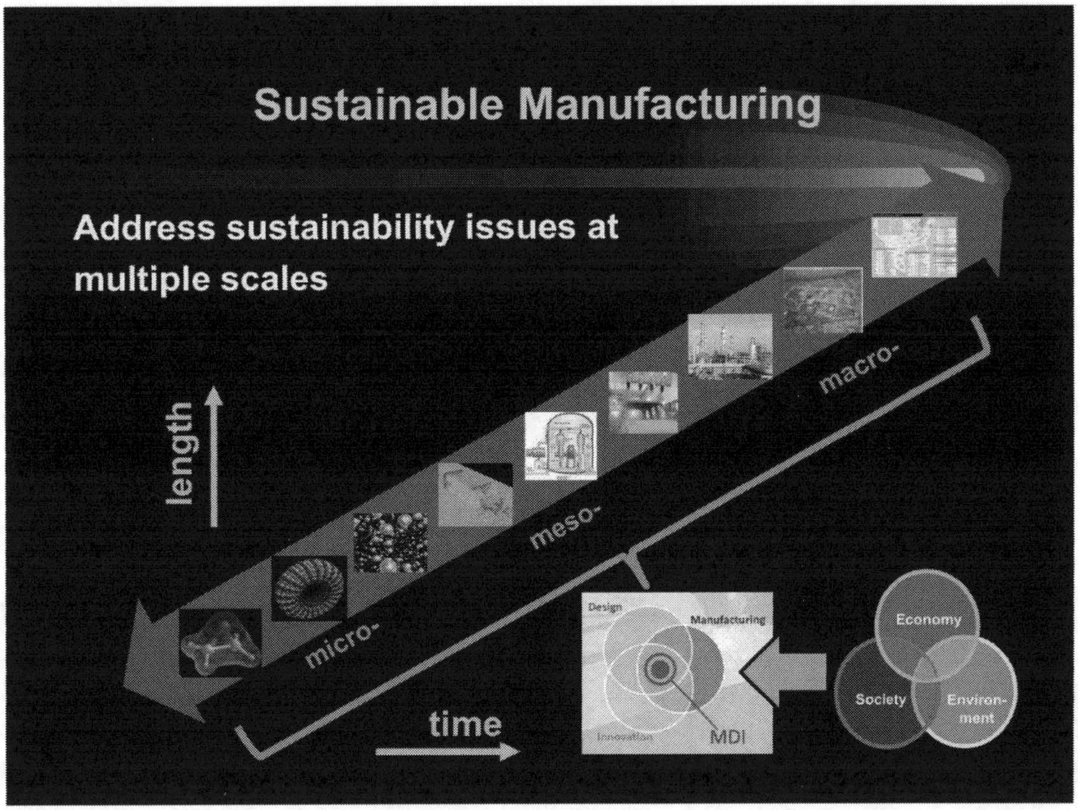
Sustainable Manufacturing
Address sustainability issues at multiple scales
length
time
micro-
meso-
macro-
Design
Manufacturing
Innovation
MDI
Economy
Society
Environ-ment

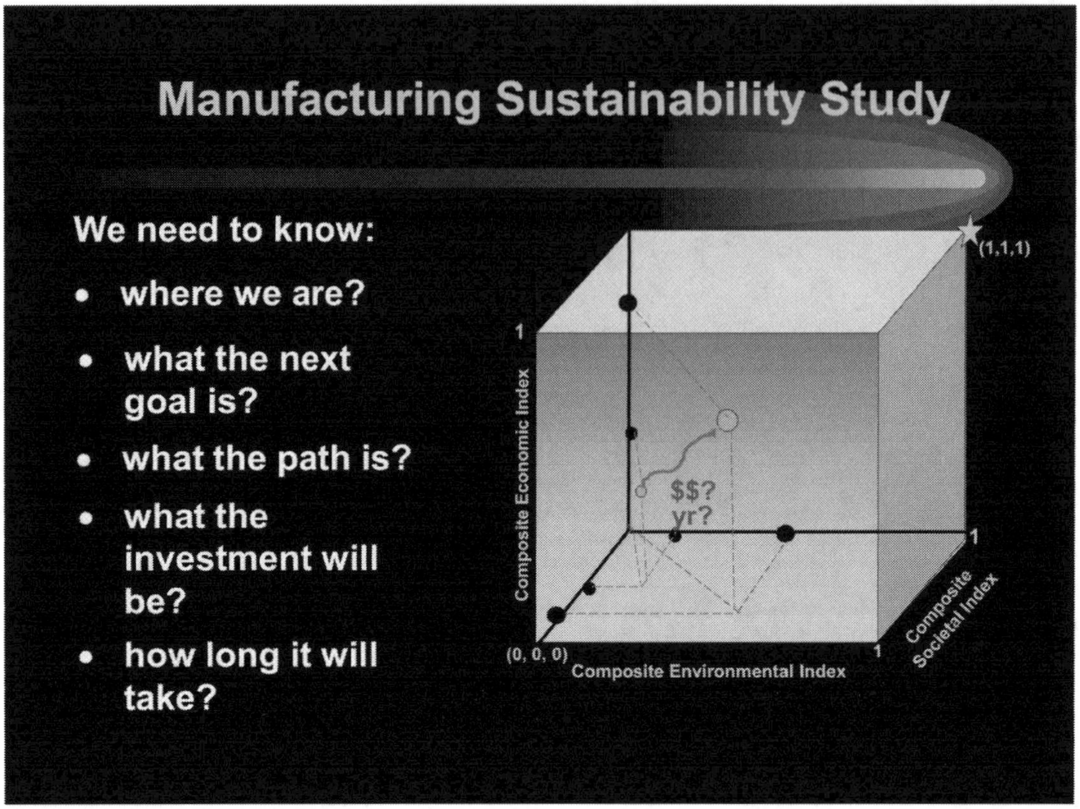
Manufacturing Sustainability Study
We need to know:
• where we are?
• what the next goal is?
• what the path is?
• what the investment will be?
• how long it will take?
(1,1,1)
1
Composite Economic Index
$$?
yr?
1
Composite Societal Index
(0, 0, 0)
1
Composite Environmental Index

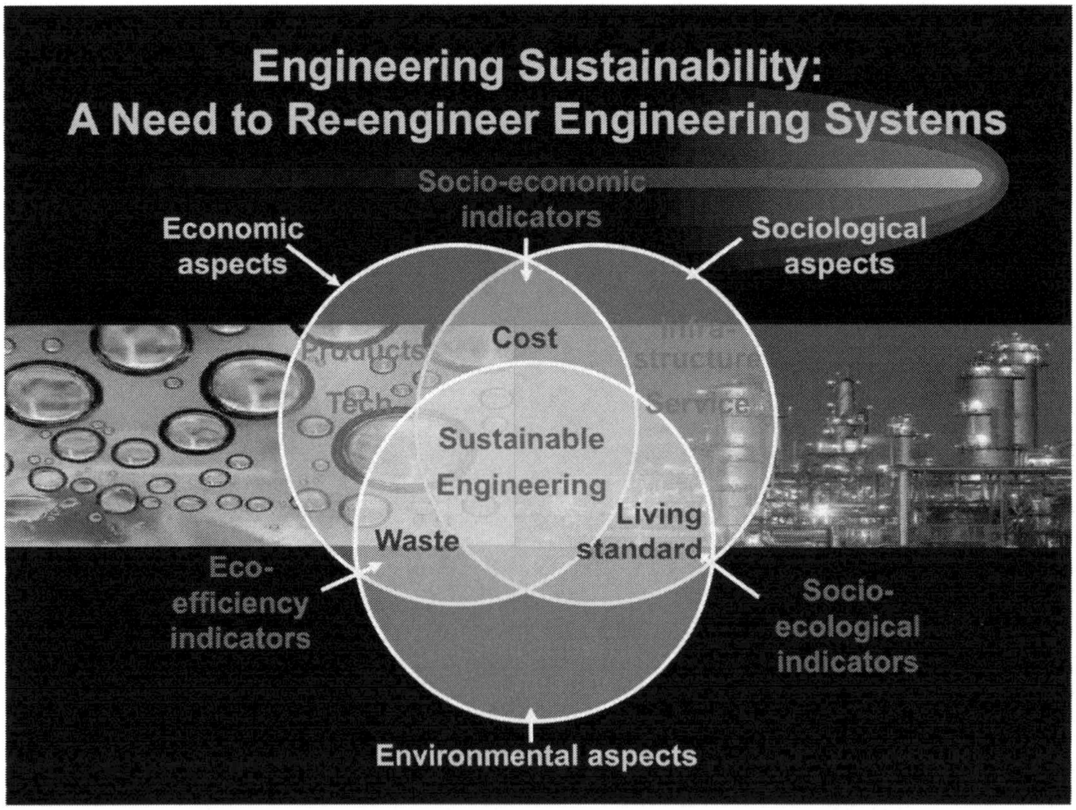
Engineering Sustainability:
A Need to Re-engineer Engineering Systems
Socio-economic indicators
Economic aspects
Sociological aspects
Products
Cost
Infra-structure
Tech
Service
Sustainable Engineering
Living standard
Waste
Eco-efficiency indicators
Socio-ecological indicators
Environmental aspects

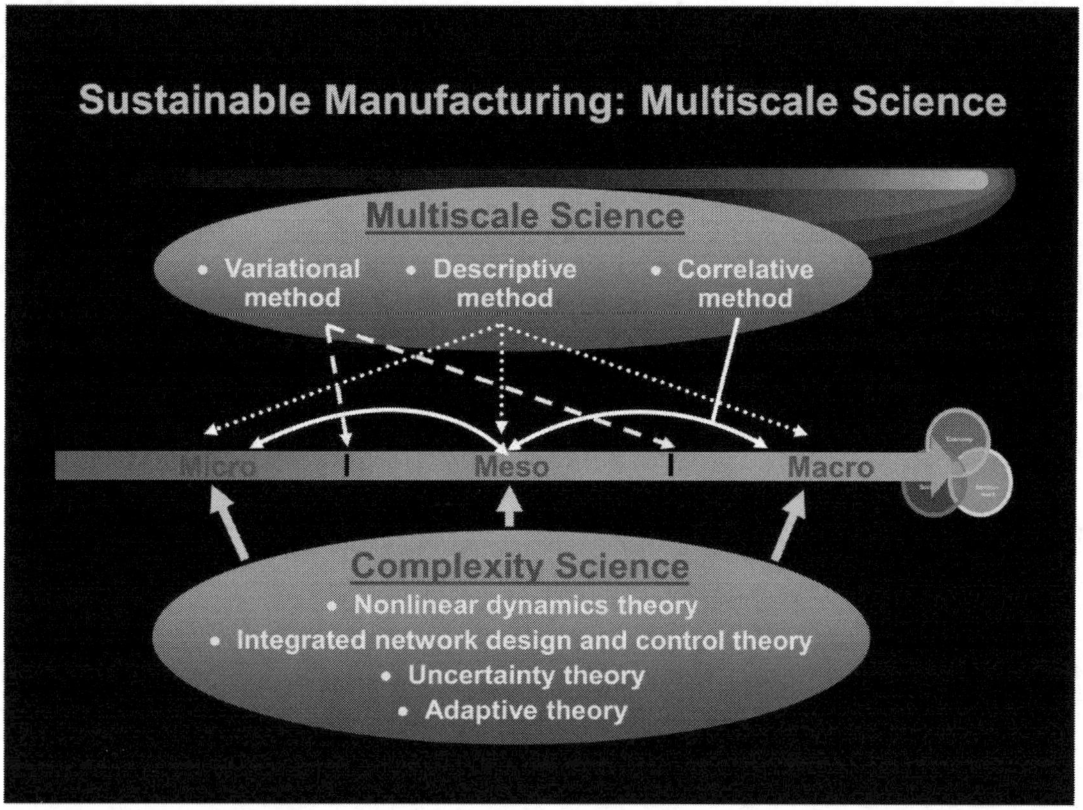
Sustainable Manufacturing: Multiscale Science
Multiscale Science
• Variational method
• Descriptive method
• Correlative method
Micro
Meso
Macro
Complexity Science
• Nonlinear dynamics theory
• Integrated network design and control theory
• Uncertainty theory
• Adaptive theory

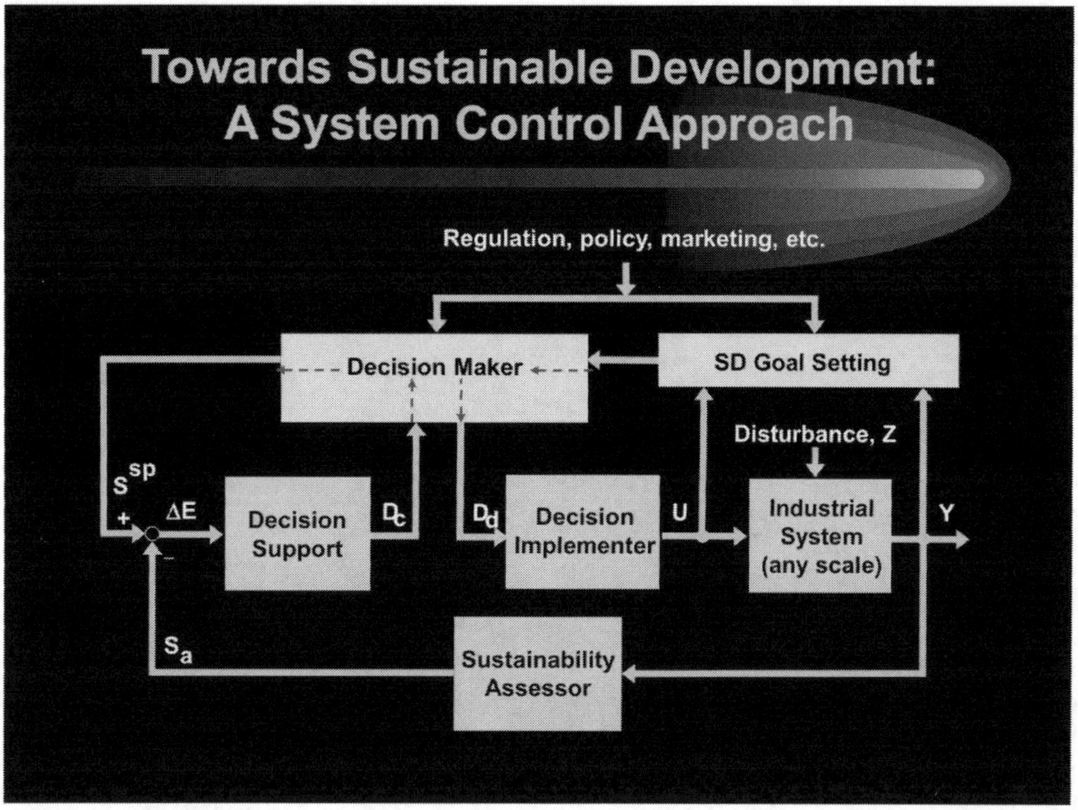
Towards Sustainable Development: A System Control Approach
Regulation, policy, marketing, etc.
Decision Maker
SD Goal Setting
Disturbance, Z
S^{sp}
+
ΔE
–
Decision Support
D_c
D_d
Decision Implementer
U
Industrial System (any scale)
Y
S_a
Sustainability Assessor

I - Sustainable Nanocoating Material Development

- Development of functional materials for super scratch resistance and anti corrosion
- Life-cycle-based sustainability assessment and decision making

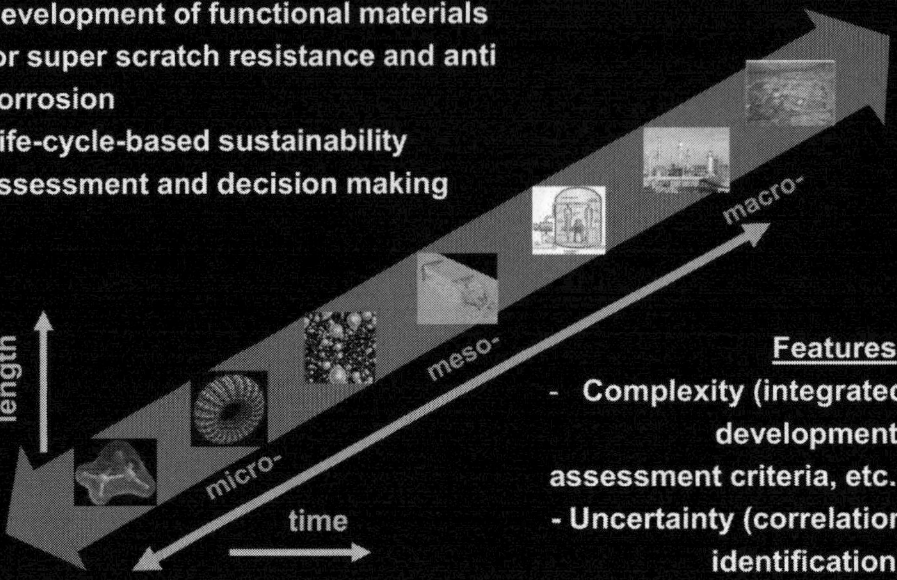

Features:
- Complexity (integrated development, assessment criteria, etc.)
- Uncertainty (correlation identification)

Fundamental Challenges

- Poor understanding of material formulation-structure-processing-property-performance correlations
- Design complexity: huge design space, multiscale nature (nano to macro)
- Lack of deep understanding of engineering sustainability (definition, quantification, goal setting, pathway towards goal achievement)

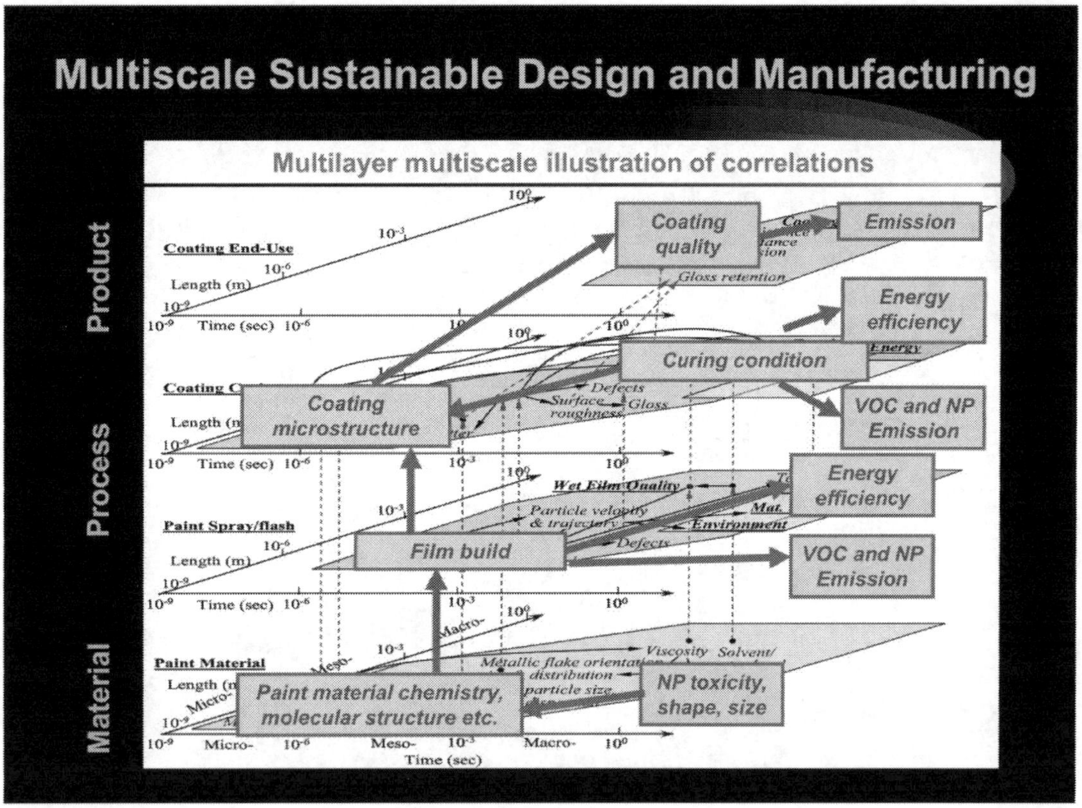
Multiscale Sustainable Design and Manufacturing
Multilayer multiscale illustration of correlations
Product
Process
Material
Coating End-Use
Length (m)
Time (sec)
Coating quality
Emission
Gloss retention
Energy efficiency
Curing condition
Coating microstructure
Defects
Surface roughness
Gloss
VOC and NP Emission
Energy efficiency
Wet Film Quality
Particle velocity & trajectory
Mat.
Environment
Paint Spray/flash
Film build
Defects
VOC and NP Emission
Paint Material
Macro-
Meso-
Micro-
Metallic flake orientation distribution
particle size
Viscosity
Solvent/
NP toxicity, shape, size
Paint material chemistry, molecular structure etc.

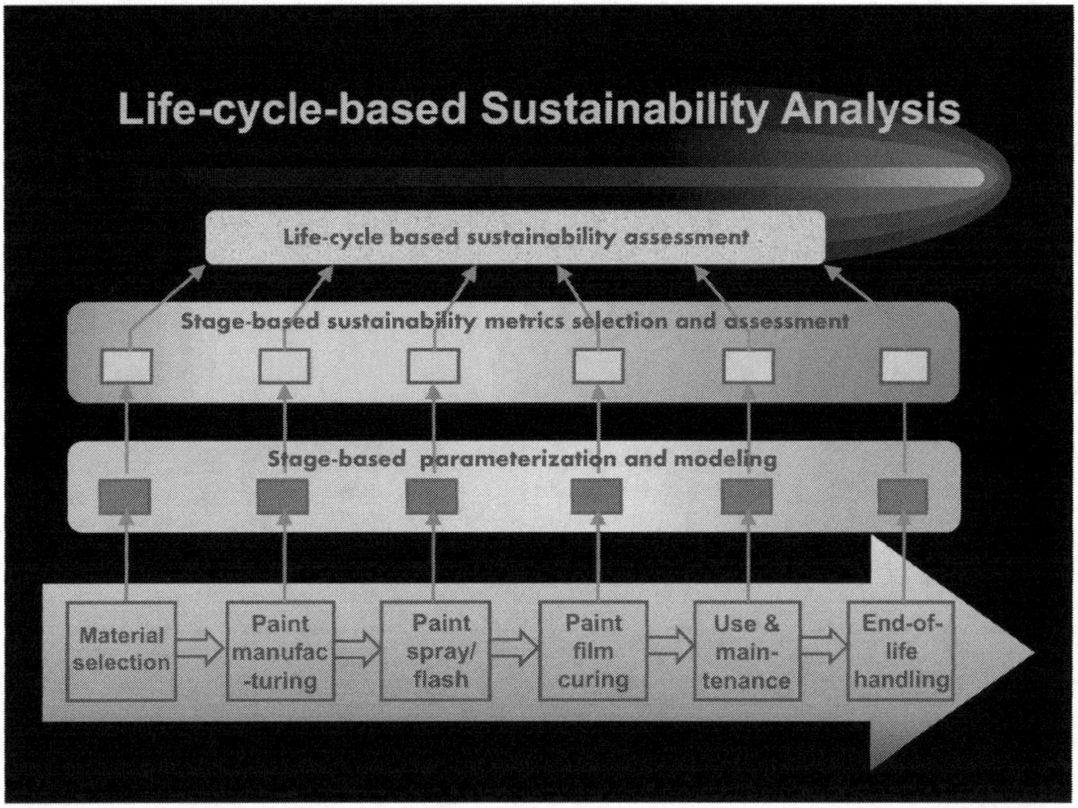
Life-cycle-based Sustainability Analysis
Life-cycle based sustainability assessment
Stage-based sustainability metrics selection and assessment
Stage-based parameterization and modeling
Material selection
Paint manufac -turing
Paint spray/ flash
Paint film curing
Use & main-tenance
End-of-life handling

Life-cycle-based Sustainability Assessment

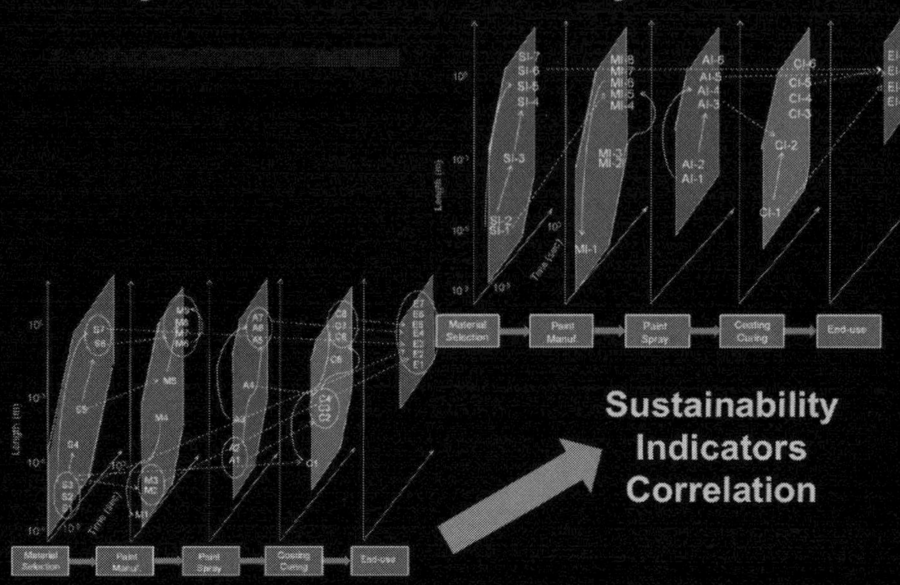

Sustainability Indicators Correlation

Parameterization

Design Optimality

Objective function: $J = f$ (nanocoating performance, functionality, processing efficiency, environmental cleanness, cost)

s.t.

- Multiscale multi-physics models for sample development and testing
- System performance evaluation models
- Product development cycle and budget constraints
- Product cost constraint
- Environmental regulation (toxicity, emissions, etc.)
- etc.

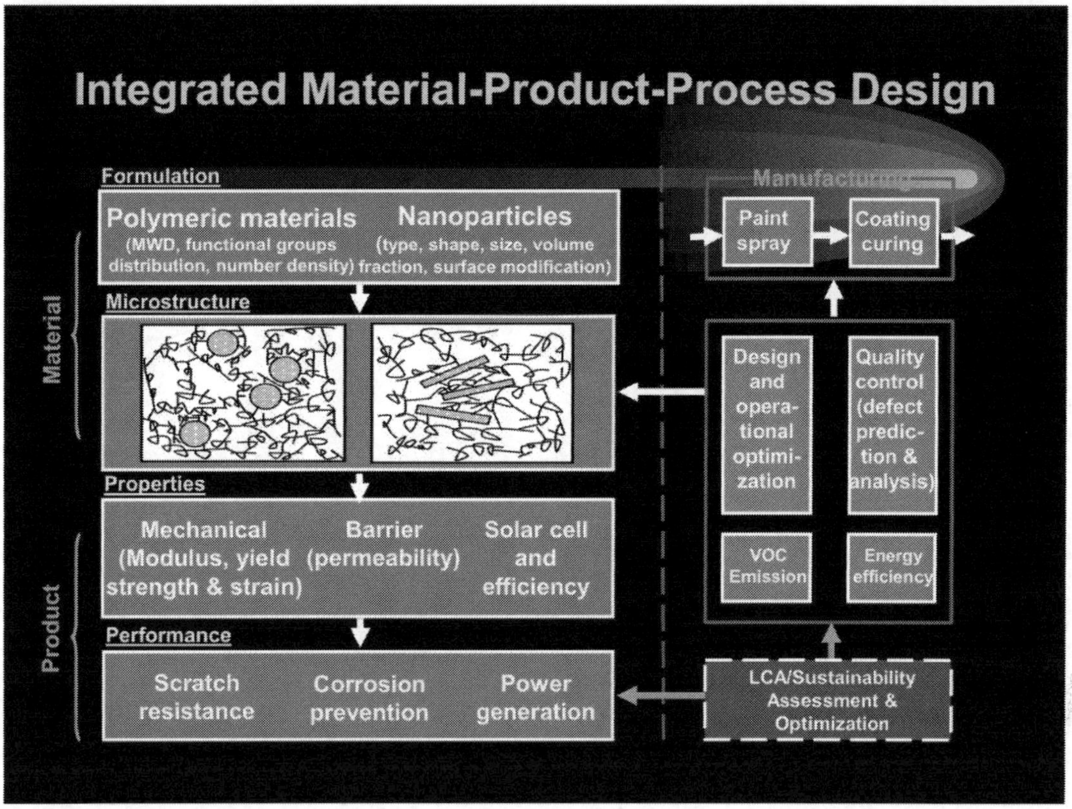
Integrated Material-Product-Process Design
Formulation
Polymeric materials
(MWD, functional groups distribution, number density)
Nanoparticles
(type, shape, size, volume fraction, surface modification)
Microstructure
Properties
Mechanical (Modulus, yield strength & strain)
Barrier (permeability)
Solar cell and efficiency
Performance
Scratch resistance
Corrosion prevention
Power generation
Material
Product
Manufacturing
Paint spray
Coating curing
Design and opera-tional optimi-zation
Quality control (defect predic-tion & analysis)
VOC Emission
Energy efficiency
LCA/Sustainability Assessment & Optimization

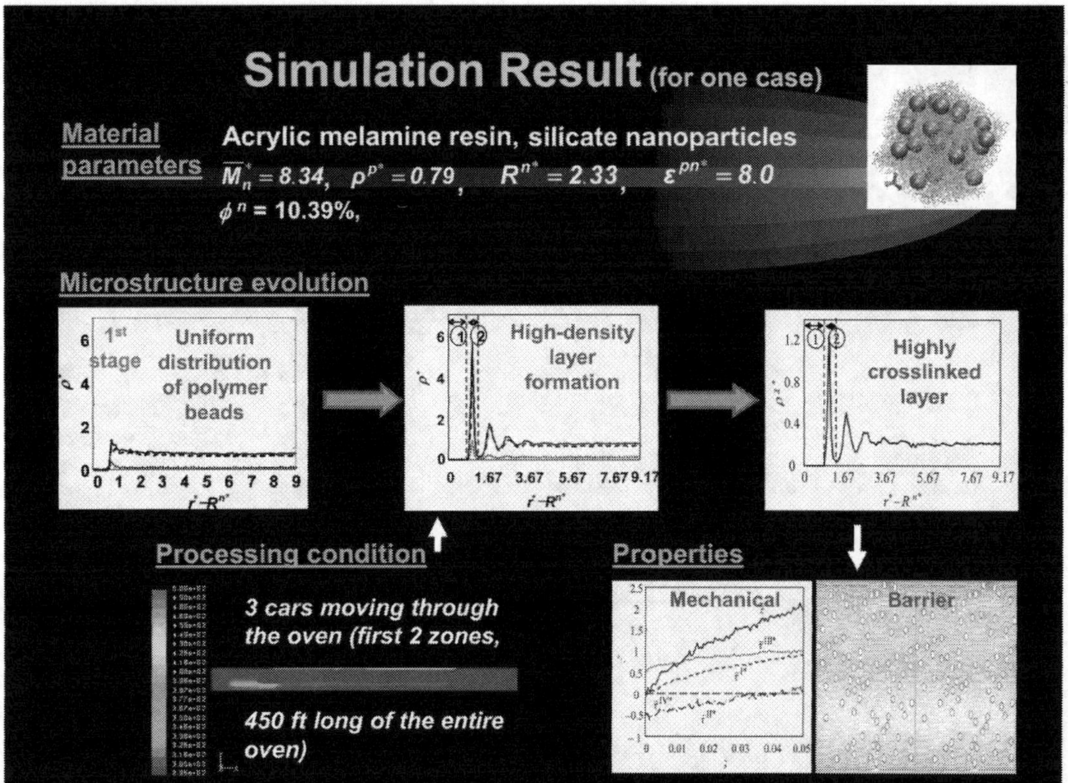
Simulation Result (for one case)
Material parameters
Acrylic melamine resin, silicate nanoparticles
$\overline{M}_n^* = 8.34$, $\rho^{p*} = 0.79$, $R^{n*} = 2.33$, $\varepsilon^{pn*} = 8.0$
ϕ^n = 10.39%,
Microstructure evolution
1st stage
Uniform distribution of polymer beads
High-density layer formation
Highly crosslinked layer
Processing condition
3 cars moving through the oven (first 2 zones, 450 ft long of the entire oven)
Properties
Mechanical
Barrier

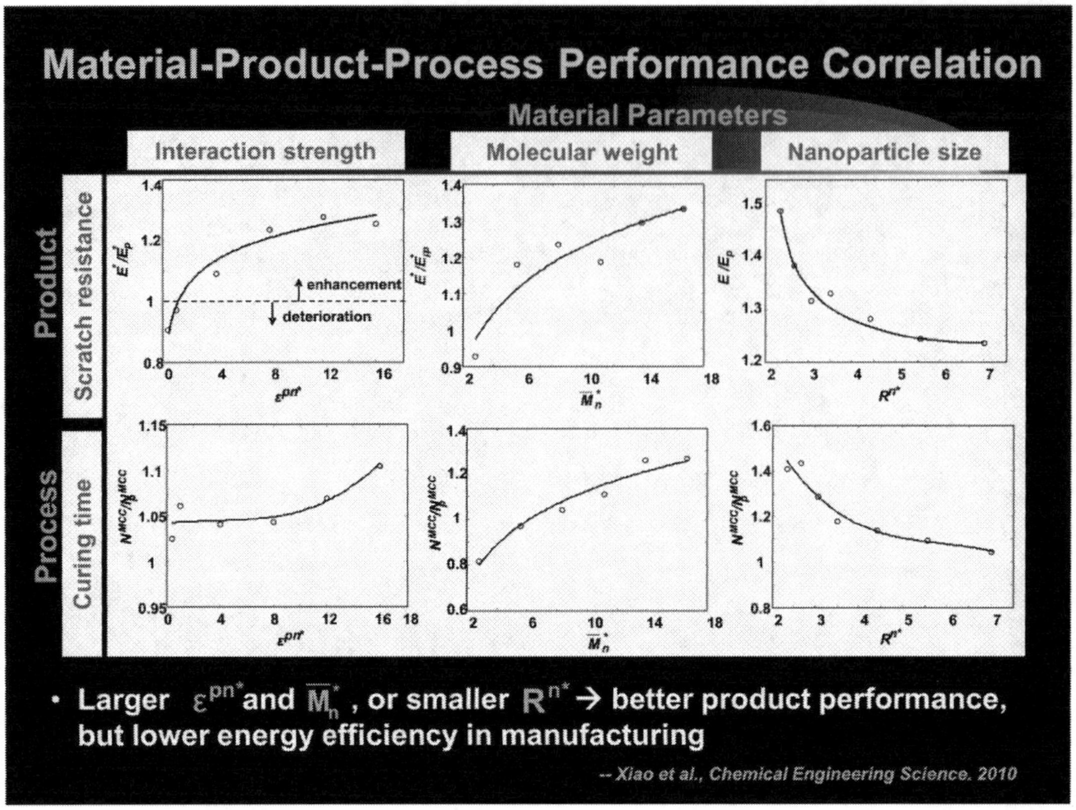
Material-Product-Process Performance Correlation
Material Parameters
Interaction strength
Molecular weight
Nanoparticle size
Product
Scratch resistance
Process
Curing time
enhancement
deterioration
• Larger ε^{pn*} and $\overline{M}_n^*$, or smaller R^{n*} → better product performance, but lower energy efficiency in manufacturing
-- Xiao et al., Chemical Engineering Science. 2010

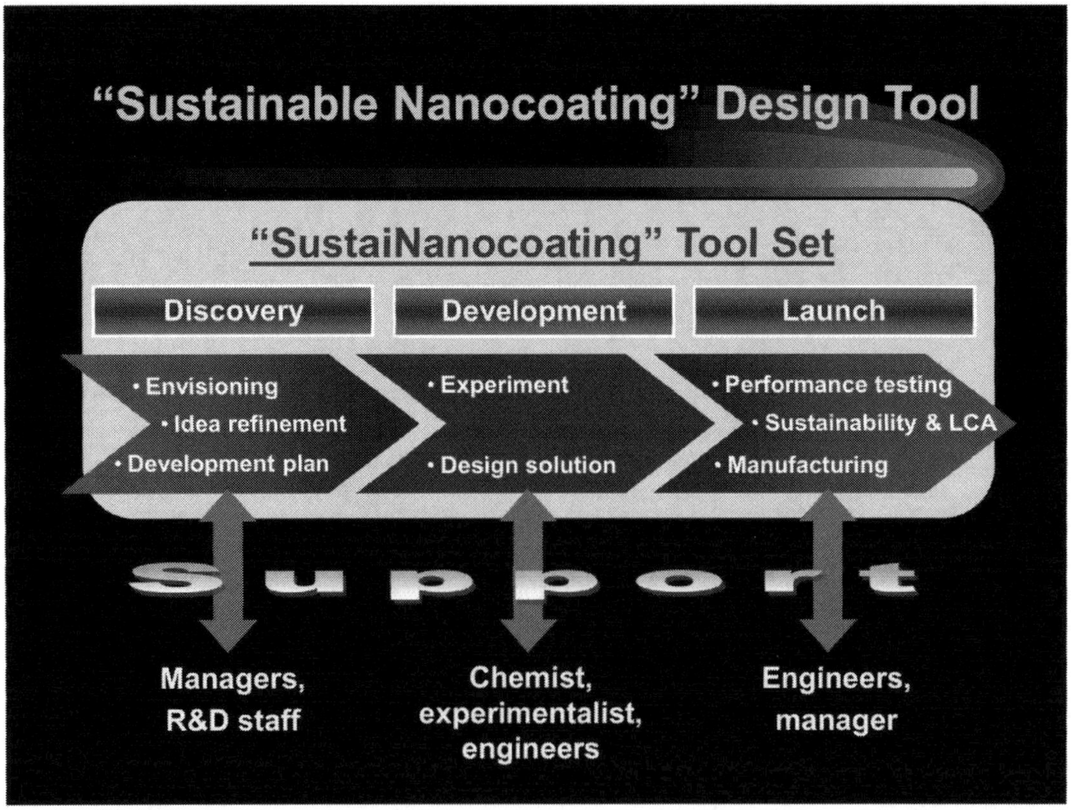
"Sustainable Nanocoating" Design Tool
"SustaiNanocoating" Tool Set
Discovery
Development
Launch
• Envisioning
• Idea refinement
• Development plan
• Experiment
• Design solution
• Performance testing
• Sustainability & LCA
• Manufacturing
Support
Managers, R&D staff
Chemist, experimentalist, engineers
Engineers, manager

II - Technology Evaluation for Sustainable Biodiesel Manufacturing

- Energy/renewable energy manufacturing
- Complex polygeneration system design and deployment

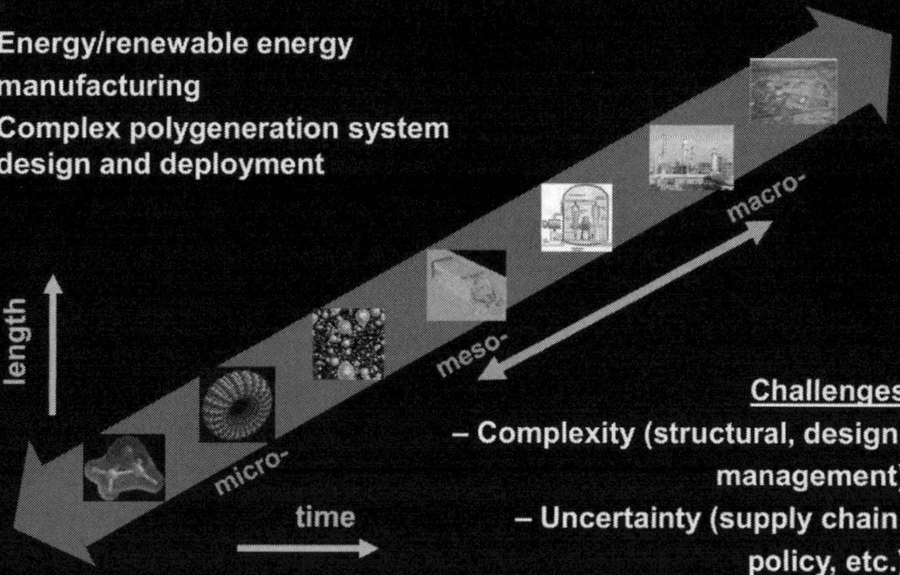

Challenges

– Complexity (structural, design, management)

– Uncertainty (supply chain, policy, etc.)

Two-Stage Sustainability Assessment

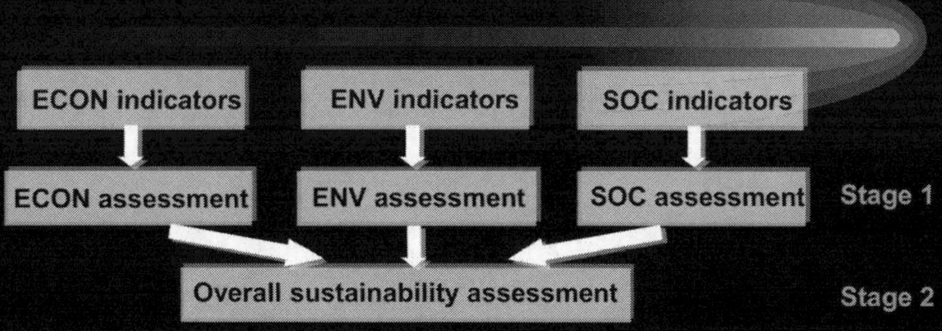

$S = \{E, \ V, \ L\}$

$E = \{E_i \mid i = 1, 2, \cdots, F\}$ the set of economic sustainability indices

$V = \{V_i \mid i = 1, 2, \cdots, G\}$ the set of environmental sustainability indices

$L = \{L_i \mid i = 1, 2, \cdots, H\}$ the set of social sustainability indices

Note: All values are internal numbers

Categorized Sustainability (1st Stage)

- **Indicator Availability**
  - IChemE (2002) and AIChE (2009) sustainability metrics, and/or other engineering criteria
- **Multi-Criteria Combination for Composite Economic, Environmental, and Social Sustainability**

$$E(P)=\frac{\sum_{i=1}^{F} a_i E_i(P)}{\sum_{i=1}^{F} a_i} \qquad V(P)=\frac{\sum_{i=1}^{G} b_i V_i(P)}{\sum_{i=1}^{G} b_i} \qquad L(P)=\frac{\sum_{i=1}^{H} c_i L_i(P)}{\sum_{i=1}^{H} c_i}$$

$$E(T)=\frac{\sum_{i=1}^{F} a_i E_i(T)}{\sum_{i=1}^{F} a_i} \qquad V(T)=\frac{\sum_{i=1}^{G} b_i V_i(T)}{\sum_{i=1}^{G} b_i} \qquad L(T)=\frac{\sum_{i=1}^{H} c_i L_i(T)}{\sum_{i=1}^{H} c_i}$$

Overall Sustainability (2nd Stage)

$$S(P)=\frac{\left\|(\alpha E(P),\ \beta V(P),\ \gamma L(P))\right\|}{\left\|(\alpha,\ \beta,\ \gamma)\right\|}, \qquad S(T;P)=\frac{\left\|(\alpha E(T;P),\ \beta V(T;P),\ \gamma L(T;P))\right\|}{\left\|(\alpha,\ \beta,\ \gamma)\right\|}$$

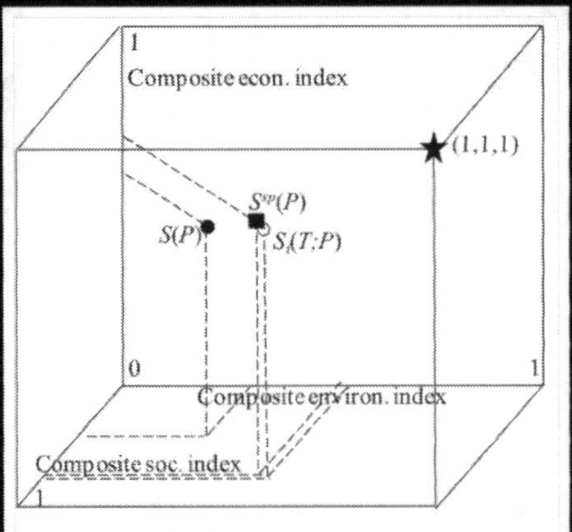

Figure 1. Sustainability cube representation.

Assessment for Technology Combination

- **For Economic Sustainability Assessment**

$$\Delta E_i(T_j;P) = E_i(T_j) - E_i(P), \quad i = 1, 2, \cdots, F, \quad j = 1, 2, \cdots, N$$

$$\Delta E(T_j;P) = E(T_j) - E(P) = \frac{\sum_{i=1}^{F} a_i \Delta E_i(T_j;P)}{\sum_{i=1}^{F} a_i}, \quad i = 1, 2, \cdots, F, \quad j = 1, 2, \cdots, N$$

$$E(T;P) = \sum_{j=1}^{m} \Delta E(T_j;P) + E(P)$$

$$E_i^{imp}(T;P)(\%) = \frac{E_i(T;P) - E(P)}{E(P)}$$

- **Total Cost of Multiple Technologies:** $B(T;P) = \sum_{j=1}^{m} B(T_j;P)$

(Same method for environmental and social sustainability assessment)

Technology Capacity in Sustainability Performance Improvement

No.	Tech. set	Cost for tech. set	Achievable categorized sustainability			Overall sust. by tech. set
			Econ.	Enviro.	Soc.	
1	$\{\omega_{1,1}, \cdots, \omega_{m,1}\}$	$B_1^c(T;P)$	$\bar{E}_1^c(T;P)$	$\bar{V}_1^c(T;P)$	$\bar{L}_1^c(T;P)$	$\bar{S}_1^c(T;P)$
…	…	…	…	…	…	…
G_{tot}	$\{\omega_{1,G_1}, \cdots, \omega_{m,G_m}\}$	$B_{G_{tot}}^c(T;P)$	$\bar{E}_{G_{tot}}^c(T;P)$	$\bar{V}_{G_{tot}}^c(T;P)$	$\bar{L}_{G_{tot}}^c(T;P)$	$\bar{S}_{G_{tot}}^c(T;P)$

- **Tech set removal based on cost checking:** $B_i^c(T;P) \le B_{tot}^{lim}(T;P)$

Case Study on Biodiesel Manufacturing

- **Sustainability performance improvement through source waste reduction and process and product quality improvement**

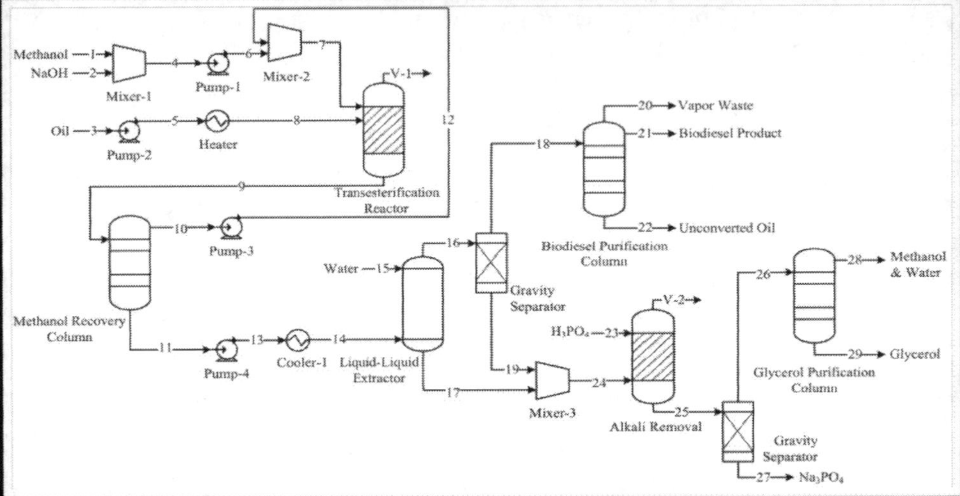

Sustainability Assessment

Category	Index	System	Technologies in Group 1				Technology in Group 2					
		P	$T_{1,1}$	$T_{1,2}$	$T_{1,3}$	$T_{1,4}$	$T_{2,1}$	$T_{2,2}$	$T_{2,3}$	$T_{2,4}$	$T_{2,5}$	$T_{2,6}$
Econ. (E)	E_1	[0.550, 0.570]	0.620	[0.620, 0.640]	0.580	[0.600, 0.610]	[0.600, 0.610]	[0.580, 0.600]	0.610	[0.620, 0.630]	0.580	[0.590, 0.600]
	E_2	0.450	[0.500, 0.530]	[0.480, 0.490]	[0.460, 0.480]	[0.490, 0.510]	0.510	[0.470, 0.480]	[0.460, 0.470]	0.460	[0.520, 0.530]	[0.460, 0.480]
Environ. (V)	V_1	0.400	0.430	0.450	[0.410, 0.420]	[0.420, 0.430]	[0.440, 0.450]	0.420	0.430	[0.460, 0.470]	[0.410, 0.420]	[0.460, 0.470]
	V_2	[0.350, 0.380]	0.400	0.360	[0.390, 0.400]	0.370	[0.360, 0.380]	[0.380, 0.400]	[0.360, 0.370]	0.350	0.400	[0.410, 0.420]
	V_3	0.420	[0.410, 0.420]	[0.450, 0.460]	0.400	0.430	0.450	[0.400, 0.410]	[0.430, 0.440]	[0.430, 0.440]	[0.420, 0.430]	0.410
Soc. (L)	L_1	[0.335, 0.340]	[0.355, 0.360]	[0.380, 0.390]	0.330	0.350	[0.310, 0.315]	[0.370, 0.380]	0.330	0.440	[0.390, 0.400]	[0.350, 0.365]
	L_2	[0.370, 0.380]	0.400	0.380	[0.378, 0.380]	[0.380, 0.385]	0.400	[0.390, 0.410]	[0.380, 0.390]	[0.360, 0.370]	[0.400, 0.410]	0.370

Technology Set Identificaiton

No.	Tech. set	Cost for tech. set $B_c^l(T;P)$	System's achievable categorized sustainability $\tilde{E}_c^l(T;P)$	$\tilde{V}_c^l(T;P)$	$\tilde{L}_c^l(T;P)$	Overall sust. by tech. set $\tilde{S}_c^l(T;P)$
1	$\{T_1, T_2\}\{T_9, T_{10}\}$	$390 K	[0.645, 0.745]	[0.483, 0.540]	[0.434, 0.501]	[0.528, 0.605]
2	$\{T_1, T_2\}\{T_5, T_8, T_9\}$	$520 K	Set deleted (cost concern)			
...	...	...	...			
9	$\{T_1, T_2\}\{T_8, T_9, T_{10}\}$	$480 K	Set deleted (cost concern)			
10	$\{T_1, T_2, T_3\}\{T_9, T_{10}\}$	$440 K	[0.655, 0.775]	[0.485, 0.558]	[0.426, 0.500]	[0.531, 0.622]
11	$\{T_1, T_2, T_3\}\{T_5, T_8, T_9\}$	$570 K	Set deleted (cost concern)			
...	...	...	...			
18	$\{T_1, T_2, T_3\}\{T_8, T_9, T_{10}\}$	$530 K	Set deleted (cost concern)			
19	$\{T_2, T_3, T_4\}\{T_9, T_{10}\}$	$420 K	[0.640, 0.760]	[0.478, 0.553]	[0.421, 0.489]	[0.521, 0.611]
20	$\{T_2, T_3, T_4\}\{T_5, T_8, T_9\}$	$550 K	Set deleted (cost concern)			
...	...	...	...			
27	$\{T_2, T_3, T_4\}\{T_8, T_9, T_{10}\}$	$510 K	Set deleted (cost concern)			

Performance Comparison

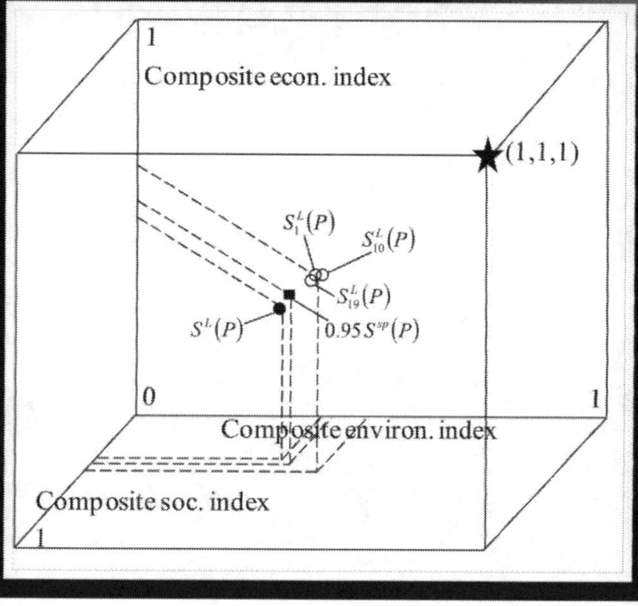

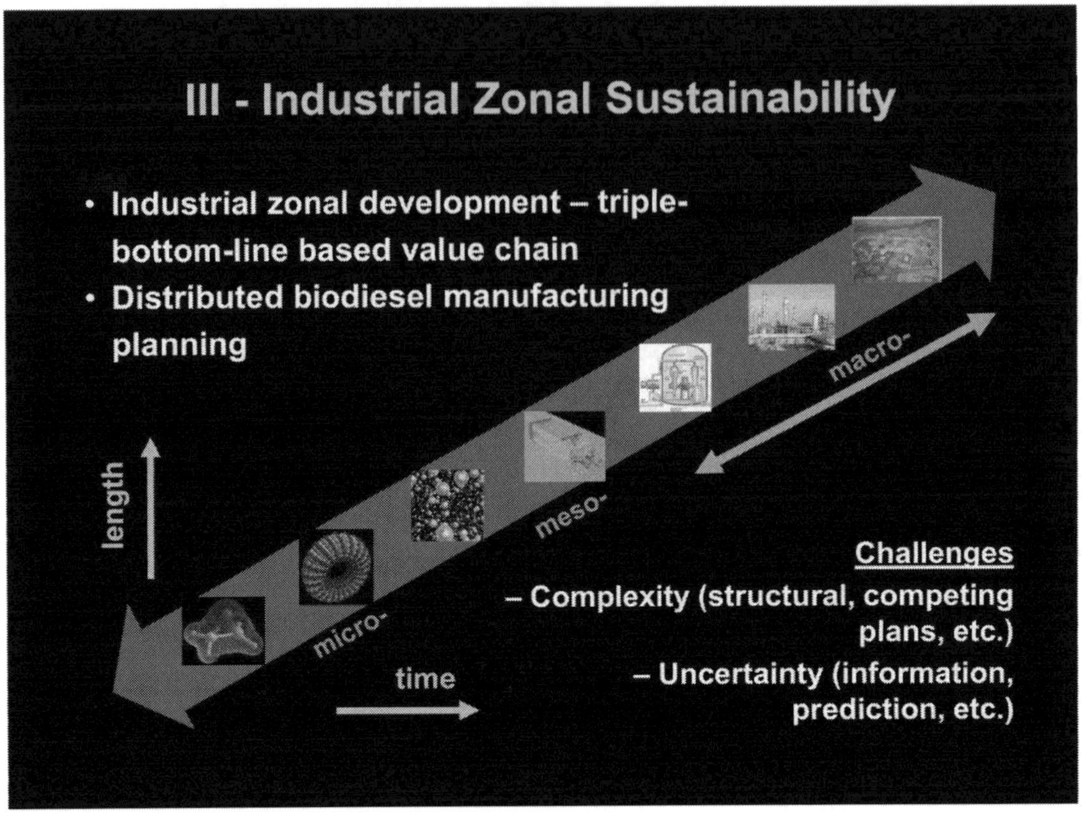
III - Industrial Zonal Sustainability
• Industrial zonal development – triple-bottom-line based value chain
• Distributed biodiesel manufacturing planning
length
time
micro-
meso-
macro-
Challenges
– Complexity (structural, competing plans, etc.)
– Uncertainty (information, prediction, etc.)

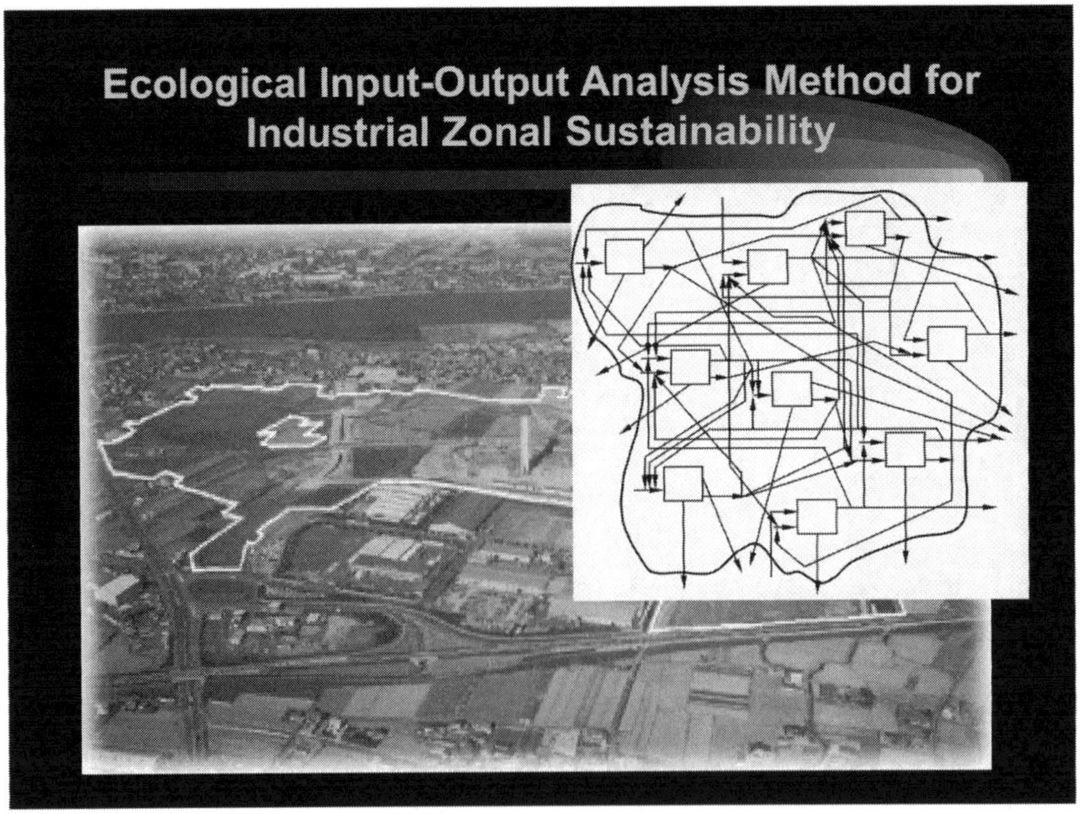
Ecological Input-Output Analysis Method for Industrial Zonal Sustainability

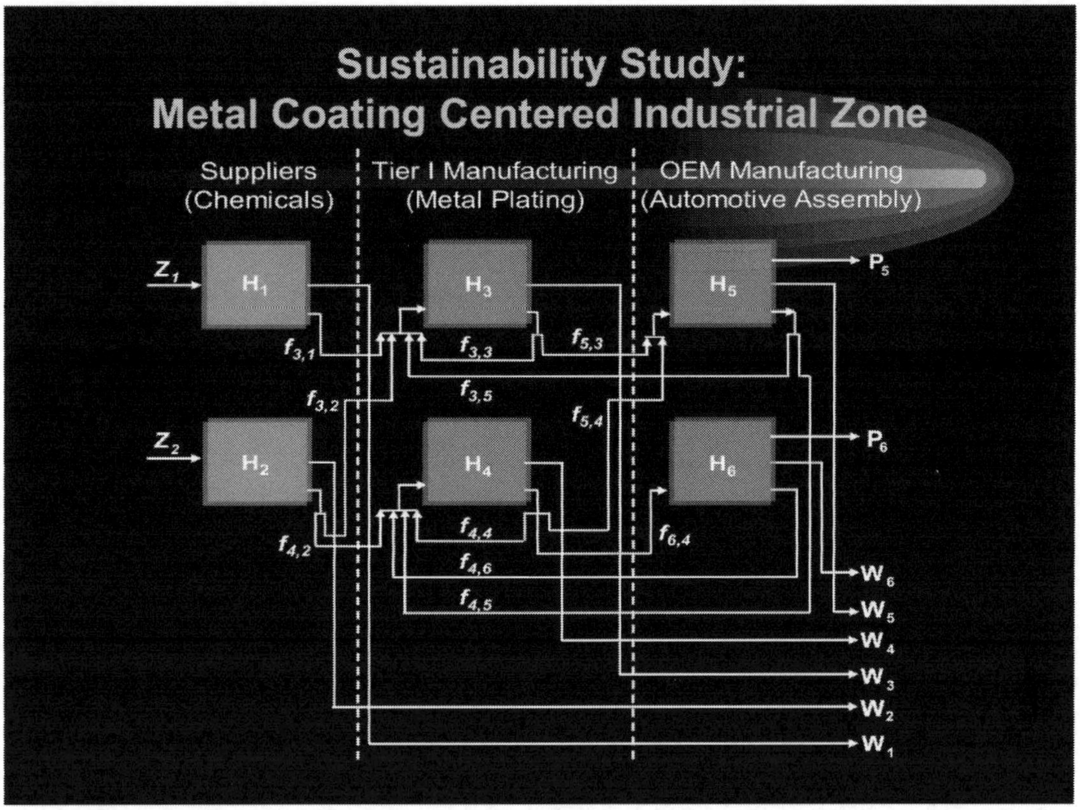

Sustainability Improvement Plans A and B

Improvement Focus	Current (Year 0)	Short-term (Year 3)	Mid-term (Year 6)
Plan A			
Main plan for environmental sustainability improvement			
• Fraction of raw materials recycled within a company ($x_{v,2}$)	0.08	0.22	0.30
• Fraction of raw materials recycled from consumers ($x_{v,3}$)	0.02	0.15	0.25
• Hazardous solid waste per unit value added ($x_{v,4}$)	3.7	1.5	1.4
Main plan for social sustainability improvement			
• Lost time accident frequency ($x_{l,1}$)	11.4	7.8	6.2
• Number of complaints per unit value added ($x_{l,3}$)	30.6	20	12
Plan B			
Main plan for environmental sustainability improvement			
• Fraction of raw materials recycled within a company ($x_{v,2}$)	0.08	0.15	0.35
• Fraction of raw materials recycled from consumers ($x_{v,3}$)	0.02	0.10	0.32
• Hazardous solid waste per unit value added ($x_{v,4}$)	3.7	3.2	1.2
Main plan for social sustainability improvement			
• Lost time accident frequency ($x_{l,1}$)	11.4	9.8	3.0
• Number of stakeholder meetings per unit value added ($x_{l,2}$)	2.2	2.2	5.4
• Number of complaints per unit value added ($x_{l,3}$)	30.6	25	6

Sustainable Development Pathway Study (Based on Short-to-Mid Term Predictions)

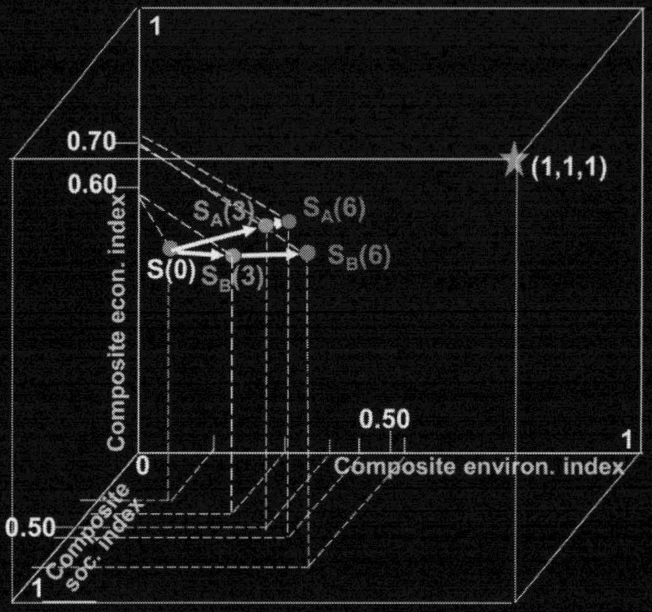

NSF RCN-SEES: Sustainable Manufacturing Advances in Research and Technology (SMART) Coordination Network

- PI: Y. Huang (Wayne State U.)
- Co-PIs: T. Edgar (U. Texas), M. El-Halwagi (TAMU), C. Davidson (Syracuse), M. Eden (Auburn)
- Steering Committee Members: L. Achenie (VPI), D.Allen (U. Texas), B. Bhakshi (OST), Burton English (U. Tennessee), D. Fasenfest (WSU), I. Grossmann (Carnegie Mellon), K. High (Oklahoma State U.), I. Jawahir (U. Kentucky), C. Maravelias (U. Wisconsin), K. Ogden (U. Arizona), M. Rezac (Kansas State U.), F. Shadman (U. Arizona)
- Foreign Universities

  Denmark (TU Denmark), Germany (TU Berlin), Norway (NUST), China (Tsinghua U., SCUT), Japan (Waseda U.), and Singapore (NUS)

Coordinated Research on Sustainable Manufacturing

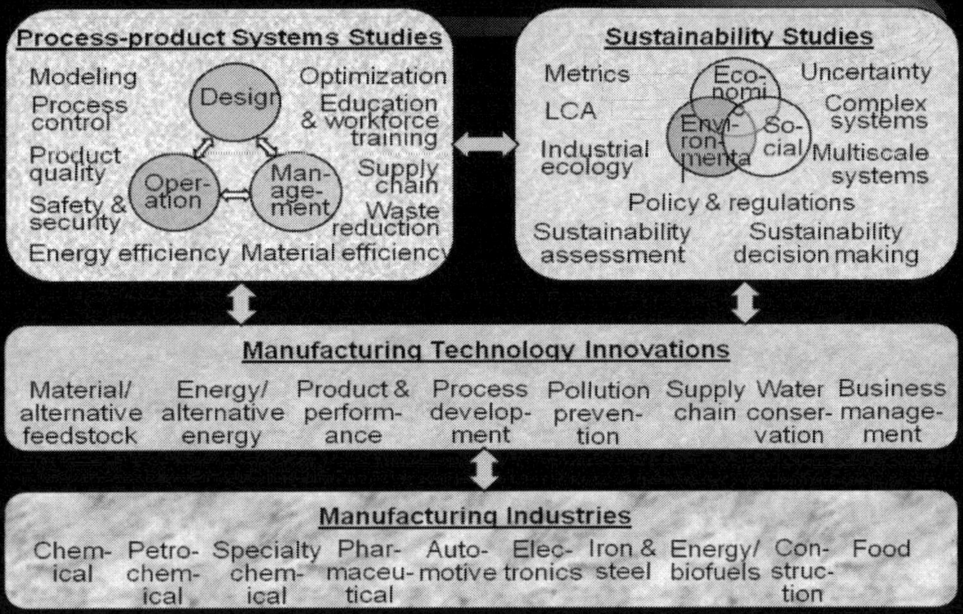

Concluding Remarks

- Manufacturing sustainability: essential to the success of manufacturing advancement
- Multiscale sustainability: ranging from nanoscale material design to large-scale industrial system management
- Sustainability assessment and decision making tools: valuable for manufacturing industries
- Sustainability study: multi-disciplinary collaboration needed

Acknowledgments

- NSF
- EPA
- DOE
- US Army and NASA
- ACS PRF and AESF Foundation
- Industries (DuPont, BASF, Ford, GM, Chrysler, Dow, Marathon-Ashland, ...)
- All students and associates

Solid-State Recycling of AZ31 Mg Alloy Using Equal Channel Angular Pressing

- **Conference:** International Congress on Sustainability Science Engineering ICOSSE
- **Year:** 2015
- **Proceeding:** ICOSSE 2015
- **Group:** Abstract Submissions
- **Session:**

Session 9: Sustainable Product and Process Design II

- **Time:**

Thursday, May 28, 2015 - 1:40pm-2:00pm

Authors:

Al-Maharbi, M.

The aim of this study is to solid-state recycle of AZ31 magnesium alloys chips through consolidation utilizing severe plastic deformation technique. The chips are consolidated using Equal Channel Angular Extrusion (ECAE) die. This severe plastic deformation technique has been used to consolidate powder successfully as well as to process bulk materials. In addition to improvement of mechanical properties of the recycled products over those recycled using conventional remelting process, starting with chips different orientation will result in more random crystallographic texture of the material. This random texture will lead to lower anisotropy that wrought magnesium alloys usually exhibit due to their hexagonal close packed (hcp) structure. A metallugraphical and mechanical investigation as well as texture measurements were carried out on the consolidated billets. Experimental parameters like processing temperatures, speed and number of passes can be varied to improve the consolidation and to engineer the crystallographic texture.

Design of Sustainable Value Creation for Production Systems through Principles of Industrial Engineering

Pinar Bilge, Fazleena Badurdeen, Günther Seliger, I.S. Jawahir

28.05.2015

4th International Congress on Sustainability Science & Engineering ICOSSE '15

Session 9: Sustainable Product and Process Design II

Content

- **Sustainable Value Creation**
- Design Conditions and Requirements
- Principles of Industrial Engineering

Breadth from Systemic Reference and Depth from Production Technology

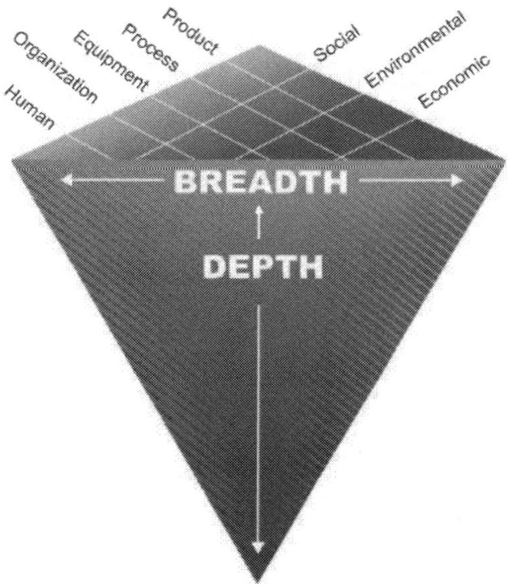

TU IWF
Technische Universität Berlin
Department of Machine Tools and Factory Management
Univ.-Prof. Dr.-Ing. Günther Seliger

Source: [CRC 1026, 2013]

Page 3

Mapping Life-Cycles with Management for Value Creation

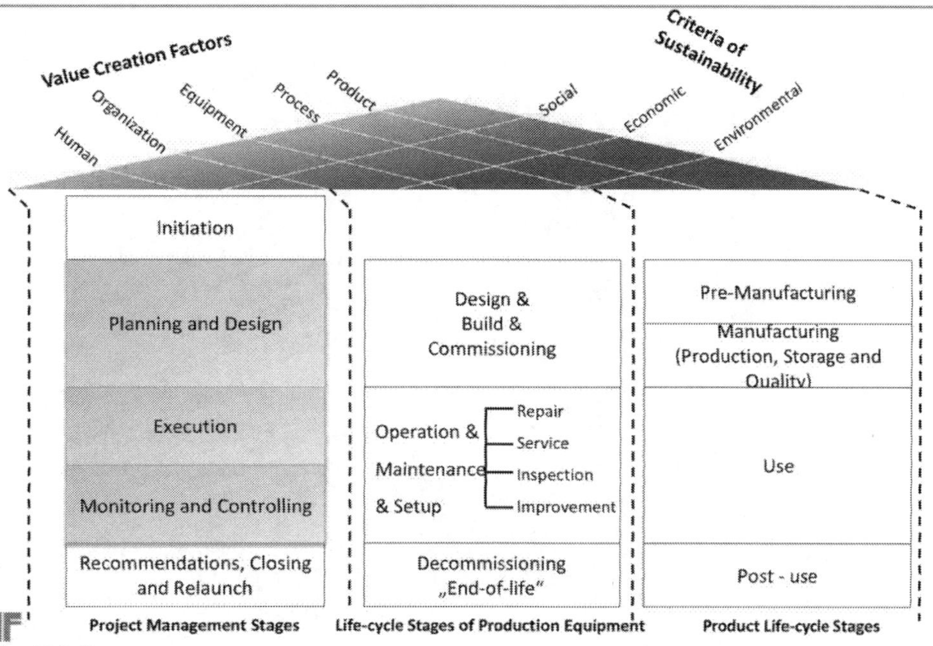

TU IWF
Technische Universität Berlin
Department of Machine Tools and Factory Management
Univ.-Prof. Dr.-Ing. Günther Seliger

Page 9

Value Creation Module

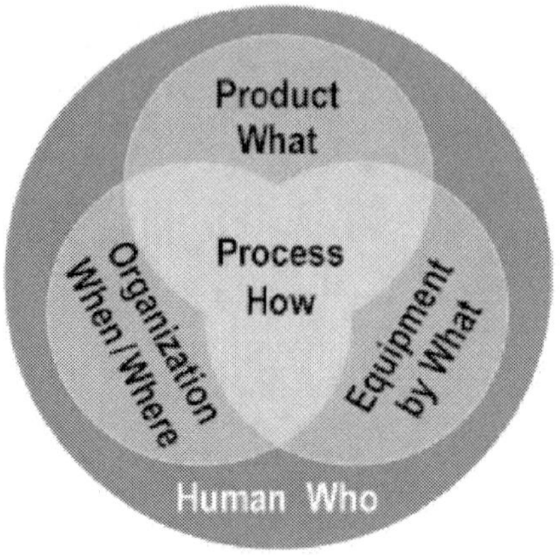

Technische Universität Berlin
Department of Machine Tools and Factory Management
Univ.-Prof. Dr.-Ing. Günther Seliger

Source: [CRC 1026, 2013]

Page 10

Value Creation Architecture

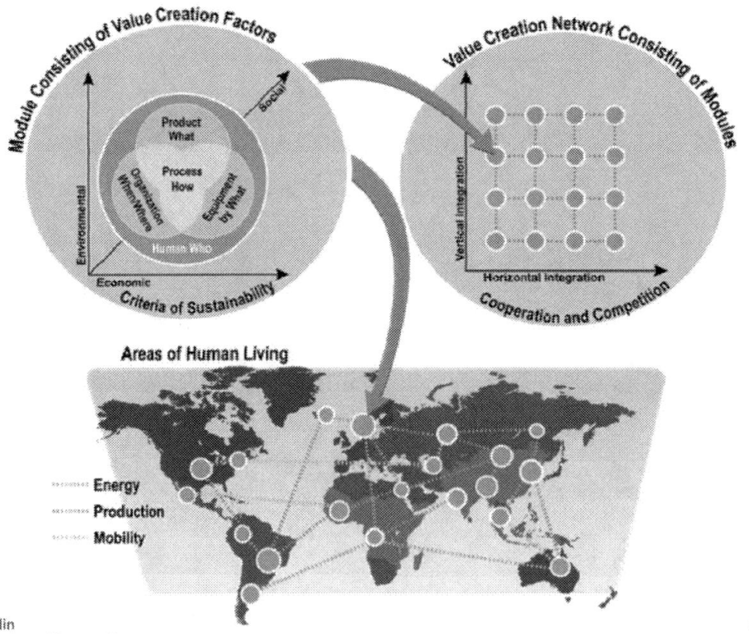

Technische Universität Berlin
Department of Machine Tools and Factory Management
Univ.-Prof. Dr.-Ing. Günther Seliger

Source: [CRC 1026, 2013]

Page 11

Model of a Production System

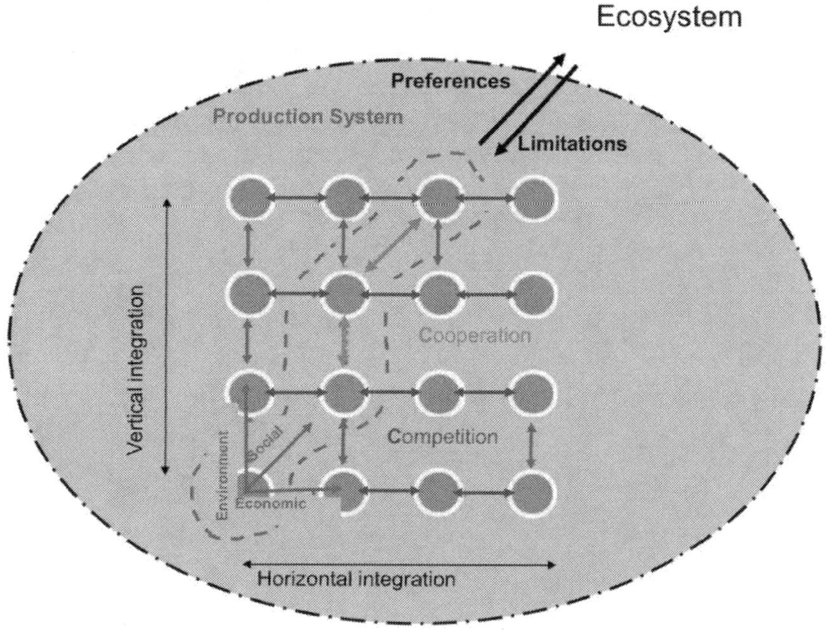

Technische Universität Berlin
Department of Machine Tools and Factory Management
Univ.-Prof. Dr.-Ing. Günther Seliger

Source: [CTEP, 2015]

Content

- Sustainable Value Creation
- **Design Conditions and Requirements**
- Principles of Industrial Engineering

Technische Universität Berlin
Department of Machine Tools and Factory Management
Univ.-Prof. Dr.-Ing. Günther Seliger

Design Conditions for „Eco-efficiency" and „Eco-effectiveness" in Material and Energy Flows

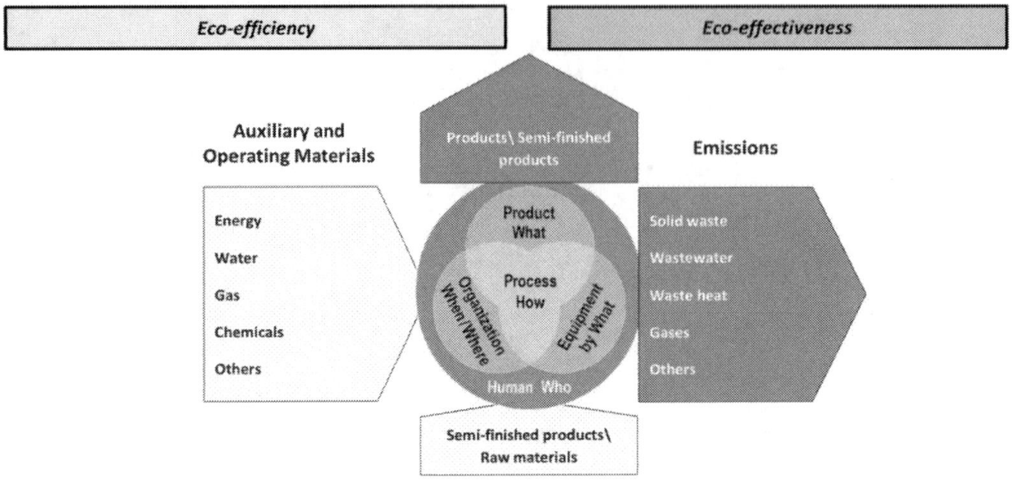

Design Conditions for „Changeability"

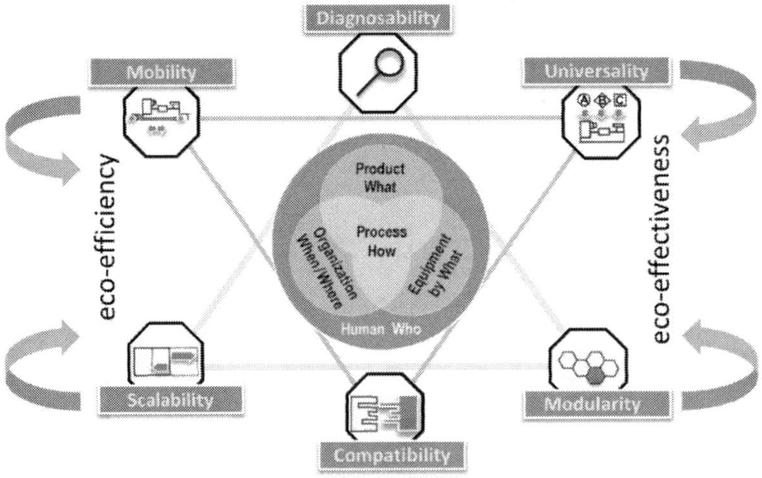

Requirements of Sustainable Improvement

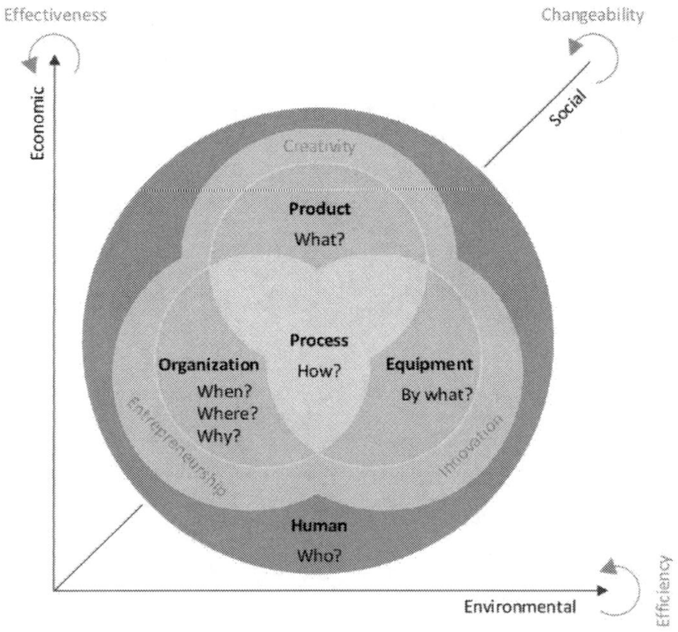

Technische Universität Berlin
Department of Machine Tools and Factory Management
Univ.-Prof. Dr.-Ing. Günther Seliger

Content

- Sustainable Value Creation
- Design Conditions and Requirements
- **Principles of Industrial Engineering**

Technische Universität Berlin
Department of Machine Tools and Factory Management
Univ.-Prof. Dr.-Ing. Günther Seliger

Education Cycle of Kolb and European Qualification Framework

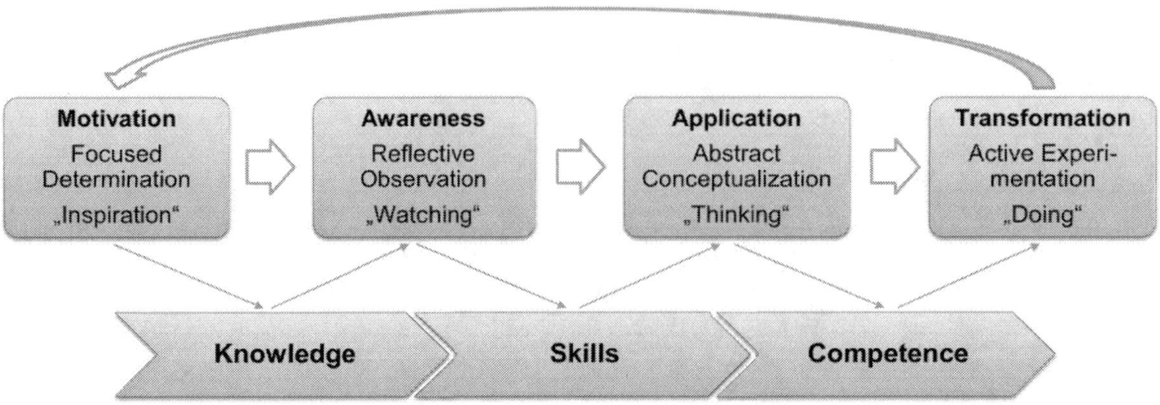

TU IWF
Technische Universität Berlin
Department of Machine Tools and Factory Management
Univ.-Prof. Dr.-Ing. Günther Seliger

Major Principles of Industrial Engineering

- Transformative
- Transdisciplinary
- International
- Sustainability-based and
- Project-based

TU IWF
Technische Universität Berlin
Department of Machine Tools and Factory Management
Univ.-Prof. Dr.-Ing. Günther Seliger

Integration of Sustainability Challenges into Engineering Education

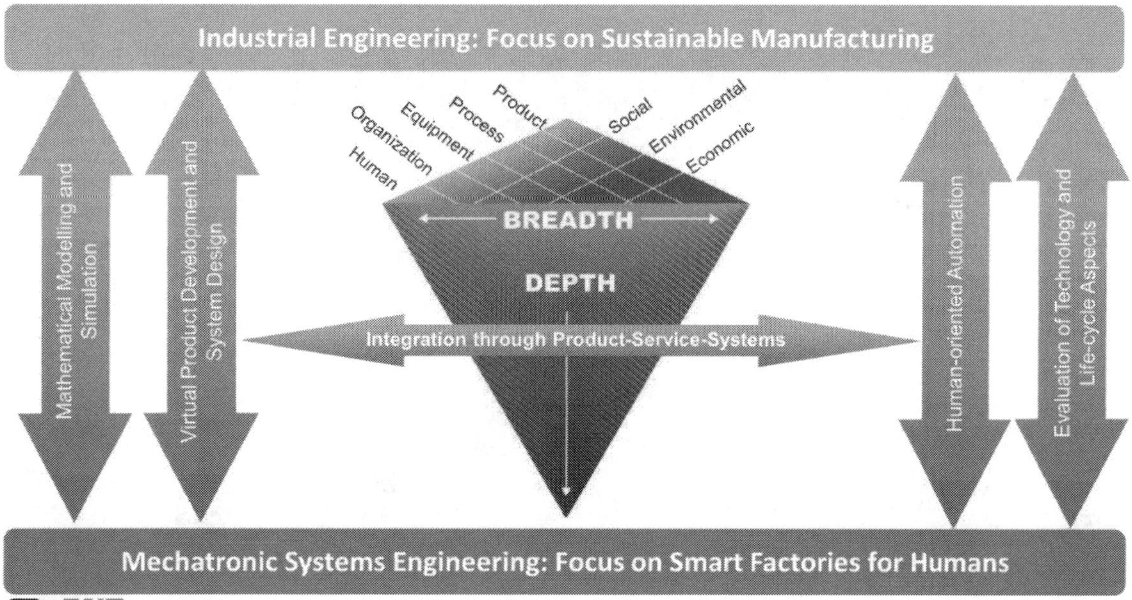

Technische Universität Berlin
Department of Machine Tools and Factory Management
Univ.-Prof. Dr.-Ing. Günther Seliger

Source: [TDU, 2015]

Page 20

Context: Sustainable Manufacturing for the Integration of Economic and Engineering Sciences

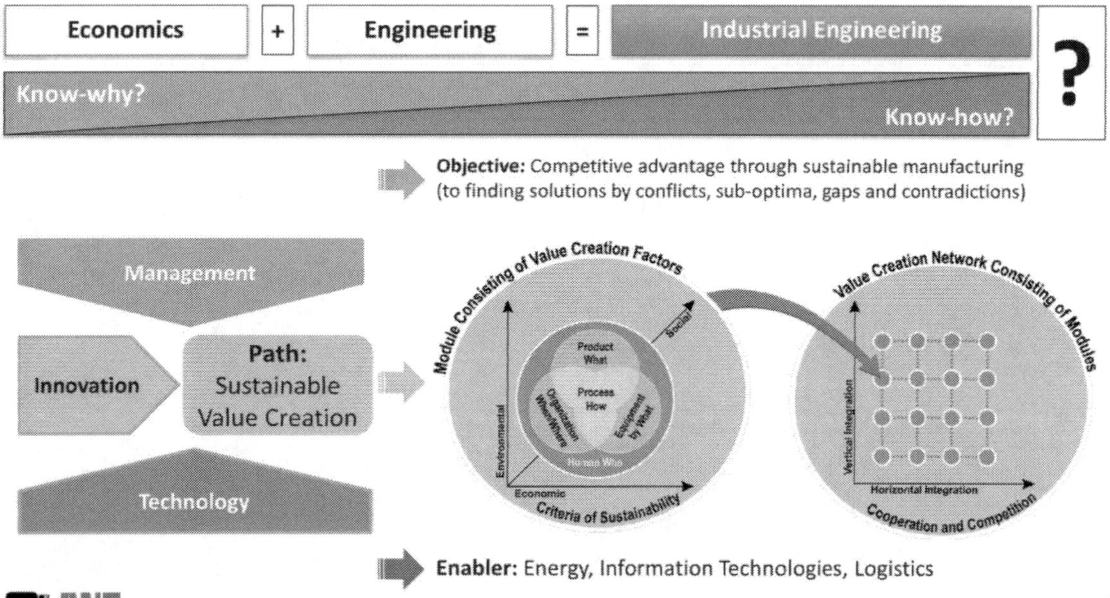

Technische Universität Berlin
Department of Machine Tools and Factory Management
Univ.-Prof. Dr.-Ing. Günther Seliger

Source: [CRC 1026, 2013]; [TDU, 2015]; [TU Berlin, 2014]

Page 21

Structure: Newly designed Industrial Engineering Education Program with focus on Sustainable Manufacturing

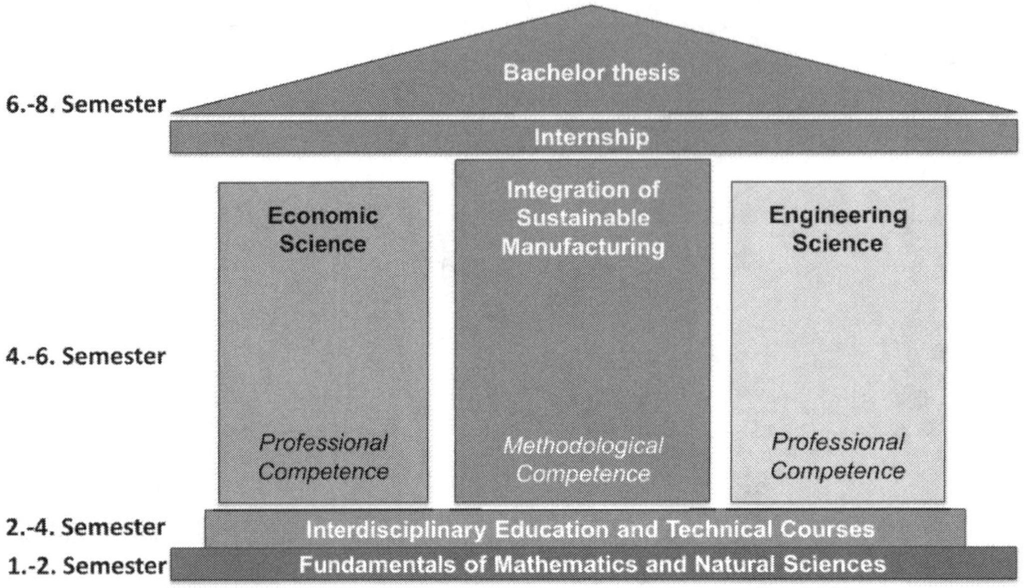

Technische Universität Berlin
Department of Machine Tools and Factory Management
Univ.-Prof. Dr.-Ing. Günther Seliger

Source: [TU Berlin, 2014]

Thank You for Your Attention!

Questions...

Pinar Bilge

bilge@mf.tu-berlin.de

Technische Universität Berlin
Department of Machine Tools and Factory Management
Univ.-Prof. Dr.-Ing. Günther Seliger

Compact Heat Exchangers to Increase Energy Sustainability

Dr. Olga P. Arsenyeva,

Professor Leonid L. Tovazhnyanskyy

Professor Petro O. Kapustenko*

AO Spivdruzhnist-T

Department of Integrated Technology and Chemical Engineering,

National Technical University "Kharkiv Polytechnic Institute", Ukraine

ICOSSE-2015

The sustainable development of mankind require steady increase in need for energy.

BP World Energy Outlook (2013):

- By 2030 more 1.3 billion people will need energy;
- World primary energy consumption will grow from 2011 to 2030 on 36 %;
- Share of fossil fuels now 82 %, same as 25 years ago. The strong rise of renewables will only reduce this to around 75% in 2035;
- Compare 2011 increase in fossil fuels use by 25 %.

To combat hazardous environmental effects require to increase efficiency of energy usage

ICOSSE-2015

NTU "KhPI"

Report for US Department of Energy BCS Incorporated, (BCS Incorporated, 2008) :

during manufacturing processes
20 to 50% of the energy consumed is
ultimately lost
via
waste heat contained

- in streams of hot exhaust gases and liquids
- heated product streams.

ICOSSE-2015

NTU "KhPI"

- Low grade waste heat is the heat that is not feasible to utilize for heat recovery inside a specific process.
- Majority of waste heat losses, about ***60-70%***, are in the low temperature range
- Low grade heat we define to be in the temperature range of ambient up to around ***260°C***
- In the temperature range up to 200°C takes place 75–85% of all process heating, cooling and inter-process heat transfer applications.

ICOSSE-2015

NTU "KhPI"

Requirements for heat transfer equipment when utilizing low grade heat

- Small temperature differences
- Close temperature approach
- Fouling mitigation
- Compactness and limited cost when using expensive materials for heat transfer surface

The Oxford dictionary provides one of the meanings of the word "compact" –
"small and economically designed".

ICOSSE-2015

NTU "KhPI"

Basic principles and terminology of compactness

ICOSSE-2015

Surface area density is characterised by two parameters:

First one is the ratio of heat transfer surface area for one side, occupied by one fluid in heat exchanger (hot or cold), to the volume on that side:

$$\sigma_i = \frac{F_i}{V_i}\text{, m}^2/\text{m}^3 \qquad (1)$$

where i=1 for hot fluid, i=2 for cold fluid.

Another parameter is the ratio of heat transfer surface area for one side to the total volume occupied by heat exchanger:

$$\psi_i = \frac{F_i}{V}\text{, m}^2/\text{m}^3 \qquad (2)$$

Hydraulic diameter for channel at one side:

$$D_h = \frac{4 \cdot f_{ch}}{\Pi_{ch}} = \frac{4 \cdot f_{ch} \cdot L}{\Pi_{ch} \cdot L} = \frac{4 \cdot V_{ch}}{F_{ch}} = \frac{4}{\sigma}\text{, m} \qquad (3)$$

For gas side of gas-liquid HE is compact if

For liquid side $\sigma_i \geq 700$

$\sigma_i \geq 400$

Total compactness

$\psi_i \geq 200$

$D_h \leq 10\_mm$

Thermal and hydraulic performance of different heat transfer surfaces

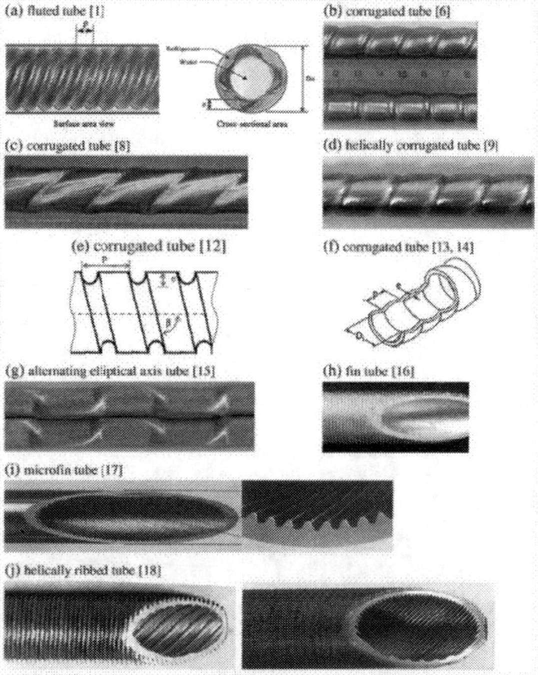

Tubes with artificial roughness, after Pethkool et al. (2011)

Plate heat exchanger channels

The friction factor dependence from Re number

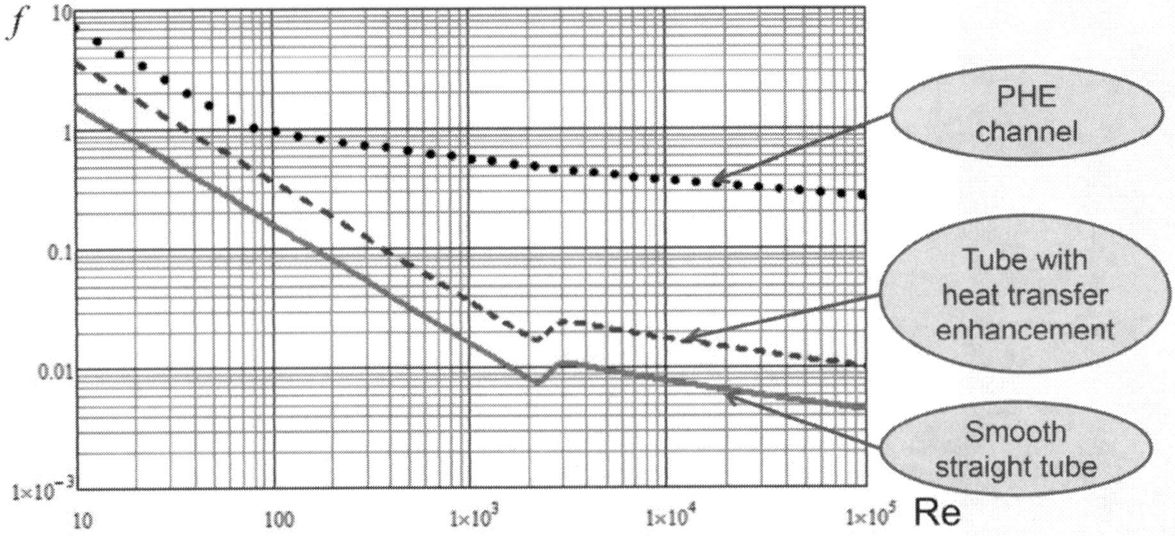

NTU "KhPI"

Nusselt number dependence from Re number for compared heat transfer surfaces

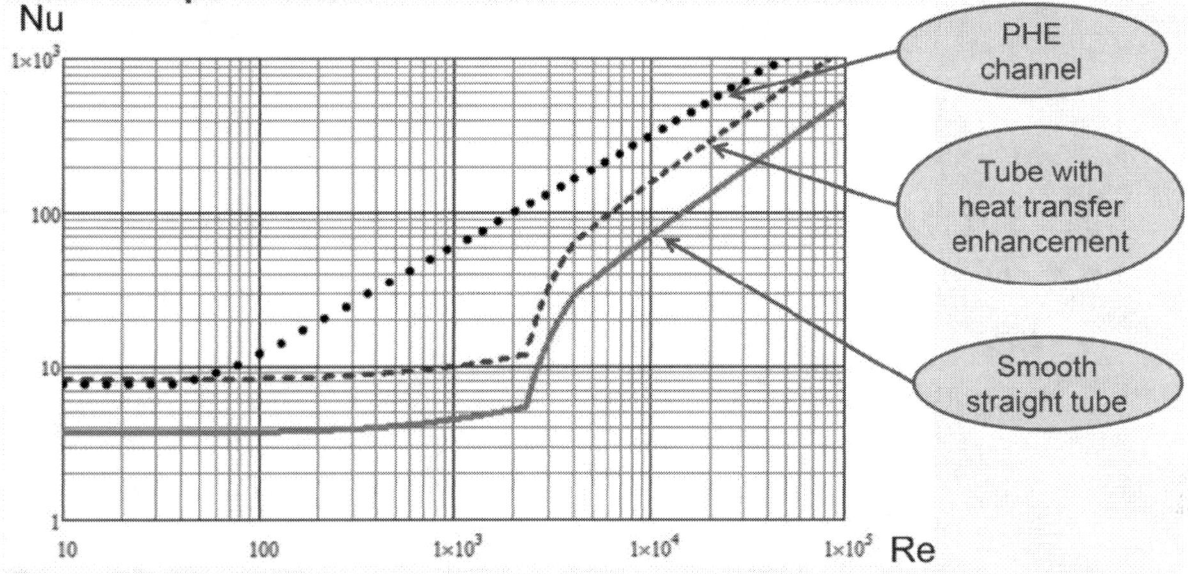

ICOSSE-2015

NTU "KhPI"

Colburn j factor dependence from Re number for compared heat transfer surfaces

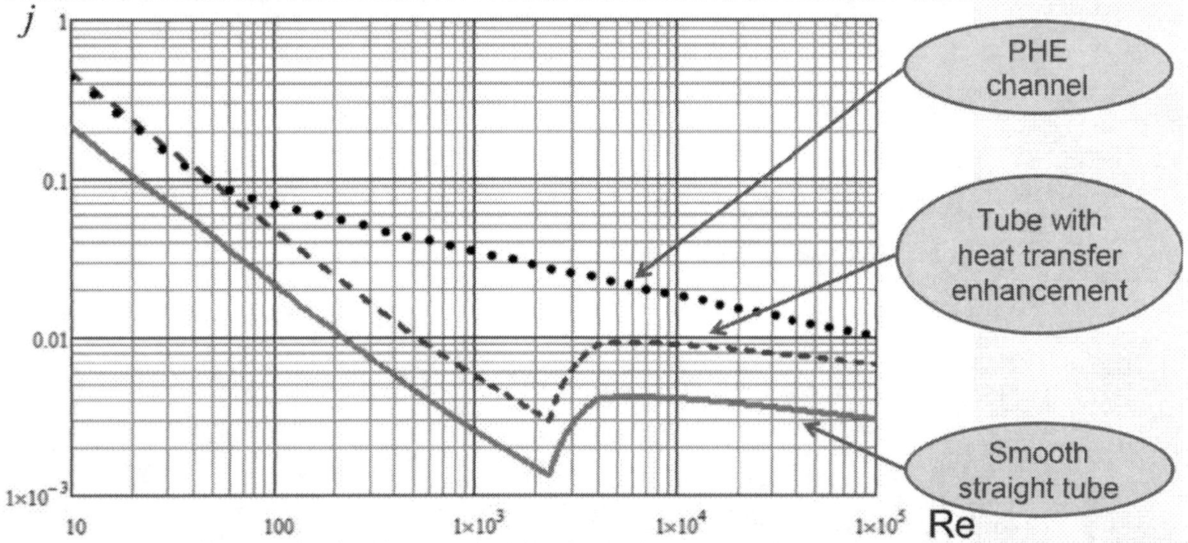

ICOSSE-2015

Factor j/f dependence from Re number for compared heat transfer surfaces

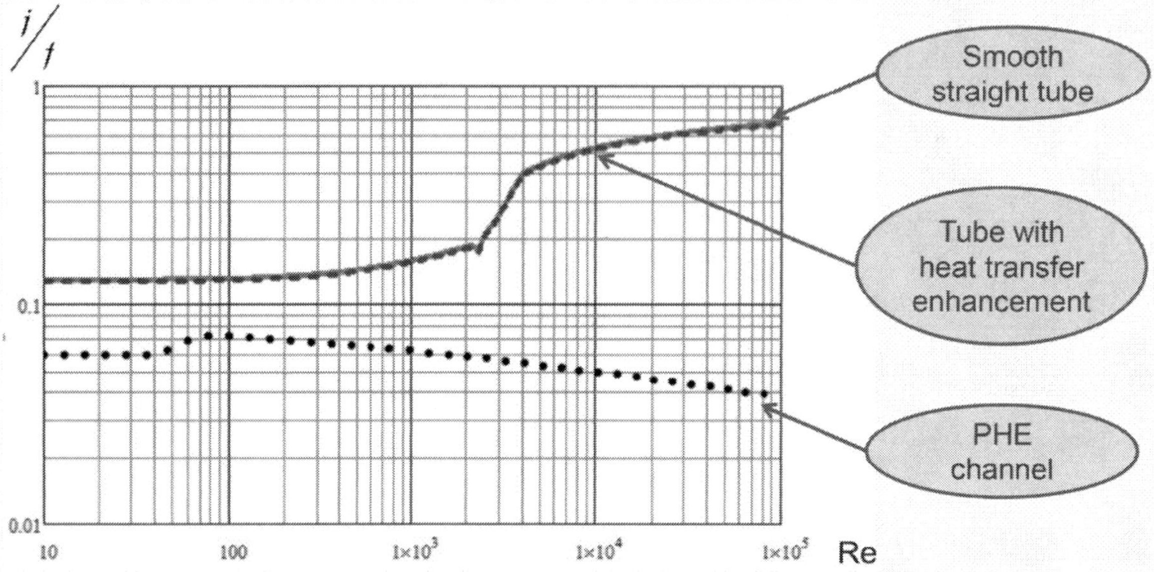

Core velocity Equation can be rewritten:

$$\frac{W_2^2 \cdot \rho_2}{2 \cdot \Delta P_2} = \frac{\dfrac{h_2 \cdot D_{h2}}{\lambda_2}}{\dfrac{W_2 \cdot D_{h2}}{\nu_2} \cdot \dfrac{c_{p2} \cdot \rho_2 \cdot \nu_2}{\lambda_2} \cdot f_2 \cdot NTU} \quad (14)$$

After rearranging:

$$W_2 = \sqrt[3]{\frac{2 \cdot \Delta P_2 \cdot h_2(W_2)}{c_{p2} \cdot \rho_2^2 \cdot NTU \cdot f_2(W_2)}} \quad (15)$$

The required heat transfer surface area:

$$F = \frac{Q}{\Delta T_{ln} h_2}, \text{ m}^2. \quad (16)$$

The ratio of heat transfer surface length to hydraulic diameter L/D_{h2}:

$$\frac{L}{D_{h2}} = \frac{\Delta P_2}{2 \cdot W_2^2 \cdot \rho_2 \cdot f_2} \quad (17)$$

This value is also satisfying thermal conditions.

The solution for specified conditions:

1. Heating of water from T_{21}= *5 °C* to T_{22}=*55 °C*
2. by some vapour with constant temperature $T_{11}=T_{12}$= *60 °C.*
3. The thermo-physical properties of water are taken for its average temperature *30 °C*:
   - ρ_2=996 kg/m^3; ν_2=0.805×10^{-6} m^2/s;
   - λ_2=9.618 W/(mK); c_{p2}=4174 J/(kgK);
   - Prandtl number is Pr=5.43.
4. The required heat load ***Q=600 kW*** and
5. The allowable pressure drop is ***ΔP₂= 20,000 Pa.***
6. The number of heat transfer units required by this process conditions: ***NTU=2.40.***

ICOSSE-2015

The change of required Re number with hydraulic diameter D_h for compared heat transfer surfaces (*NTU=2.40, ΔP$_2$=20 kPa*)

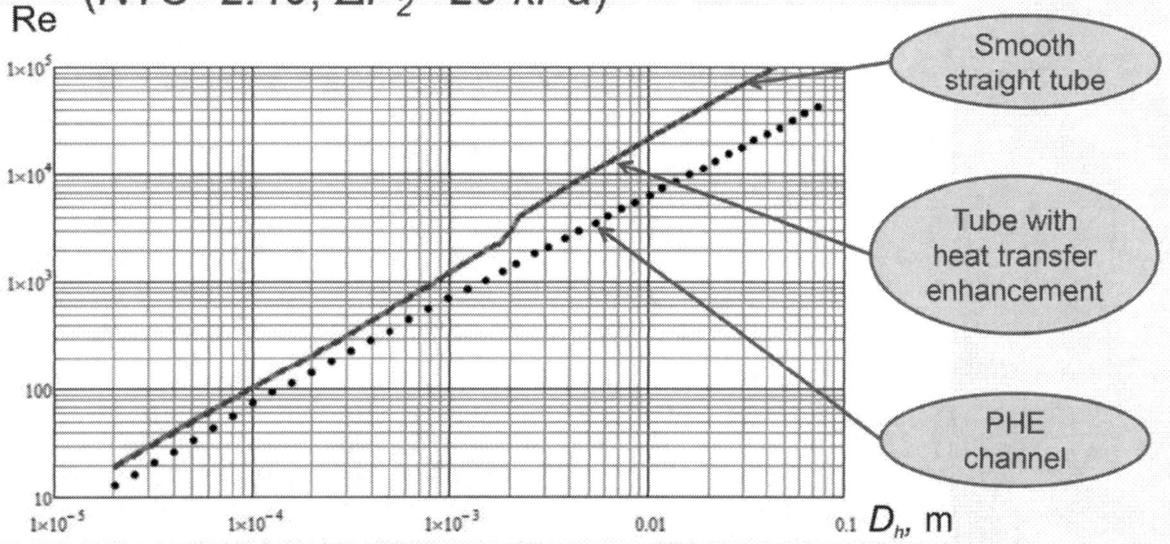

ICOSSE-2015

The change of required heat transfer surface area F with hydraulic diameter D_h for compared heat transfer surfaces($NTU=2.40$, $\Delta P_2=20$ kPa)

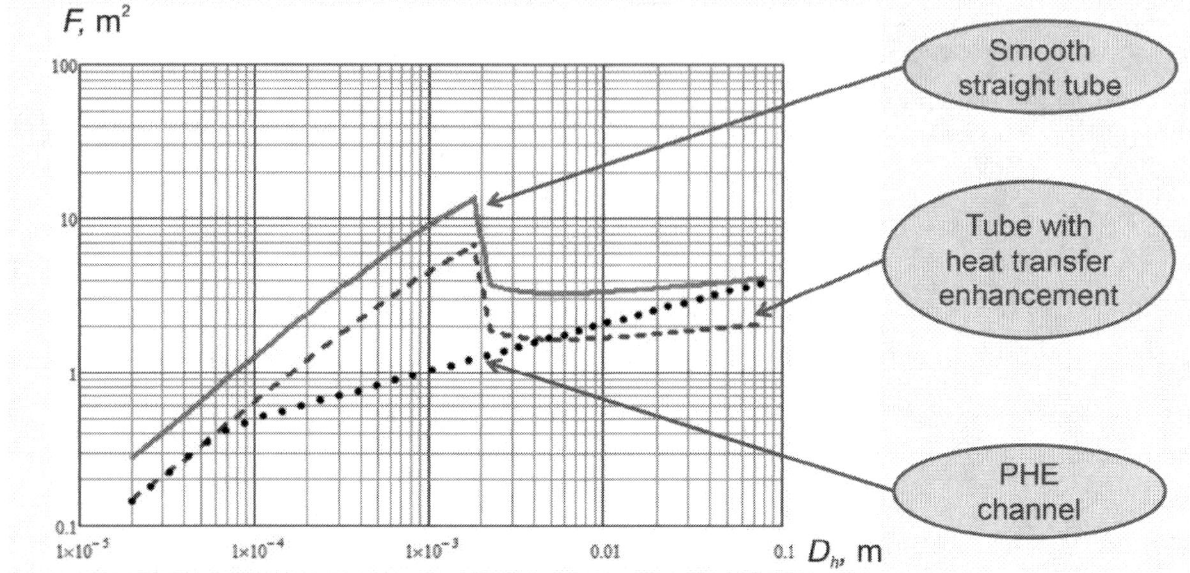

The required length of compared heat transfer surfaces ($NTU=2.40$, $\Delta P_2=20$ kPa)

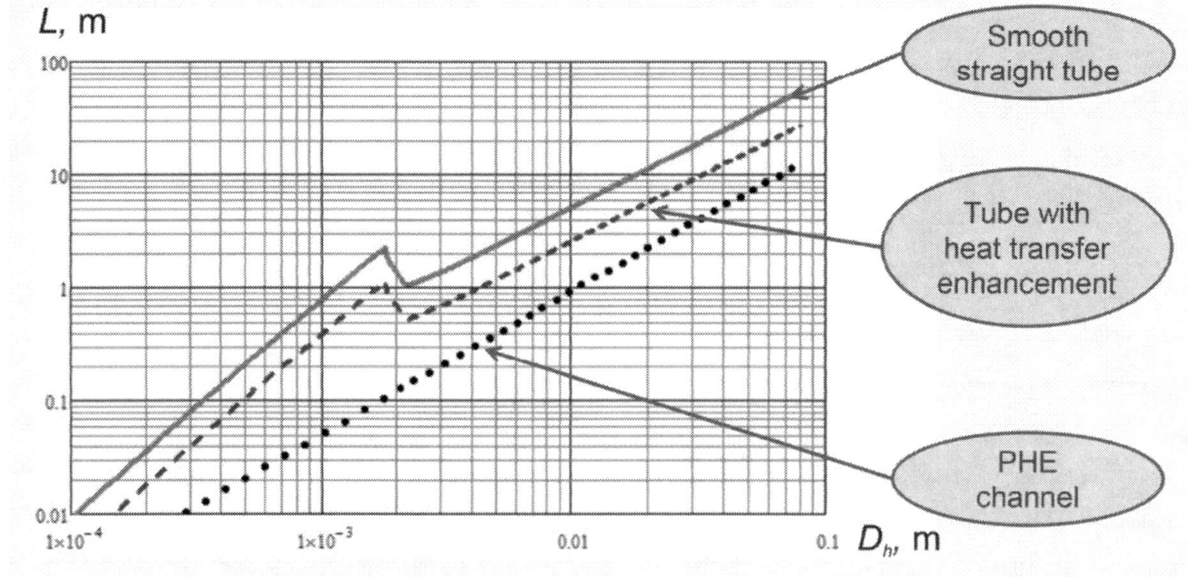

Fouling mitigation

NTU "KhPI"

The surface shear stress on compared heat transfer surfaces ($NTU=2.40$, $\Delta P_2=20$ kPa)

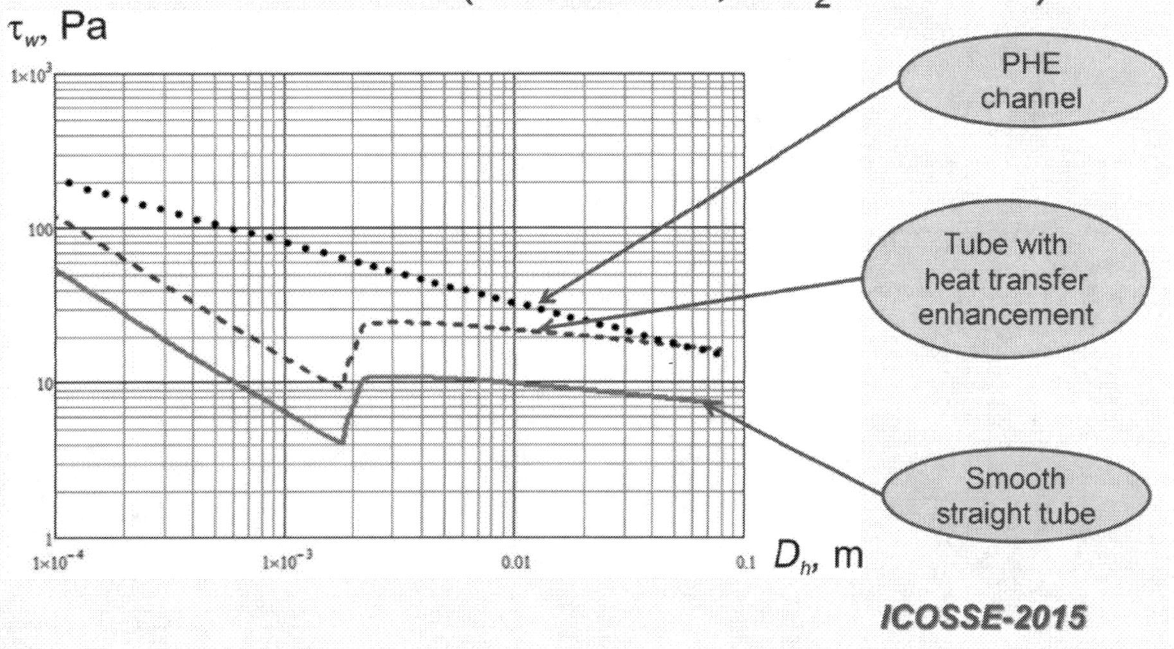

ICOSSE-2015

Types of Compact Heat Exchangers

According to hydraulic diameter

- Conventional channels >3mm;
- Minichannels 3mm≥D_h>200μm;
- Microchannels 200μm≥D_h>10μm;
- Transitional Microchannels 10μm≥D_h>1μm;
- Transitional Nanochannels 1μm≥D_h>0.1μm;
- Nanochannels 0.1μm≥D_h.

NTU "KhPI"

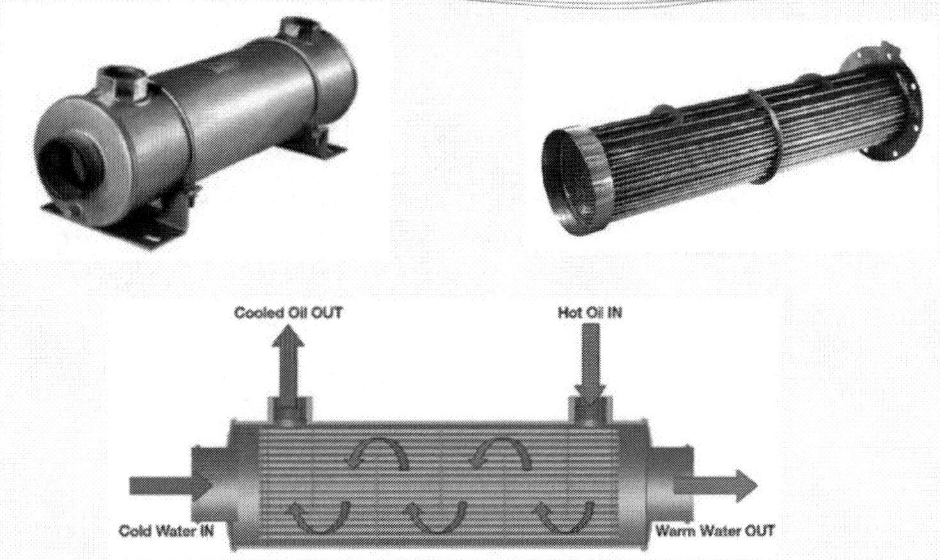

Compact brazed shell and tube heat exchanger: a) assembled HE; b) tube bundle; c) the movement of heat exchanging streams in HE (after Alfa Laval, 2014a).

ICOSSE-2015

NTU "KhPI"

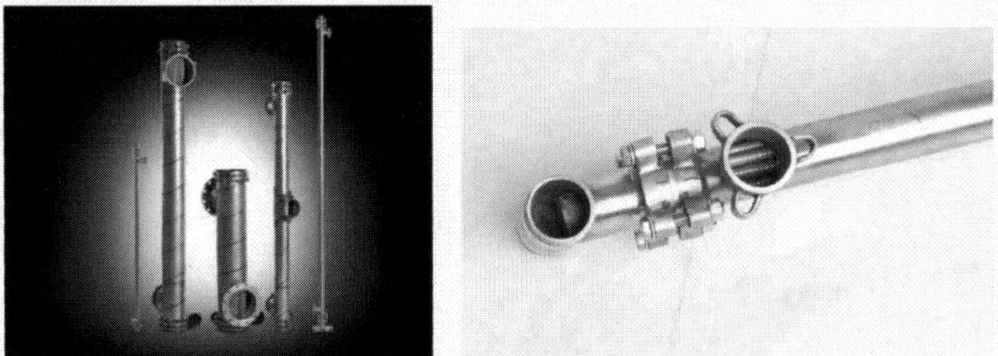

Compact stainless steel heat exchangers TTAI produced by Teploobmen LLC in Sevastopol, Ukraine (after Teploobmen, 2014)

ICOSSE-2015

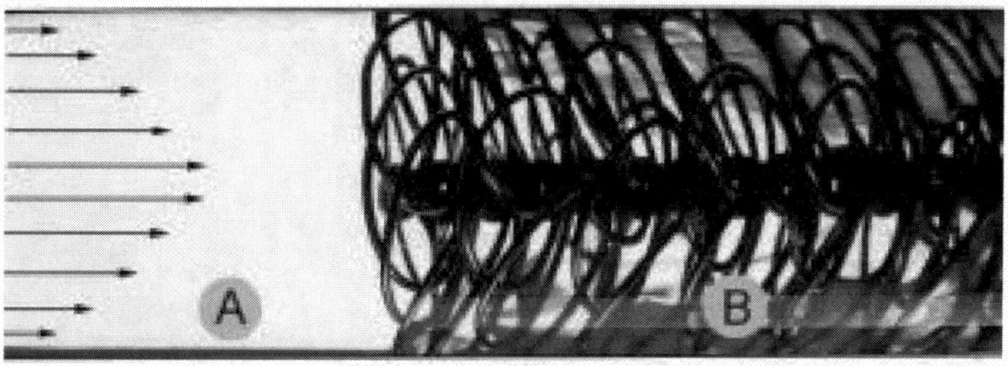

HiTRAN Matrix Element tube insert promoting turbulence and enhancement of heat transfer. (After Calgavin, 2014).

NTU "KhPI"

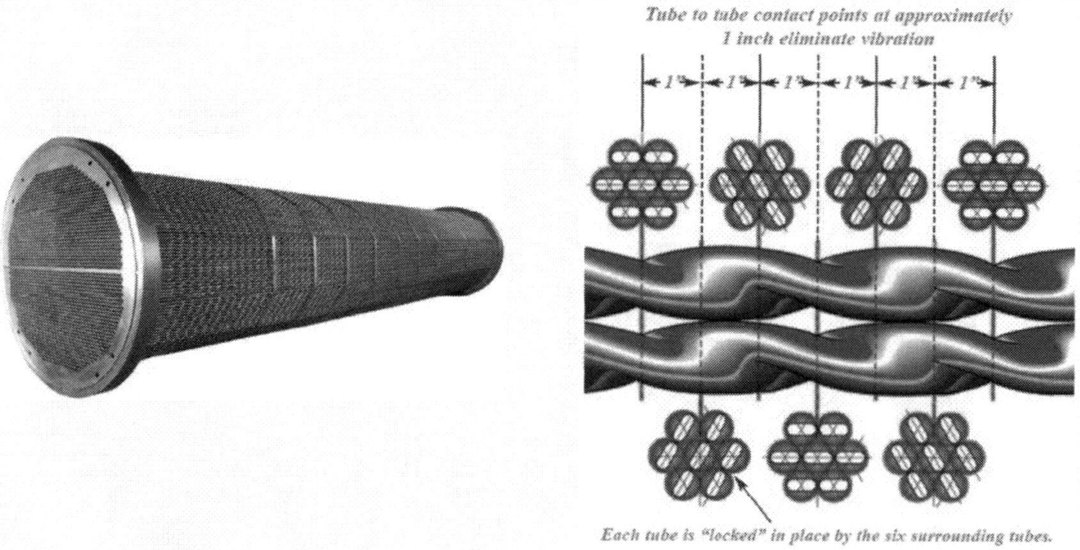

Koch Heat Transfer Company's TWISTED TUBE® bundle (a) and scheme of tubes arrangement (b) (see Koch Heat Transfer Company, 2014)

ICOSSE-2015

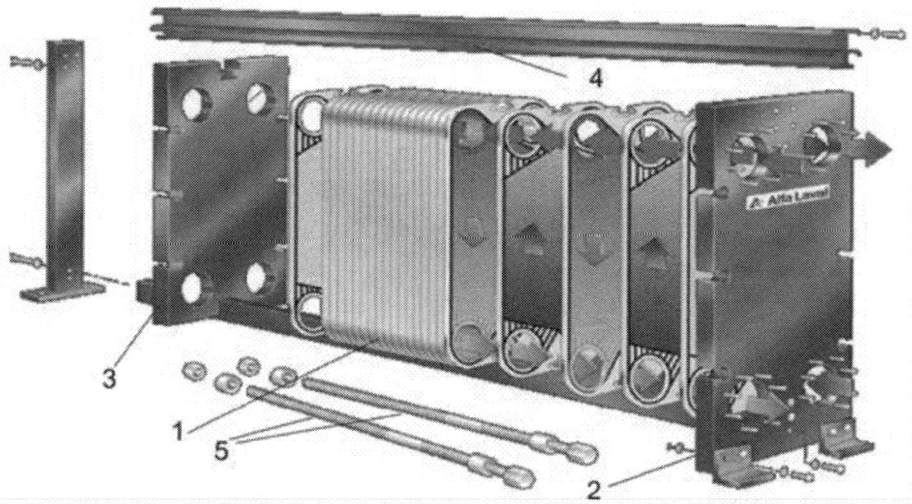

Schematic drawing of one pass frame-and-plate PHE:
1-pack of plates with gaskets; 2- fixed frame plate;
3- movable frame plate; 4- carrying bar; 5- tightening bolts.

ICOSSE-2015

NTU "KhPI"

Corrugated fields with different angles of corrugations for plates that can be assembled in one single pass PHE. Selecting exact numbers of different plates enables close satisfaction to required duty.

ICOSSE-2015

NTU "KhPI"

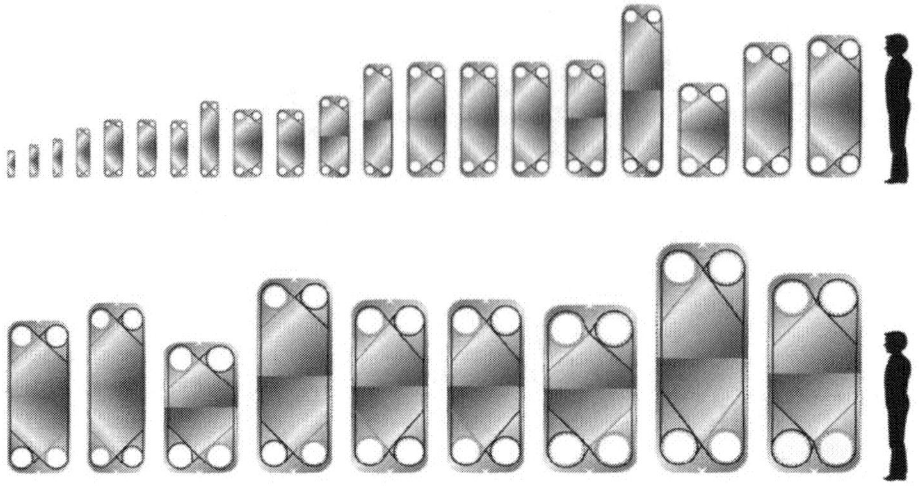

Range of plates manufactured for plate-and-frame PHEs by AlfaLaval, Sweden (see www.alfalaval.com).

ICOSSE-2015

NTU "KhPI"

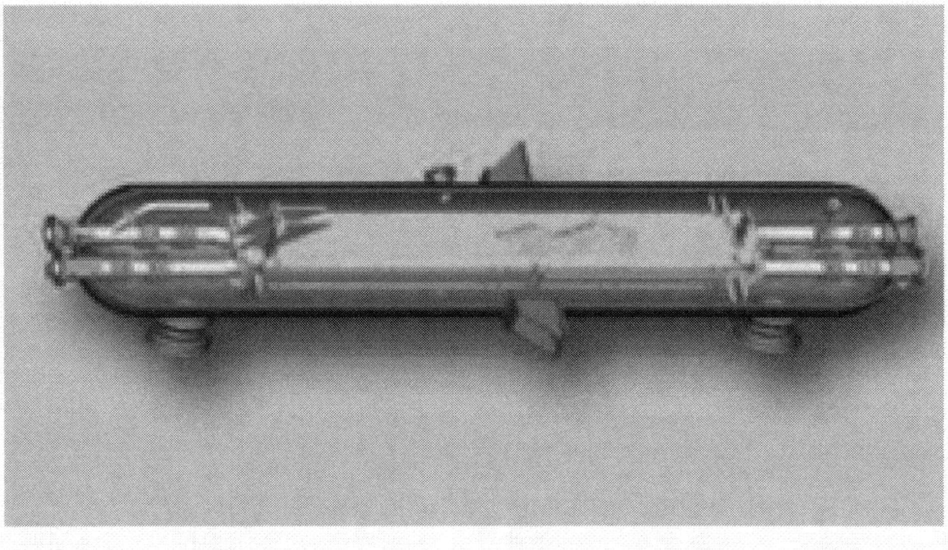

Drawing of Alfa Laval Packinox welded PHE (after Alfa Laval, 2014b).

ICOSSE-2015

NTU "KhPI"

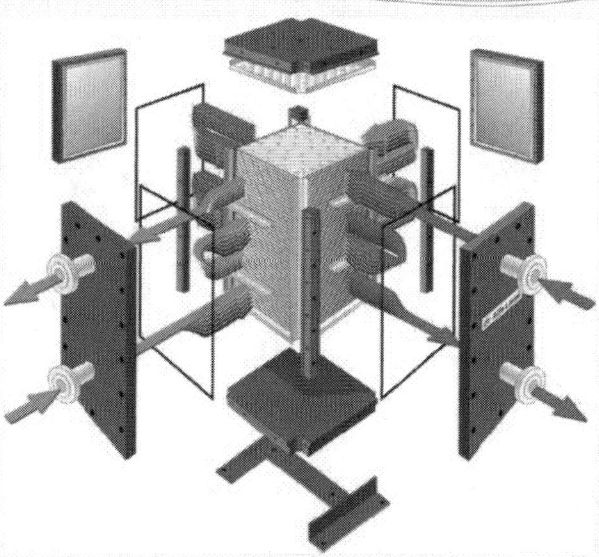

CompablocTM welded PHE of AlfaLaval production: exploded view showing cross-counter current arrangement (after Alfa Laval, 2014).

ICOSSE-2015

NTU "KhPI"

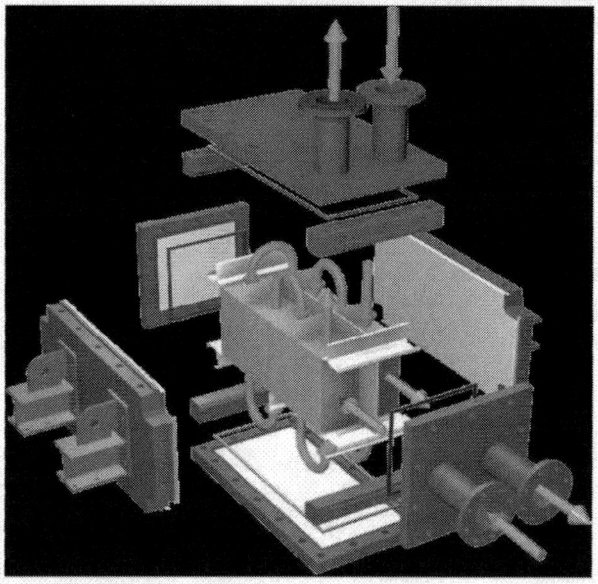

APV Hybrid welded PHE: exploded view showing cross-counter current arrangement (after SPX, 2012).

ICOSSE-2015

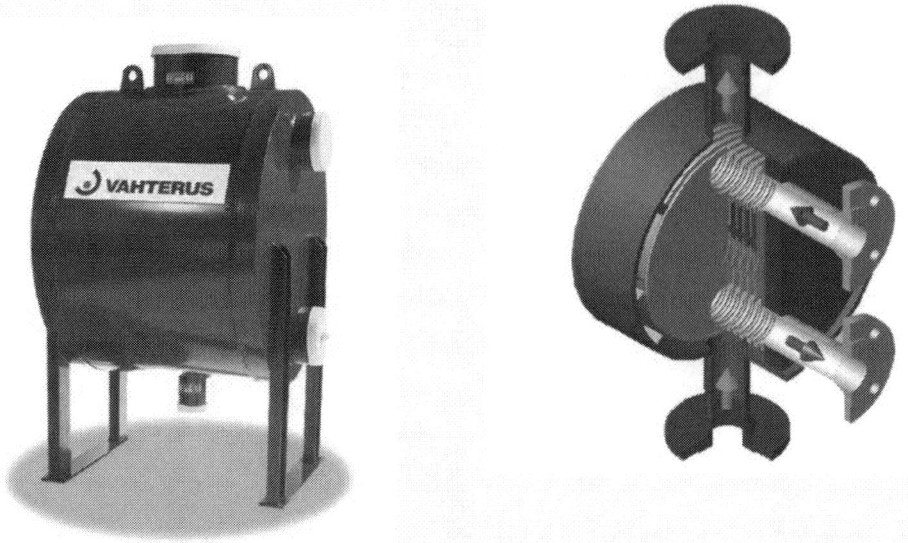

Plate & Shell ® heat exchanger:
a) assembled PSHE;
b) flow arrangement in one pass unit (after Vahterus Oy, 2013).

ICOSSE-2015

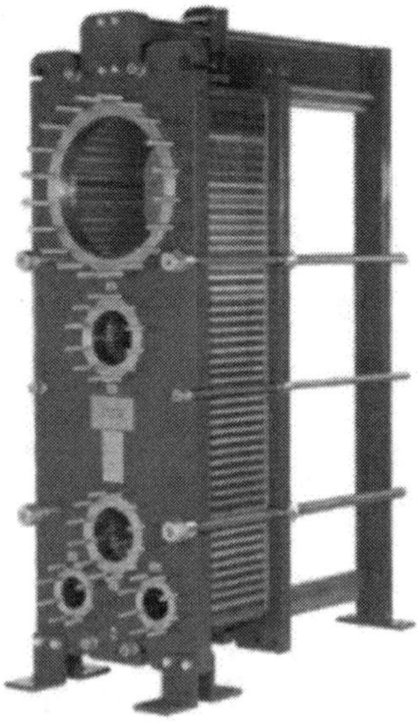

Semi-welded PHE AlfaCond manufactured by AlfaLaval for condensing duties (after Alfa Laval, 2014).

ICOSSE-2015

NTU "KhPI"

Evaporator semi-welded PHE AlfaVap:
a) AlfaVap-500;
b) AlfaVap-700 (after Alfa Laval, 2014).

ICOSSE-2015

NTU "KhPI"

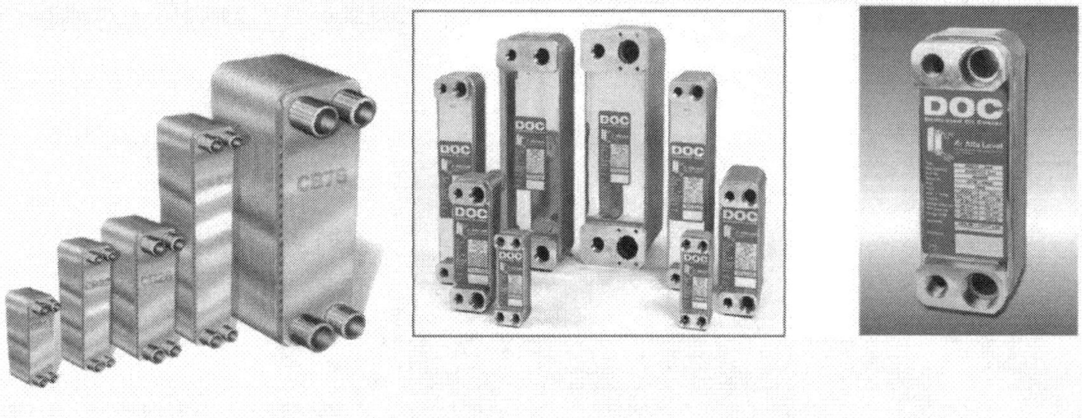

Some of BPHEs produced by AlfaLaval:
a) general purpose BPHEs;
b) special design for oil cooling (after Alfa Laval, 2014).

ICOSSE-2015

Some of fusion bonded AlfaNova PHEs produced by AlfaLaval (after Alfa Laval, 2014).

ICOSSE-2015

Plate-Fin Heat Exchanger

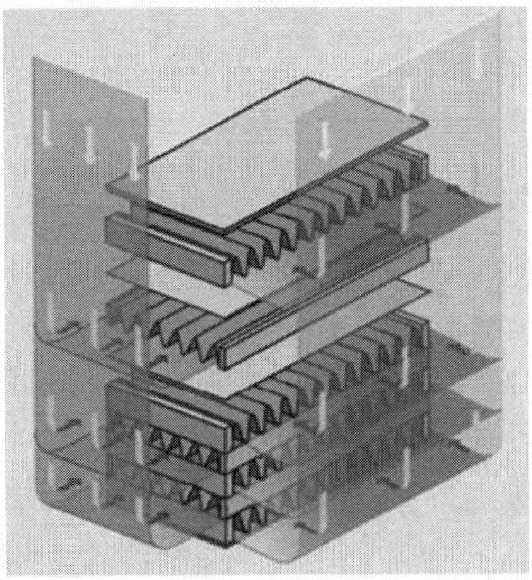

ICOSSE-2015

NTU "KhPI"

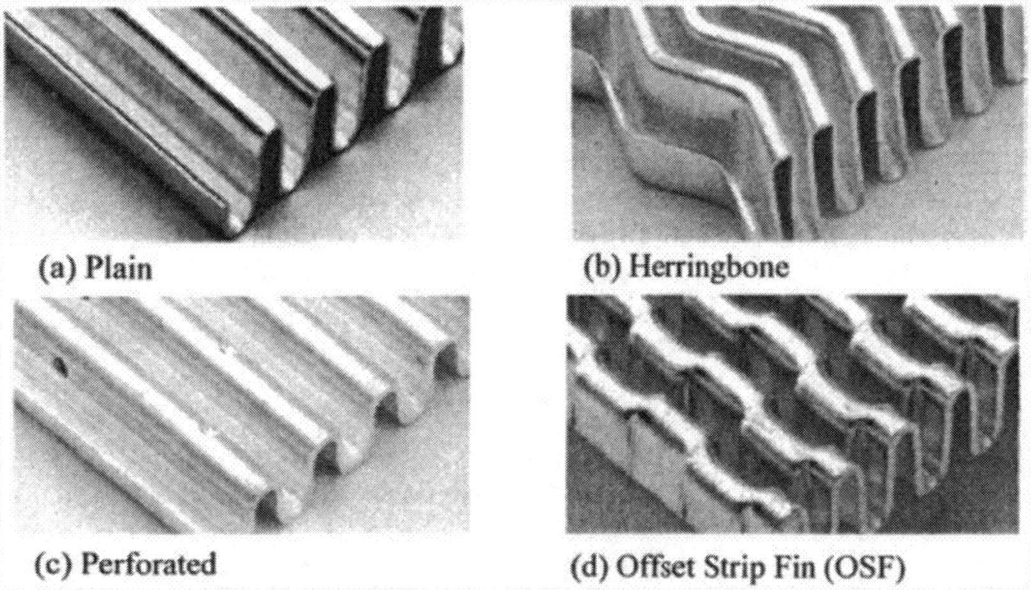

PFHE Fin types (courtesy Chart Heat Exchangers)

ICOSSE-2015

NTU "KhPI"

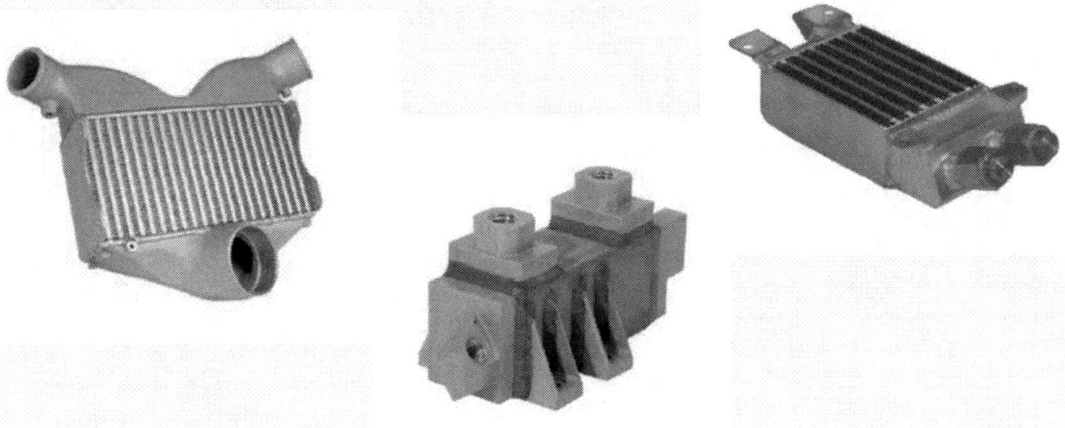

Custom designed PFHE produced by Lytron, USA:
a) PFHE for air-to-air cooling;
b) PFHE for liquid cooling by air in an airplane;
c) liquid-to-liquid oil cooler for commercial aircraft (after www.lytron.com/Heat-Exchangers) .

ICOSSE-2015

NTU "KhPI"

Chart's Brazed Aluminum PFHE (After Chart, 2014).

ICOSSE-2015

NTU "KhPI"

Example of finned tube (a) and TFHE made of such tubes (manufactured by Defon Heat Exchanger Co., Ltd. In China, see Defon, 2013).

ICOSSE-2015

NTU "KhPI"

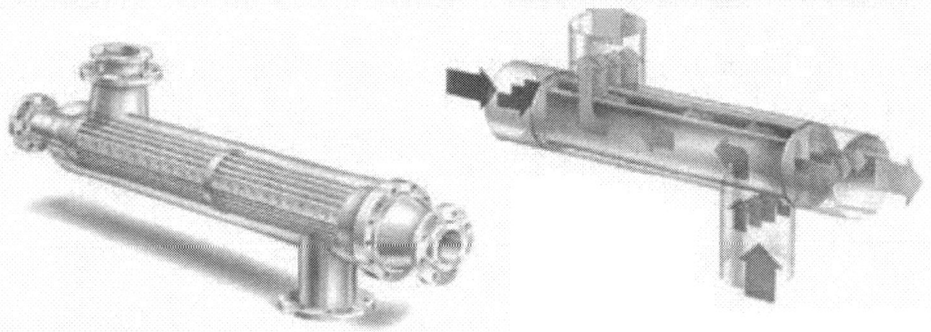

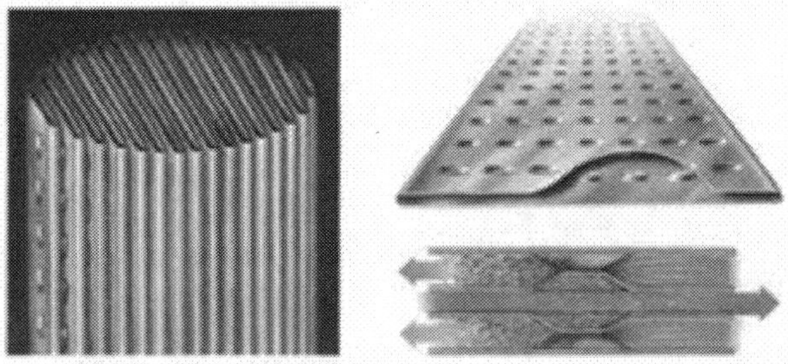

Lamella heat exchanger:
a) drawing of fluids flow;
b) lamellas bundle (after Sondex Tapiro OY AB, 2011).

ICOSSE-2015

NTU "KhPI"

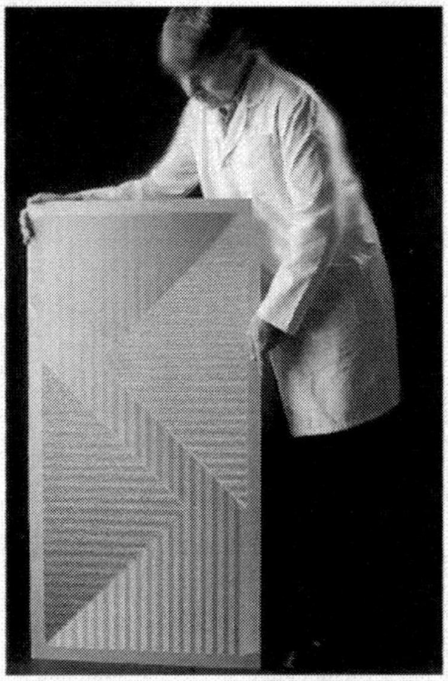

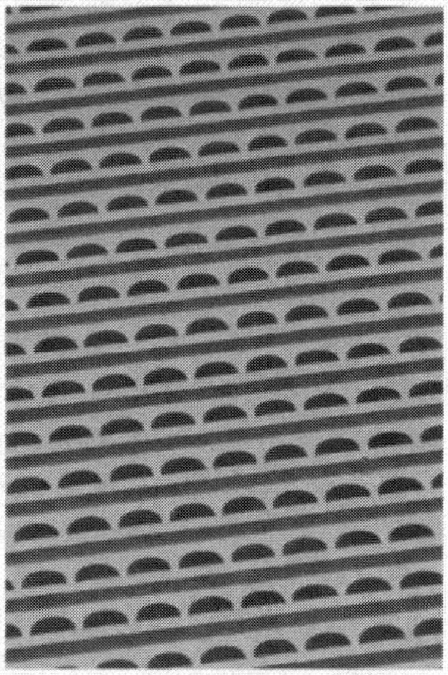

Example of Printed Circuit PCHE Heatric™ plate (a) and block cross section (b), see Heatric, 2014.

ICOSSE-2015

NTU "KhPI"

Example of PCHE Heatric™ for oil and gas processing industry, see Heatric, 2014.

ICOSSE-2015

NTU "KhPI"

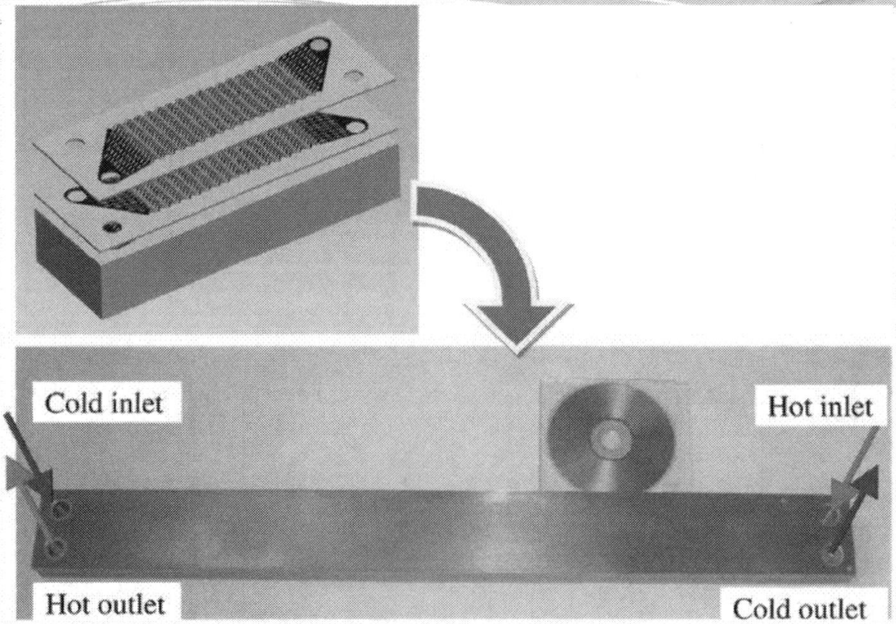

PCHE for water heating by CO_2 as heat pump working fluid (after Tsuzuki et al., 2009).

ICOSSE-2015

NTU "KhPI"

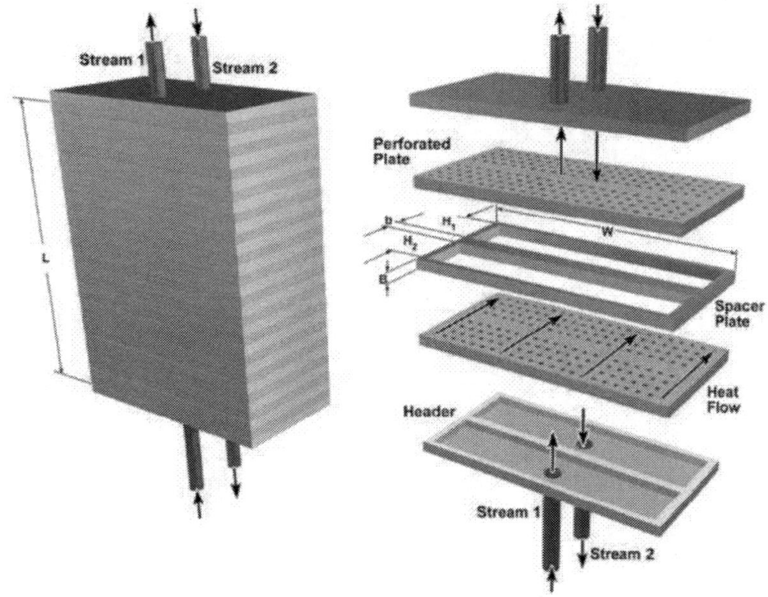

The construction of matrix MCHE (after Moheisen, 2009).

ICOSSE-2015

NTU "KhPI"

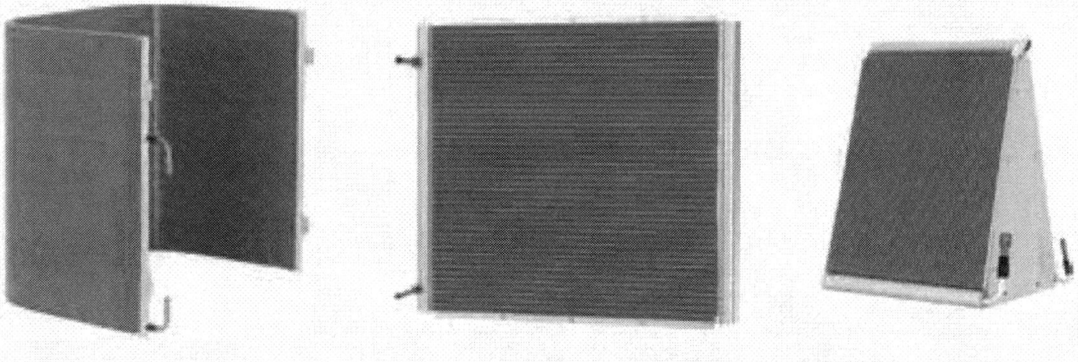

Plate-fin MCHEs for refrigeration and air conditioning produced by Danfoss (after Danfoss, 2013).

ICOSSE-2015

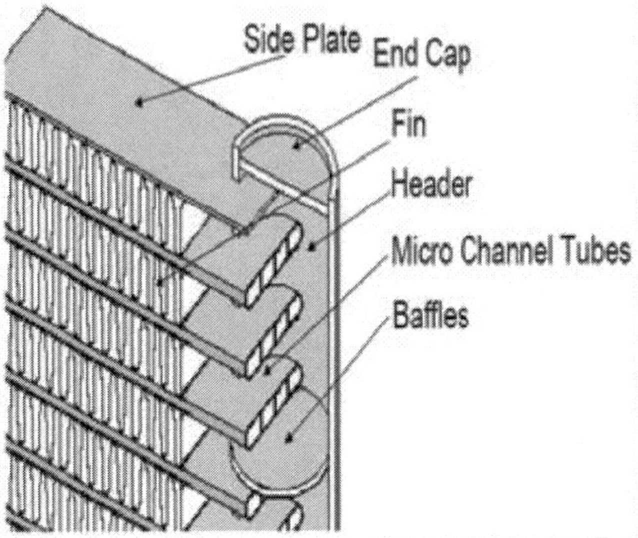

The construction principle of Plate-fin MCHEs for refrigeration and air conditioning (after Popescu et al., 2012).

Non-metal Ametek shell and tube Teflon heat exchanger (After Ametek, 2010).

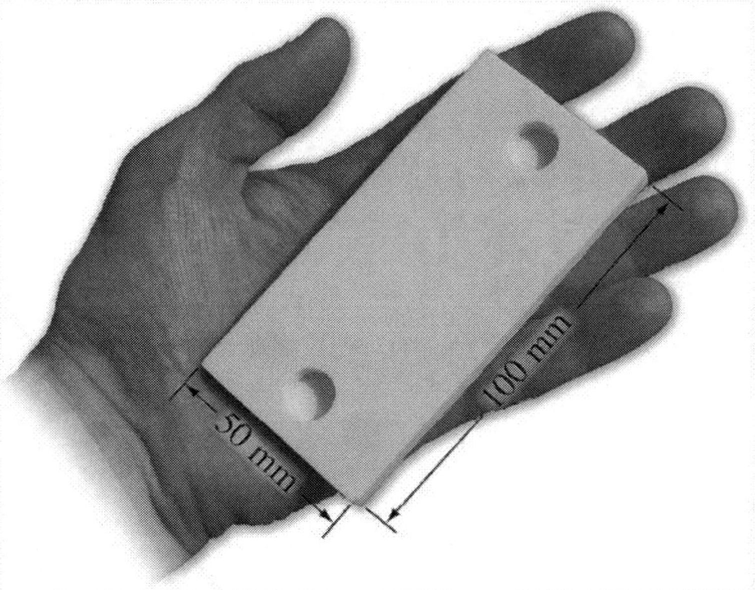

Photograph of a compact ceramic microchannel counter-flow heat exchanger.

NTU "KhPI"

MonoblockTM heat exchanger produced by AB Segerfrojd (/www.segerfrojd.com/ after Zaheed and Jachuck ,2004).

ICOSSE-2015

NTU "KhPI"

CONCLUSIONS

1. Compact heat exchangers are able to significantly improve energy efficiency and contribute to energy sustainability.
2. The correct selection of heat transfer surface for compact heat exchanger must be made based on thermo-hydraulic design for which correlations for heat transfer and friction factor should be known.
3. As a rule the decreasing of hydraulic diameter is leading not only to increase in geometrical compactness, but also to smaller required heat transfer surface area
4. Heat transfer intensification can significantly improve efficiency of compact heat exchangers.
5. The fouling tendencies of heat exchanger streams are significantly influencing its performance. For most types of fouling mechanisms the fouling tendencies on enhanced surfaces are significantly lower than on smooth surfaces because of higher levels of shear stress on the wall.

ICOSSE-2015

NTU "KhPI"

Thank you for your attention!

ICOSSE-2015

Pollution Reduction in Croatian Food Industry Via Total Site Heat Recovery

- **Conference:** International Congress on Sustainability Science Engineering ICOSSE
- **Year:** 2015
- **Proceeding:** ICOSSE 2015
- **Group:** Abstract Submissions
- **Session:**

Session 9: Sustainable Product and Process Design II
- **Time:**

Thursday, May 28, 2015 - 2:40pm-3:00pm

Authors:

Dominkovic, D. F.
Boldyryev, S.
Cosic, B., University of Zagreb, Facuty of Mechanical Engineering and Naval Architecture
Krajacic, G.
Stefanec, K.
Duic, N., University of Zagreb, Facuty of Mechanical Engineering and Naval Architecture

The statistic indicates the energy consumption in food industry: to produce 1 J of food energy 10 J of primary energy is required. The population growth and its food supplying are causing the annual energy consumption rise on 24 – 40%. On the other hand it leads to fast deterioration of surrounding environment owing to CO2, NOx, SOx, dust, soot and other industrial emissions. The steep rise of energy prices in recent years, as well as the necessity to curb carbon dioxide emissions, have intensified the need for solutions that are increasing energy efficiency in the food industry.

This paper deals with selection of optimum amount heat to be recovered during Total Site integration and utility targets for external heating, cooling, refrigerating etc. The methodology provides the calculation of minimum capital investment during heat integration of site. The results may be used for minimisation of investments during Site retrofit.

Strategy to Achieve Sustainability of Green Integrated Biorefineries

- **Conference:** International Congress on Sustainability Science Engineering ICOSSE
- **Year:** 2015
- **Proceeding:** ICOSSE 2015
- **Group:** Abstract Submissions
- **Session:**

Session 11: Sustainable Supply Chains

- **Time:**

Thursday, May 28, 2015 - 3:20pm-3:50pm

Authors:

Kafarov, V.

A biorefinery is a facility to produce bio-based products and bioenergy.

The PSE, Process Integration, LCA, exergy and economic analysis tools were applied at all steps of integrated biorefinery design, integration and evaluation for selected case study and allowing significant reductions of emissions and improve global energy balance. This strategy permit to achieve sustainability of green integrated biorefinery based on optimal design of multicomponent feedstock and biorefinery topology.

With energy and material integration of Sugarcane and Multi Species Algae biorefineries, it is possible to approach a zero emission system due to CO2 sequestration in phototrophic algae cultivation systems, and vinasse utilization in algal mixotrophic systems, and a positive energy balance due to utilization of energy stored in sugarcane bagasse or lignin in an integrated topology, numerous components in multicomponent feedstock, producing several low-volume, but high-value, chemical biochemical or biological products, increasing the general profitability and assuring the economic sustainability.

Quantitative Modeling to Assess Total Life-Cycle Risk Implications Across the Supply Chain for Sustainable Manufacturing Decision Making

- **Conference:** International Congress on Sustainability Science Engineering ICOSSE
- **Year:** 2015
- **Proceeding:** ICOSSE 2015
- **Group:** Abstract Submissions
- **Session:**

Session 11: Sustainable Supply Chains

- **Time:**

Thursday, May 28, 2015 - 3:50pm-4:10pm

Authors:

Badurdeen, F., University of Kentucky
Amundson, J.
Brown, A.,

Globalization of supply chains increasingly exposes organizations to unforeseen events, underscoring the need for better risk modeling tools. While promoting sustainable manufacturing requires exploring the use of new materials, technologies and partners, these can also be sources of potential risk exposure in supply chains. When such changes are introduced, it is imperative that relevant risks are assessed from a total life-cycle perspective, from pre-manufacturing to post-use. For example, using carbon composites instead of titanium is advantageous due, particularly to, the energy efficiencies possible. However, future outlook for carbon composite recycling or take-back legislations is uncertain, and the potential impact of new legislation is unclear. Strategic sustainable manufacturing decisions must be evaluated for potential long-term risks such decisions can bring about for the company as well as the supply chain. This presentation will share quantitative, total life-cycle risk models to evaluate sustainable manufacturing decisions and its application to a case study.

The University of Nottingham
UNITED KINGDOM · CHINA · MALAYSIA

GREEN Supply Chain towards Sustainable Development

Hon Loong LAM
The University of Nottingham Malaysia Campus

1

The University of Nottingham
UNITED KINGDOM · CHINA · MALAYSIA

OUTLINE

- Introduction
- Previous works
- New Developments in our team
- Future work
- Conclusions

2

The University of Nottingham
UNITED KINGDOM · CHINA · MALAYSIA

Question 1:
Is Biomass Energy = GREEN ?

3

The University of Nottingham
UNITED KINGDOM · CHINA · MALAYSIA

Carbon Cycle in Biomass

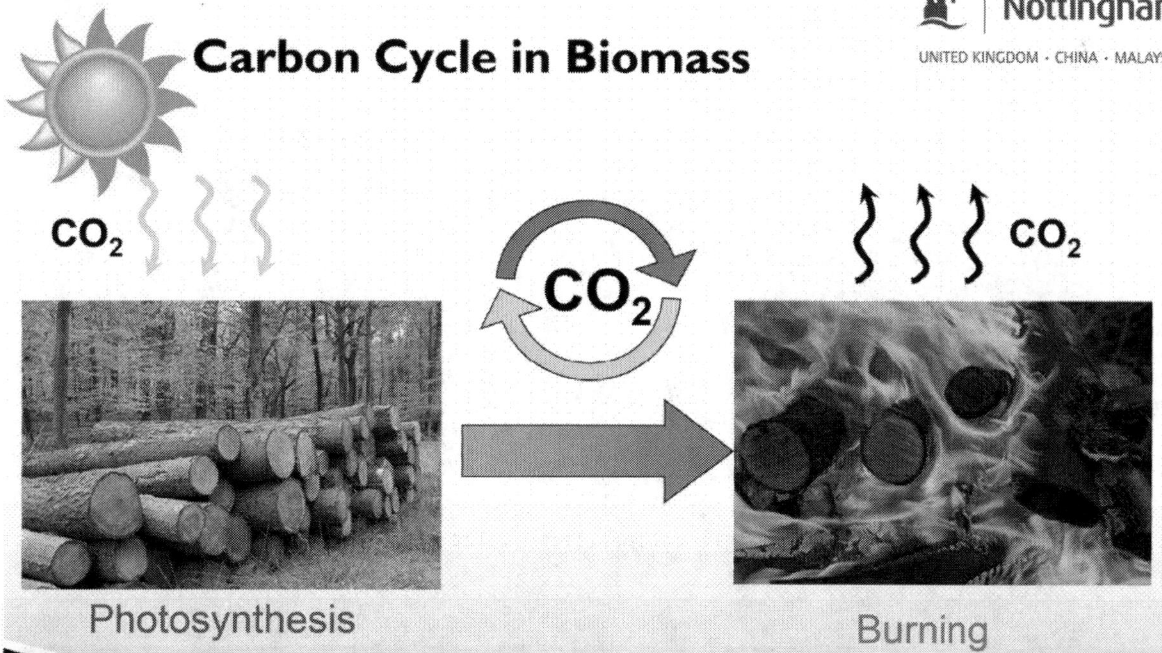

4

The University of Nottingham
UNITED KINGDOM · CHINA · MALAYSIA

BIOMASS
is not entirely **CARBON**-neutral

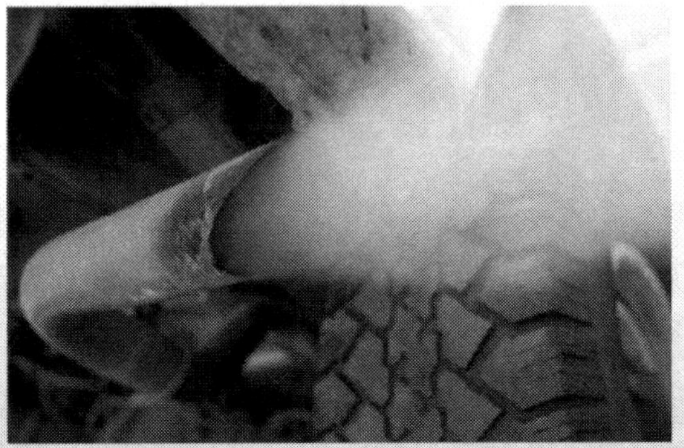

5

The University of Nottingham
UNITED KINGDOM · CHINA · MALAYSIA

CARBON FOOTPRINT (CFP)

6

The University of Nottingham
UNITED KINGDOM · CHINA · MALAYSIA

Question 2:

Where are those GREEN Sources?

7

The University of Nottingham
UNITED KINGDOM · CHINA · MALAYSIA

POTENTIAL BIOMASS

8

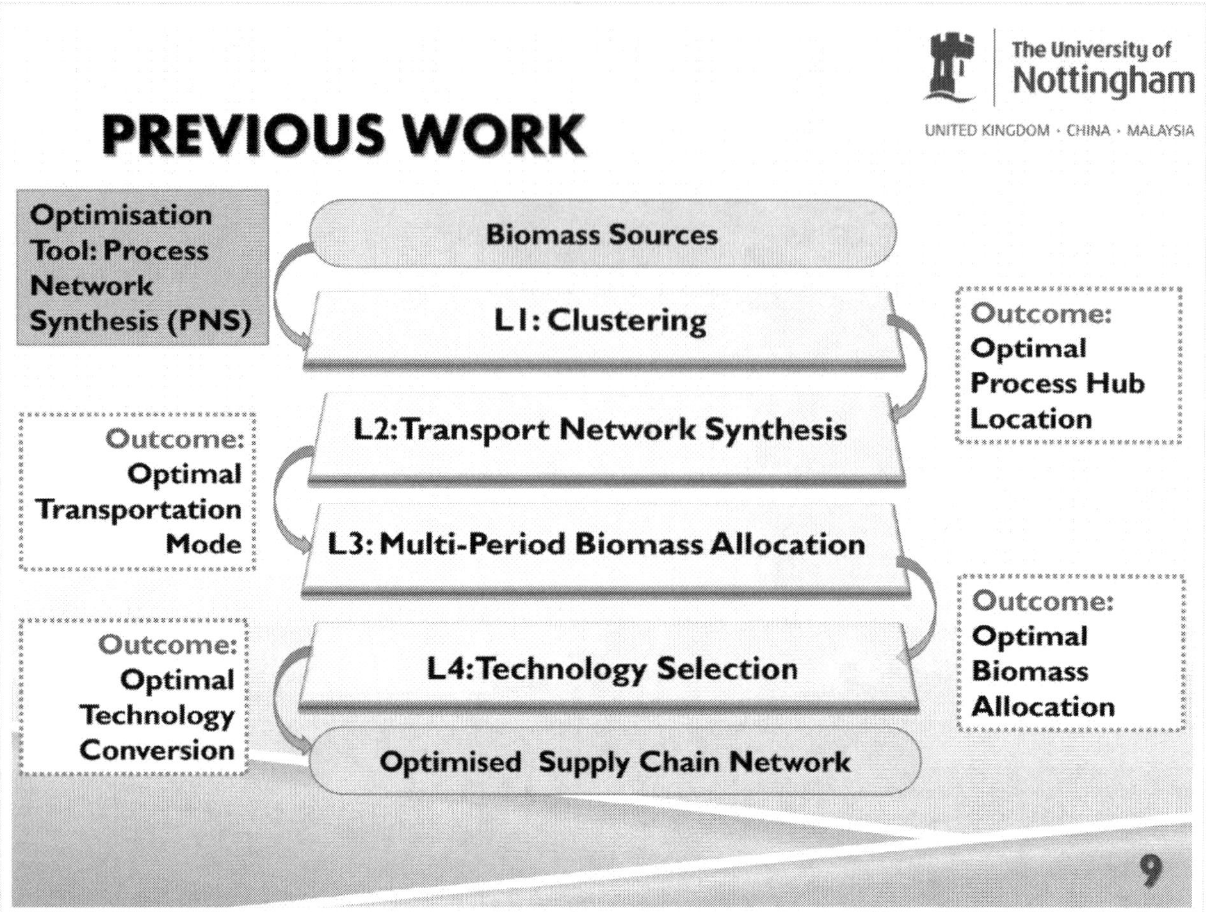
The University of Nottingham
UNITED KINGDOM · CHINA · MALAYSIA
PREVIOUS WORK
Optimisation Tool: Process Network Synthesis (PNS)
Biomass Sources
L1: Clustering
Outcome: Optimal Process Hub Location
L2:Transport Network Synthesis
Outcome: Optimal Transportation Mode
L3: Multi-Period Biomass Allocation
Outcome: Optimal Biomass Allocation
L4:Technology Selection
Outcome: Optimal Technology Conversion
Optimised Supply Chain Network
9

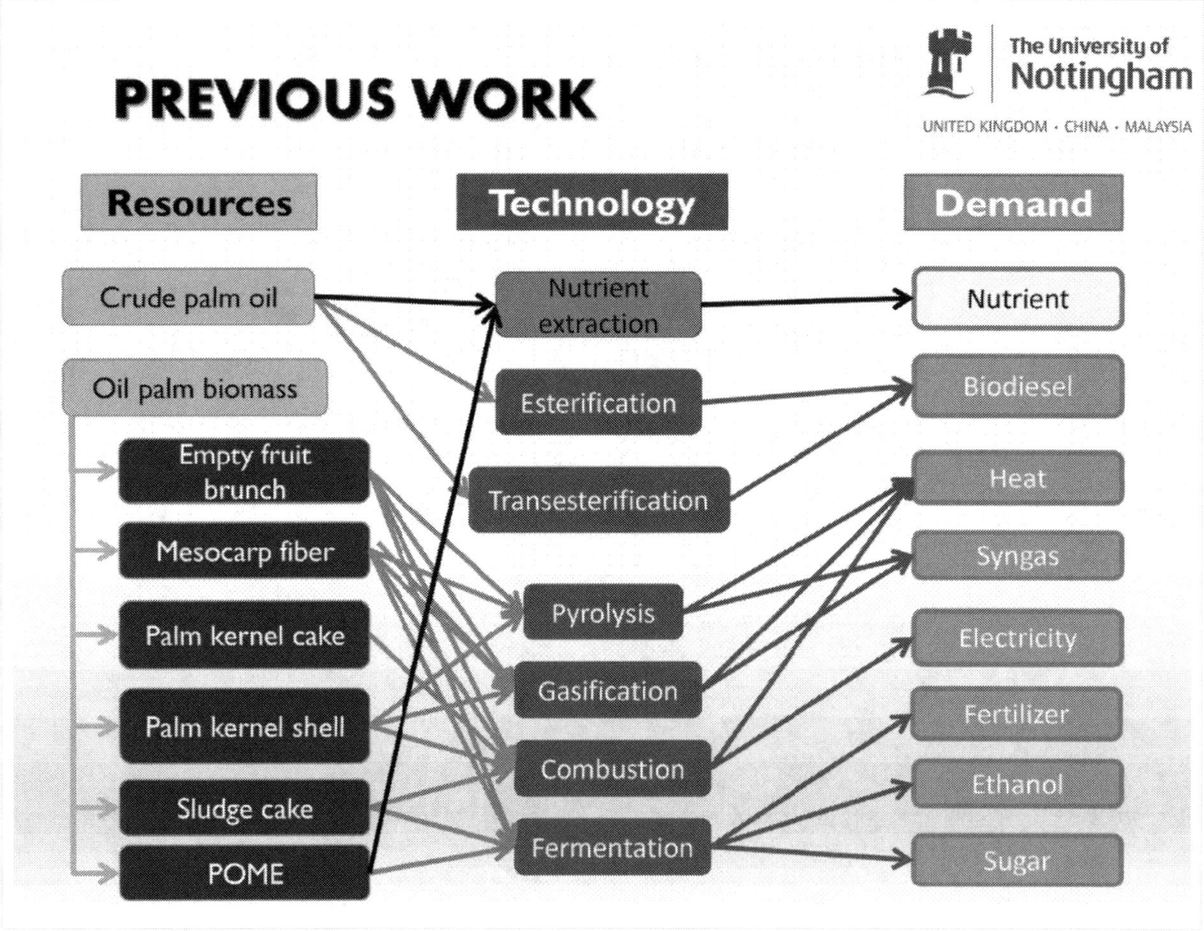
The University of Nottingham
UNITED KINGDOM · CHINA · MALAYSIA
PREVIOUS WORK
Resources
Technology
Demand
Crude palm oil
Oil palm biomass
Empty fruit brunch
Mesocarp fiber
Palm kernel cake
Palm kernel shell
Sludge cake
POME
Nutrient extraction
Esterification
Transesterification
Pyrolysis
Gasification
Combustion
Fermentation
Nutrient
Biodiesel
Heat
Syngas
Electricity
Fertilizer
Ethanol
Sugar

SUPERSTRUCTURE

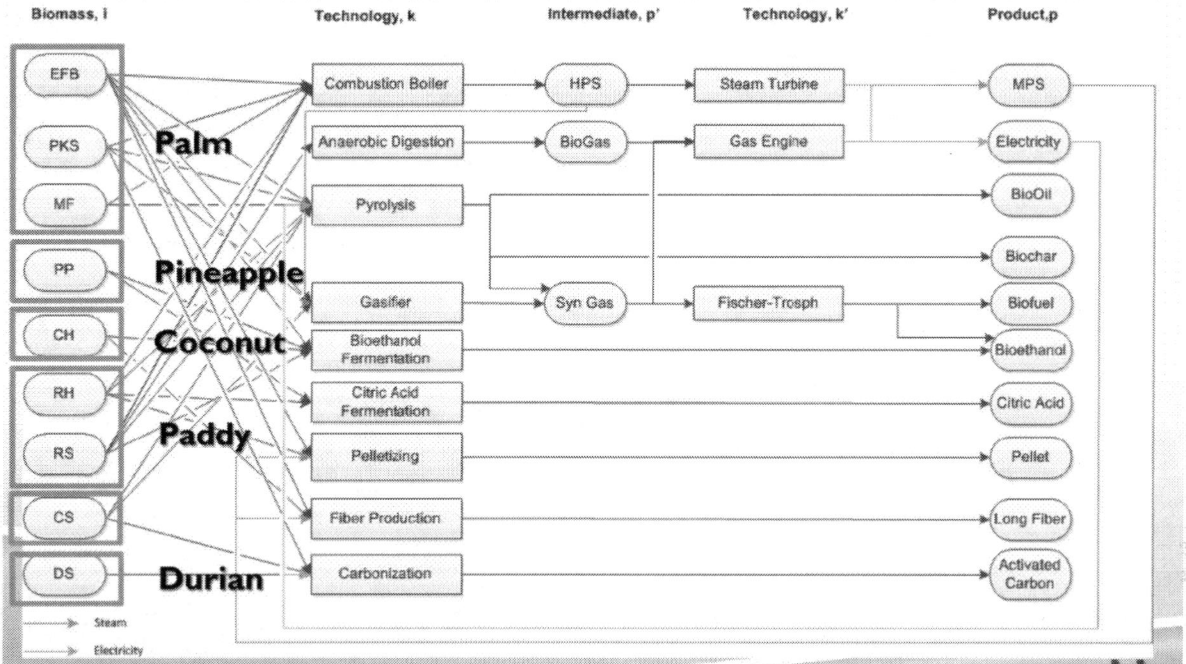

11

P-GRAPH APPROACH

The University of Nottingham
UNITED KINGDOM · CHINA · MALAYSIA

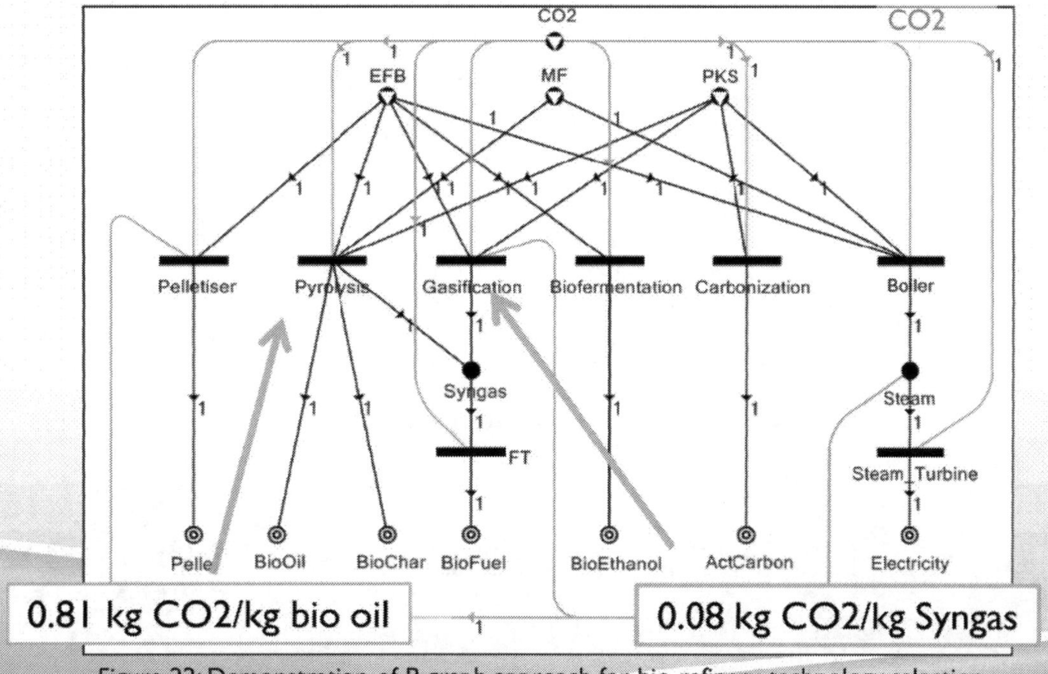

Figure 22: Demonstration of P-graph approach for bio-refinery technology selection.

12

MULTI-PERIOD BIOMASS ALLOCATION

Cluster	JAN	FEB	MAR	APR	MAY	JUN	JUL	AUG	SEP	OCT	NOV	DEC
1	L	L	L	L	H	H	H	H	M	M	M	M
	H	H	H	H	H		L	H	H	H		L
						H	H	M	L			
2	L	L	L	L	H	H	H	H	M	M	M	M
						H	H	M				
3	L	L	L	L	H	H	H	H	M	M	M	M
						H	H	M				
4	L	L	L	L	H	H	H	H	M	M	M	M
						H	H	M				

Legends: Palm Biomass | Paddy | Durian

13

The University of Nottingham
UNITED KINGDOM · CHINA · MALAYSIA

MULTI-PERIOD BIOMASS ALLOCATION P-GRAPH APPROACH

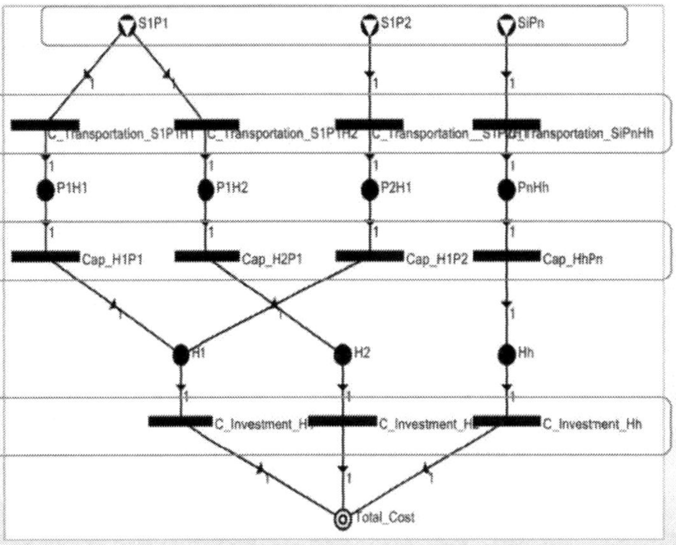

Figure 16: Multi-period Optimisation for Biomass Allocation

Biomass Sources of Period *n*

Transportation Cost (vary by distance)

Capacity Constraints of Processing Hub

Investment Cost of Processing Hub

Total Transportation and Investment Cost

14

The University of Nottingham
UNITED KINGDOM · CHINA · MALAYSIA

DEMAND-RESOURCES VALUE TARGETING

Scope: To optimise the biomass value

15

The University of Nottingham
UNITED KINGDOM · CHINA · MALAYSIA

Value Targeting

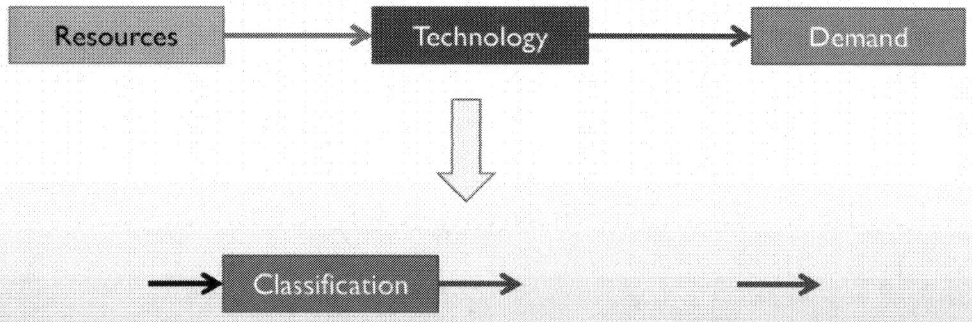

The University of Nottingham
UNITED KINGDOM · CHINA · MALAYSIA

Value Targeting

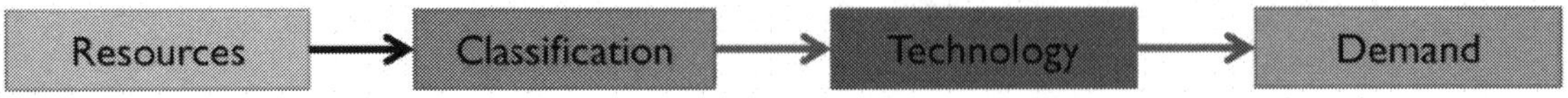

How does it work?

Classification of bio-resources based on the **study** of its **elements** and link to its **potential processing technology.**

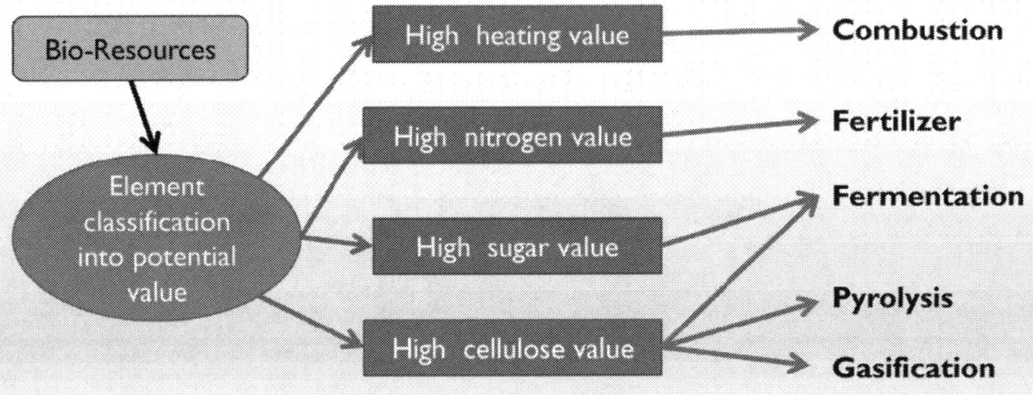

Element Targeting

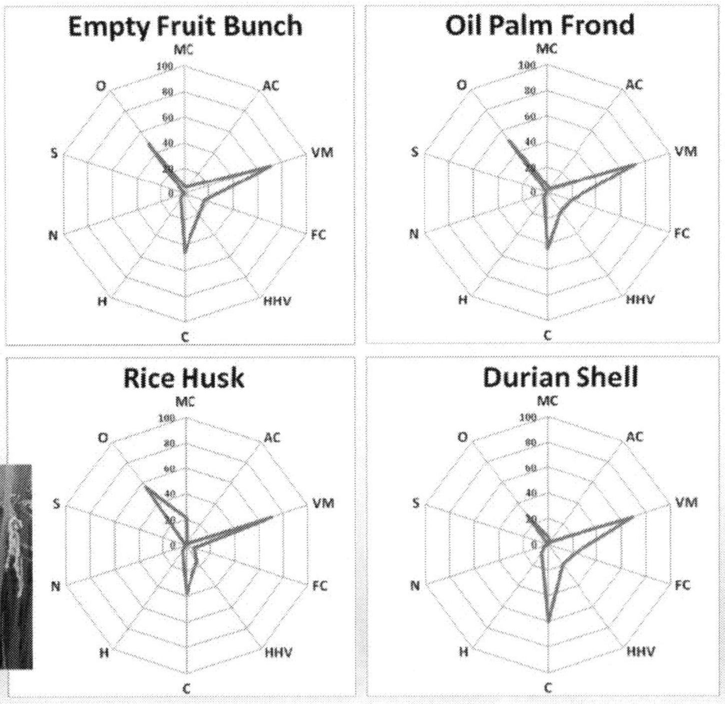

18

Element Targeting

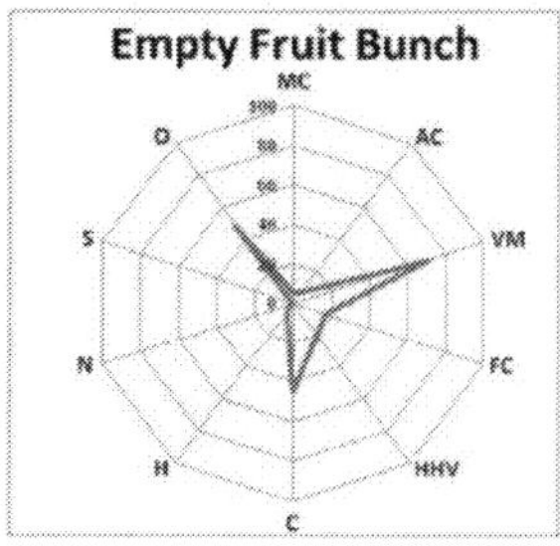

100% EFB
Yield: 1st run- **51.64%**
2nd run- **49.34%**

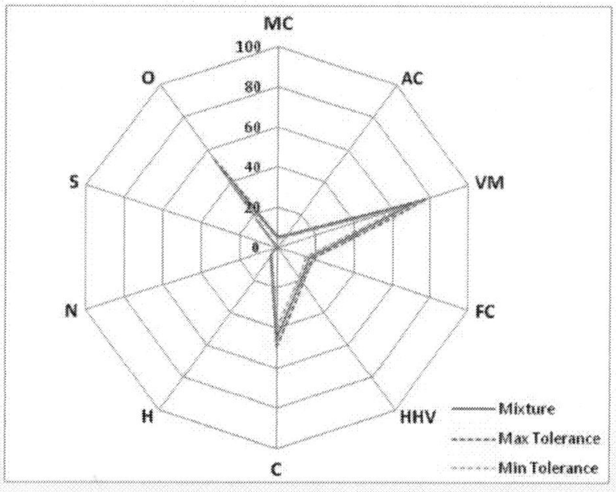

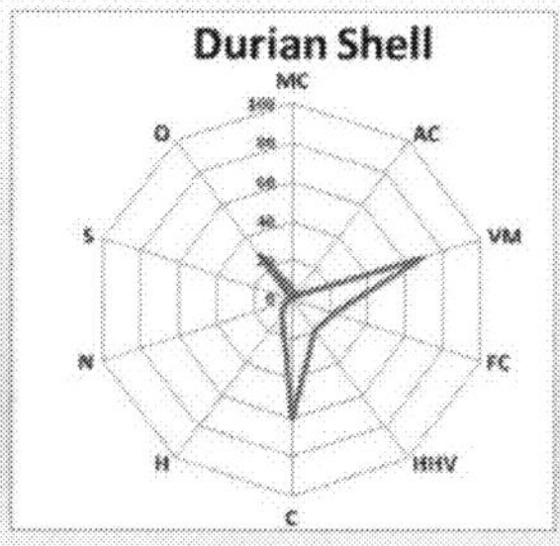

100% DS
Yield: 38.53%

96.6% EFB + 3.4% DS

Yield: 1st run- 38.97%
2nd run- 42.00%

19

The University of Nottingham
UNITED KINGDOM · CHINA · MALAYSIA

CLUSTERING

Scope: To determine the optimal process hub locations.

20

CLUSTERING

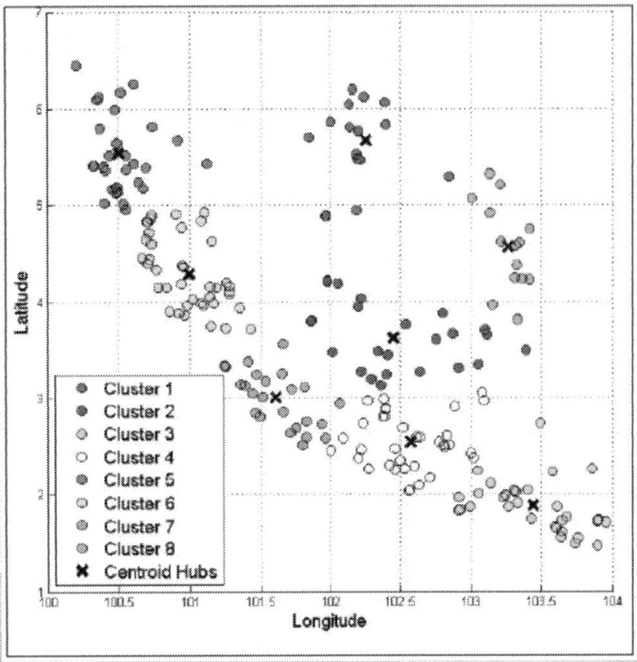

Figure 4: Clustered Sources through Kmeans

- Issues with Kmeans algorithm

lues or nature

  i. Based on Euclidean Distance calculation.
  ii. Does not consider the availabilities of sources as well as demand of destination points

cluster groups

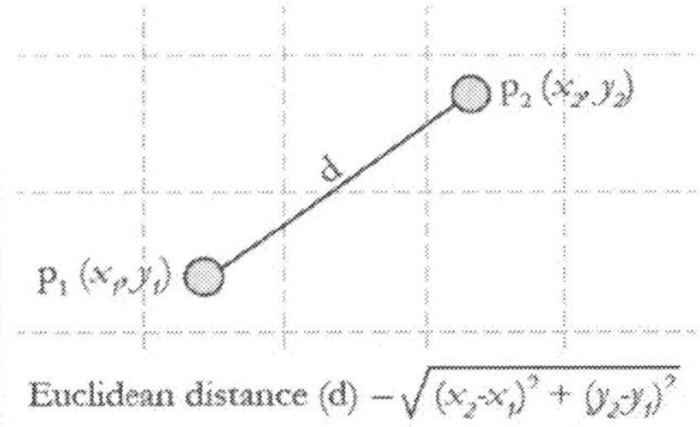

Figure 5: Euclidean Distance

21

The University of Nottingham
UNITED KINGDOM · CHINA · MALAYSIA

- Function
  i. Locate and optimise centroids based on cluster groups determined by Kmeans with consideration of availabilities and distances
- Features
  i. Utilises Google Distance Matrix API
  ii. Grid Reducing Algorithm to locate optimal centroids
  iii. Availability normalisation to range of 0 to 1

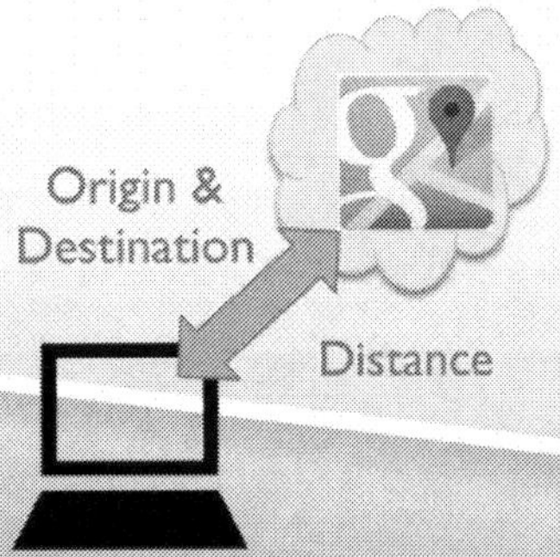

Figure 6: Google API Usage

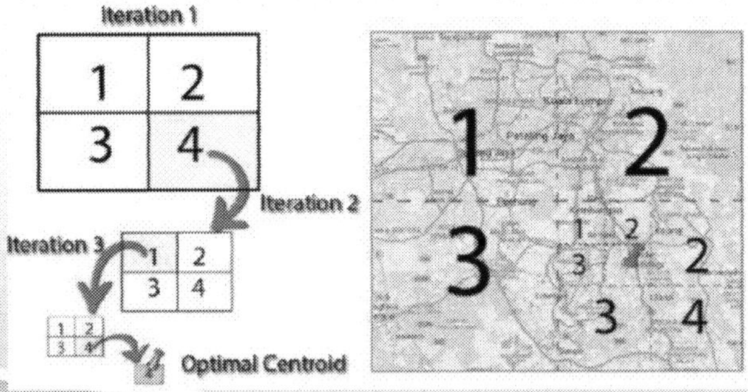

Figure 7: Grid Reducing Algorithm

22

The University of Nottingham
UNITED KINGDOM · CHINA · MALAYSIA

SCENARIO STUDY

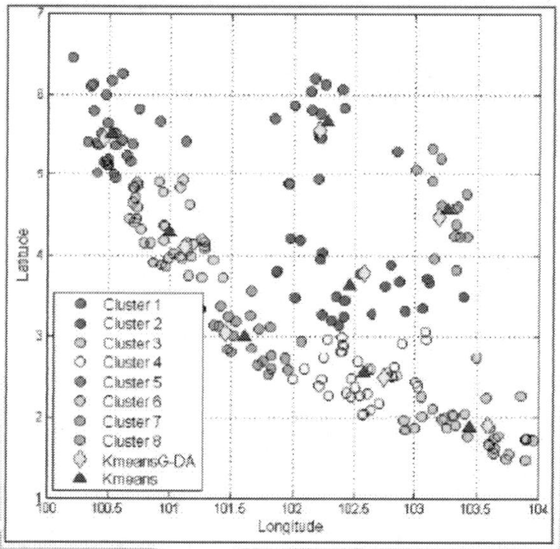

Figure 9a: Scenario 1a – KmeansG without Port Demands

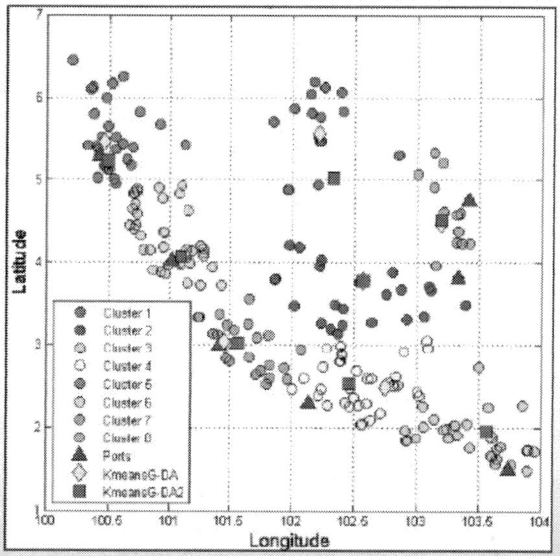

Figure 9b: Scenario 1b – KmeansG with Port Demands

23

The University of Nottingham
UNITED KINGDOM · CHINA · MALAYSIA

TRANSPORT NETWORK SYNTHESIS

Scope: To minimise total transportation cost for supply chain.

24

TRANSPORTATION COST

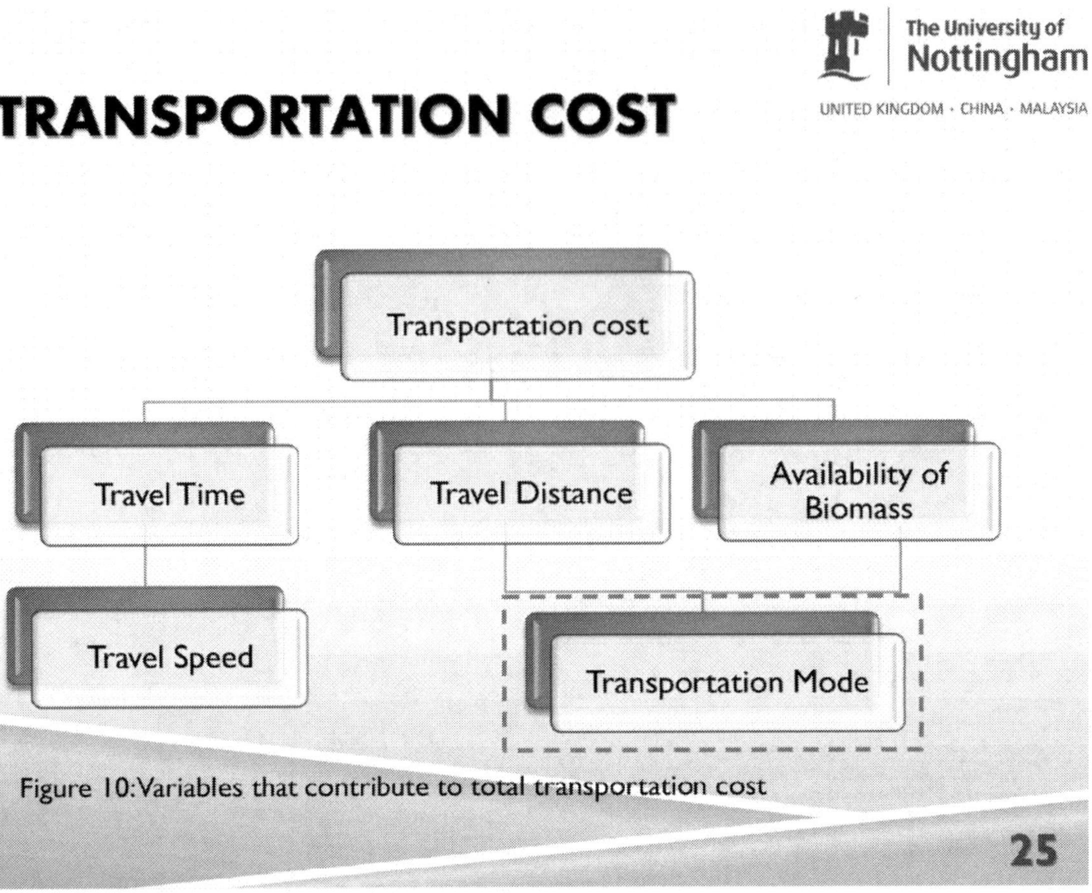

Figure 10: Variables that contribute to total transportation cost

TRANSPORT MODE: Biomass load vs. Distance

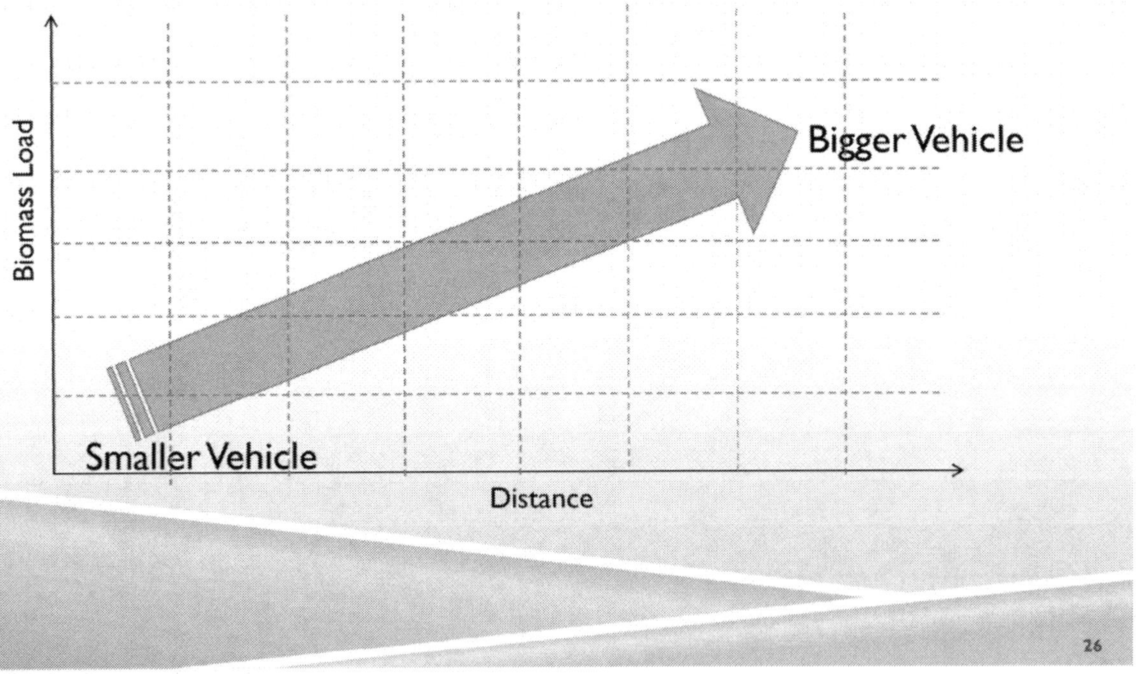

The University of Nottingham
UNITED KINGDOM · CHINA · MALAYSIA

SMART VEHICLE SELECTION (SVS) DIAGRAM

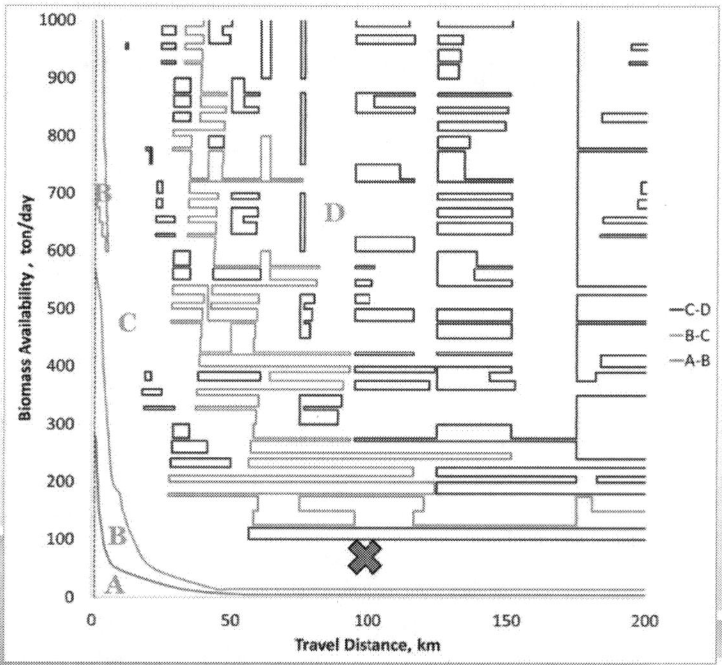

Figure 12: SVS Diagram

SVS Diagram is a smart tool generated to enable easy access to optimal transportation mode selection.

A – 5-ton lorry
B – 15-ton truck
C – 25-ton truck
D – 30-ton trailer

27

The University of Nottingham
UNITED KINGDOM · CHINA · MALAYSIA

UNCERTAINTY SCENARIO

- Four uncertainty are considered to investigate their effects on total transportation cost and vehicle type selection:
  i. Fuel price
  ii. Geographical condition (slope)
  iii. Weather
  iv. Traffic condition

Figure 13: Four major transport uncertainties considered in this research

28

The University of Nottingham
UNITED KINGDOM · CHINA · MALAYSIA

CONCLUSIONS & FUTURE WORK

- An integrated biomass supply chain towards Sustainable Development is very important .
- The uncertainties considered can help to improve the reliability and sensitivity of the biomass supply chain.
- Multivariate will be the keywords for the future

29

PRES'15 is the sixteenth in the series. It was initiated in collaboration with the late Professor Zdeněk Burianec in the framework of CHISA congresses. The events in the series are as follows:

Event	Year	Place	Country
1	1998	Prague	Czech Republic
2	1999	Budapest	Hungary
3	2000	Prague	Czech Republic
4	2001	Florence	Italy
5	2002	Prague	Czech Republic
6	2003	Hamilton, Ontario	Canada
7	2004	Prague	Czech Republic
8	2005	Giardini Naxos	Italy
9	2006	Prague	Czech Republic
10	2007	Ischia, Naples	Italy
11	2008	Prague	Czech Republic
12	2009	Rome	Italy
13	2010	Prague	Czech Republic
14	2011	Florence	Italy
15	2012	Prague	Czech Republic
16	2013	Rhodas	Greece
17	2014	Prague	Czech Republic

THE AIM

The aim of the PRES conference is to review the latest development and applications of process integration for energy conservation, pollution reduction and related topics. Industrial experience of the application of any available method is also welcome.

TOPICS

* Process integration for sustainable development
* Energy saving technology
* CO_2 minimisation and mitigation
* Combined heat & power, Combined cycles
* Heat exchangers as equipment and integrated items
* Integration of renewables, biomass and energy conversion technologies
* Integrated and multifunctional operations
* Operational research, supply chain and technology management
* Pulp & paper
* Clean technologies - Low emissions technologies
* Sustainable processing and production
* Waste minimisation, processing and management
* Thermal treatment of waste including waste to energy
* Dynamic, flexible and sustainable plant operation
* Industrial & experimental studies
* Industrial application & optimal design
* e-learning, e-teaching and e-knowledge
* CFD and numerical heat transfer
* Sustainable biofuel production
* Sustainable Green Palm Oil Transformation
* Green Asia Move

ABSTRACTS SUBMISSION

Authors wishing to present a contribution are asked to prepare a 300 words abstract (one page). Please indicate the topic, which the abstract belongs to and submit it electronically to the Conference Secretariat.

FORMAT

The Conference technical program is structured in two parallel sessions. Besides invited lectures; papers submitted will be considered for oral/poster presentations.

PROCEEDINGS

The full set of accepted papers (in the form of short papers of 6 pages) will be published in a Proceedings volume in Chemical Engineering Transactions **www.aidic.it/cet** (indexed by SCOPUS and TSI), which will be distributed at the PRES' 15 Conference desk.

PUBLICATION POLICY

The selected contribution will be developed and published in special thematic issues as propose:

* **Clean Technologies and Environmental Policy**
  Editor-in-Chief: Dr. Subhas K. Sikdar, National Risk Management Research Laboratory, United States Environment al Protection Agency, Cincinnati, USA

* **Applied Thermal Engineering**
  Editor-in-Chief: Prof. David A Reay, DRA Ltd, Tyne and Wear, UK

* **Journal of Cleaner Production**
  General Editor: Prof. Don Huisingh, Center for Cleaner Products and Clean Technologies, University of Tennessee, Knoxville, USA

* **ENERGY - The International Journal**
  Editor-in-Chief: Prof. Henrik Lund, Aalborg University, DK

IMPORTNT DATES

Abstract submission	**: 30 Nov 2014**
Full manuscript	**: 28 Feb 2015**
Early bird fees	**: 31 May 2015**

ORGANISER & SECRETARIAT

General Enquiry and Secretariat
Email: PRES15@PRESconference.com

Dr Hon Loong LAM (Organization Chair)
Centre of Excellence for Green Technologies,
University of Nottingham Malaysia, Broga Road,
43500 Semenyih, Selangor, MALAYSIA
Phone: +60-3-89248716
Fax: +60-3-89248017
Email: Lam@PRESconference.com

Dr Petar VARBANOV
(Scientific Programme Executive Secretary)
Phone: +36-88-421664
Fax: +44 871 244 774
Email: PRES_Secretary@cpi.uni-pannon.hu

The University of Nottingham

UNITED KINGDOM · CHINA · MALAYSIA

THANK YOU

35

Environmental Sustainability Assessment of a Multifunctional Process: Municipal Solid Waste Incineration in the Iberian Peninsula

Margallo M, Aldaco R, Irabien A

Department of Chemical and Biomolecular Engineering
University of Cantabria

Departamento de Ingenierías Química y Biomolecular

UC UNIVERSIDAD DE CANTABRIA

Balantonfüred, Hungary
May 2015

Contents

1 INTRODUCTION

2 METHODOLOGY AND RESULTS

3 CONCLUSIONS

1. Introduction: Waste generation and management

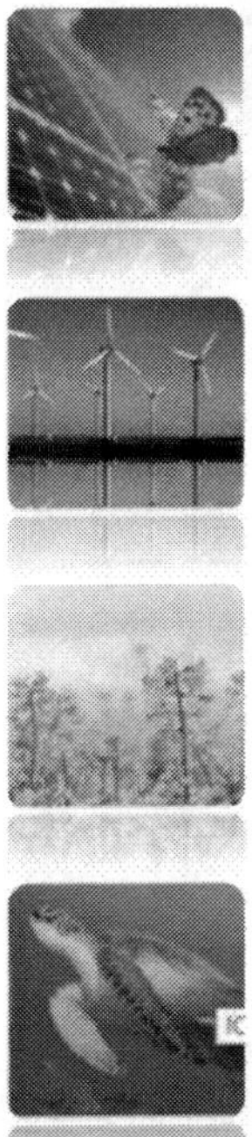

Waste generation in Europe has increased regularly in recent years, amounting to more than 2 billion t waste in 2012. Municipal Solid Wastes (MSW) make up 10 % of total waste production in Europe, representing a generation of 492 kg MSW person^{-1}

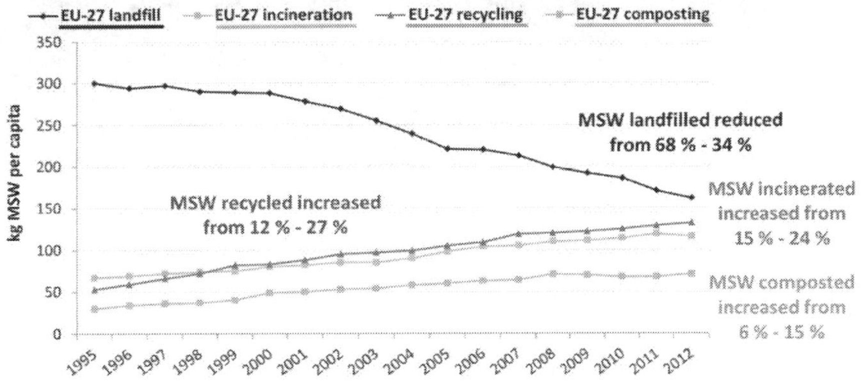

Landfill reduction

- Directive packaging and packaging waste sets recovery of packaging of 60% by 2008 [1]
- Directive landfill requires to reduce amount biodegradable MSW landfilled to 35 % by 2016 [2]

EUROSTAT 2013

[1] *Directive 94/64/EC* [2] *Directive 1999/31/EC*

3

1. Introduction: Waste generation and management

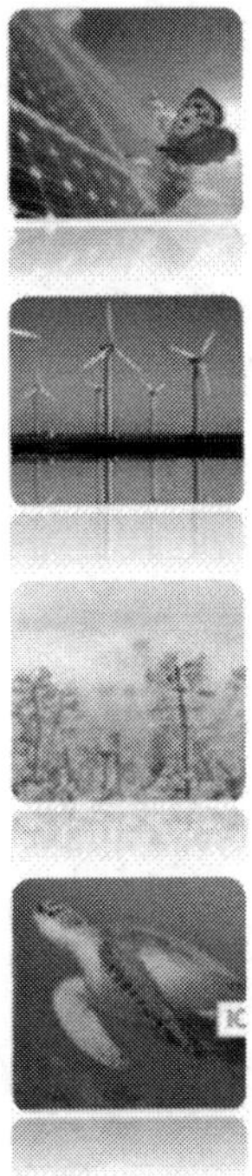

Waste treatment: incineration

Any technical unit and equipment used to burn waste and recover the combustion heat generated [1]

Incineration is used as a treatment for a wide range of wastes

Rapid technological development incineration sector over the last 10 to 15 years driven by specific legislation to reduce emissions to air

[1] *Directive 2000/76/EC waste incineration*

4

1. Introduction: Waste generation and management

Advantages

Reduction of mass and volume of residues and the recovery of the energy content in that waste with a significant heating value

Disadvantages

It has gained a bad reputation owing to the environmental impact specifically emissions of ultratoxic pollutants

5

1. Introduction: Waste generation and management

To assess the advantages, drawbacks, and environmental performance of waste incineration the use of environmental tools is essential

Life cycle assessment (LCA) is a powerfull methodology to assess the environmetal impacts from cradle to grave

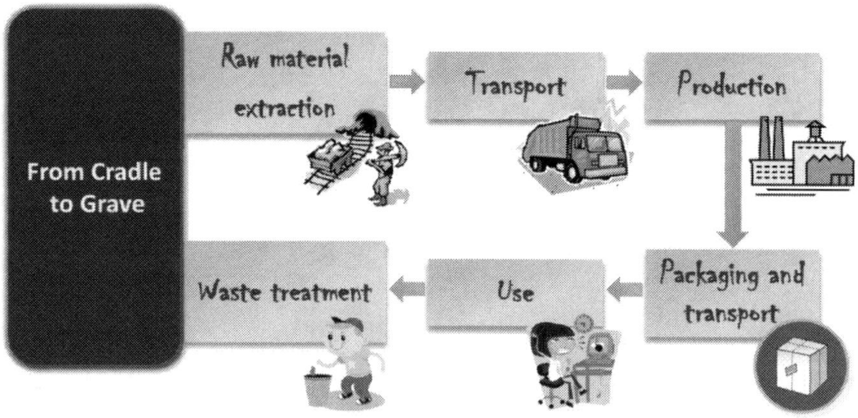

ISO 14040 2006

6

1. Introduction: Objetives

Main Objective

Assess and compare the environmental impacts WtE process in the Iberian Peninsula from a life cycle approach

Specific Objectives

- Review the **state of the art** of **WtE process** in **Spain** and **Portugal**
- Develop a **life cycle model** of **WtE** in the **Iberian Peninsula** to assess and compare the environemtal impacts WtE process
- Development of **environmental sustainability assessment (ESA)** methodology
- **Application** of the life cyle model and ESA methodology to WtE plants

7

Contents

INTRODUCTION

METHODOLOGY AND RESULTS

2.1 State of the art of WtE process

2.2 Life cycle model of WtE process

2.3 Environmental assessment of WtE plants in Spain

CONCLUSIONS

2.1 State of the art of WtE process

WtE plant structure

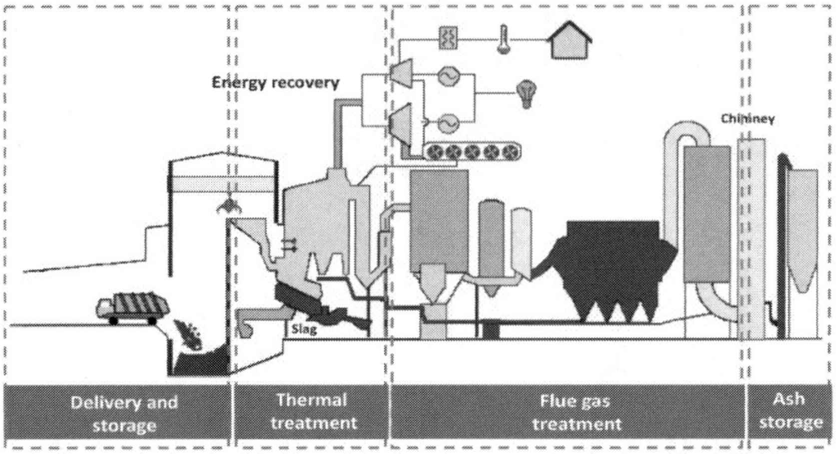

9

2.1 State of the art of WtE process

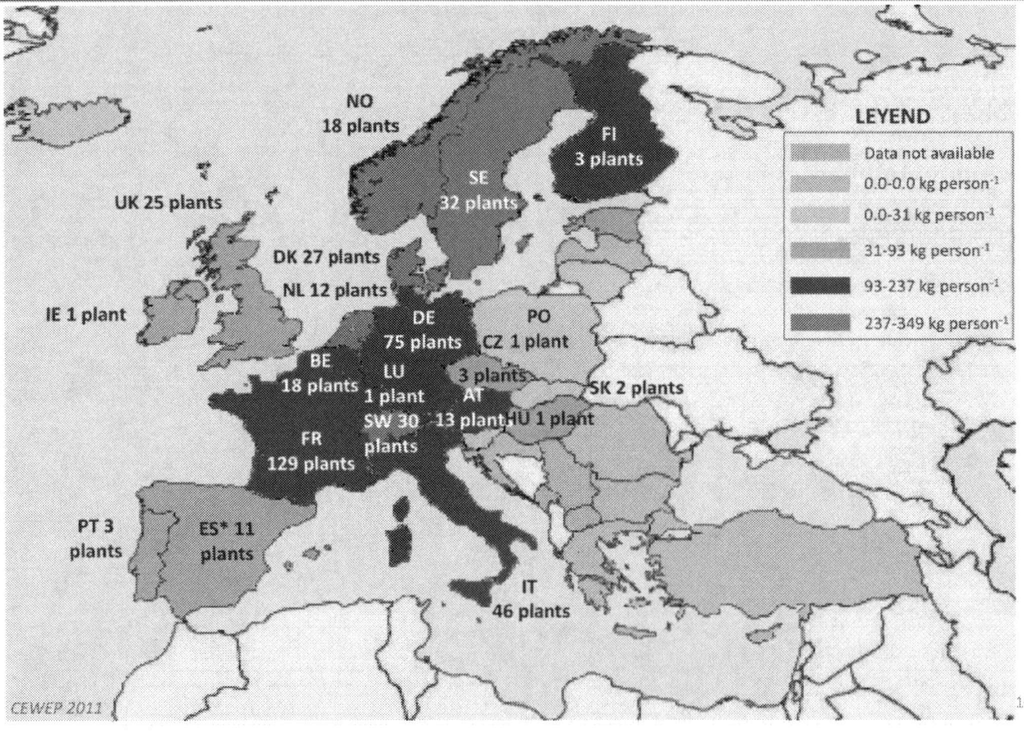

10

2.1 State of the art of WtE process

2.1 State of the art of WtE process

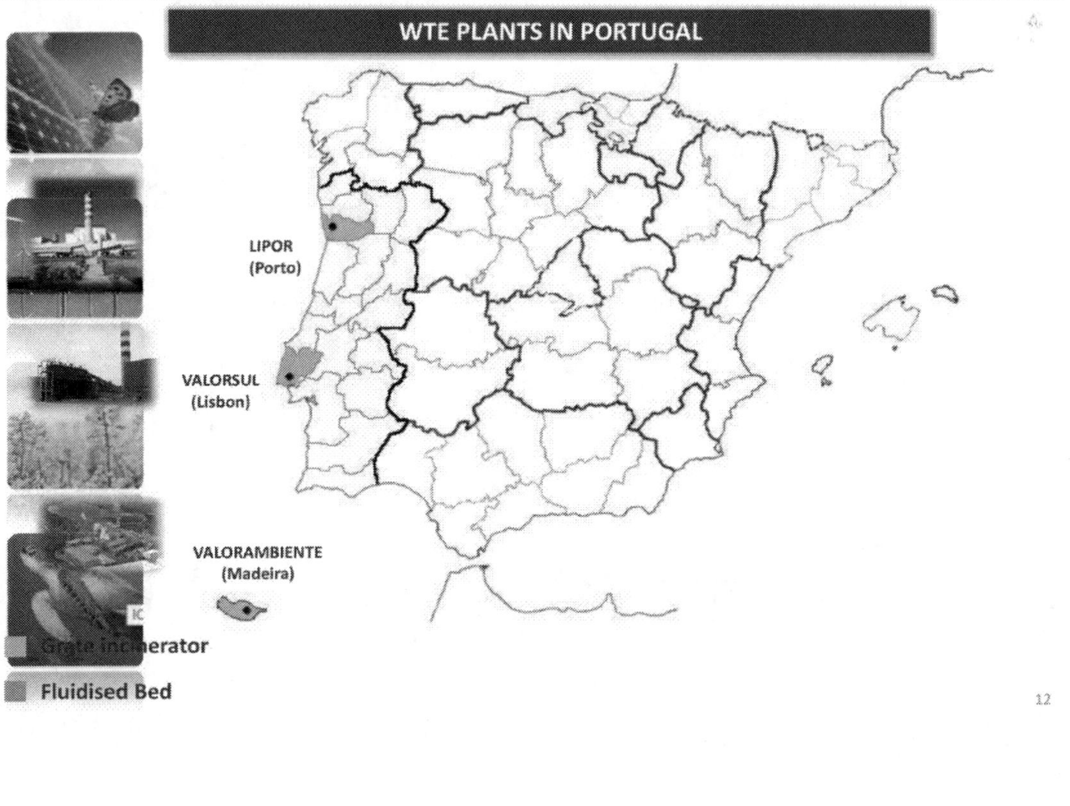

2.2 Life Cycle model of WtE process

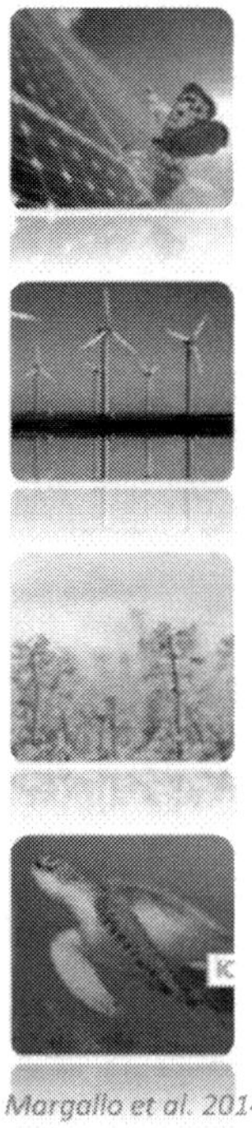

Develop a life cycle model representative of WtE plants in Spain and Portugal

Incineration is a multifunctional process where several inputs -the different input waste fraction- and outputs - energy recovery, ash, slag and scrap generation and emissions- coexist

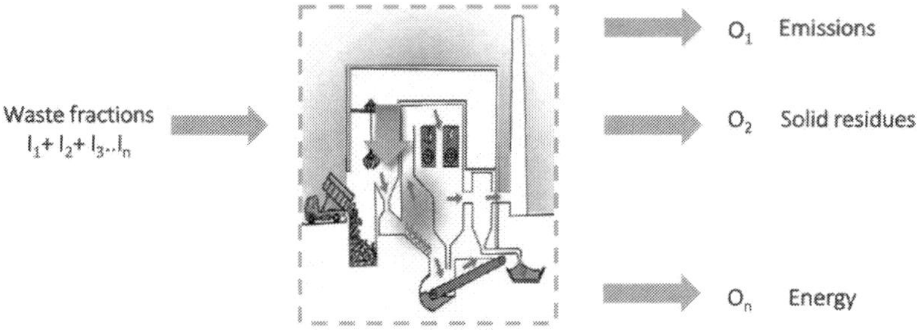

Margallo et al. 2014. Waste Manage Res

12

2.2 Life Cycle model of WtE process

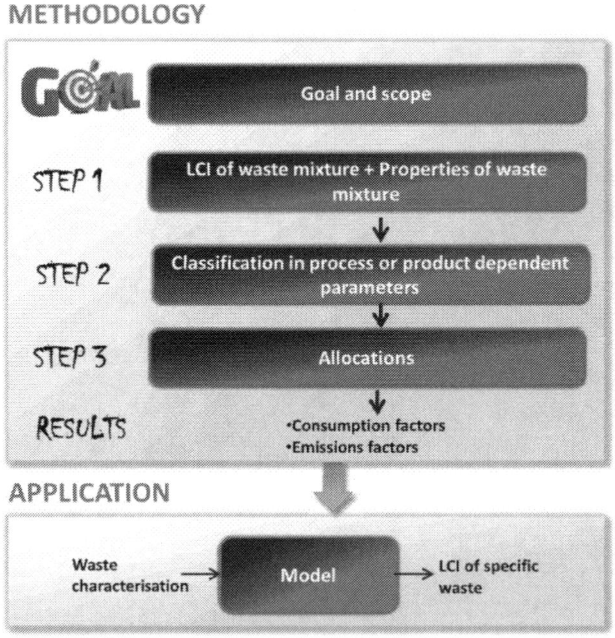

Margallo et al. 2014. Waste Manage Res

12

2.2 Life Cycle model of WtE process

GOAL AND SCOPE

System description

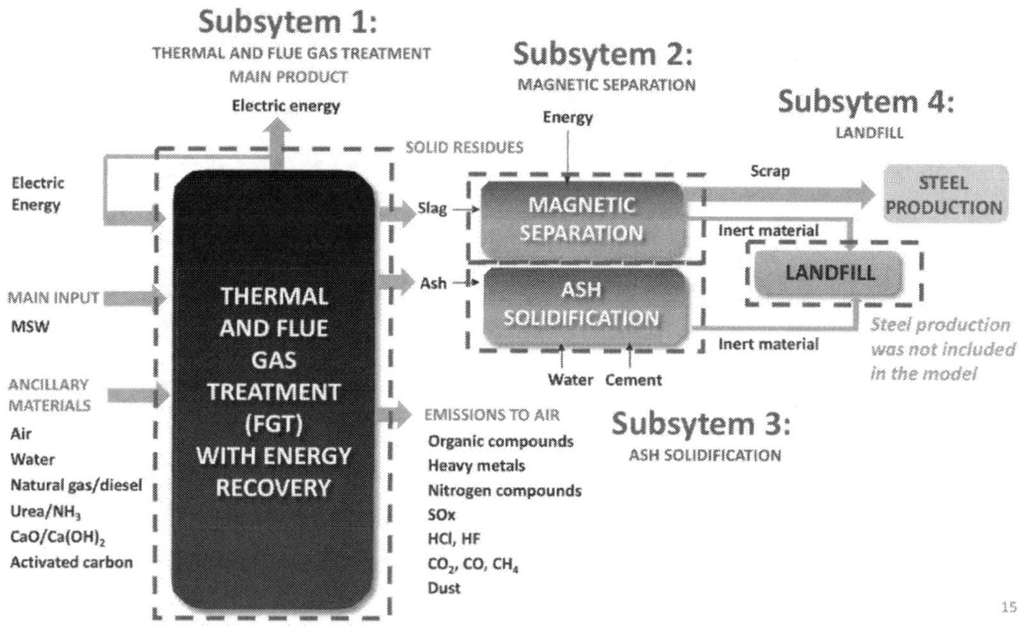

15

2.2 Life Cycle model of WtE process

GOAL AND SCOPE

Functional unit: 1 tonne of MSW incinerated in Spain and Portugal in 2009. The 18 waste fractions that composed MSW were considered

Assumptions

Thermal treatment and FGT were modelled a single process → there was not information composition and flow of FG before its treatment

FB was modeled as if it were a grate: lack information of FB and they only represent the 15 % in the Iberian Peninsula

Due to the lack of significance, emissions of some pollutants were excluded

Only 1 WtE plant had emissions of **hexachlorobenzene, endrin, heptachlor, naphthalene, pentachlorobenzene, tetrachloroethylene, tetrachloromethane, trichlorobenzene, benzene ethylene oxide, vinyl oxide and HCN**

Heavy metals are significant for toxicity assessment, most representative emissions were considered: **As, Cd, Cr, Cu, Mn, Hg, Ni, Pb and Zn**

16

2.2 Life Cycle model of WtE process

GOAL AND SCOPE: Data Source and LCI

	DATA SOURCE	TIMEFRAME	GEOGRAPHY
ECOEMBES	Tons of waste incinerated	2008-2009	Spain
lipor valorsul	Tons of waste incinerated	N.A	Portugal
AEVERSU	HHV, moisture, S content	N.A	Europe
	Energy production, generation of ash and slag	2008-2009	Spain
IPPC	Consumption of energy, fuels, ancillary materials and reagents	2004-2006	Spain/Portugal
PRTR	Emissions to air	2008-2009	Spain/Portugal
[1]	Total carbon and fossil carbon content of waste fraction	2009	Europe
ecoinvent [2]	Content of heavy metals, Cl_2, F_2 of waste fraction	2000	Switzerland
NORDTEST [3]	Consumption of energy, water, cement ash solidification and energy slag separation	2000-2001	Europe

[1] Riber et al. 2009; [2] Doka 2003; [3] Bjarnadóttir et al. 2002

17

2.2 Life Cycle model of WtE process

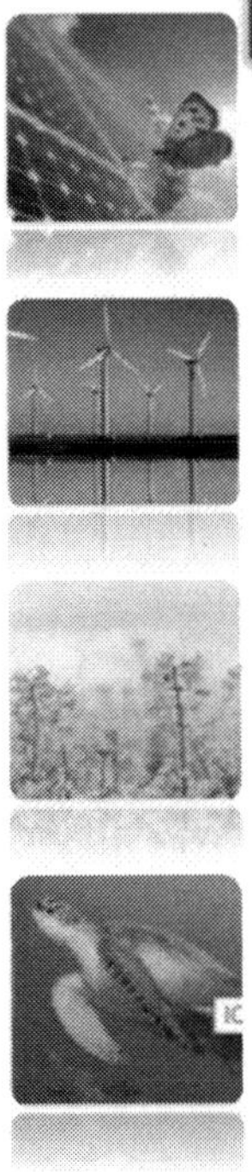

MODEL DESCRIPTION: Type allocation

Process dependent

Ancillary materials, fuels, reagents for FGT →
- Diesel, natural gas
- Air, water
- Urea, NH_3, CaO, Ca(OH)2, activated C

Solid residues →
- Bottom ash, scrap
- Fly ash

Products →
- Energy

Process dependent parameters

Emissions to air →
- NO_x, PCDD/F, TSP, PM_{10}
- CO_2
- CO, CH_4, NMVOC, TOC, PAHs
- HCl
- HF
- Metals and metalloids
- SO_x

Product dependent parameters

Product dependent

15

2.2 Life Cycle model of WtE process

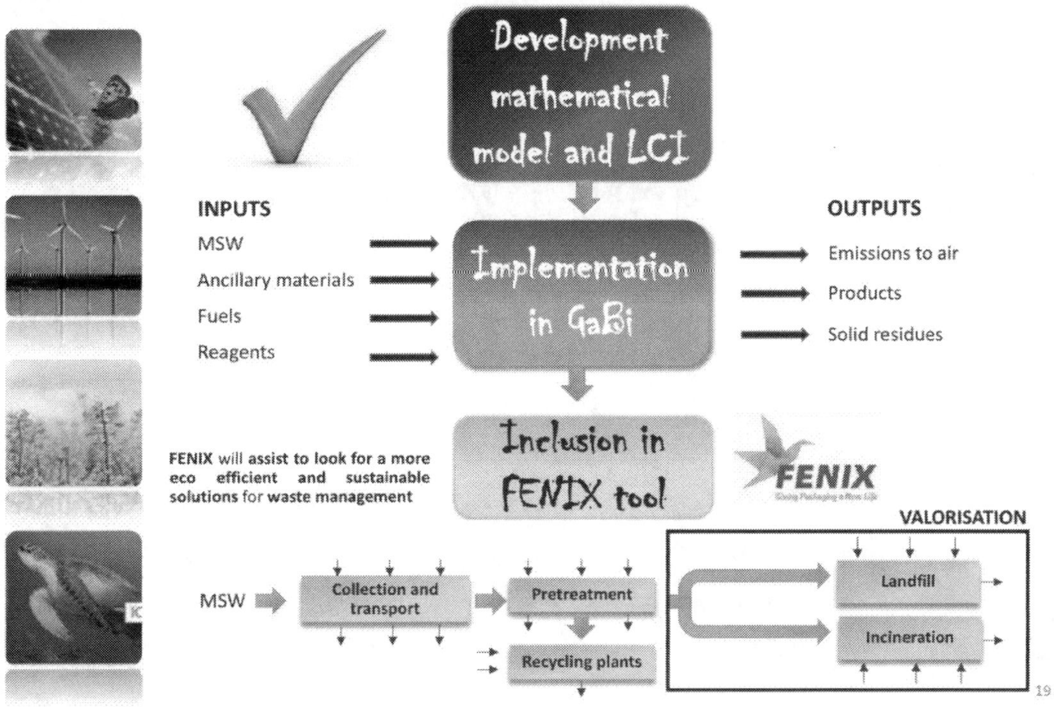

19

2.3 Environmental assessment of WtE plants in Spain

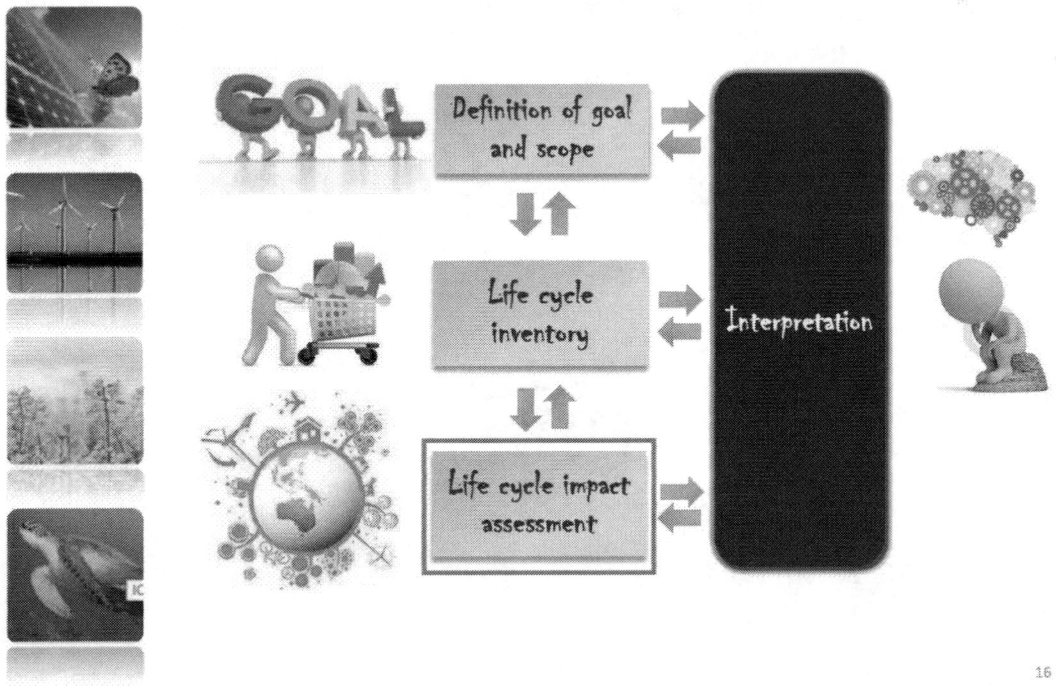

16

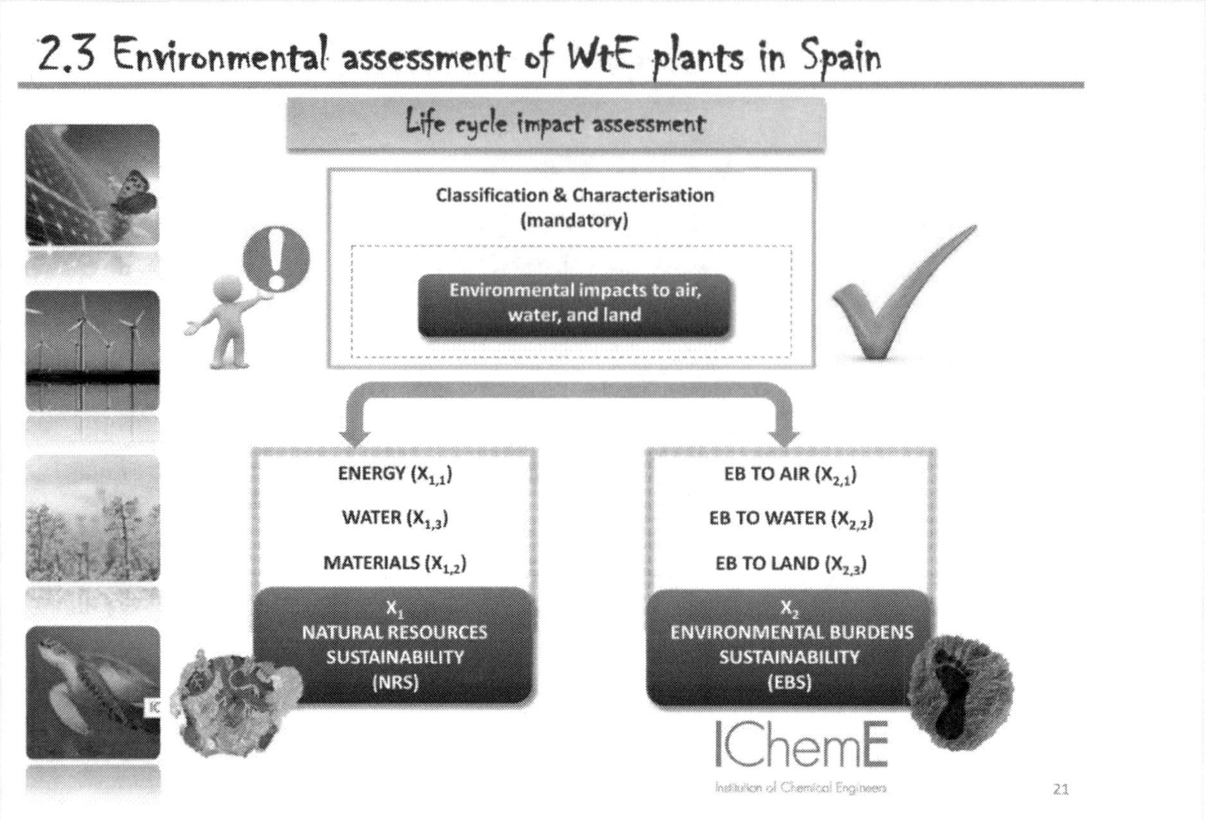
2.3 Environmental assessment of WtE plants in Spain
Life cycle impact assessment
Classification & Characterisation (mandatory)
Environmental impacts to air, water, and land
ENERGY ($X_{1,1}$)
WATER ($X_{1,3}$)
MATERIALS ($X_{1,2}$)
X_1 NATURAL RESOURCES SUSTAINABILITY (NRS)
EB TO AIR ($X_{2,1}$)
EB TO WATER ($X_{2,2}$)
EB TO LAND ($X_{2,3}$)
X_2 ENVIRONMENTAL BURDENS SUSTAINABILITY (EBS)
IChemE
Institution of Chemical Engineers
21

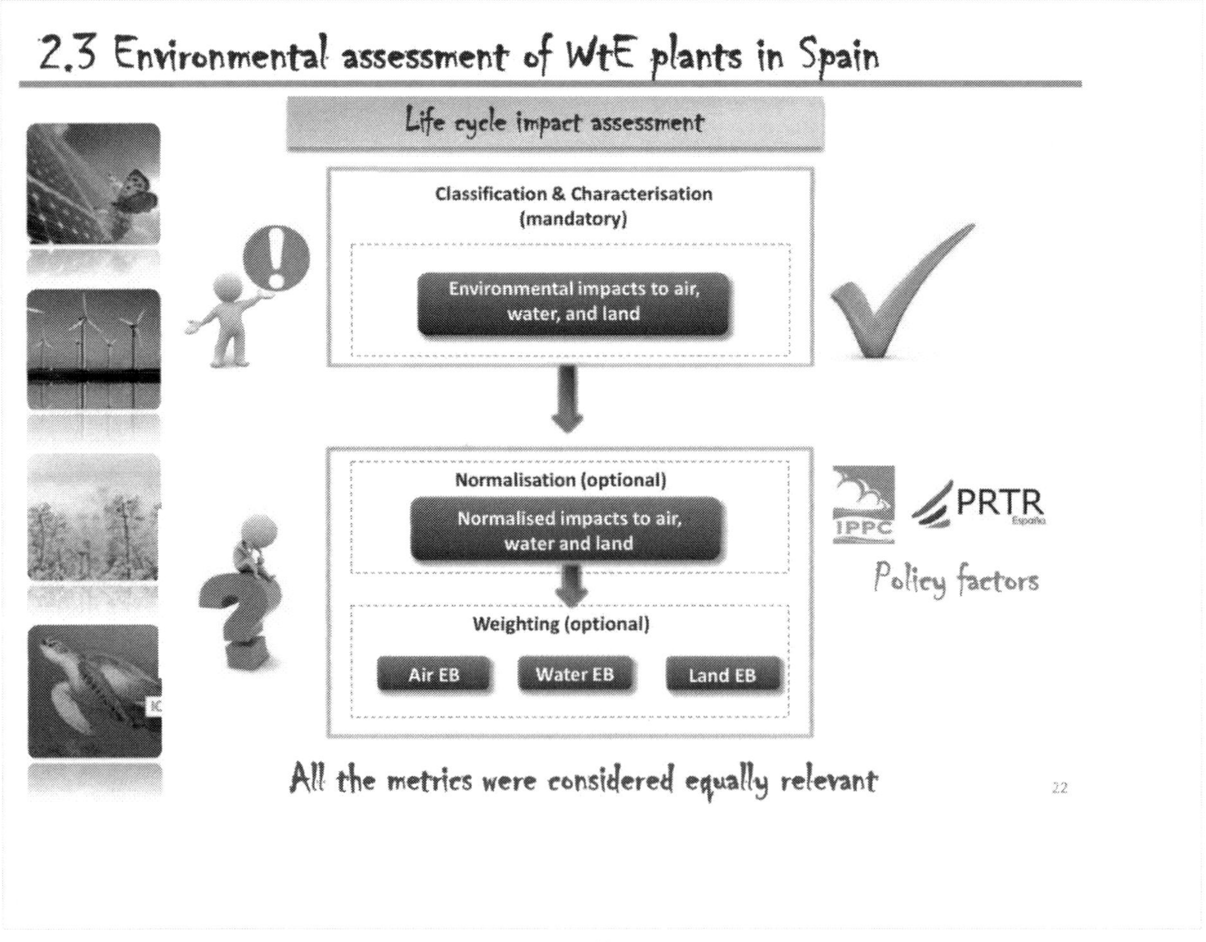
2.3 Environmental assessment of WtE plants in Spain
Life cycle impact assessment
Classification & Characterisation (mandatory)
Environmental impacts to air, water, and land
Normalisation (optional)
Normalised impacts to air, water and land
Weighting (optional)
Air EB
Water EB
Land EB
IPPC
PRTR España
Policy factors
All the metrics were considered equally relevant
22

2.3 Environmental assessment of WtE plants in Spain

NORMALISATION AND WEIGHTING

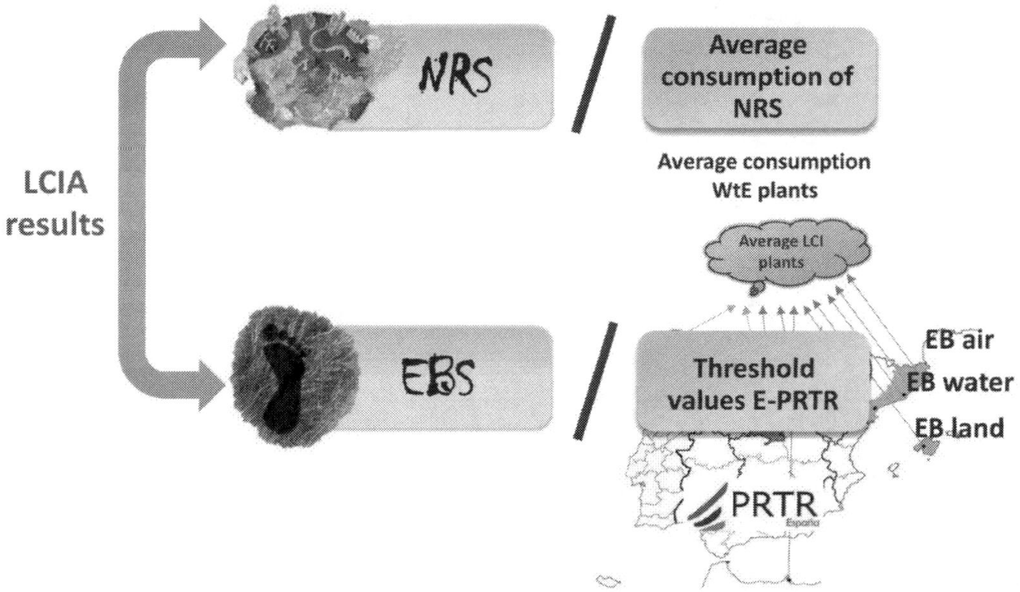

23

2.3 Environmental assessment of WtE plants in Spain

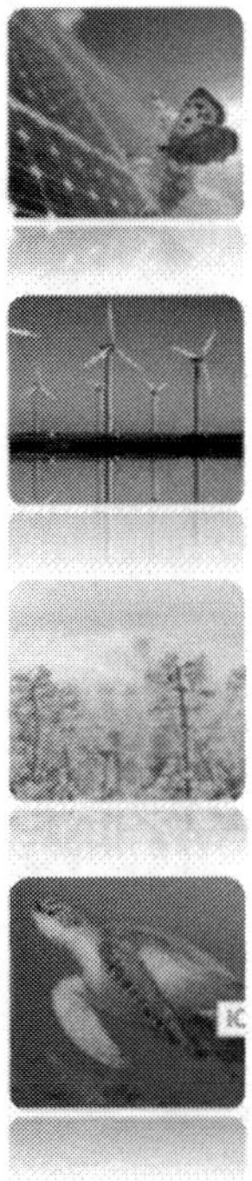

ESA was applied to 4 case studies, I_{S1}, I_{S2}, I_{S3}, I_{S4}, in order to assess the environmental performance of organic matter incineration in the plants

Data were the most representative from all the Spanish plants

		I_{S1}	I_{S2}	I_{S3}	I_{S4}
Start up year		1975	1994	1984	1991
Incineration capacity (t h^{-1})		14.5	10.0	2.50	9.60
Nº lines		3	2	2	2
Thermal treatment		Roller GI	Travelling GI	Reverse acting GI	Roller GI
Combustion temperature (ºC)		950	1,000-1,100	1,050	900-1,500
FGT treatment	NO_X	SNCR	SNCR	SNCR	SNCR
	Particles	Electro-filter	Bag filter	Electrofilter / bag filter	Bag filter
	Acid gases	Semidry	Semidry/dry	Dry	Semidry
FGT reagents	Activated C	✓	✓	✓	✓
	Urea	✓	✓	✓	
	NH_3				✓
	CaO	✓			✓
	$Ca(OH)_2$		✓	✓	

Same thermal treatment: GI

Organic matter was the most abundant waste flow in MSW

24

2.3 Environmental assessment of WtE plants in Spain

EBS RESULTS

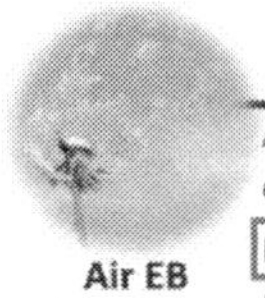
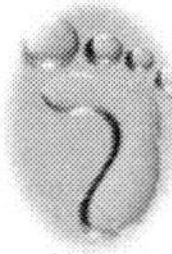

		Threshold value (kg y^{-1})	Factor	Normalised results (FU^{-1})				
				I_{S1}	I_{S2}	I_{S3}	I_{S4}	Spain average
Air EB	AA $X^*_{2,1,1}$	150,000	10^{-6}	-6.59	-7.58	-0.036	-2.38	-8.90
	GW $X^*_{2,1,2}$	100,000,000	10^{-6}	-1.48	-1.51	0.41	-0.58	-1.95
	HHE $X^*_{2,1,3}$	1,000	10^{-6}	9.86	281	44.7	4.26	-15.0
	POF $X^*_{2,1,4}$	1,000	10^{-6}	-68.9	-75.0	-10.4	-33.0	-96.8
	SOD $X^*_{2,1,5}$	1.00	10^{-6}	35.5	-41.8	-8.47	-18.0	-49.6
	Total EB to air ($X_{2,1}$)		**10^{-6}**	**-103**	**155**	**25.5**	**-49.6**	**-172**
Water EB	AOD $X^*_{2,2,1}$	50,000	10^{-6}	0.26	0.83	0.12	0.05	0.34
	AqA $X^*_{2,2,2}$	100	10^{-6}	0.01	0.02	0.01	0.01	0.02
	MEco $X^*_{2,2,3,1}$	50	10^{-6}	-0.21	-0.24	-0.03	-0.10	-0.28
	NMEco $X^*_{2,2,3,2}$	50	10^{-6}	-0.07	-0.08	-0.003	-0.03	-0.10
	EU $X^*_{2,2,4}$	5,000	10^{-6}	-18.1	-17.8	-3.35	-7.94	-26.1
	Total EB to water ($X_{2,2}$)		**10^{-6}**	**-0.18**	**0.36**	**0.07**	**-0.16**	**-0.28**
Land EB	HW $X^*_{2,3,1}$	2,000						
	NHW $X^*_{2,3,2}$	2,000,000	10^{-6}	113	208	123	136	154
	Total EB to land ($X_{2,3}$)		**10^{-6}**	**113**	**208**	**123**	**136**	**154**
Total EB	X_2		**10^{-6}**	**10.6**	**364**	**148**	**86.2**	**-18.5**

HHE showed a **positive value** due to great influence **emissions PCDD/F, heavy metals, dust**

AOD and AqA positive values. MEco control negatives values

I_{S2} **highest EB** due to larger amount of ash and slag generated

I_{S2} highest total EB

25

2.3 Environmental assessment of WtE plants in Spain

EBS RESULTS

There is a **great interdependence** between **NRS** and **EBS.** I_{S2} presented the **greatest consumption of NR and** the **highest total EB,** whereas I_{S1} displayed the **best environmental performance**

	Environmental Burden intensity (+ → -)					**Natural Resources intensity** (+ → -)			
Air $X_{2,1}$	I_{S2}	I_{S3}	I_{S4}	I_{S1}	**Energy $X_{1,1}$**	I_{S3}	I_{S4}	I_{S1}	I_{S2}
Water $X_{2,2}$	I_{S2}	I_{S3}	I_{S4}	I_{S1}	**Materials $X_{1,2}$**	I_{S2}	I_{S3}	I_{S1}	I_{S4}
Land $X_{2,3}$	I_{S2}	I_{S4}	I_{S3}	I_{S1}	**Water $X_{1,3}$**	I_{S4}	I_{S2}	I_{S1}	I_{S3}
Total X_2	I_{S2}	I_{S3}	I_{S4}	I_{S1}	**Total X_1**	I_{S2}	I_{S3}	I_{S4}	I_{S1}

26

2.3 Environmental assessment of WtE plants in Spain

NRS RESULTS

Negatives values, the plants are able to export much more energy than amount obtained from fuels

Normalised NRS

Dimensionless NRS variables		I_{S1}	I_{S2}	I_{S3}	I_{S4}	Spanish average
Energy	$X^*_{1,1}$	(-)0.74	(-)0.88	(-)0.14	(-)0.36	(-)1.00
Materials	$X^*_{1,2}$	0.83	1.48	0.72	0.58	1.00
Water	$X^*_{1,3}$	0.89	1.27	0.46	1.45	1.00
TOTAL ($X_1 = \frac{\gamma X^*_{1,1}}{3} + \frac{X^*_{1,2}}{3} + \frac{X^*_{1,3}}{3}$)	X_1	0.33	0.62	0.34	0.56	0.33

I_{S1} showed the lowest consumption of NR and I_{S2} the highest consumption

2.3 Environmental assessment of WtE plants in Spain

EBS RESULTS

There is a **great interdependence** between **NRS** and **EBS.** I_{S2} presented the **greatest consumption of NR and the highest total EB,** whereas I_{S1} displayed the **best environmental performance**

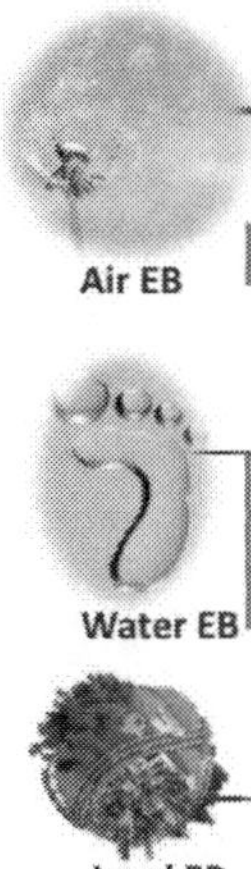

Air EB	AA $X^*_{2,1,1}$	150,000	10^{-6}	-6.59	-7.58	-0.036	-2.38	-8.90
	GW $X^*_{2,1,2}$	100,000,000	10^{-6}	-1.48	-1.51	0.41	-0.58	-1.95
	HHE $X^*_{2,1,3}$	1,000	10^{-6}	9.86	281	44.7	4.26	-15.0
	POF $X^*_{2,1,4}$	1,000	10^{-6}	-68.9	-75.0	-10.4	-33.0	-96.8
	SOD $X^*_{2,1,5}$	1.00	10^{-6}	35.5	-41.8	-8.47	-18.0	-49.6
	Total EB to air ($X_{2,1}$)		10^{-6}	**-103**	**155**	**25.5**	**-49.6**	**-172**
Water EB	AOD $X^*_{2,2,1}$	50,000	10^{-6}	0.26	0.83	0.12	0.05	0.34
	AqA $X^*_{2,2,2}$	[illegible]	[illegible]	[illegible]01	0.02	0.01	0.01	[illegible]
	MEco $X^*_{2,2,3,1}$	[illegible]	[illegible]	[illegible]21	-0.24	-0.03	-0.10	[illegible]
	NMEco $X^*_{2,2,3,2}$	[illegible]	[illegible]	[illegible]07	-0.08	-0.003	-0.03	[illegible]
	EU $X^*_{2,2,4}$	[illegible]	[illegible]	[illegible]8.1	-17.8	-3.35	-7.94	[illegible]
	Total EB to water ($X_{2,2}$)		[illegible]	[illegible].18	**0.36**	**0.07**	**-0.16**	[illegible]
Land EB	HW $X^*_{2,3,1}$	[illegible]	[illegible]	[illegible]	[illegible]	[illegible]	[illegible]	[illegible]
	NHW $X^*_{2,3,2}$	2,000,000	10^{-6}	113	208	123	136	154
	Total EB to land ($X_{2,3}$)		[illegible]	[illegible]13	**208**	**123**	**136**	[illegible]
Total EB	X_2		10^{-6}	10.6	364	148	86.2	-18.5

HHE showed a **positive value** due to great influence **emissions PCDD/F, heavy metals, dust**

Highest air impact I_{S2} and I_{S3}. EB air I_{S2} 6 times higher than I_{S3}

...ve values. MEco control negatives values ...st EB. ...gher than I_{S3} due to the greatest consumption of ... in I_{S2}

... highest amount of ... generated

Environmental Burden intensity					Natural Resources intensity				
Air $X_{2,1}$	I_{S2}	I_{S3}	I_{S4}	I_{S1}	Energy $X_{1,1}$	I_{S3}	I_{S4}	I_{S1}	I_{S2}
Water $X_{2,2}$	I_{S2}	I_{S3}	I_{S4}	I_{S1}	Materials $X_{1,2}$	I_{S2}	I_{S3}	I_{S1}	I_{S4}
Land $X_{2,3}$	I_{S2}	I_{S4}	I_{S3}	I_{S1}	Water $X_{1,3}$	I_{S4}	I_{S2}	I_{S1}	I_{S3}
Total X_2	I_{S2}	I_{S3}	I_{S4}	I_{S1}	Total X_1	I_{S2}	I_{S3}	I_{S4}	I_{S1}

I_{S2} highest total EB

Contents

1 INTRODUCTION

2 METHODOLOGY AND RESULTS

3 CONCLUSIONS

Conclusions

1 The **LCI** and **life cycle model** of WtE process in the Iberian Peninsula **were developed.** The **model** was **included** in an **ad-hoc software** and the **LCI** is part of a **specific database** on waste treatments **compatible with ILCD**

High heating value, water content, waste composition were critical factors

An **ESA methodology was developed** to analyse WtE plants. This procedure **allows to :** 2

- Reduce LCA complexity and assist the decision making process
- Help WtE plants to optimised the use of NR and reduce EB

3 The generation of **solid residues** in incineration supposed **the highest environmental impact,** therefore **its management is a critical issue**

23

Holistic Implementation of a Sustainable Energy Management Programme

- **Conference:** International Congress on Sustainability Science Engineering ICOSSE
- **Year:** 2015
- **Proceeding:** ICOSSE 2015
- **Group:** Abstract Submissions
- **Session:**

Session 12: Renewable Energy Technologies

- **Time:**

Thursday, May 28, 2015 - 3:20pm-3:50pm

Authors:

Manan, Z., Universiti Teknologi Malaysia
Wan Alwi, S. R., Universiti Teknologi Malaysia

Energy efficiency measures employed by organisations are typically focused on energy-saving measures that are largely technical by nature; aimed at containing the effects of rising costs. In the absence of holistic and systemic changes, these technical measures may not be sustainable. As a result, the benefits that can be derived from the piece-meal implementation energy conservation measures can be marginal. This paper highlights a holistic approach for implementing effective and sustainable energy management programme (SEMP) that is multi-disciplinary by nature and considers aspects of managerial, capacity building as well as technical. This entails the integration of the SEMP into an organisation's structure and practice in order to sustain energy efficiency practices and hence, cost savings. Holistic implementation of the SEMP in Universiti Teknologi Malaysia (UTM) that comprises of 22,000 students and 3000 staff enabled UTM to sustain an average of 15% energy savings in the last five years.

Hydrothermal carbonization of cow manure

Presented by Chuck Coronella, Ph.D., P.E.
With coauthors:
M. Toufiq, Reza, Sage Hiibel, Hongfei Lin, Mi Lu, Tianlin Song, of the University of Nevada, Reno
Alireza Shekarriz of Hydrotorr, LLC

ICOSSE, 28 May 2015, Balatonfüred, Hungary

HydroTORR
Agriculture | Environment | Energy

University of Nevada, Reno

University of Nevada, Reno

Comprehensive university- wide range of academic disciplines

Traditional practices of manure management need updating for contemporary dairies

- Sustainable practices suitable for large, modern dairies are needed
- Average dairy has herd of 100 cows (up from 20 in 1970)
- Numerous dairies have > 1000 cows

Chuck Coronella   Hydrothermal Carbonization of Cow Manure   3

Beneficial uses of manure

- "Cow pies" have been used for fuel for centuries
  - Intrinsic fuel value
- Macronutrients (especially nitrogen and phosphorous)
  - Valuable fertilizer (when applied in the right quantities)

Chuck Coronella   Hydrothermal Carbonization of Cow Manure   4

The challenge:

- Modern dairies produce a lot of manure (up to 50 kg per cow per day)
  - More than 20 million ton/year (dry basis) in the U.S.
  - Moisture > 90%
- Traditional methods of spreading on fields
  - Large amounts must be spread over large distances- high costs
  - Over application → Eutrophication
- "Nutrient management" is a headache for many dairymen
  - Associated regulatory and compliance issues
  - Lagoon management is especially unwelcome

Our approach: Hydrothermal carbonization

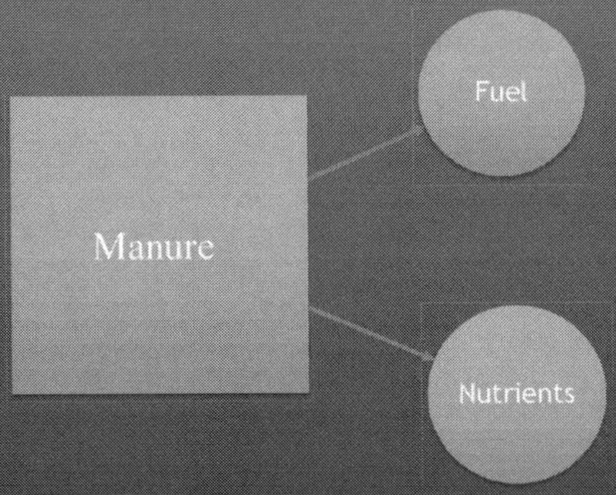

Hydrothermal conditions

Properties of liquid water:

Biomass undergoes chemical reactions:

- Hydrolysis
- Decarboxylation
- Dehydration
- Deammination
- etc.

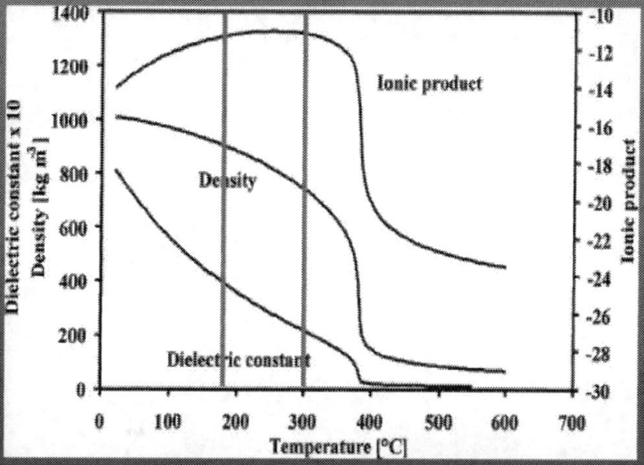

Figure taken from Peterson et al. *Energy Environ. Sci.*, 2008, 1, 32–65

Hydrothermal carbonization

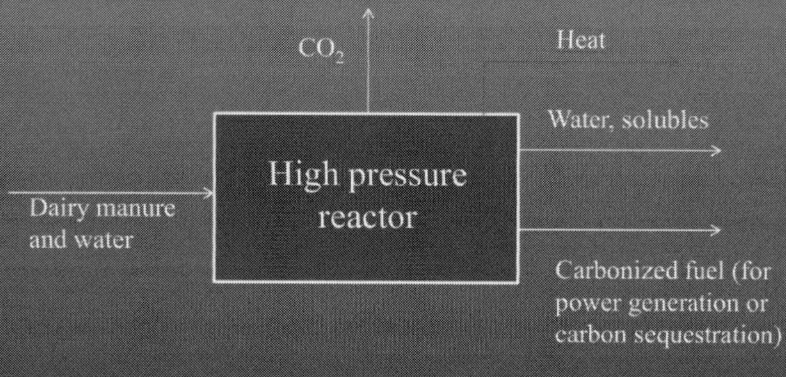

Reaction conditions:
180 – 280 °C
$P > P^{sat}$ (10 – 64 bar)

Manure → products

Dried manure

→

Solid "hydrochar"

Very hydrophobic

Solid fuel for on-farm power generation

\+

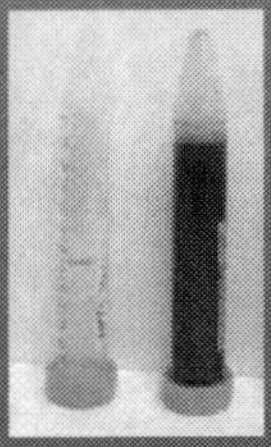

Water and soluble chemicals

Experimental methods

- T is between 180 and 260 °C
- Reaction time 5 or 30 min
- Biomass : water = 1:5

100 mL Parr reactor

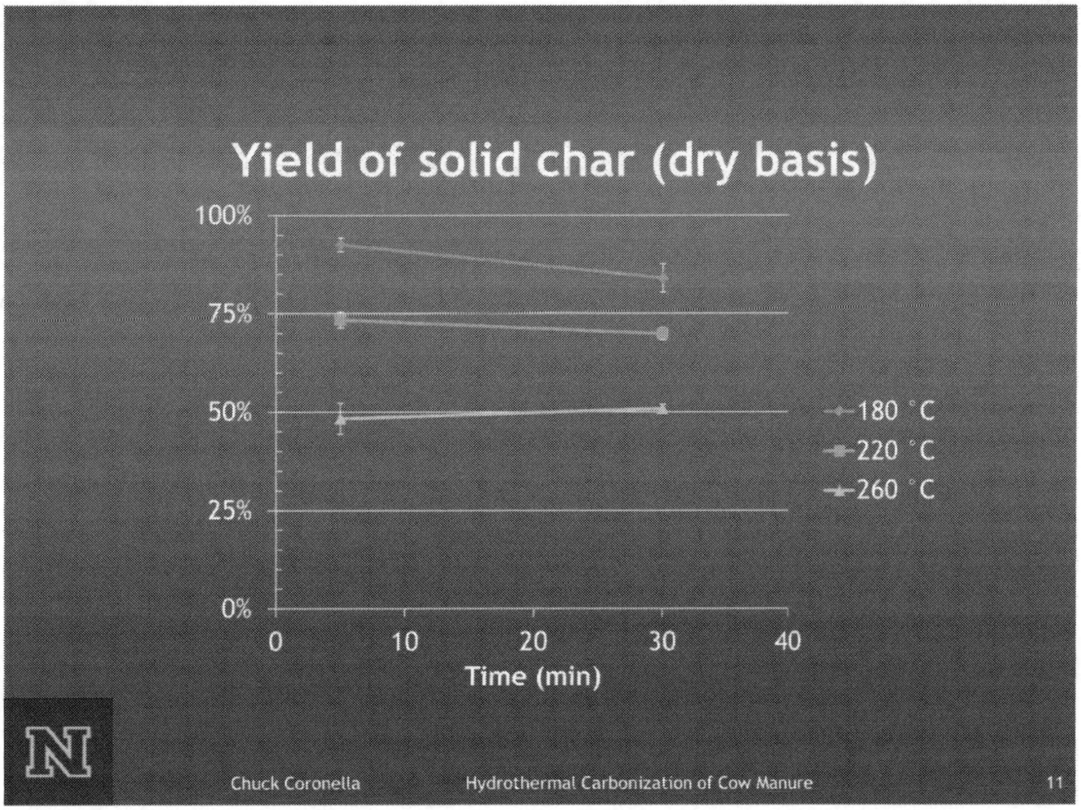

Yield of solid char (dry basis)
100%
75%
50%
25%
0%
0
10
20
30
40
Time (min)
180 °C
220 °C
260 °C
Chuck Coronella
Hydrothermal Carbonization of Cow Manure
11

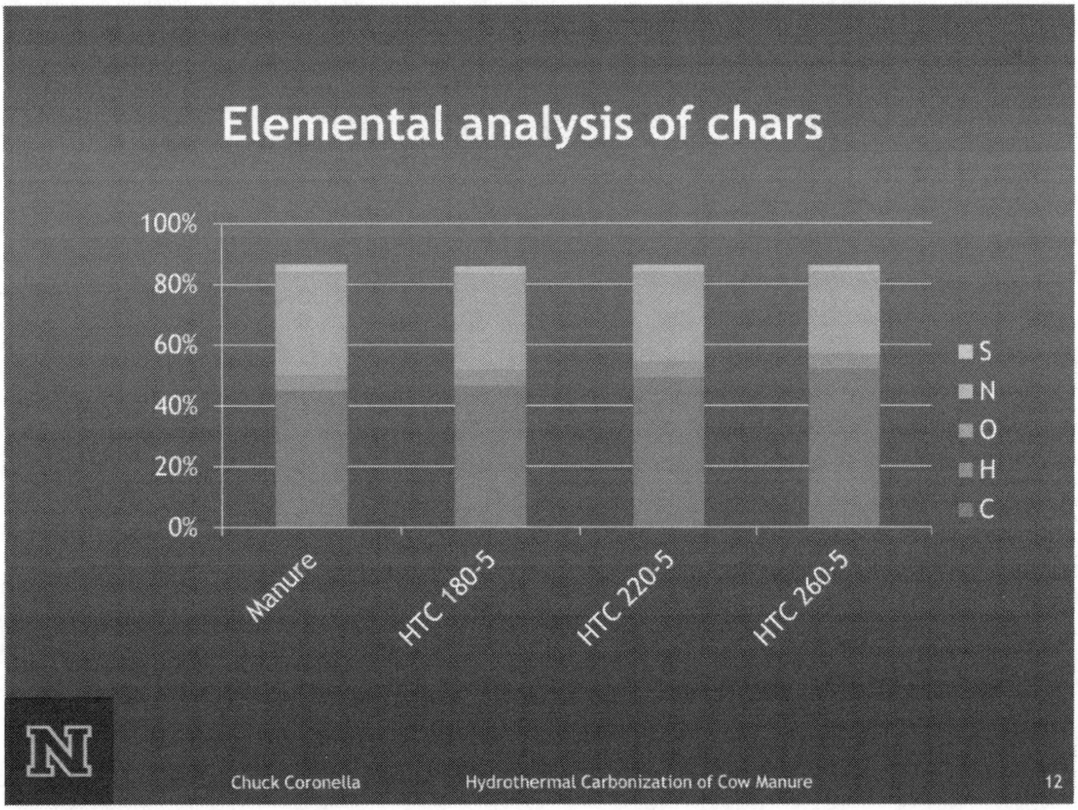

Elemental analysis of chars
100%
80%
60%
40%
20%
0%
Manure
HTC 180-5
HTC 220-5
HTC 260-5
S
N
O
H
C
Chuck Coronella
Hydrothermal Carbonization of Cow Manure
12

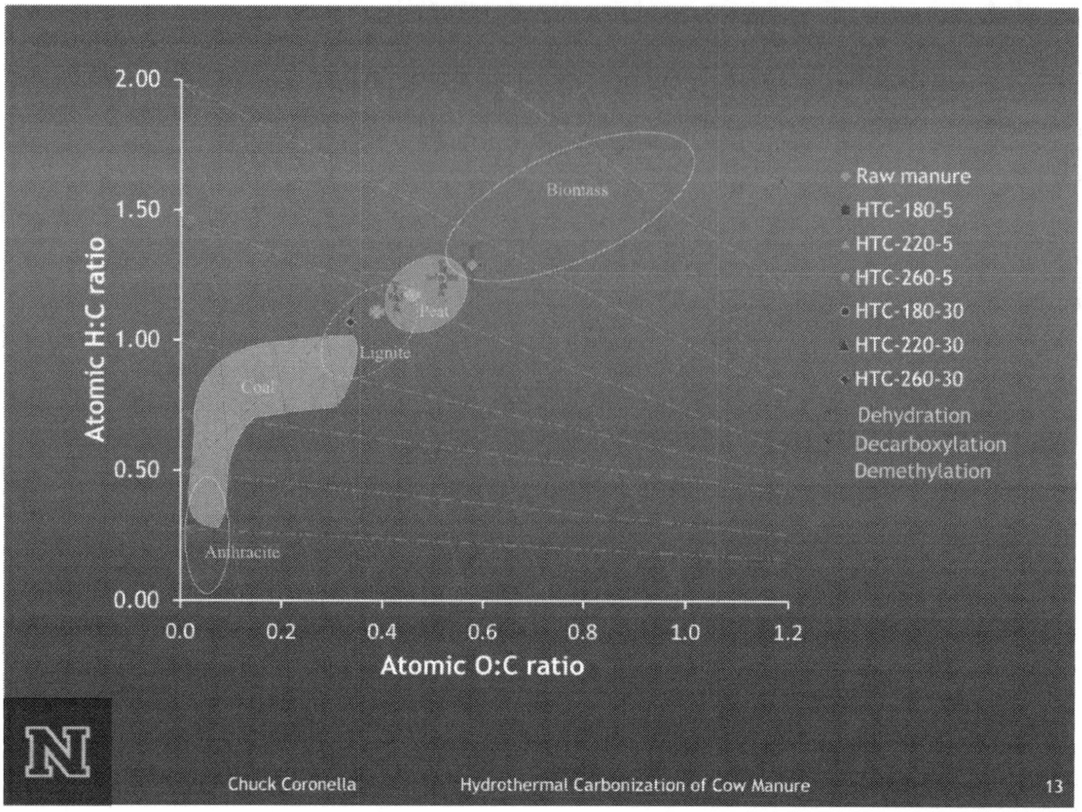
2.00
1.50
1.00
0.50
0.00
Atomic H:C ratio
0.0
0.2
0.4
0.6
0.8
1.0
1.2
Atomic O:C ratio
Biomass
Peat
Lignite
Coal
Anthracite
Raw manure
HTC-180-5
HTC-220-5
HTC-260-5
HTC-180-30
HTC-220-30
HTC-260-30
Dehydration
Decarboxylation
Demethylation
Chuck Coronella
Hydrothermal Carbonization of Cow Manure
13

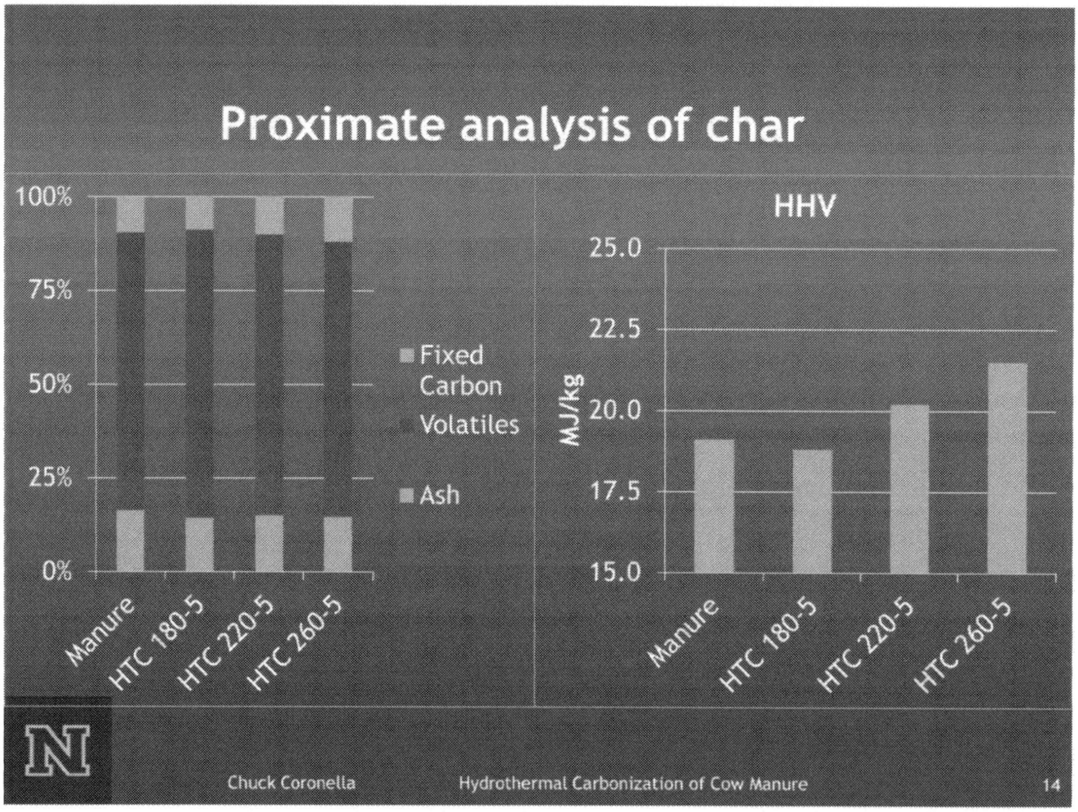
Proximate analysis of char
100%
75%
50%
25%
0%
Manure
HTC 180-5
HTC 220-5
HTC 260-5
Fixed Carbon
Volatiles
Ash
HHV
25.0
22.5
20.0
17.5
15.0
MJ/kg
Manure
HTC 180-5
HTC 220-5
HTC 260-5
Chuck Coronella
Hydrothermal Carbonization of Cow Manure
14

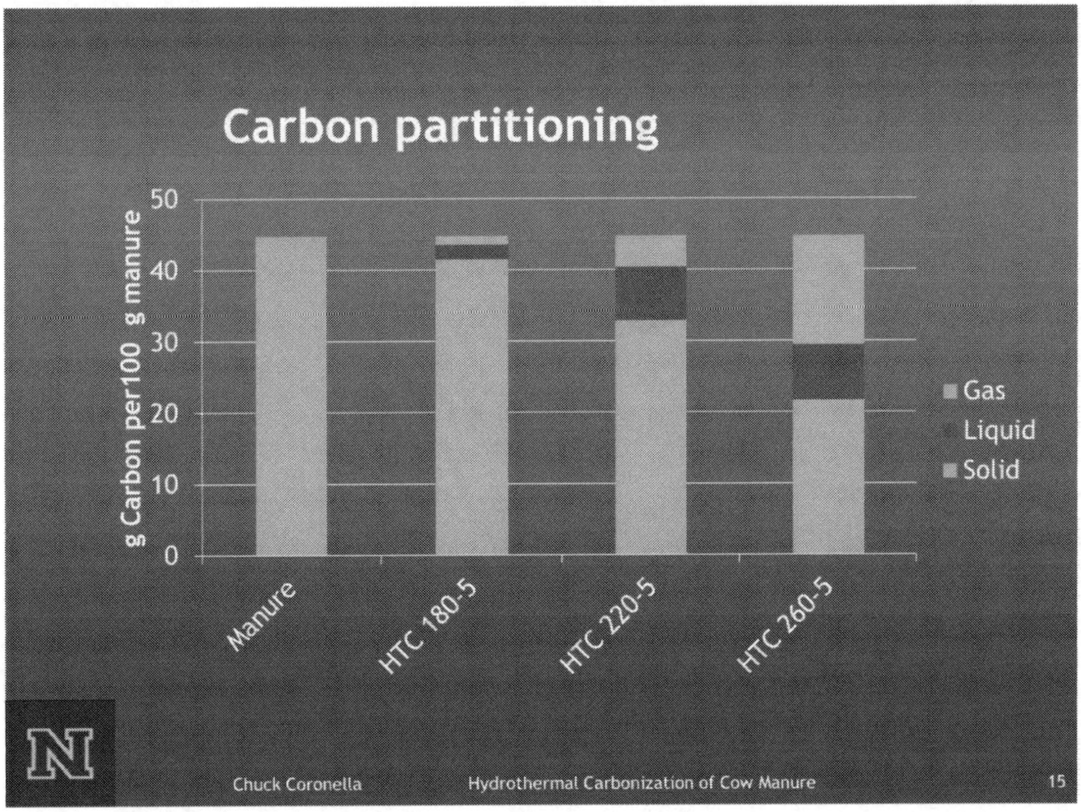

Carbon partitioning
g Carbon per100 g manure
50
40
30
20
10
0
Manure
HTC 180-5
HTC 220-5
HTC 260-5
Gas
Liquid
Solid
Chuck Coronella
Hydrothermal Carbonization of Cow Manure
15

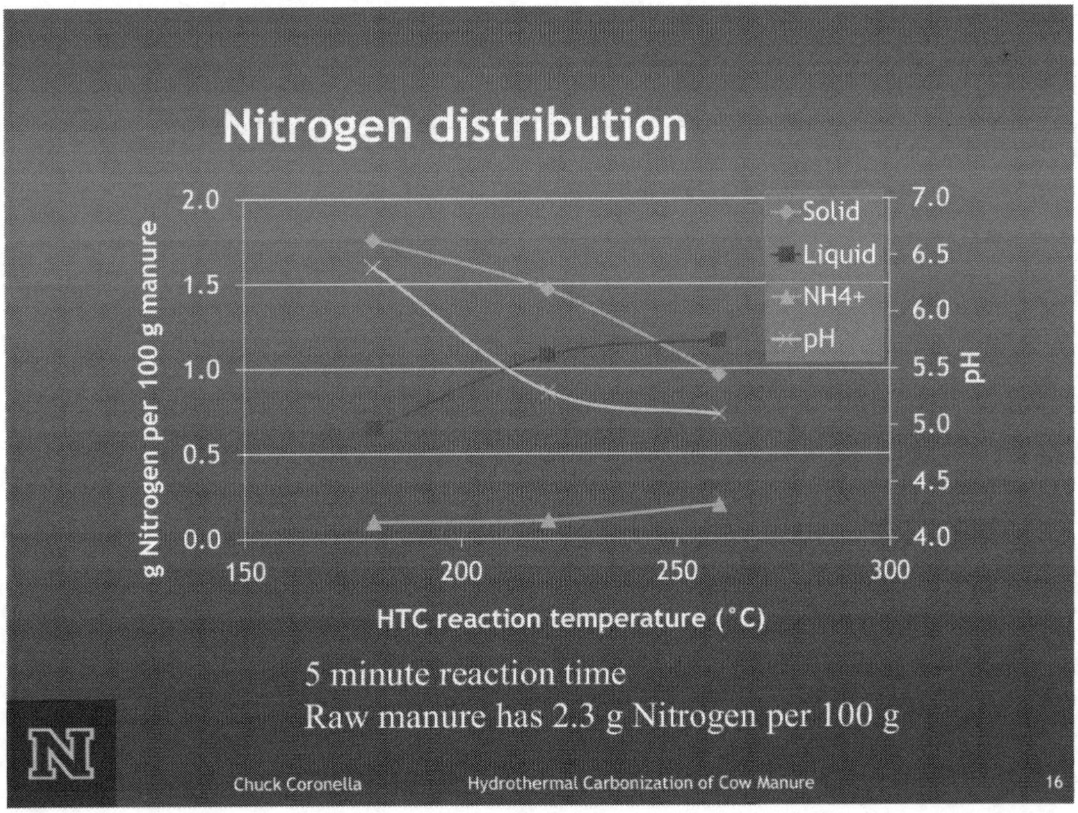

Nitrogen distribution
g Nitrogen per 100 g manure
2.0
1.5
1.0
0.5
0.0
150
200
250
300
HTC reaction temperature (°C)
Solid
Liquid
NH4+
pH
7.0
6.5
6.0
5.5
5.0
4.5
4.0
pH
5 minute reaction time
Raw manure has 2.3 g Nitrogen per 100 g
Chuck Coronella
Hydrothermal Carbonization of Cow Manure
16

Macronutrients in hydrochar

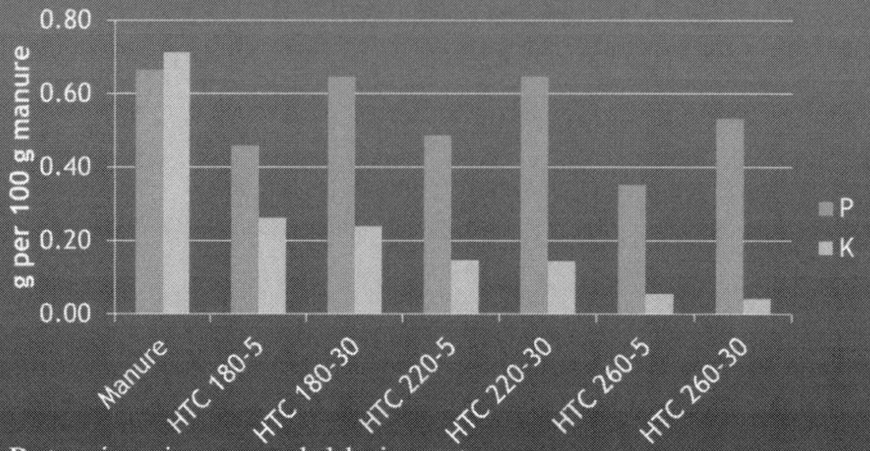

- Potassium is very soluble in water;
- Phosphorous seems to be absorbed into the char matrix with time

Chuck Coronella Hydrothermal Carbonization of Cow Manure 17

Ash in char

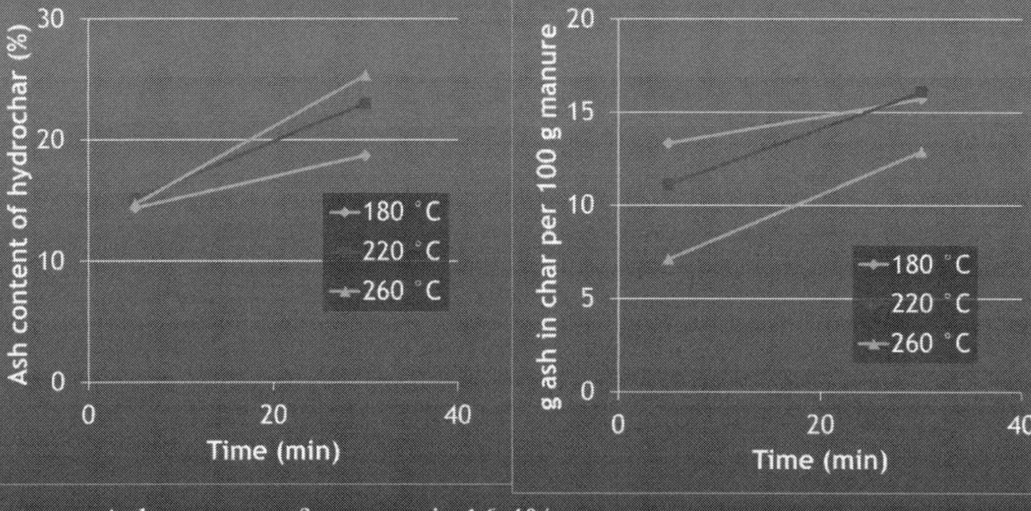

Ash content of manure is 16.4%

Chuck Coronella Hydrothermal Carbonization of Cow Manure 18

Aqueous product characterization

- Detailed analysis is underway
- Simple sugars, simple organic acids, and furfurals are significant

Chuck Coronella Hydrothermal Carbonization of Cow Manure 19

Remaining steps

- Optimize conditions for nutrient capture
- Couple with membrane distillation
- Demonstrate gasification of hydrochar pellets
- Life cycle analysis

- Construction of a continuous reactor
  - Nominal capacity of 5 gal/h ~ 0.3 lpm
  - Design complete, fabrication is taking place now
  - Initial hot experiments expected around July 1

Chuck Coronella Hydrothermal Carbonization of Cow Manure 20

Other markets:

- Municipal solid waste
  - Wastewater sludge “biosolids”
  - Food wastes
  - Green wastes
- Other high-moisture wastes
  - Corn ethanol wastes, food processing, etc.
- Microalgae processing

Chuck Coronella | Hydrothermal Carbonization of Cow Manure | 21

Acknowledgements

Collaborators at UNR:

Chemical engineering: Dr. Toufiq Reza, Dr. Hongfei Lin, Mi Lu, Tianlin Song

CEE: Dr. Sage Hiibel

Cooperative extension: Jay Davison

Undergraduate students:

Sean Sullivan, Akbar Saba, Jasmine Miller, Keenan Conrad-Williams

Hydrotorr, LLC:

Dr. Alireza Shekarriz

Desert Research Institute:

S. Kent Hoekman, Curt Robbins, Vivian Liu

APL, Bear Kaufmann

Financial support from the USDA through the Western Sun Grant Initiative, grant number 2010-38502-21839

Chuck Coronella | Hydrothermal Carbonization of Cow Manure | 22

LIFE CYCLE ASSESSMENT OF DIMETHYL CARBONATE PRODUCTION PROCESSES: OXIDATIVE CARBONYLATION VS. DIRECT SYNTHESIS FROM CO_2

I. Garcia-Herrero[a], R. Cuéllar-Franca[b], V. M. Enríquez-Gutiérrez[b], M. Alvarez-Guerra[a], A. Irabien[a] and A. Azapagic[b]

[a]Department of Chemical and Biomolecular Engineering
University of Cantabria (SPAIN)

[b]School of Chemical Engineering and Analytical Science University of Manchester (UK)

Outline

1. Motivation

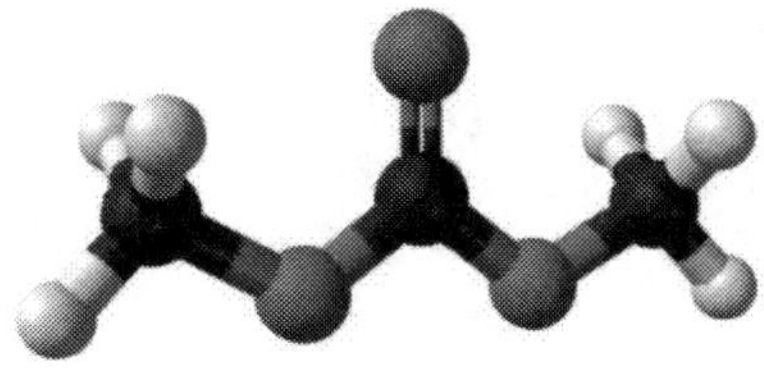

2. Methodology

3. Results
   - 3.1. Global Warming Potential
   - 3.2 Other environmental impacts
   - 3.3 Sensitivity analysis

4. Final remarks

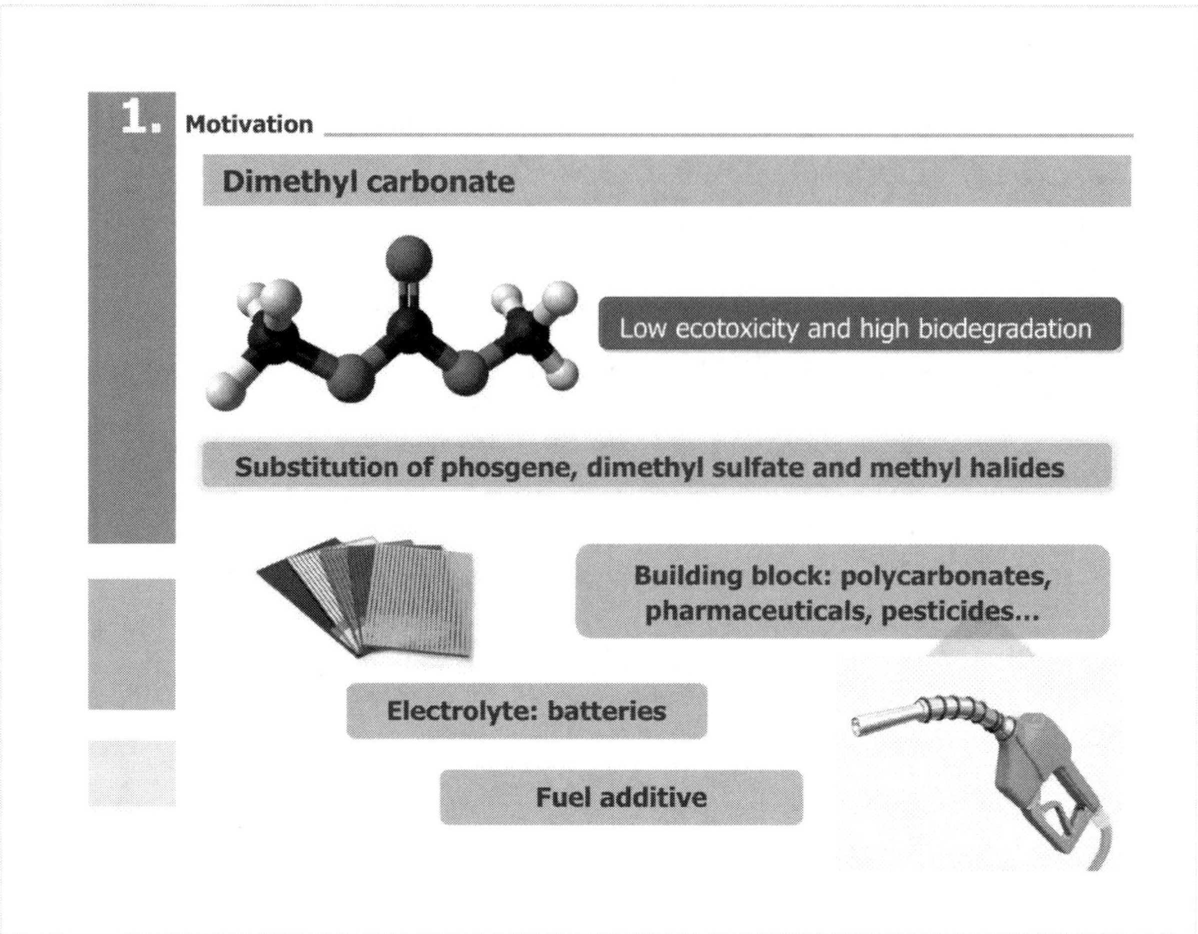
1. Motivation
Dimethyl carbonate
Low ecotoxicity and high biodegradation
Substitution of phosgene, dimethyl sulfate and methyl halides
Building block: polycarbonates, pharmaceuticals, pesticides...
Electrolyte: batteries
Fuel additive

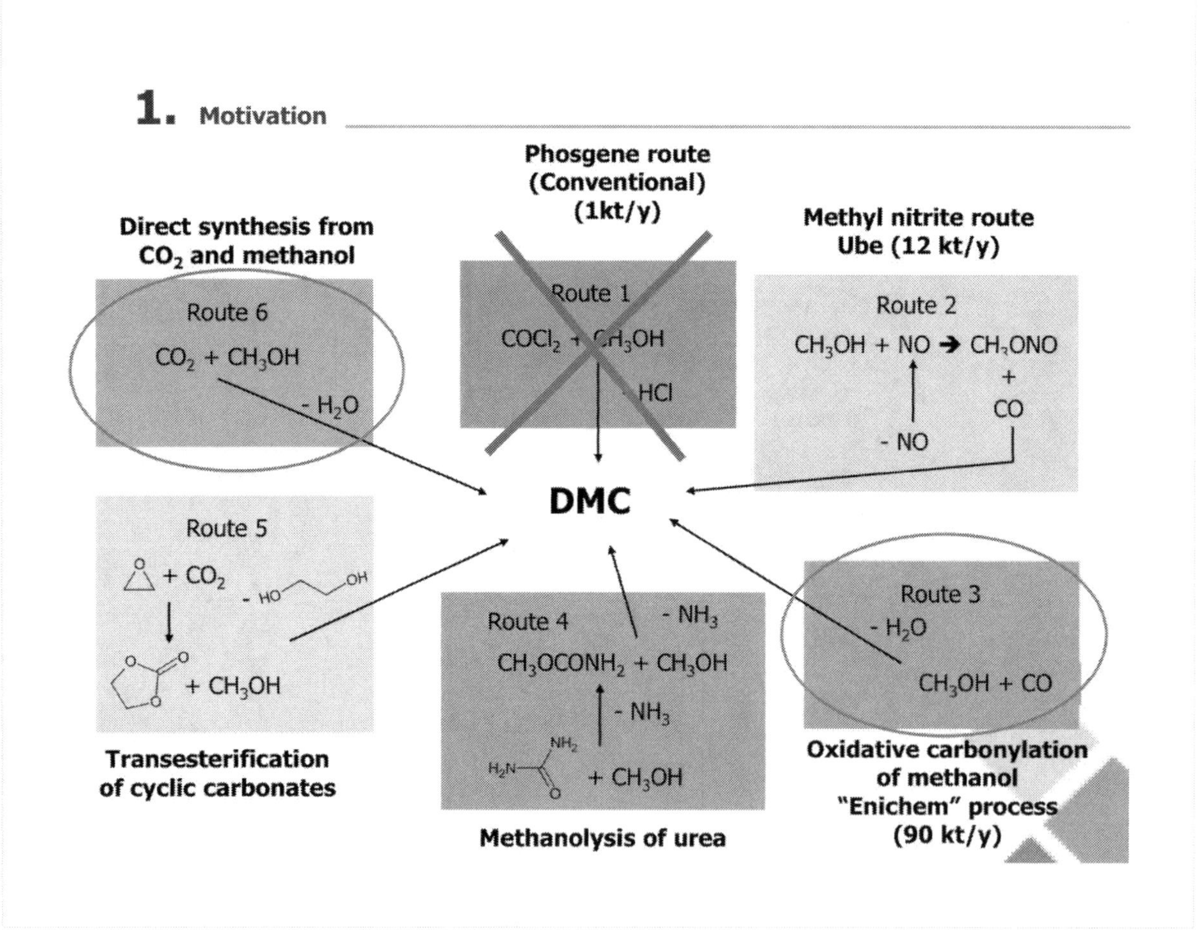
1. Motivation
Phosgene route (Conventional) (1kt/y)
Route 1
$COCl_2$ + CH_3OH
- HCl
Methyl nitrite route Ube (12 kt/y)
Route 2
CH_3OH + NO → CH_3ONO + CO
- NO
Direct synthesis from CO_2 and methanol
Route 6
CO_2 + CH_3OH
- H_2O
DMC
Route 5
+ CO_2
+ CH_3OH
Transesterification of cyclic carbonates
Route 4
- NH_3
CH_3OCONH_2 + CH_3OH
- NH_3
+ CH_3OH
Methanolysis of urea
Route 3
- H_2O
CH_3OH + CO
Oxidative carbonylation of methanol "Enichem" process (90 kt/y)

1. Motivation

However...

High thermodynamic stability of CO_2

Electrochemical synthesis of DMC from CO_2

- Particularly attractive
- Storing intermittent renewable energy
- reuse of captured CO_2

Electrochemical activation to carry out the DMC synthesis, especially using **Ionic Liquids (ILs)** ➡ **Great interest**, but previous work... ... added carcinogenic alkylating agents (e.g. CH_3I)

1. Motivation

Objective

Assess and compare the environmental sustainability of the commercial oxidative carbonylation of methanol and the innovative electrochemical process for the synthesis of DMC from CO_2 and methanol

Quantify and compare the life cycle environmental impacts

Identify improvement opportunities

2. Methodology

Scope

- **Functional unit:**
'Production of 1 kg of dimethyl carbonate with 99.8% puritiy'

- **'Cradle to gate' approach**

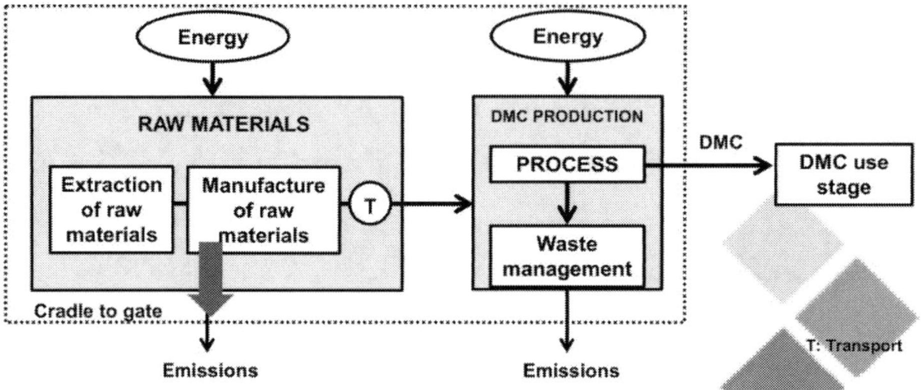

2. Methodology

System description

Eni process (Reference case)

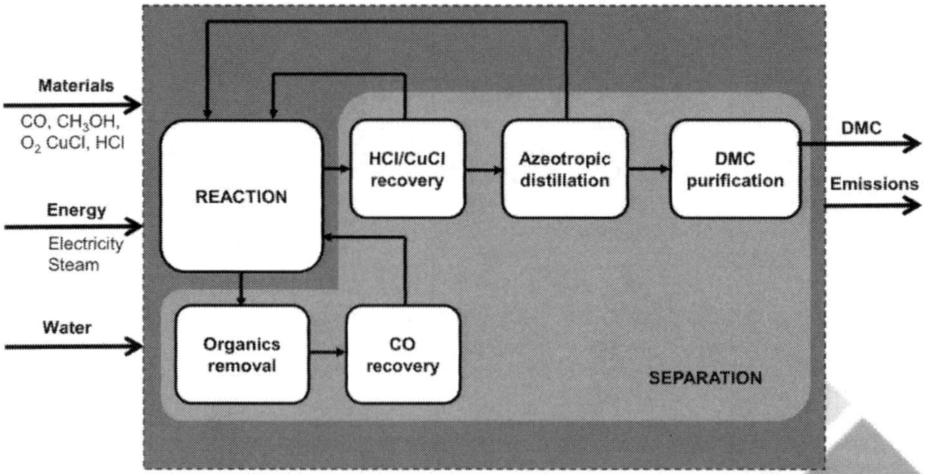

2. Methodology

System description

Innovative electrochemical processes (Alternative process 1 and Alternative process 2)

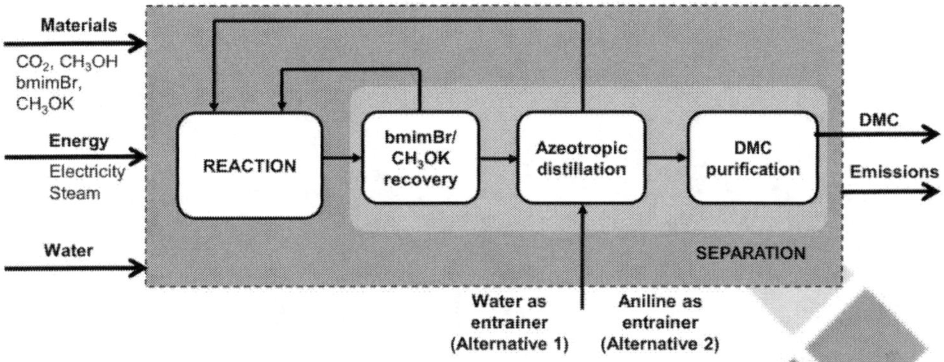

2. Methodology

Life cycle inventory and assumptions

- LCA modelling in Gabi V4.4
- Ecoinvent database used

Reference case

- Inventory sourced from available commercial data
- CO assumed to be sourced from coal gasification using CO_2
- Organic purged assumed to be incinerated

GaBi Software

ecoinvent Centre

2. Methodology

Life cycle inventory and assumptions

Alternative processes

- Separation were simulated using Aspen HYSYS
- Kinetics modelled based on experimental[1,2] data using Matlab
- Aspen Energy Analyzer used for heat integration
- Solar photovoltaic energy assumed for the ER reactor

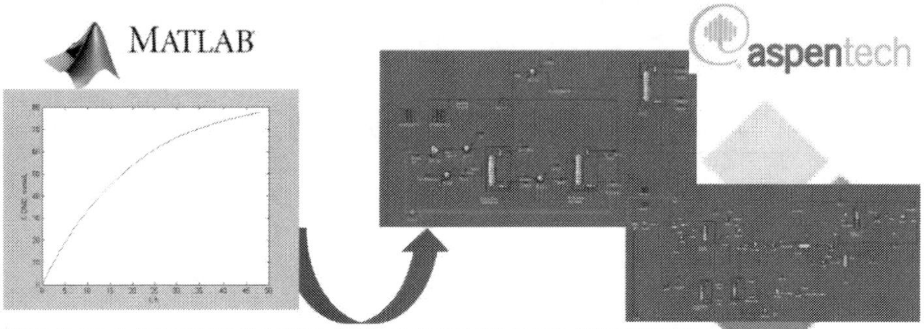

[1]Garcia-Herrero, I., Alvarez-Guerra, M. & Irabien. A. J. Chem. Technol. Biot. In press. DOI: 10.1002/jctb.4445
[2]Garcia-Herrero, I., Alvarez-Guerra, M. & Irabien. A. J. Chem. Technol. Biot. In press. DOI: 10.1002/jctb.4605

3. Results

3.1. Global Warming Potential

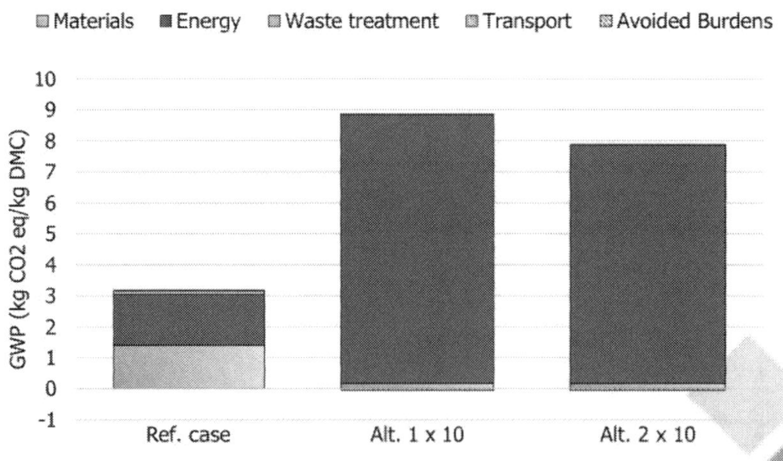

3. Results

3.2. Other environmental impacts

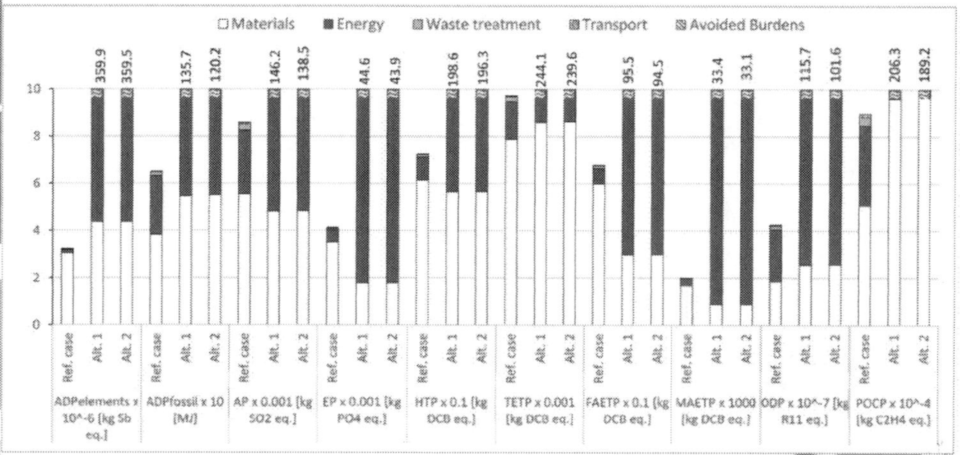

3. Results

3.2. Sensitivity analysis

Global Warming Potential

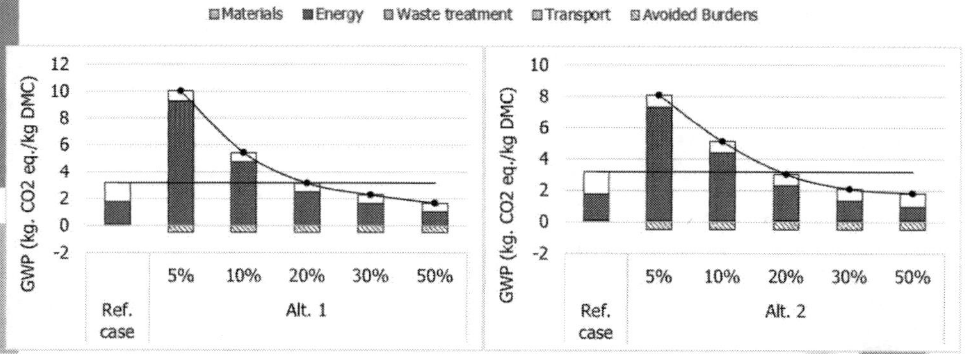

3. Results

3.2. Sensitivity analysis

Influence of 20% in other environmental impacts

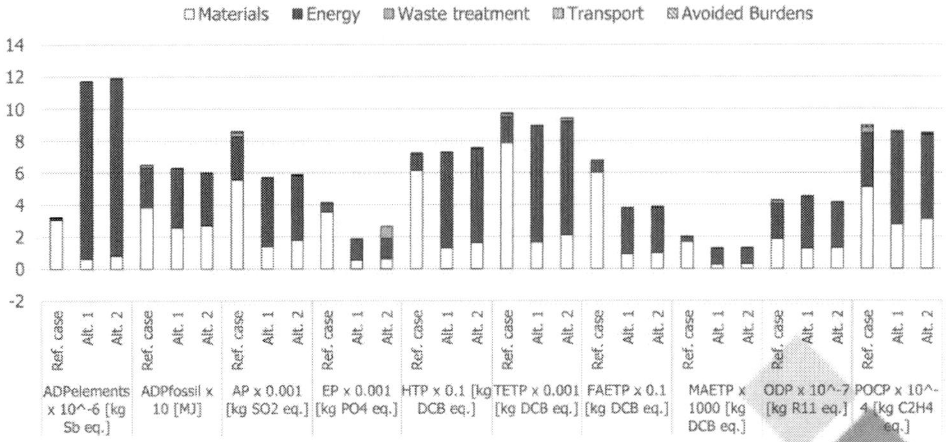

4. Final remarks

- Energy consumption has been identified as the main hotspot for the 3 cases.

  The low conversion achieved with the current electrochemical design results in high separation requirements.

- Environmental impacts for the electrochemical alternatives are higher than for the reference case.

- A sensitivity analysis suggests that the production yields in alternative 1 and 2 must be increased by 20% to reduce their GWP to a comparable level to the reference case.

- Further research must be done to overcome the current limitations.

- LCA plays a key role in the delivery of more sustainable processes at the design stage by combining experimental and simulation work.

Acknowledgements

- This work was conducted under the framework of the Spanish Ministry of Science and Innovation Project CTQ2013-48280-C3-1-R. M.I. Garcia-Herrero thanks the FPI postgraduate research grant (BES-2011-047906).
- M.I. Garcia-Herrero is also grateful for the financial help of the Spanish Ministry of Economy and Competitiveness (EEBB-I-14-08917) to carry out a research stay in the University of Manchester.

ICOSSE

International Congress on Sustainability Science & Engineering

LIFE CYCLE ASSESSMENT OF DIMETHYL CARBONATE PRODUCTION PROCESSES: OXIDATIVE CARBONYLATION VS. DIRECT SYNTHESIS FROM CO_2

I. Garcia-Herrero[a], R. Cuéllar-Franca[b], V. M. Enríquez-Gutiérrez[b], M. Alvarez-Guerra[a], and A. Irabien[a] and A. Azapagic[b]

[a]Department of Chemical and Biomolecular Engineering University of Cantabria (SPAIN)

[b]School of Chemical Engineering and Analytical Science University of Manchester (UK)

Departamento de Ingenierías Química y Biomolecular
Universidad de Cantabria (SPAIN)

Grupo Desarrollo de Procesos Químicos y Control de Contaminantes

Departamento de Ingenierías Química y Bimolecular
ETSIIyT, Avda. de los Castros s/n 39005
Santander, Cantabria. SPAIN
E-mail: lrabienj@unican.es Tel. +34 942 201597 Fax +34 942 201591

«We know what we have to do to achieve the sustainability, now it's time to do it »

Environmental Sustainability Applied to Energy Systems and Energy Use

- **Conference:** International Congress on Sustainability Science Engineering ICOSSE
- **Year:** 2015
- **Proceeding:** ICOSSE 2015
- **Group:** Abstract Submissions
- **Session:**

Session 12: Renewable Energy Technologies

- **Time:**

Thursday, May 28, 2015 - 4:30pm-4:50pm

Authors:

Cucek, L., University of Pannonia
Klemes, J. J., University of Pannonia

Environmental sustainability has become very important issue in the last decades, especially during the last years. Many different concepts, methods and tools have been developed for measuring and monitoring environmental sustainability, and also in combination with other pillars of sustainability. Different concepts and methods will be presented for measuring environmental sustainability and sustainability in general with special focus on new developments. High potential for improved sustainability lays in sustainable use of energy, such as by using renewable energy, within transport sector, when selecting environmentally more sustainable travel options, within industrial plants, and industrial and regional Total Sites. Several examples will be shown in order to illustrate recent developments in different concepts and methods that could be related to more sustainable use of energy within various sectors.

Thermal-Hydraulic Performance and Optimization of Printed Circuit Heat Exchangers for High-Temperature/High-Pressure Applications

- **Conference:** International Congress on Sustainability Science Engineering ICOSSE
- **Year:** 2015
- **Proceeding:** ICOSSE 2015
- **Group:** Abstract Submissions
- **Session:**

Session 13: Closing Session: Some Ideas in Practical Sustaiability

- **Time:**

Friday, May 29, 2015 - 9:10am-9:40am

Authors:

Wang, Q.
Zeng, M.

Recently, the printed circuit heat exchanger (PCHE) is considered as one of the most potential candidates for the intermediate heat exchanger in the helium and supercritical carbon dioxide (SCO2) Brayton cycles in the nuclear application. The operating temperature in the helium Brayton cycle is up to 900°C, while the operating pressure in the supercritical carbon dioxide Brayton cycle is up to 8 MPa. The ultra high temperature and high pressure give a significant challenge to the design and fabrication of PCHE. In the first part, the PCHE using helium as the working fluid is studied in the laminar flow region. The effect of fluid-solid coupled thermal boundary on the local thermal-hydraulic performance is analyzed and compared to that of the constant temperature boundary. A two-sided etched zigzag-type PCHE is proposed to reduce the thermal stress produced at high temperature. In the second part, the PCHE using SCO2 as the working...

Participatory Sustainability Assessment Tool for Wood-based Bioenergy Industry in Upper Michigan, USA

Ashma Vaidya[a] and Audrey L. Mayer[a,b]
[a]Department of Social Sciences
[b]School of Natural Resources and Environmental Science
Michigan Technological University

4th International Congress on Sustainability Science and Engineering
29 May 2015
Balatonfured, Hungary

Background

- USA is the largest producer of biofuel (corn ethanol and soy biodiesel) in the world
- Issues with crop-based biofuel production
  - Food-versus-fuel
    - Food prices
    - Competition for land, labor, and other resources
  - Direct and indirect land use changes and its subsequent impact on GHGs emissions
  - Long-term environmental and economic sustainability not very clear

MichiganTech

Alternatives

- Advanced biofuels
  - cellulose-rich materials like wood, grasses, agricultural residues and non-grain materials
- Have relatively smaller environmental and socioeconomic costs than fossil fuels and crop-based biofuels
- **Long-term sustainability** yet to be adequately explored

MichiganTech

Assessment Techniques

- ***Cost Benefit Analysis***

includes cost & benefits that do not have market prices (e.g. health risk, pollution, land degradation)

- ***Life Cycle Assessment***

Typically uses GHGs, energy balance, resources (water usage) as indicators

Assessment Techniques

- *Environmental Impact Assessment/ Strategic Environmental Assessment*

soil contamination, air pollution, health effects, ecology impacts, endangered species, geological hazards and water pollution

- *Issues*
  - Mostly expert-driven with no or limited input from local stakeholders
  - Social dimensions such as indigenous rights, labor protection, poverty, local access to resources are often not included (Fredericks 2012; Ribeiro 2013)

Bioenergy SAFs: Examples

- Global Bioenergy Partnership Sustainability Indicators for Bioenergy (GBEP)
- Roundtable on Sustainable Biomaterials (RSB), RTRS, RSPO
- Environmental and socio-economic indicators for sustainable bioenergy systems (Oak Ridge Environmental Laboratory)
- Sustainable Forest Initiatives
- Forest Stewardship Council
- BONSUCRO (for sugarcane ethanol)

MichiganTech

Limitations

- Largely driven by multilateral trade obligations, market-oriented.
- Conflicting and minority's concerns are often sidelined. Generated mostly through consensus-based multi-stakeholder decision-making processes (Schouten et al. 2010)
- Too many indicators, too broad and too costly to implement at local levels–GBEP, RSB (Dale et al. 2013)
- Often lack 'hard' and quantitative indicators for local-level implementations. (Lewandowski & Faaij 2006, Solomon 2010)

MichiganTech

Research Objectives

- Understanding sustainability goals of the community
- Defining 'sustainability' for forest-based bioenergy production
- Develop criteria and indicators to assess sustainability of forest-based bioenergy industry, which also overlaps with the sustainability goals of the community

MichiganTech

Methodology

- Qualitative case study research method
- Study site:
  - Western Upper Peninsula, Michigan
  - Wood-based bioenergy development is being planned and researched

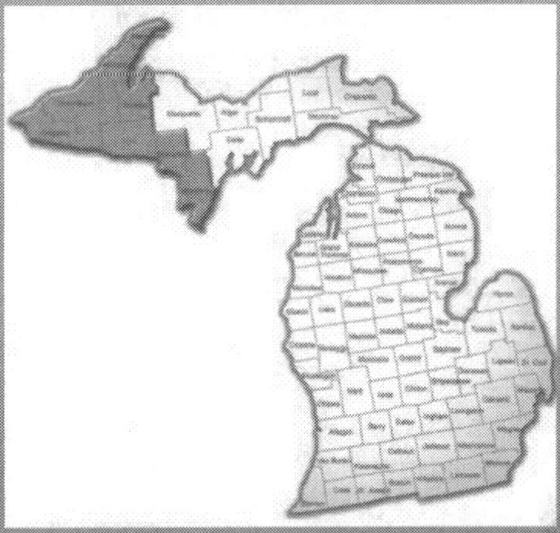

Photo source: http://store.milwaukeemap.com

MichiganTech

Sustainable Energy Pathways: Forest-Based Cellulosic Ethanol

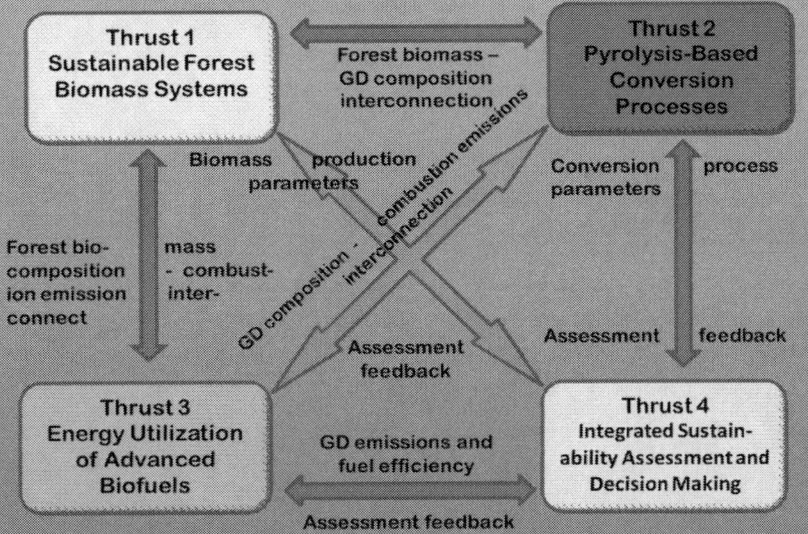

Michigan Tech

Sustainable Energy Pathways: Forest-Based Cellulosic Ethanol

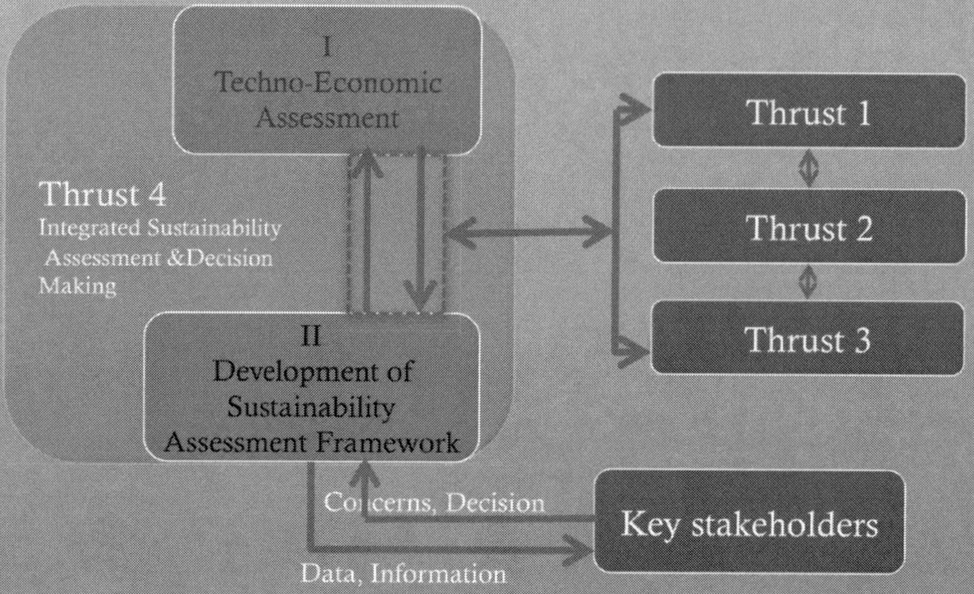

Michigan Tech

Research Steps

- Task 1: Stakeholder selection using stakeholder analysis
- Task 2: Data Collection
  - Focus group meetings (7)
  - Individual semi-structured interviews (29)
- Task 3: Stakeholder Workshop
- Task 4: Framework development based on workshop results and other works.

Focus Group Meetings

Telephone Interviews

S.N.	Components of bioenergy production cycle	Stakeholders
1	Feedstock production	Forest land owners (public, private, industrial, nonindustrial), tribal members, business owners; foresters and consultants; regional natural resource managers
2	Production plant	Venture capitalists; Timberland management companies, Subsidy providers (e.g., state agencies)
3	Energy use	Local consumers of wood products and wood pellets, potential users
4	Decision making and sustainability assessment	GOs, Interest groups, forestland management agencies, land owners, Potential users, Researchers

Participants	# of participants
Farmers	3
Landowners	12
Tribal community representatives	2
Timberland Management Companies	4
Venture capitalists	4
Foresters	2
Biologist and Ecologist	2
USDA Rural Development Program	1
DNR	2
Township Officials/Local government	8
NGOs	3
Current/potential users of wood-based energy	7+

MichiganTech

Task III: Stakeholder workshop

- Representation from each component was ensured
- A mix of scientific as well as non-scientific experts from the region were brought together
- Solicited feedback on preliminary list of criteria and indicators generated through focus group meetings and interviews based on their importance to them and relevance to bioenergy production system.

MichiganTech

Results

Focus group meetings and Interviews

- Environmental concerns
- Economic concerns
- Social concerns
- Policy and regulations
- Institutional capacity and strength

Workshop

- Feedstock production
- Biofuel Production
- End Use
- Decision making

Michigan Tech

Task IV: Framework development

- Criteria and indicators generated from the participatory approach will be evaluated further based on their measurability, scientific relevance, practicality, sensitivity to changes, biasedness
- Based on literature review, local experts

Michigan Tech

Conclusion

✓ SAF more reflective of the concerns and interests of local stakeholders, and more relevant to the regional context which is expected to make it more likely to be accepted by local practitioners and implementers

✓ This work provided an opportunity for both scientific and non-scientific stakeholders to collaborate and engage in mutual learning.

✓ Transparency was maintained throughout the research process thereby ensuring validity, reliability and credibility.

Thank you!

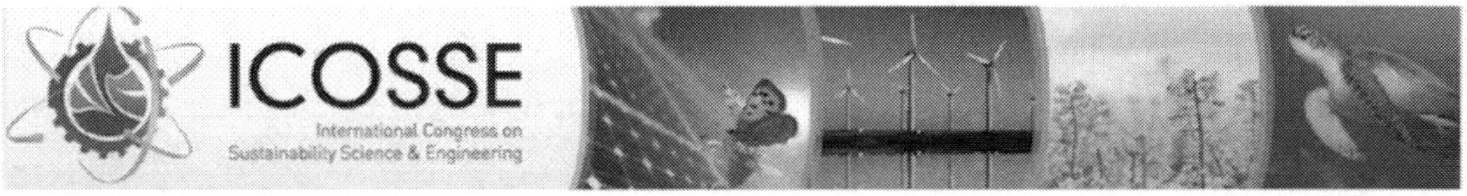

Sustainable Engineering Education: New Community Information

Miriam Heller and Cliff I. Davidson

May 29, 2015
Balantonfüred, Hungary
ICOSSE '15

Acknowledgements

- Brad Allenby, Liv Haselbach, Bill Kelly; Syracuse staff/students; ASEE Staff; 35 f2f participants; and 27 virtual participants of the SEEDR workshop held on August 5-6, 2014.
- 250 survey respondents
- The authors acknowledge the support of the U.S. National Science Foundation through grant CBET-1440208.

5/29/2015 Heller - Engineering Ethics for Sustainable Water

Topics

- The Center For Sustainable Engineering
- NSF Workshop in August 2014: A Digital Repository To Serve The Sustainable Engineering Community
- Survey of Current Sustainable Engineering Educational Content, Practice, and Needs

Background:

Need for Sustainable Engineering Education

ASCE: Sustainability is a set of environmental, economic and social conditions in which all of society has the capacity and opportunity to maintain and improve its quality of life indefinitely without degrading the quantity, quality or the availability of natural, economic and social resources.

→

Demands practicing engineers who can make sustainable engineering decisions:

- Improve quality of life for an ever-increasing global population
- Manage use of resources and control of wastes
- Move society toward economic and social sustainability.

Home Institutions of CSE Workshop Participants

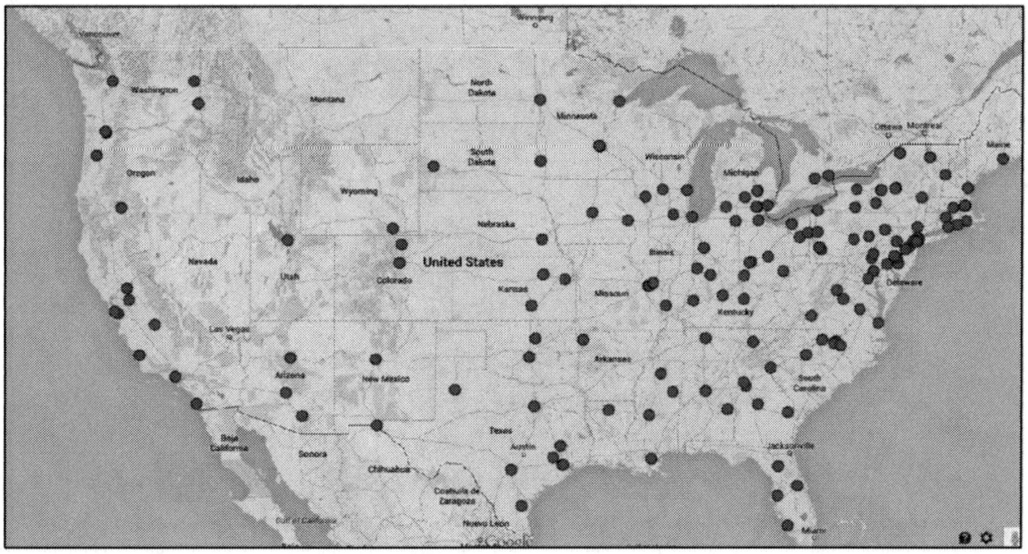

We Have Made Progress, But Not Enough

- Departments develop their own courses, curricula, and programs in sustainable engineering
- Lack of coordination in developing new educational materials and assessing quality of those materials
- Little to no coordinated resource sharing among sustainable engineering educators
- Result: many opportunities to make teaching SE more efficient and effective

Role for State-of-the-Art Digital Repository / Web Platform for SE Educational Materials?

Goals of the NSF Sustainable Engineering Workshop:

- Expose SE Education Community to benefits of a repository to improve teaching efficiency
- Illustrate state-of-the-art online educational resources and instructional design and assessment methods to improve teaching effectiveness
- Solicit and assess the community's requirements and preferences and identify barriers to adoption

Sustainable Engineering Workshop Motivation: *Ubiquitous Online Content and Learning*

- 7.1 million students (33.5 %) total students enrolled took at least one online course in 2011-12 ay

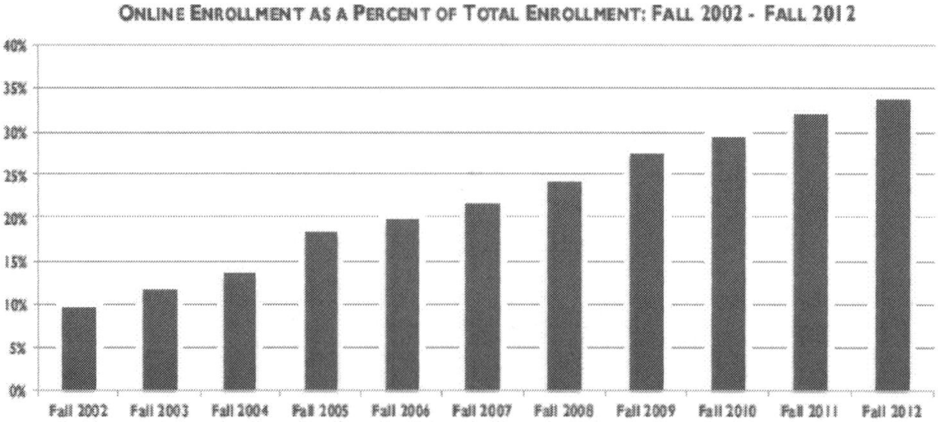

Babsen Survey, 2014

Sustainable Engineering Workshop Motivation:

Game Changing by MOOCs

- Tens of thousand of students per course
- 8 – 20 minute lecture video segments
- Consortia of top universities (Coursera, Udacity, edX, Canvas Network, Novoed)
  - As of January 2014 > 1000 US courses
  - As of June 2014 > 2,625 EC courses (⇧327% from '13)
- "Only" 5 – 10% students complete course
  - ML: 58K students started / 13K completed course
  - ~ 43 (430) sections of 300 (30) students/section

Sustainable Engineering Workshop Motivation:

Sustainable Engineering MOOCs

Course Title
Introduction to Sustainability
Global Sustainable Energy: Past, Present and Future
Climate Change in Four Dimensions
Wheels of Metals: The Global Metals Challenge
How Green is this Product? An Introduction to Life Cycle Environmental Assessment
Sustainable Product Development and Manufacturing
Sustainable Energy Innovation
Sustainability in Practice
The Age of Sustainable Development
Sustainability of Food Systems: A Global Life Cycle Perspective
Sustainable Agricultural Land Management
Supply Chain Management: A Learning Perspective

Sustainable Engineering Workshop Motivation:

Instructional Design Opportunities

- Little to no training in instructional design
- SE educators don't regularly read learning science literature
- SE educators don't exploit Engineering Education Research and Teaching Centers
- Instructional strategies research is growing
- Sustainable engineering- specific instructional strategies are emerging
  - Int'l Journal Engineering Education – special issues
  - Int'l Journal of Sustainability in Higher Education

Sustainable Engineering (SE) Community Survey Educational Content, Practice and Needs

- Survey informed by 2008 Benchmark survey (Allen
- Administered June-July 2014 mainly in US
- Five sections:
  - Your Information
  - Courses and Topics You Teach
  - Sources for the Content You Use
  - Your Instructional Practices
  - Sharing Community Resources

ASCE | INTERNATIONAL CONFERENCE ON SUSTAINABLE INFRASTRUCTURE 2014
Long Beach, California | November 6 – 8, 2014

SE Community Survey
Cohort Demographics

Cohort	Size
In-person Workshop Participants *(16 CSE Alumni; 4 Kelly WS Alumni; 1 SEES Fellow)*	39
CSE Alumni	245
ASEE Annual 2013 – Kelly Workshop Alumni	60
Others recommended by any of the above	217
SEES Grantees	4
ASEE Divisions	
Civil Engineering Division	584
Architectural Engineering Division	537
Chemical Engineering Division	273
Total	1959

5/29/2015 Heller - Engineering Ethics for Sustainable Water

ASCE | INTERNATIONAL CONFERENCE ON SUSTAINABLE INFRASTRUCTURE 2014
Long Beach, California | November 6 – 8, 2014

SE Community Survey
Respondent Demographics

- 150 respondents with valid surveys
- 98 U.S. academic institutions; 2 U.S. NGOs
- 93 U.S. cities
- 42 U.S. states
- 8 International academic institutions:
  - Canada: 3
  - Hong Kong: 2
  - Netherlands: 1
  - Nepal: 1
  - Ecuador: 1

5/29/2015 Heller - Engineering Ethics for Sustainable Water

Geography of Survey Responses

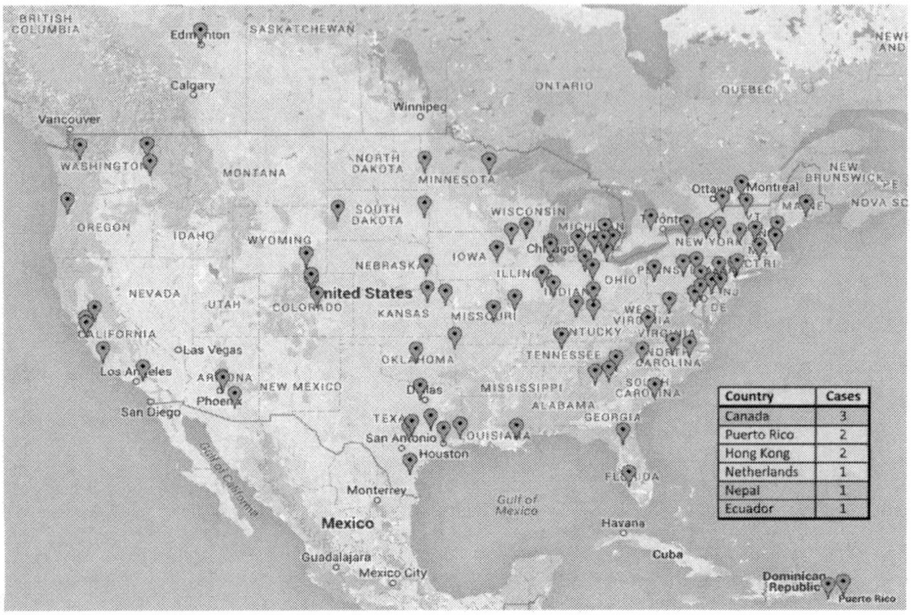

Country	Cases
Canada	3
Puerto Rico	2
Hong Kong	2
Netherlands	1
Nepal	1
Ecuador	1

SE Community Survey

Departments / Disciplines Represented

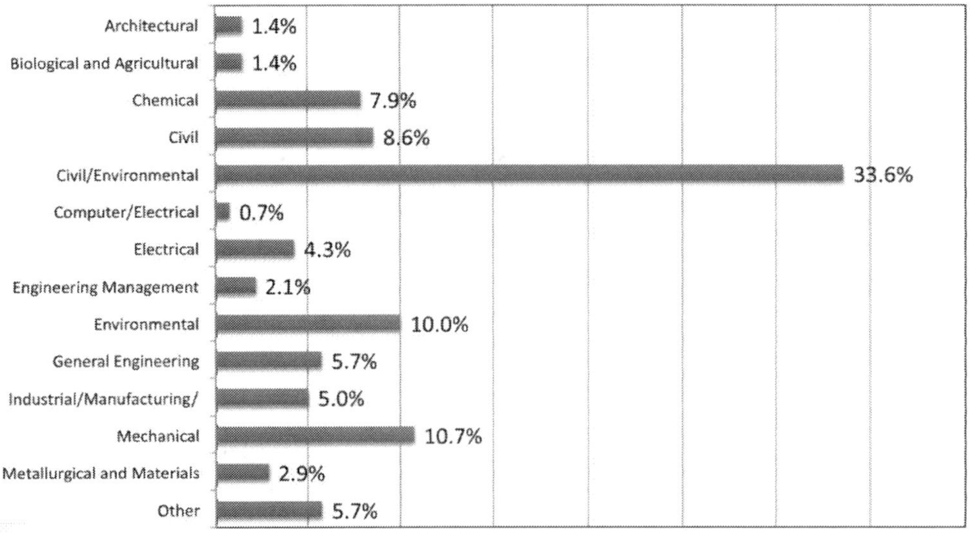

SE Community Survey
Courses and Content Taught

For each of the 21 topics provided, plus up to 3 more, indicate whether the topic:

- Is taught as a full (standalone) course or part of a course
- As taught is in a course required for a major or minor degree or certificate program
- Targets which of the following student level:
  - Freshman/sophomore
  - Junior/senior
  - Graduate

SE Community Survey
Topics: How They Fit in the Curriculum

Topic	Full Course*	Part of Course
Agriculture / Land Use	--	23.0%
Biogeochemical Systems (incl. Ecology)	--	17.8%
Business / Economics	--	27.0%
Climate Change	--	50.0%
Design	13.2%	49.3%
End of Life / Waste Management	--	38.2%
Energy / Power Generation	8.6%	48.7%
Green Buildings / Infrastructure / Construction	5.9%	36.8%
Human Health	--	22.4%
Humanities (Philosophy, Ethics, History)	--	21.1%

* "-" means < 5.0%

SE Community Survey

Topics: How They Fit in the Curriculum

Topic	Full Course*	Part of Course
Industrial Processes	--	22.4%
Life-cycle Assessment (LCA)	--	52.0%
Materials Flow Analysis (MFA)	--	25.7%
Matierals	--	33.6%
Natural Resources	--	30.3%
Policy	--	34.2%
Pollution Prevention/ Fate & Transport	--	32.2%
Systems / Metrics/ Information Management	--	23.0%
Transportation	--	27.6%
Urbanism / Urban Systems	--	19.7%
Water	--	42.8%

Conclusions:
- Few (3) of 21 SE topics are taught as standalone courses
- 17% of respondents teach some or all topics
- 40% of respondents teach some or all of 4 topics (LCA is #1)

5/29/2015 Heller - Engineering Ethics for Sustainable Water

SE Community Survey

Source of Content: Importance

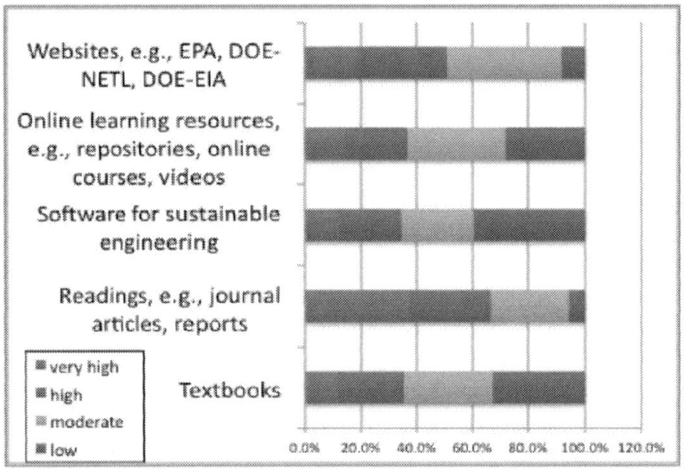

- Readings, e.g., journal articles, reports are most important
- Textbooks, SW, online learning are least important (purple)

5/29/2015 Heller - Engineering Ethics for Sustainable Water

SE Community Survey

Source of Content: Ease of Use

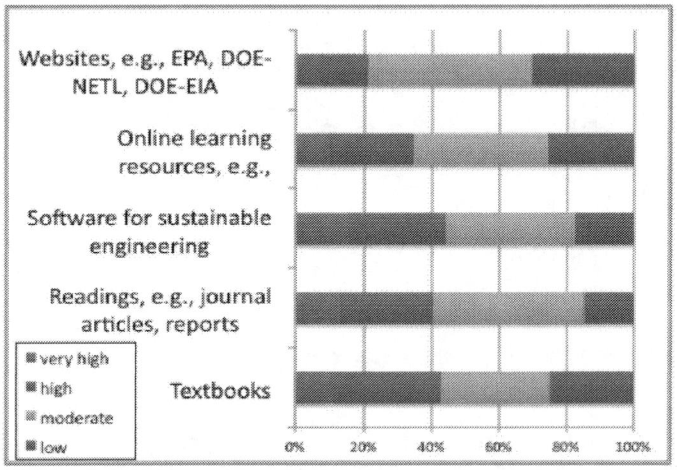

- No one source of content is easier to find, discover, locate, access and translate into course content

SE Community Survey

Community Instructional Practice

- Instructional design elements used in community SE classes
  - Course Objectives
  - Teaching goals inventory
  - Learning outcomes and objectives
  - Instructional strategies, e.g., service learning
  - Assessment methods: formative and summative
- Do you receive assistance preparing courses with SE content?
  - From a national, regional or campus education center
- Which design elements does the center help you prepare?

SE Community Survey

Community Instructional Practice

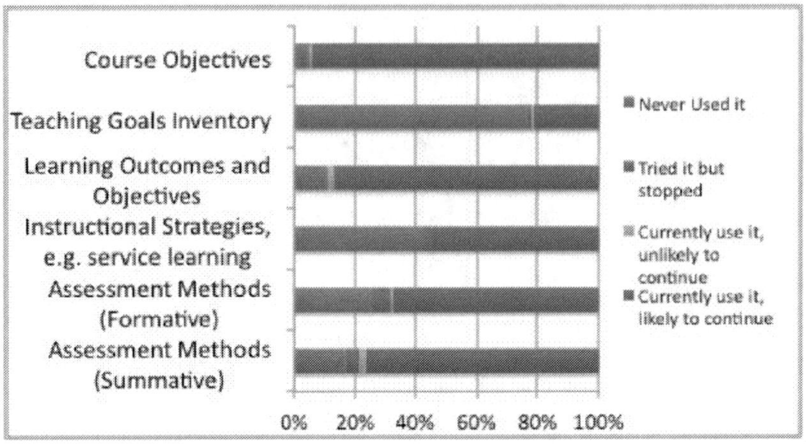

- Course/learning objects and summative assessment are widely used
- Faculty development opportunities exist to use instructional strategies, teaching goals inventory, and formative assessment to improve teaching

SE Community Survey

Course Development Assistance

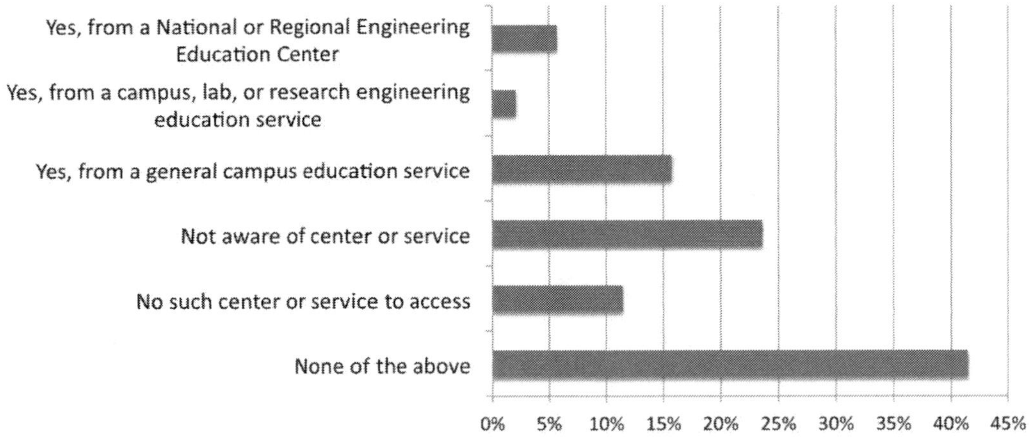

Conclusions

Based on preliminary survey results:

- There is a well-defined set of Sustainable Engineering topic areas that are relatively consistent throughout the community
- Journal articles and readings are more important sources of SE content than textbooks
- Course objectives and learning objectives are widely used, but other instructional practices are yet to be tapped.

5/29/2015 Heller - Engineering Ethics for Sustainable Water

ASCE | INTERNATIONAL CONFERENCE ON SUSTAINABLE INFRASTRUCTURE 2014
Long Beach, California | November 6 – 8, 2014

SE Community Survey
Courses with SE Content

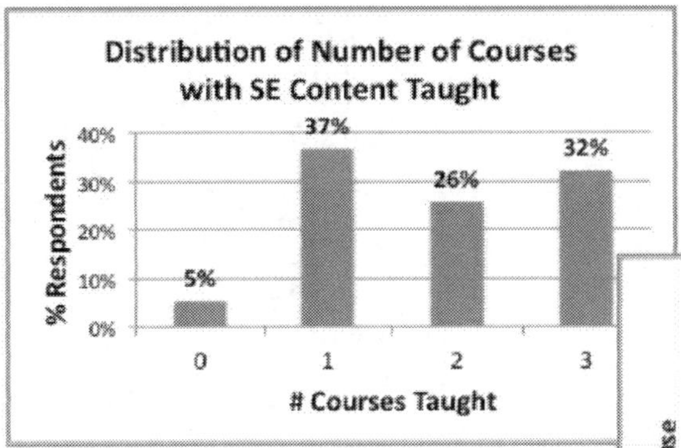

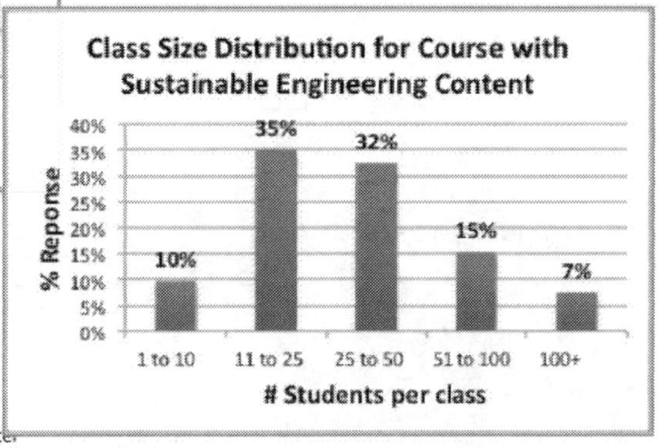

5/29/2015 Heller - Engineering Ethics for Sustainable Water

SE Community Survey
Courses and Content Taught

For each of the 21 topics provided, plus up to 3 more, indicate whether the topic:

- Is taught as a full (standalone) course or part of a course
- As taught is in a course required for a major or minor degree or certificate program
- Targets which of the following student level:
  - Freshman/sophomore
  - Junior/senior
  - Graduate

ASCE | INTERNATIONAL CONFERENCE ON SUSTAINABLE INFRASTRUCTURE 2014
Long Beach, California | November 6 – 8, 2014

SE Community Survey
Topics: How They Fit in the Curriculum

Topic	Full Course*	Part of Course
Agriculture / Land Use	--	23.0%
Biogeochemical Systems (incl. Ecology)	--	17.8%
Business / Economics	--	27.0%
Climate Change	--	50.0%
Design	13.2%	49.3%
End of Life / Waste Management	--	38.2%
Energy / Power Generation	8.6%	48.7%
Green Buildings / Infrastructure / Construction	5.9%	36.8%
Human Health	--	22.4%
Humanities (Philosophy, Ethics, History)	--	21.1%

* "-" means < 5.0%

5/29/2015 Heller - Engineering Ethics for Sustainable Water

SE Community Survey

Topics: How They Fit in the Curriculum

Topic	Full Course*	Part of Course
Industrial Processes	--	22.4%
Life-cycle Assessment (LCA)	--	52.0%
Materials Flow Analysis (MFA)	--	25.7%
Matierals	--	33.6%
Natural Resources	--	30.3%
Policy	--	34.2%
Pollution Prevention/ Fate & Transport	--	32.2%
Systems / Metrics/ Information Management	--	23.0%
Transportation	--	27.6%
Urbanism / Urban Systems	--	19.7%
Water	--	42.8%

Conclusions:
- Few (3) of 21 SE topics are taught as standalone courses
- 17% of respondents teach some or all topics
- 40% of respondents teach some or all of 4 topics (LCA is #1)

ASCE | INTERNATIONAL CONFERENCE ON SUSTAINABLE INFRASTRUCTURE 2014
Long Beach, California | November 6 – 8, 2014

SE Community Survey

Topics: Are They Required ?

Courses containing the following topics are required for a Major Degree for at least 10% of respondents:

- Climate Change (13.2%)
- Design (23.7%)
- Energy Systems and Power Generation (19.1%)
- Green Buildings / Infrastructure/ Construction (11.2%)
- Life-cycle Assessment (13.8%)
- Materials (10.5%)
- Pollution Prevent / Fate and Transport (11.8%)
- Water (13.2%)

SE Community Survey
Topics: Are They Required ?

Courses containing the following topics are required for a Minor Degree for at least 5% of respondents:

- End of Life / Waste Management (5.9%)
- Energy Systems and Power Generation (5.3%)
- Life-cycle Assessment (6.6%)

No courses with any of the 21 or added topics were required for Certificate programs by ≥ 5% respondents

SE Community Survey
Topics: : To Whom are They Taught?

Courses containing the following topics are taught at the Freshman or Sophomore levels for at least 10% of respondents:

- Climate change (11.2%)
- Design (17.1%)
- Energy and power generation (13.8%)
- Water (11.8%)

ASCE | INTERNATIONAL CONFERENCE ON SUSTAINABLE INFRASTRUCTURE 2014
Long Beach, California | November 6 – 8, 2014

SE Community Survey
Topics: : To Whom are They Taught?

Courses containing the following topics are taught at the Junior or Senior levels for at least 20% of respondents:

- Business / Economics (20.4%)
- Climate Change (28.3%)
- Design (41.4%)
- End of Life / Waste Management (27.0%)
- Energy Systems and Power Generation (32.4%)
- Green Buildings / Infrastructure/ Construction (28.3%)
- Life-cycle Assessment (LCA) (35.5%)
- Materials (24.3%)
- Policy (23.0%)
- Pollution Prevention/ Fate & Transport (24.3%)
- Water (28.3%)

ASCE | INTERNATIONAL CONFERENCE ON SUSTAINABLE INFRASTRUCTURE 2014
Long Beach, California | November 6 – 8, 2014

SE Community Survey
Source of Content: Importance

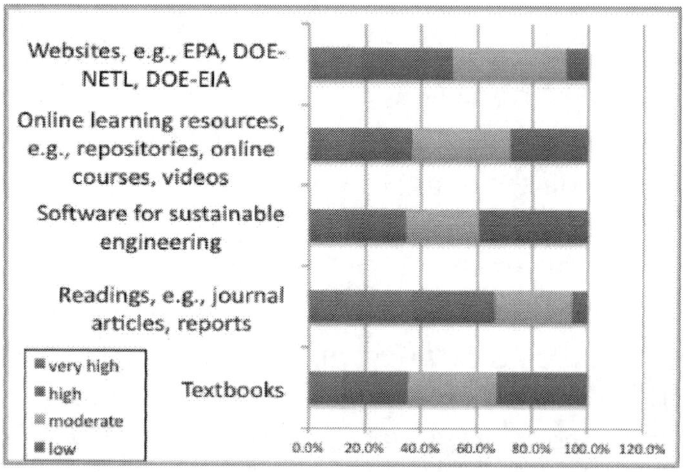

- ➢ Readings, e.g., journal articles, reports are most important
- ➢ Textbooks, SW, online learning are least important (purple)

ASCE | INTERNATIONAL CONFERENCE ON SUSTAINABLE INFRASTRUCTURE 2014
Long Beach, California | November 6 – 8, 2014

SE Community Survey
Source of Content: Ease of Use

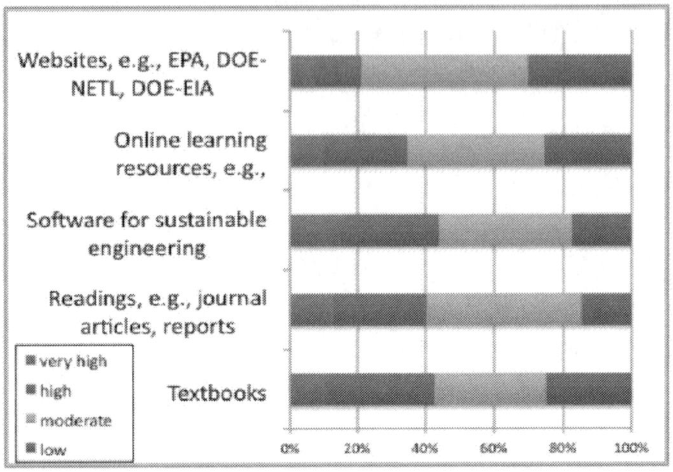

- No one source of content is easier to find, discover, locate, access and translate into course content

5/29/2015 Heller - Engineering Ethics for Sustainable Water

ASCE | INTERNATIONAL CONFERENCE ON SUSTAINABLE INFRASTRUCTURE 2014
Long Beach, California | November 6 – 8, 2014

SE Community Survey
Community Instructional Practice

- Instructional design elements used in community SE classes
  - Course Objectives
  - Teaching goals inventory
  - Learning outcomes and objectives
  - Instructional strategies, e.g., service learning
  - Assessment methods: formative and summative
- Do you receive assistance preparing courses with SE content?
  - From a national, regional or campus education center
- Which design elements does the center help you prepare?

5/29/2015 Heller - Engineering Ethics for Sustainable Water

ASCE | INTERNATIONAL CONFERENCE ON SUSTAINABLE INFRASTRUCTURE 2014
Long Beach, California | November 6 – 8, 2014

SE Community Survey
Community Instructional Practice

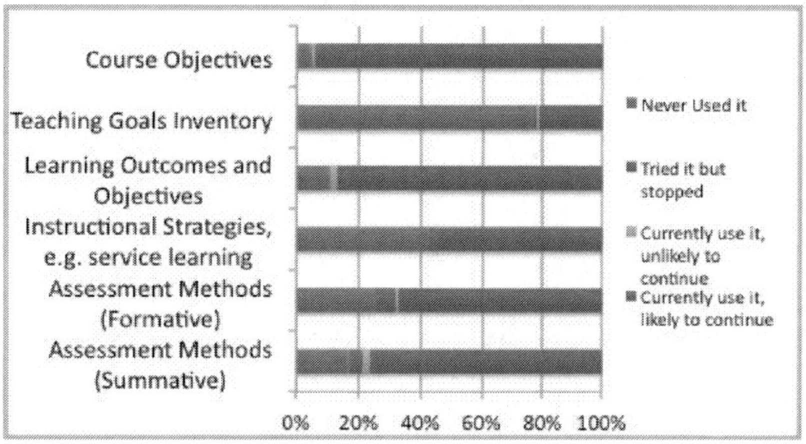

- Course/learning objects and summative assessment are widely used
- Faculty development opportunities exist to use instructional strategies, teaching goals inventory, and formative assessment to improve teaching

ASCE | INTERNATIONAL CONFERENCE ON SUSTAINABLE INFRASTRUCTURE 2014
Long Beach, California | November 6 – 8, 2014

SE Community Survey
Assistance in Preparing?

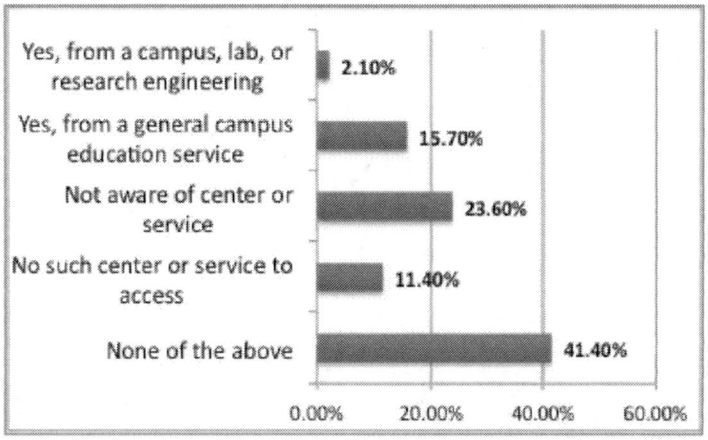

- 83% of respondents did not exploit instructional practice and hel;[
- Faculty development opportunities exist to use instructional strategies, teaching goals inventory, and formative assessment to improve teaching

ASCE | INTERNATIONAL CONFERENCE ON SUSTAINABLE INFRASTRUCTURE 2014
Long Beach, California | November 6 – 8, 2014

Conclusions

5/29/2015 Heller - Engineering Ethics for Sustainable Water

ASCE | INTERNATIONAL CONFERENCE ON SUSTAINABLE INFRASTRUCTURE 2014
Long Beach, California | November 6 – 8, 2014

References

5/29/2015 Heller - Engineering Ethics for Sustainable Water

Thank you for your attention.

Sustainable Engineering Workshop: *Open Educational Resources (OER)*

- MOOCs, contrary to their name, are not all open
- *Sustainability: a Comprehensive Foundation,* foundation for Coursera's *Introduction to Sustainability* MOOC is open
- Khan Academy, MIT OpenCourseware, and Washington's Open Course Library are all open
- creative commons has devised licensing options
- How will we permit reuse/remix of the content that is used and contributed by our community

Sustainable Engineering Workshop: *Instructional Design Opportunities*

- Little to no training in instructional design
- Don't regularly read learning science literature
- Reach out to Engineering Education Research and Teaching Centers
- Instructional strategies research growing
- Sustainable engineering-specific instructional strategies are emerging

  - Int'l Journal Engineering Education – special issues
  - Int'l Journal of Sustainability in Higher Education

Optimizing Control of an Improved Single-Column Chromatographic Process for the Separation of Enantiomers: Experimental Implementation

MOHAMMAD AMANULLAH, KAZI MONZURE KHODA
Department of Chemical Engineering, College of Engineering, Qatar University 2713, Doha, Qatar

4th International Congress on Sustainability Science & Engineering (ICOSSE '15), May 26-29, 2015, Balatonfüred, Hungary.

جامعة قطر
QATAR UNIVERSITY
College of Engineering – كلية الهندسة
Department of Chemical Engineering – قسم الهندسة الكيميائية

جامعة قطر
QATAR UNIVERSITY

Outline

1. Introduction
2. ISCC process and monitoring system
3. Results and Discussions
4. Conclusions

Introduction

Enantiomers/optical isomers

- Identical physical properties
- Identical chemical properties
- Biologically distinguishable

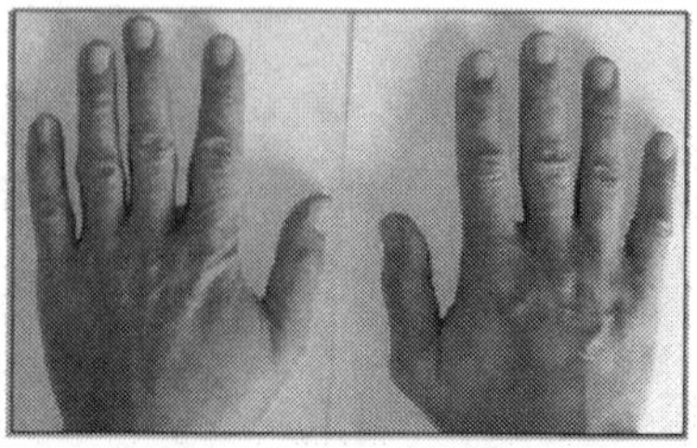

Year	Single enantiomer	Total drug
	sales ($ Billions)	sales ($ Billions)
2001	139	368
2002	159	387
2005	172	-
2007	210	-
2012	>250	-

C&EN (Tech. Cat. Int.)

Introduction…

Single enantiomers production routes

Criteria	Chemical	Biological	Physical
Mechanism	Asymmetric hydrogenation	Enzyme	Chromatography
Productivity	High	Low	Moderate
Selectivity	Varies	High	High
Processing	Varies	Complex	Complex
Development time	3-5 years	2-3 years	Months
Development cost	$3-5 million	$2-3 million	$0.5-0.7 million

C&EN

Needs:

- Speed
- Reliability
- Cost efficiency

200 metric tons/year

Currently, Chromatography is the method of choice for single enantiomer drug development

Objectives

- ❑ Experimental Implementation of the ISCC Process
  - ❑ Open-loop experiments
  - ❑ Closed loop experiment

Performance indicators:

- Productivity, and
- Desorbent requirement

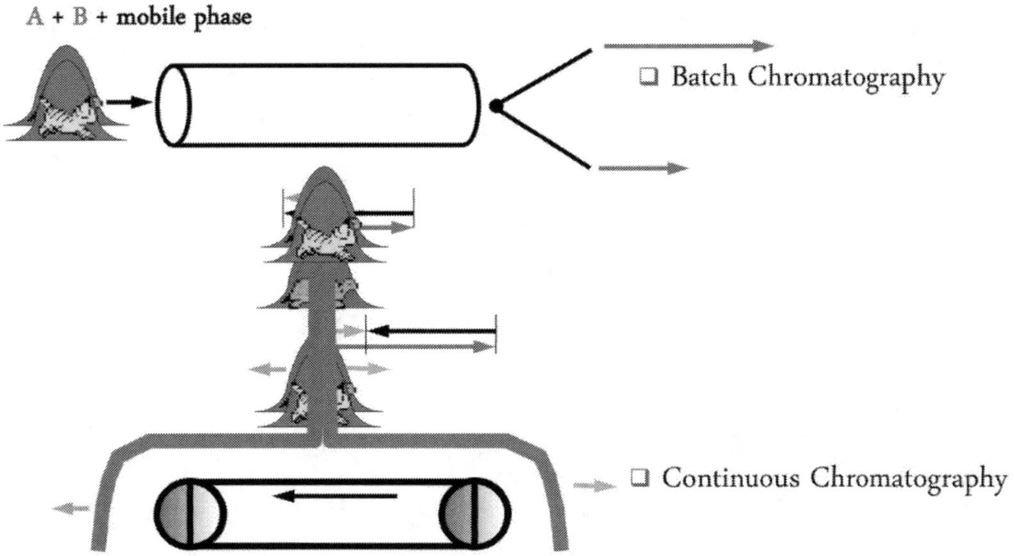

ISCC Process

- **Improved Single-Column Chromatographic (ISCC) Process**
  - (Semi-) Continuous
  - Cost-effective
  - Flexible in changing feed and optimization demands
- **Improvements**
  - Conceptual
    - Fraction collection scheme
    - Overlapped peaks from adjacent cycles
  - Physical
    - Fraction collector
    - Online monitoring system

Improvements

- Overlapped peaks
  - Same cycle
  - Adjacent cycles

- New fraction collection scheme
  - First fraction: B
  - Second fraction: B+A
  - Third fraction: A
  - Fourth fraction: A+B

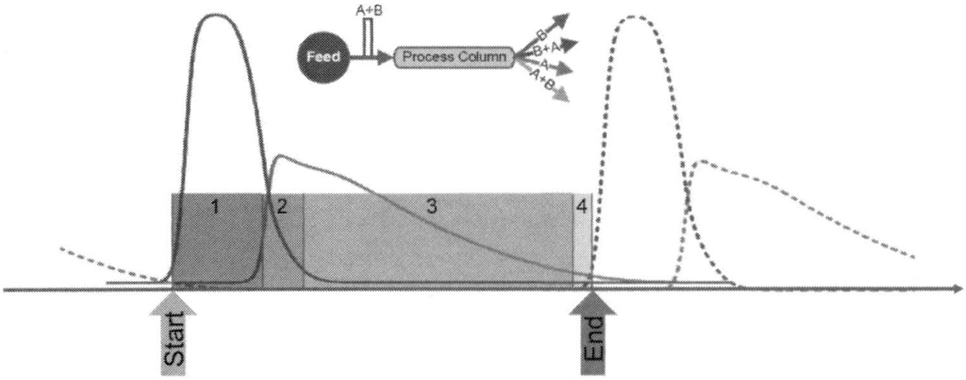

Process design

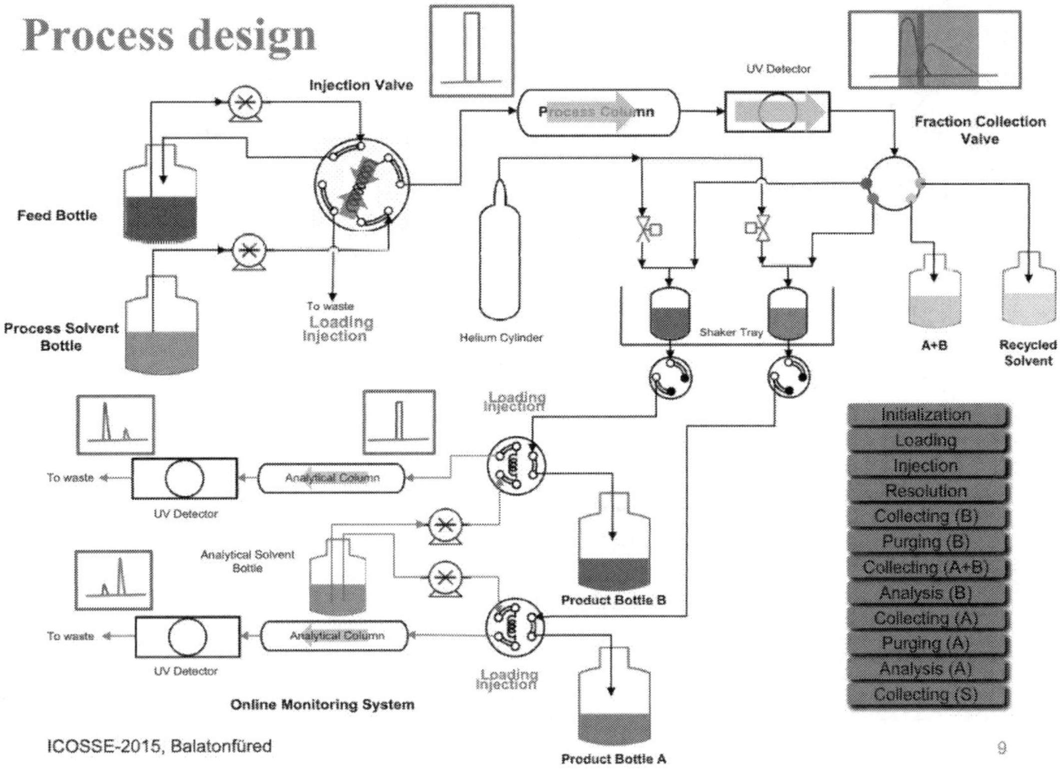

Laboratory set-up

Improved Single-Column Chromatographic (ISCC) Process

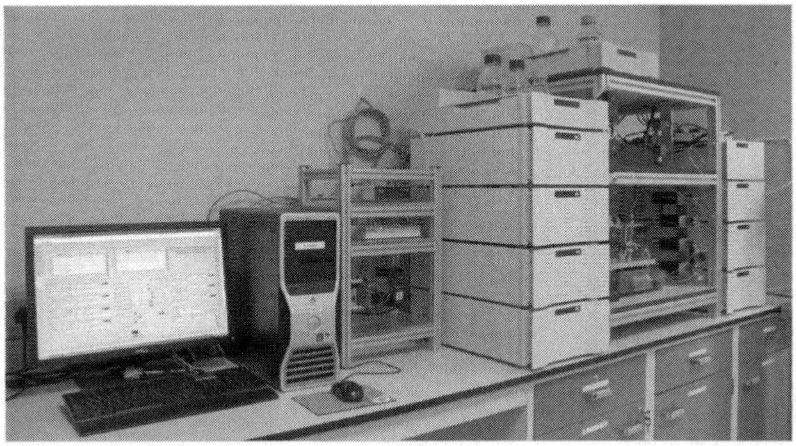

Experiments- Model Chiral Compound

- Chiral model compound
  - Guaifenesin

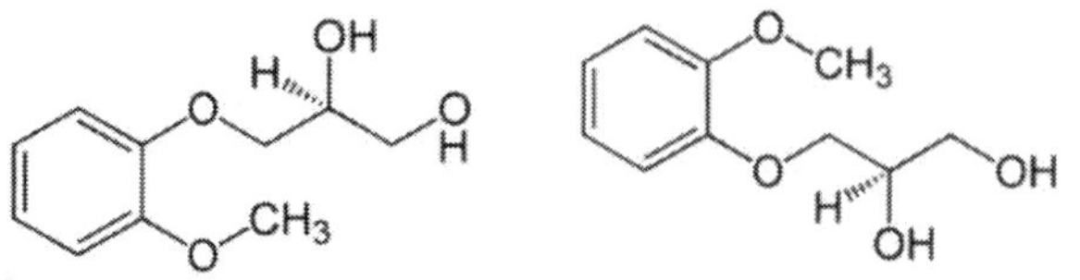

- Widely used in racemic form
- Important ingredients for cough remedy formulation
- One enantiomer has greater physiological activity and fewer undesired side effects

Experiments: Set-up

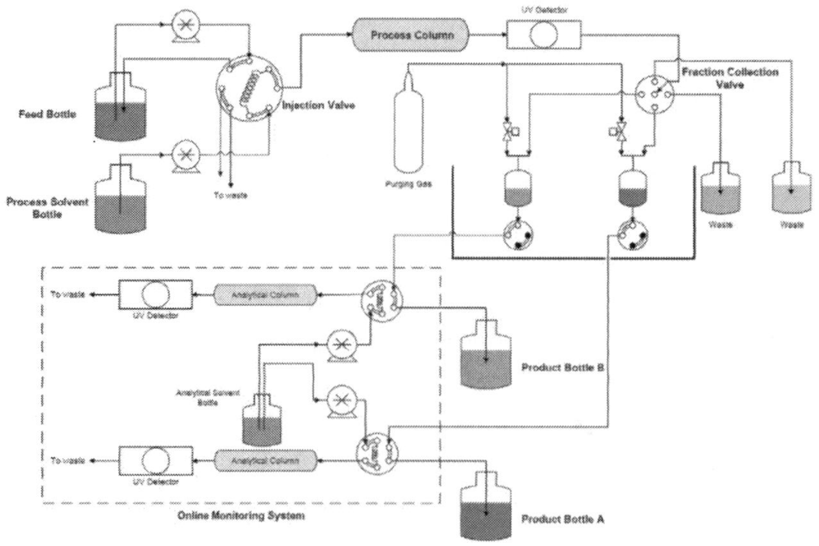

Experiments- Characterization

Overall bed void fraction (ε_t)

The overall bed void fraction was calculated using:

$$t_0 = t_D + \frac{V\varepsilon_t}{Q}$$

t_0 = Residence time
t_D = Extra-column dead volume
V = Column volume
Q = Volumetric flow rate

The value of t_D of the system was obtained by injecting the tracer with a zero dead volume column instead of real column

Overall void fractions of the columns

Column name	Type	Specification	Particle size	ε_t	% error
Chiralcel OD	Preparative	10 cm × 1 cm	20 µm	0.71	< 5%
Chiralcel OD-H	Analytical	15 cm × 0.46 cm	5 µm	0.68	< 5%

Experiments- Open-loop operation

To test and verify the separation performance of the ISCC process without the controller

- To check compatibility of all customized parts

Purity requirement constraints for two case studies

Case	Purity of A	Purity of B
A	98%	98%
B	99.9%	99.9%

Operating parameters

Parameter	Case study A	Case study B
c_T^F	10 g/L	10g/L
t_{cy}	170 s	170 s
dt_{c1}	70 s	75 s
dt_{c2}	10 s	5 s
dt_{c3}	80 s	90 s
V_{inj}	2 ml	2 ml
V_{inj}^s	20 µL	20 µL
V_H	2 ml	2 ml
ω	150 RPM	150 RPM

Experiments- Open-loop operation…

Case study A ($P_A = 98\%$. $P_B = 98\%$)

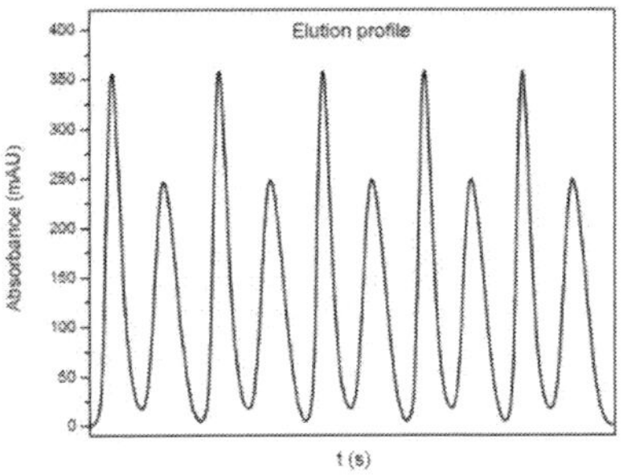

Elution profile

Analytical response of product in vial 1 ⇨

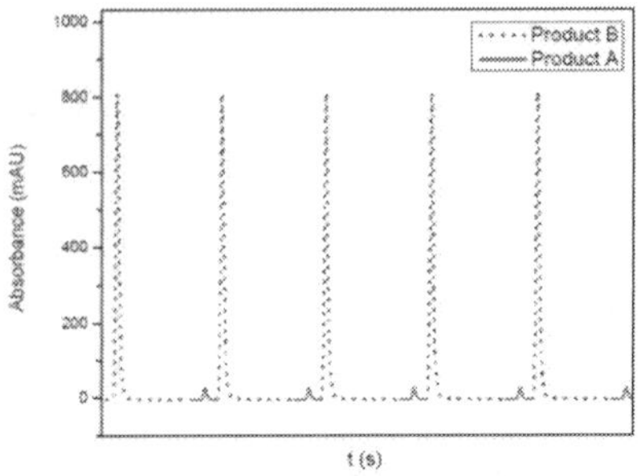

Analytical response of product in vial 2 ⇨

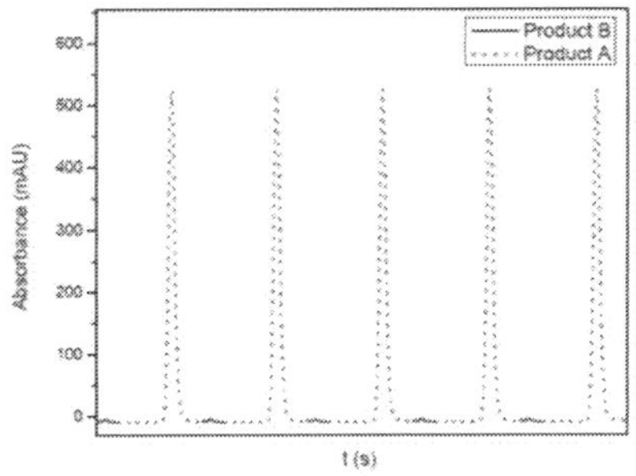

ICOSSE-2015, Balatonfüred

15

Experiments- Open-loop operation…

Case study B ($P_A = 99.9\%$. $P_B = 99.9\%$)

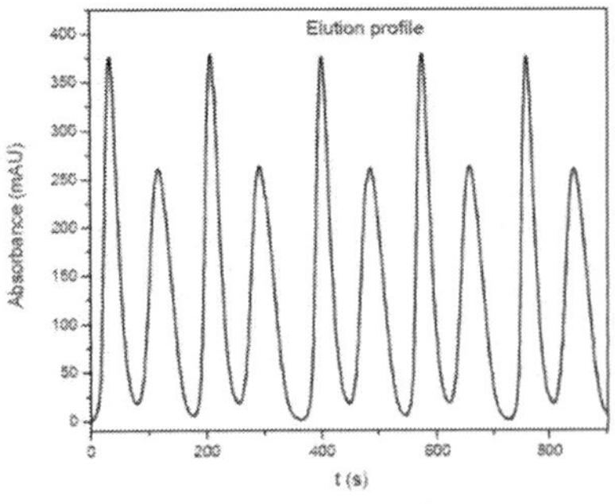

Elution profile

Analytical response of product in vial 1 ⇨

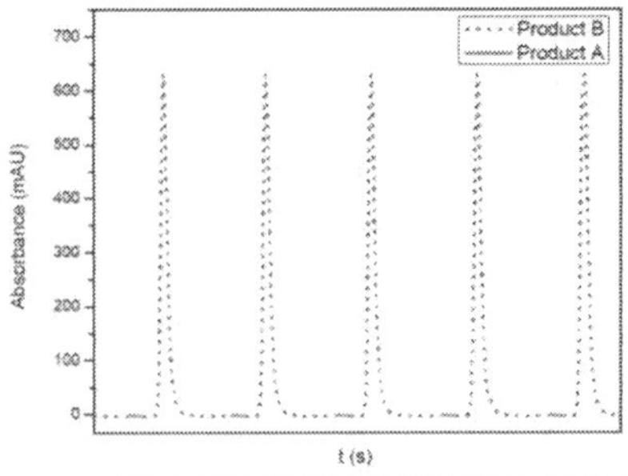

Analytical response of product in vial 2 ⇨

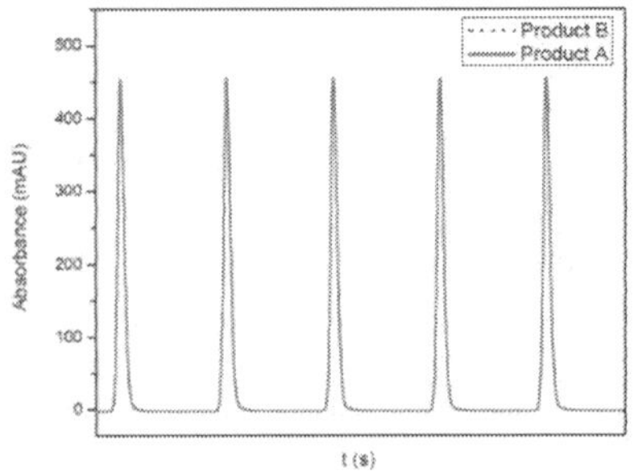

ICOSSE-2015, Balatonfüred

16

Experiments- closed loop operation

Schematic diagram of the implementation of model predictive control (MPC) concept on the process

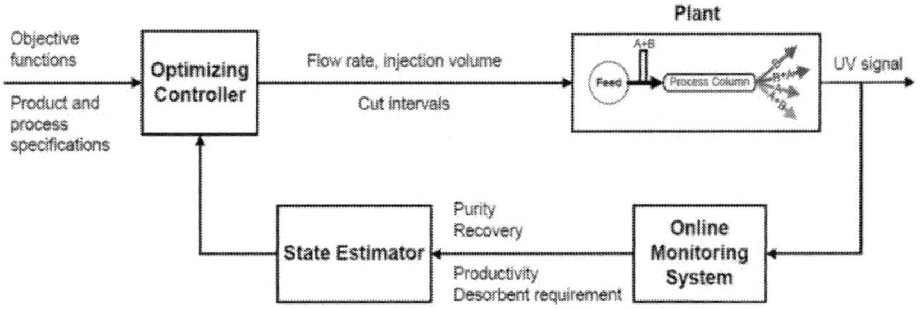

Performance function: $F_{PD} = \lambda_1 (Pr - Pr_s) - \lambda_2 (Dr - Dr_s)$

Experiments- closed loop operation…

Close-loop operation of ISCC process

Set point tracking

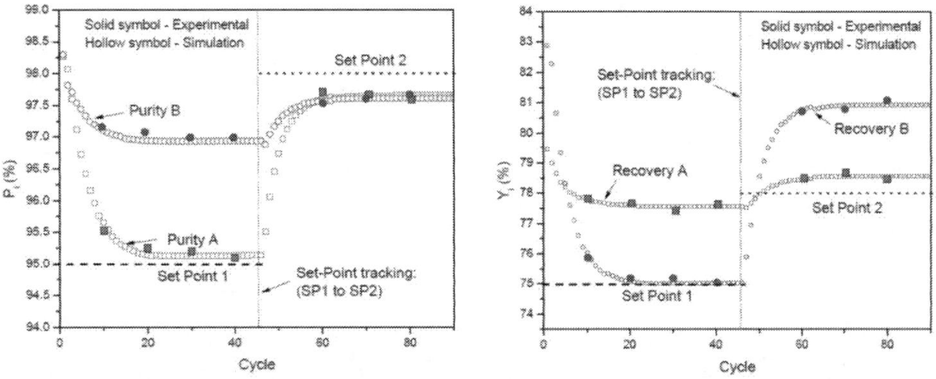

The results of set-point tracking; after 45th injection the set-point values of purity (*P*) and recovery (*Y*) for both products are changed from SP_1 ($P_i = 95\%$, $Y_i = 75\%$) to SP_2 ($P_i = 98\%$, $Y_i = 78\%$): (a) purities; (b) recoveries

Conclusions and Outlook

- **ISCC: superior to stacked injection**
  - Higher productivity at small cost of loss in recovery
- **Overlapped peaks and new cut-time strategy**
  - Minor improvement for high purity and recovery
  - Promising for lower purity/recovery
- **Efficient online monitoring system**
  - Faster than analytical HPLC system
  - Analyze both products simultaneously without interruption
- **Future works/extension**
  - Implement optimizing controller in online manner
  - Steady state recycling

Questions, comments, suggestions

Pore Surface Functionalization of Metal Organic Frameworks for Carbon Dioxide Capture Applications

- **Conference:** International Congress on Sustainability Science Engineering ICOSSE
- **Year:** 2015
- **Proceeding:** ICOSSE 2015
- **Group:** Poster Submissions
- **Session:**

Rapid Fire Session

- **Time:**

Wednesday, May 27, 2015 - 5:00pm-5:02pm

Authors:

Zhao, X.

Metal organic frameworks belong to a class of inorganic-organic crystalline materials where metal constituting units are connected by organic molecules to form 3-D nanoporous structures. Ultra high surface area, high pore volume, tunable chemical and physical properties of those materials have led to investigation of MOFs as promising adsorbents in carbon dioxide capture from coal- burning power plants. While promising, the hydrophilic nature of MOFs causes loss of crystallinity and competitive adsorption of H_2O inhibiting an effective overall CO_2 capture. In this work, the pore space of Mg/DOBDC MOF has been modified with amine functionalities to preserve material structure under humid environment, which is characteristic to the practical conditions such as flue gas (5-7% of H_2O by volume). Such modification showed structural preservation of the material characterized by XRD and surface area analyser and further a fully regenerable CO_2 capacity upon material exposure to high temperature steam.

An Appropriate Technology Based Multifunctional Processor for Sustainable Production of Bio-Based Products

- **Conference:** International Congress on Sustainability Science Engineering ICOSSE
- **Year:** 2015
- **Proceeding:** ICOSSE 2015
- **Group:** Poster Submissions
- **Session:**

Rapid Fire Session

- **Time:**

Wednesday, May 27, 2015 - 5:02pm-5:04pm

Authors:

Joshi, C., University of Kentucky College of Engineering

As underdeveloped countries seek advancement, the use of appropriate technology will play a critical role. The University of Kentucky Appropriate Technology and Sustainability (UKATS) research team has developed a multi-function processor that can be used to produce biodiesel, biochar, kerosene, and a non-synthetic pesticide from locally sourced biomass and waste plastic in rural regions, including sub-Saharan Africa. The processor, built using the principles of appropriate technology and sustainability fundamentals, is a simple, non-automated processor that uses non-standard construction materials typically present in under-developed regions, while keeping costs at a minimum. This contribution details the processor design and its operation in the production of biodiesel, biochar, kerosene and a non-synthetic pesticide. This research is developed in collaboration with the African Center for Renewable Energy and Sustainable Technology (ACREST) in Bangang, Cameroon.

Sustainability Assessment Methodology for Appropriate Technology Based Products in Developing Regions

- **Conference:** International Congress on Sustainability Science Engineering ICOSSE
- **Year:** 2015
- **Proceeding:** ICOSSE 2015
- **Group:** Poster Submissions
- **Session:**

Rapid Fire Session

- **Time:**

Wednesday, May 27, 2015 - 5:04pm-5:06pm

Authors:

Croft, W. R., University of Kentucky
Seay, Ph.D., P.E., J. R., University of Kentucky

Assessing the sustainability of products is particularly difficult in underdeveloped or remote regions. Products based on the principles of appropriate technology are intended to be beneficial to developing communities. A widely accepted methodology capable of assessing societal, economic and environmental impacts of products at a regional level is currently unavailable.

This contribution demonstrates a novel approach for assessing the sustainability of appropriate technology based products in developing regions. This proposed methodology introduces a system for quantifying the impact of products based on a series of rubrics to assess regionally appropriate indicators. The magnitudes of the impact variables are evaluated utilizing an absolute scale. Current sustainability assessments utilized in industrialized regions are not applicable in underdeveloped or remote regions. This is particularly true when trying to assess the societal impacts of products in underdeveloped regions. The novel methodology proposed is capable of assessing the impacts of sustainable products in developing communities.

The Impact of High Frequency Flow Pulsation on Fluidisation Behaviour of Powder

- **Conference:** International Congress on Sustainability Science Engineering ICOSSE
- **Year:** 2015
- **Proceeding:** ICOSSE 2015
- **Group:** Poster Submissions
- **Session:**

Rapid Fire Session

- **Time:**

Wednesday, May 27, 2015 - 5:06pm-5:08pm

Authors:

Ireland, E. J.
Smith, R. M.
Zimmerman, W. B.

In a manufacturing world dominated by incentives to increase energy efficiency, pulsed flow fluidisation theoretically offers a simple means to significantly reduce the air velocity requirements of fluidised bed equipment. To date, promising research has shown the capability of a pulsed flow, defined as the intermittent fluidising gas stream, to reduce gas requirements by over 70% in Group C powders, and a third in Group A/B powders. In terms of energy use, the potential savings are sizeable. However, the research has been limited to a low pulsation range of approximately 0-15 Hz due to the design of common pulsation devices. Consequently the influence of a new high pulsation frequency range is presented, as an indication of the extent of the benefit from pulsed flow technology. The effect of pulsation on pressure drop, bed structure and mixing performance are investigated.

On the Weighting Factors of the Environmental Impacts in Sustainable Manufacturing

- **Conference:** International Congress on Sustainability Science Engineering ICOSSE
- **Year:** 2015
- **Proceeding:** ICOSSE 2015
- **Group:** Poster Submissions
- **Session:**

Rapid Fire Session

- **Time:**

Wednesday, May 27, 2015 - 5:08pm-5:10pm

The Integrated Prevention Pollution Control (IPPC) and the Integrated Product Policy (IPP) have been considered efficient regulations to increase the Environmental performance of the european process industries and its related products respectively. IPPC identifies 90 main pollutants of concern and the threshold to be considered to inform to the administration and the public in general (E-PRTR European-Pollutants Release and Tansfer Register).

In this contribution the pollutants thresholds have been used as weighting factors in order to account the dimensionless impacts allowing to get a quantitative assessment of the environmental sustainability. The method has been applied to the comparison among the main manufacture sectors in the different european countries.

The procedure allows a way to order the investments in the manufacture sectors and countries in order to get the higher efficiency in the use of the economic resources. Social considerations can be also included in the sustainability assessment

Assessing Effects of Supplier Relationship Management Strategies on Supply Chain Resiliency, Efficiency, and Sustainability Performance Trade-Offs

- **Conference:** International Congress on Sustainability Science Engineering ICOSSE
- **Year:** 2015
- **Proceeding:** ICOSSE 2015
- **Group:** Poster Submissions
- **Session:**

Rapid Fire Session
- **Time:**

Wednesday, May 27, 2015 - 5:10pm-5:12pm

Authors:

Brown, A.

Badurdeen, F., University of Kentucky

A major difficulty presented in modern supply chain management is the need to balance performance trade-offs such as those between operational efficiency, sustainability, and risk exposure. Of particular concern is the increasing cost and occurrence of low-likelihood, high-impact disruptions. To increase supply chain resilience, managers increasingly rely on effective Supplier Relationship Management (SRM) and supplier segmentation which identifies key suppliers on which to focus management efforts. By incorporating quantitative modeling data regarding the impact of major disruptions into the segmentation process, managers will be able to better-identify the most vulnerable areas of the supply chain. The objective of this research is to increase supply chain resilience by improving the process of supplier segmentation. Agent-based simulation (ABS) is proposed as an appropriate method because of its ability to account for autonomous behavior by supply chain entities. Expected research outcomes include an improved approach to supply chain risk management and SRM.

A Hybrid Optimization Model for Biomass Trigeneration Plant Location: The Case of the City of Petrinja, Croatia

- **Conference:** International Congress on Sustainability Science Engineering ICOSSE
- **Year:** 2015
- **Proceeding:** ICOSSE 2015
- **Group:** Poster Submissions
- **Session:**

Rapid Fire Session

- **Time:**

Wednesday, May 27, 2015 - 5:12pm-5:14pm

Authors:

Dominkovic, D. F.
Cosic, B., University of Zagreb, Facuty of Mechanical Engineering and Naval Architecture
Perkovic, L.
Ban, M.
Duic, N., University of Zagreb, Facuty of Mechanical Engineering and Naval Architecture

In this paper, optimization of the location of biomass trigeneration power plant was considered. The system combined biomass cogeneration power plant, absorbers and the seasonal pit thermal energy storage. Several factors influence merit of potential locations, which was shown in the different case studies for the City of Petrinja, Croatia. When the biomass availability, capacity of the power plant or the number of power plants changes the optimal location also change. Furthermore, economic assessment of choosing optimal and non-optimal location was performed. Case studies have shown that significant amount of yearly spending on fuel (biomass) can be avoided, if the optimal location has been chosen for the power plant location. Moreover, it was shown that while choosing the optimal location of the power plant, economic figures such as net present value (NPV) can be satisfying for the potential investor in the trigeneration power plant even in residential area.

Carbon and Water Flows Embodied in International Trade: A Review on Consumption-Based Analysis

- **Conference:** International Congress on Sustainability Science Engineering ICOSSE
- **Year:** 2015
- **Proceeding:** ICOSSE 2015
- **Group:** Poster Submissions
- **Session:**

Rapid Fire Session

- **Time:**

Wednesday, May 27, 2015 - 5:14pm-5:16pm

Authors:

Cucek, L., University of Maribor, Faculty of Chemistry and Chemical Engineering
Liu, X., South China University of Technology
Qian, Y., South China University of Technology

A comprehensive literature review of virtual carbon and water flows is provided which is based on the consumption-based inventories in the international trade, by including single-country, bi-national, and multi-country trade. It is found that e.g., there are significant CO_2 gaps between producer's and consumer's emissions, the global emissions associated with consumption in many developed countries have increased with a large share of the emissions originating in developing countries. However, by International trade it may also be possible to reduce overall environmental pressure if such imported products are consumed that are produced with lower carbon intensity and water consumption than in the domestic industry. While results across current study are robust to be used in further application, more research is needed to towards the self-sufficient regions based on more efficient processes by combining surrounding countries/regions and to develop participation mechanism, market share of virtual carbon and virtual water between international trading partners.

Design of Transportation Networks Under Uncertainties By the P-Graph Framework

- **Conference:** International Congress on Sustainability Science Engineering ICOSSE
- **Year:** 2015
- **Proceeding:** ICOSSE 2015
- **Group:** Poster Submissions
- **Session:**

Rapid Fire Session

- **Time:**

Wednesday, May 27, 2015 - 5:16pm-5:18pm

Authors:

Süle, Z., University of Pannonia
König,, University of Pannonia
Bertok, B., University of Pannonia

Rapidly changing prices and taxes, as well as emerging new products and technologies, demand that the design of any company's supply chain has to be frequently revised to stay competitive. Mathematical modeling and optimization provide solid basis for evaluating alternative designs and scenarios, reducing cost, improving efficiency, and adapting to changing requirements.

The proposed methodology is based on the combinatorial foundations of algorithmic process synthesis or more specifically on the P-graph framework. With this method it is possible to identify which is the least sensitive of all the feasible structures, if the cost or the available flow of the resources, the cost of the activity, and the required flow of the product can be stochastic.

A biodiesel supply network involving two type of transportation, such as barge and railway, serves as an illustrative example. A novel algorithm generates the mathematical model and alternative solutions to increase reliability of supply scenarios.

Optimization of Energy Networks By the Extended P-Graph Framework

- **Conference:** International Congress on Sustainability Science Engineering ICOSSE
- **Year:** 2015
- **Proceeding:** ICOSSE 2015
- **Group:** Poster Submissions
- **Session:**

Rapid Fire Session

- **Time:**

Wednesday, May 27, 2015 - 5:18pm-5:20pm

Authors:

Szlama, A., University of Pannonia

An extended P-graph methodology, with new mathematical model for the operating units, is introduced here for optimizing complex energy systems involving alternative energy sources. The energy consumption of the world has been on the rise. If the current trends continue, we can face energy shortag and serious environmental problems. One solution to this pressing issue is to complement the current energy systems with renewables whose environmental impact is significantly less damaging. The goal of this work is to determine the minimal cost for an energy production system while considering the energy demands and emission constraints.

Friedler et al. have elaborated a methodology called the P-graph framework for solving PNS (process network synthesis) problems. The representation is based on bipartite graphs and it is suitable for modeling and designing production systems and energy supply systems. This work details the extension of the mathematical model of operating units in the P-graph framework.

Branch and Bound Algorithm for Designing Supply Chains Under Uncertainties By the P-Graph Framework

- **Conference:** International Congress on Sustainability Science Engineering ICOSSE
- **Year:** 2015
- **Proceeding:** ICOSSE 2015
- **Group:** Poster Submissions
- **Session:**

Rapid Fire Session

- **Time:**

Wednesday, May 27, 2015 - 5:20pm-5:22pm

Authors:

Baumgartner, J.
Süle, Z., University of Pannonia
König,, University of Pannonia

The utilization of renewable resources poses new challenges to process design. The present paper shows a methodology based on process graphs (P-graph) for generating optimal and near-optimal supply networks where the set of activities and those reliabilities are given. A branch and bound algorithm was developed for designing supply chains under uncertainty. Such steps comprise superstructure generation, construction of the mathematical model, optimization, calculating the reliability constraints and the solution interpretation. The developed algorithm is illustrated with an optimization problem for a region that is to support a sustainable environment. This example involving the utilization of renewable feedstock, namely agricultural products and the reliability of the solutions are calculated where lower bound of the expected reliability is input parameter.

Sustainability of Atomic Layer Deposition Nanotechnology

- **Conference:** International Congress on Sustainability Science Engineering ICOSSE
- **Year:** 2015
- **Proceeding:** ICOSSE 2015
- **Group:** Poster Submissions
- **Session:**

Rapid Fire Session
- **Time:**

Wednesday, May 27, 2015 - 5:22pm-5:24pm

Authors:

Ma, L.
Pan, D., University of Wisconsin Milwaukee
Wang, F.
Xie, Y.
Yuan, Y., University of Wisconsin Milwaukee

Sustainability performance of atomic layer deposition (ALD) is significant because of ALD's high energy consumption, toxic chemical and nano-particle emissions. This project is conducted systematically using both experimentation and modeling on ALD of Al_2O_3 process. Specific energy consumption is obtained around 1.54×10^{12} J/kg, increasing with temperature and purging time. Thermodynamics and kinetics inside ALD chamber are simulated by Lattice Boltzmann Method and 3D transient numerical models. About 90% of precursors are found wasted during experimental analysis. 2×10^{4}- 8×10^{4} nano-particles (<100 nm) and about 0.5 cm^3 of CH_4 (25 °C, 1 atm) emissions are generated each cycle, while nano-particle emissions decrease with increase of purging time and gas emissions. The results would be useful for ALD system designers and technology developers to improve ALD process efficiency, minimize its environmental emissions and reduce risk of exposure.

An Innovation-Based Method for the Design of Sustainable Products

- **Conference:** International Congress on Sustainability Science Engineering ICOSSE
- **Year:** 2015
- **Proceeding:** ICOSSE 2015
- **Group:** Poster Submissions
- **Session:**

Rapid Fire Session

- **Time:**

Wednesday, May 27, 2015 - 5:24pm-5:26pm

Authors:

DuPont, B., Oregon State University
Wisthoff, A.

While methods exist to determine the environmental impact of products, these methods are only applicable after design decisions are fully ideated - no viable methods that enable the design of these products in an informed way currently exist. This work discusses the development of an advanced product design methodology that incorporates sustainable design thinking throughout an open-ended product design process, reducing sustainability-as-constraint design and enabling the development of products that are both innovative and environmentally friendly. Sustainable design tools are integrated into a new product design method, including the development and research of customer needs, product opportunity evaluation, concept generation and refinement, and prototyping. The result of this work is a new product development process that enables the design of truly innovative products that are also more environmentally friendly, which can reduce environmental impact due to consumer goods manufacturing and use.

Biomass utilization as chemical building blocks in Central Europe

- **Conference:** International Congress on Sustainability Science Engineering ICOSSE
- **Year:** 2015
- **Proceeding:** ICOSSE 2015
- **Group:** Poster Submissions
- **Session:**

Rapid Fire Session

- **Time:**

Wednesday, May 27, 2015 - 5:26pm-5:28pm

Authors:

Nady, E., University of Pannonia
Hegedüs, I.

Advanced biofuels produced from lignocellulosic biomass offer an exciting opportunity to produce renewable liquid transportation fuels, biochemicals, and electricity from locally available agriculture and forest residues. There is a great economical potential in the utilization of agricultural lignocellulosic wastes (e.g. corn stalk, straw and wood residues) in Central European countries. The estimated useable energy from these wastes can be about 40 PJ/y in Hungary (1 PJ = 10^{15}J). The possible products are bioethanol, biodiesel and numerous chemicals produced from lignocellulosic biomass and other intermediates (e.g. from lignin components) by application of green technologies. The conversion technologies from biomass to biofuels can be broadly classified in three major categories: biochemical, thermochemical, and hybrid conversions. This poster discusses the alternative possibilities of lignocellulosic biomass utilization as bioenergy and building block chemicals, focusing on Central Europe.

The National Development Agency grant TÁMOP-4.1.1.C-12/1/KONV-2012-0017 greatly acknowledged for the financial support.

Risk Assessment and Mapping of Fecal Contamination in the Ohio River Basin

- **Conference:** International Congress on Sustainability Science Engineering ICOSSE
- **Year:** 2015
- **Proceeding:** ICOSSE 2015
- **Group:** Poster Submissions
- **Session:**

Rapid Fire Session

- **Time:**

Wednesday, May 27, 2015 - 5:28pm-5:30pm

Authors:

Cabezas, A.
Yeghiazarian, L.
Morehead, D., Un
Teklitz, A.

The leading cause of surface-water impairment in the US is fecal microbial contamination, which can potentially trigger massive outbreaks of gastrointestinal disease. It is well known that the difficulty in prediction of water contamination is rooted in the stochastic variability of microbes in the environment, and in the complexity of environmental systems.

To address these issues, we employ a risk-based design format to compute the variability in microbial concentrations and the probability of exceeding the *E. coli* target in the Ohio River Basin (ORB). This probability is then mapped onto the basin's stream network within the ArcGIS environment. We demonstrate how spatial risk maps can be used in support of watershed management decisions, in particular in the assessment of best management practices for reduction of *E. coli* load in surface water.

Advancements in Carbondioxide Capture-Ionic Liquids

- **Conference:** International Congress on Sustainability Science Engineering ICOSSE
- **Year:** 2015
- **Proceeding:** ICOSSE 2015
- **Group:** Poster Submissions
- **Session:**

Poster Session

- **Time:**

Wednesday, May 27, 2015 - 5:30pm-6:30pm

Authors:

Singh, S.
Kumar, N.

<p> Advancements in Carbon dioxide capture- ionic liquids</p><p> Carbon capture and storage (CCS) is a technique which capture CO2 from industrial processes by different technologies. Liquids have shown huge potential to replace the conventional solvents used to capture CO2. In this work ionic liquids are compared by making use of ASPEN PLUS simulation for their CO2 capture tendencies and were further compared with a conventional solvent methyl ethanolamine MEA. Effect of operation parameters like temperature, pressure and solvents were modeled using selectivity analysis. Simulation results showed that about 95% CO2capture was obtained from the natural gas. Bmim BF4] showed brilliant CO2 capture tendency among the others. The loss of Ionic liquids was found to be negligible which suggested that the IL investment can be recovered in a very short period of time.

An Examination of Alternative Algae Feedstock on the Economic Feasibility of Biofuel Production

- **Conference:** International Congress on Sustainability Science Engineering ICOSSE
- **Year:** 2015
- **Proceeding:** ICOSSE 2015
- **Group:** Poster Submissions
- **Session:**

Poster Session

- **Time:**

Wednesday, May 27, 2015 - 5:30pm-6:30pm

Authors:

Nti, F.

The need for alternative feedstock in biofuel production has been increasing even as public policies supporting the industry undergoes changes around the world. One feedstock receiving increasing attention is algae in biodiesel production. The study assesses the economic feasibility of alga biodiesel production using algae from two alternative source – an open-raceway pond and naturally-harvested from the ocean using a system dynamic modeling approach. The primary differences in the two sources lie in the former's production capital cost and quality and availability uncertainty surrounding the latter. The study assumes a mass balance/unit operation system for carbon to ensure environmental sustainability of the production system. Working on the assumption of producing 50 million gallons of biodiesel per year and using 14% of the diesel plus by-products from the diesel production as the system's fuel, we find that the algal yield was a major bottleneck in the feasibility of the open-raceway system.

Applications of the Dry Disposal Process in the Arsenic Contaminated Soil Treatment

- **Conference:** International Congress on Sustainability Science Engineering ICOSSE
- **Year:** 2015
- **Proceeding:** ICOSSE 2015
- **Group:** Poster Submissions
- **Session:**

Poster Session
- **Time:**

Wednesday, May 27, 2015 - 5:30pm-6:30pm

Authors:

Yoon, J. H., .
Cheong, J. G., Huyundai Engineering & Construction
Kim, K., Huyundai Engineering & Construction
Kang, W. H., Huyundai Engineering & Construction
Chae, H., Hyundai Engineering & Construction

The wet washing method using solvent like water has been widely used in the contaminated soil cleaning process, however the wet process tends to be complicated and expensive by solvent handling requirements.

In previous tests, the arsenic concentration shows sharp variations with soil diameters, and the dry disposal process is proposed as an alternative or complementary process.

Computer simulations are carried out to determine the proper air classifier type, and the operation conditions are optimized according to the cut-size. The rotating plate type disintegration device is adopted to increase the whole system recovery ratio.

The air classification can recover more than 60% arsenic contaminated soils, based on the national standard, and, with the additional feed-back and disintegration unit, the total recovery ratio of larger than 95 % is achieved. The proposed method is proven to reduce wet by-products and be advantageous in economic and environmental aspects.

Coal Bed Methane in Pakistan: Difficulties and Prospects

- **Conference:** International Congress on Sustainability Science Engineering ICOSSE
- **Year:** 2015
- **Proceeding:** ICOSSE 2015
- **Group:** Poster Submissions
- **Session:**

Poster Session

- **Time:**

Wednesday, May 27, 2015 - 5:30pm-6:30pm

Authors:

Soomro, T. H.

The greatest challenge, the energy sector is facing today is how to meet the rising demand of energy on the other hand, the depletion of natural resources.Technological development suggests the enhancement of non-renewable energy resources. The energy crises and price escalation have led to other alternative resources to compete daily increasing energy demand. To meet the rapidly increasing demand for energy and faster depletion of conventional energy resources, Coal bed Methane is an important solution to meet the rising demands of energy considering nation growth. Pakistan with other countries is madly searching for alternate resources like coal bed methane (CBM), shale gas, and gas hydrate. CBM is considered to be the most reliable resource and Pakistan is indeed blessed with vast yet untapped resources waiting to be explored.The present paper discussed about the prospect of CBM as a clean energy source, difficulty involved in production of CBM, enhanced recovery techniques.

Comparative Characterization of Biomass

- **Conference:** International Congress on Sustainability Science Engineering ICOSSE
- **Year:** 2015
- **Proceeding:** ICOSSE 2015
- **Group:** Poster Submissions
- **Session:**

Poster Session

- **Time:**

Wednesday, May 27, 2015 - 5:30pm-6:30pm

Authors:

Saleem, H., University of the Punjab
Sajid, D.

The use of renewable energy sources is becoming increasingly necessary, if we are to achieve the changes required to address the impacts of global warming. Biomass is the most common form of renewable energy, widely used in the third world but until recently, less so in the Western world. The type of biomass required is largely determined by the energy conversion process and the form in which the energy is required.

The different industrial wastes samples collected from different industries were analyzed and their proximate and calorific analysis was carried out using muffle furnace and bomb calorimeter. The analysis was carried out under ASTM (American Society for Testing Materials).

The choice of a solid material to be used as a fuel depends upon all the factors calculated in this report, all these factors are important during the selection, combustion system, operating cost, maintenance, transportation and storage. Samples were Bagasse, wood, cow dung, rice husk and wheat straw.

Development of Acid Gas Purification for Higher CO2 Capture and Low Energy Consumption

- **Conference:** International Congress on Sustainability Science Engineering ICOSSE
- **Year:** 2015
- **Proceeding:** ICOSSE 2015
- **Group:** Poster Submissions
- **Session:**

Poster Session

- **Time:**

Wednesday, May 27, 2015 - 5:30pm-6:30pm

Authors:

Liu, X., South China University of Technology
Qian, Y., South China University of Technology

An integrated acid gas purification process with higher CO_2capture rate is proposed, which is featured by Rectisol® process heat integrated with CO_2 multistage compression. Three key factors contribute to CO_2 capture rate have been identified, which also serve as the foundation to process energy consumption and cost analysis. Compared with the conventional Rectisol® process, proposed process realize CO_2 capture rate of 80-95 %, which remarkably reduce CO_2 emission and collect highly concentrated CO_2 for further processing and utilization. Energy consumption of the proposed acid gas purification process first decrease and then increase as CO_2 capture rate increases. The energy consumption reaches the lowest point of 0.93 GJ/t CO_2 when the CO_2 capture rate is 80 %. Although CO_2capture rate of the proposed process can be as high as 95 %, a reasonable CO_2 capture rate within 80-90 % is recommended to achieve the best economic and environmental performances.

Investigation of Waste Water Cleaning Treatment Process with Biofilm

- **Conference:** International Congress on Sustainability Science Engineering ICOSSE
- **Year:** 2015
- **Proceeding:** ICOSSE 2015
- **Group:** Poster Submissions
- **Session:**

Poster Session

- **Time:**

Wednesday, May 27, 2015 - 5:30pm-6:30pm

Authors:

Nagy, R., University of Pannonia
Nagy, E. Sr., University of Pannonia

In the last 20 years, the use of biofilms has been strongly researched for biological wastewater treatment, for nitrate, bromat, perchlorate, arsenate, etc. removal. The microorganisms form a few tens/hundreds microns thick biofilm on the surface of the mobile or fixed carriers during this process. The carrier particles should have high specific surface area, and they have to move constantly with the flowing fluid or they are immobilized. The substrates (carbon, oxygen) can enter the bio-film by diffusion from the bulk fluid phase. Biofilms are agglomerations of microorganisms or bacteria, which due to their metabolic activity convert the contaminant components of the wastewaters into harmless products. Contrary to the suspended cell systems, biofilm reactors provide high mass concentration, and higher substrate conversion without need for solid separation devices, thus, the biofilm using sewage cleaning systems has several advantages comparing to the traditional activated sludge wastewater treatment.

Logistic Simulation of Sites for Short Term Scheduling in the Oil Industry

- **Conference:** International Congress on Sustainability Science Engineering ICOSSE
- **Year:** 2015
- **Proceeding:** ICOSSE 2015
- **Group:** Poster Submissions
- **Session:**

Poster Session
- **Time:**

Wednesday, May 27, 2015 - 5:30pm-6:30pm

Authors:

Halasz, L.,
Frits, M., University of Pannonia
Borbás, P. I.

A simulator framework is extended here with the logistic simulation of sites in the oil industry. The simulation system is already actively used in the MOL Plc which is a major regional oil company in Central Europe. The system assists the short term planning of the product movements in the pipelines by verifying the feasibility of the plan. Now, the simulator is extended by the inner logistics, the pipeline model of the sites. Until now sites were handled without any inner structure. It was assumed that a product from a tank always can be transported to the required main (outer) pipeline. Now the simulator systematically looks over possible the transportation routes within the site and warn us if there is a problem. It is important to highlight that though the inner pipe network is much shorter than the outer one, the inner network is significantly more complex.

Matrix Representation of the Grid Diagram for Heat Exchanger Networks

- **Conference:** International Congress on Sustainability Science Engineering ICOSSE
- **Year:** 2015
- **Proceeding:** ICOSSE 2015
- **Group:** Poster Submissions
- **Session:**

Poster Session
- **Time:**

Wednesday, May 27, 2015 - 5:30pm-6:30pm

Authors:

Yong, J. Y.

Hot and cold utility usage can be significantly reduced using heat exchangers in a chemical plant. Besides recording the specification of heat exchangers individually, the topology of the heat exchanger network (HEN) is important as well. One of the established ways to represent the network is by drawing a Grid Diagram but it can be time consuming if it is drawn manually. Manipulating the directly a Grid Diagram for design evolution and retrofit procedures can be difficult and confusing.

In this paper, a matrix representation equivalent to the Grid Diagram for HEN is presented. The novel representation clearly shows the duties, temperature differences at the cold and hot ends of each exchanger. Using the proposed table format to represent a HEN helps avoiding confusion which usually arises from complicated drawing representation. This paper also discusses its usefulness to determine the maximum amount of utilities can be saved.

Maximize Aircraft Utilization by P-graphs

- **Conference:** International Congress on Sustainability Science Engineering ICOSSE
- **Year:** 2015
- **Proceeding:** ICOSSE 2015
- **Group:** Poster Submissions
- **Session:**

Poster Session

- **Time:**

Wednesday, May 27, 2015 - 5:30pm-6:30pm

Authors:

Frits, M., University of Pannonia
Szili, L.
Bertok, B., University of Pannonia

The actual investigation aims at providing computer aid to member nations of Strategic Airlift Capability to optimally coordinate airlift requests and utilize spare airlift capacities and capabilities from other nations. From economic point of view 97% of the cost of a mission is the fuel cost. Thus, minimizing the number of flights with no useful load by more effective scheduling is crucial. Shipments are stored on aircraft pallets, and thus the availability of such pallets on the airbase is critical from the scheduling point of view.

It has been shown that the P-graph approach to process-network synthesis (PNS) originally conceived for conceptual design of chemical process provides appropriate tools generating and analyzing structural alternatives for supply scenarios Time-Constrained PNS (TCPNS) is capable to solve classical scheduling problems and vehicle scheduling problems as well. In current examinations the aircraft mission planning and scheduling is formulated as TCPNS problems.

Optimal design of multi-period process networks including storages for renewable resources

- **Conference:** International Congress on Sustainability Science Engineering ICOSSE
- **Year:** 2015
- **Proceeding:** ICOSSE 2015
- **Group:** Poster Submissions
- **Session:**

Poster Session

- **Time:**

Wednesday, May 27, 2015 - 5:30pm-6:30pm

Authors:

Bartos, A.
Bertok, B., University of Pannonia
Szlama, A., University of Pannonia

During the design of process systems seasonal effects on demands and available resources cannot be avoided. Due to this characteristic the capacity and seasonal load of an operation or operating unit has to be distinguished during modeling. In 2014 the P-graph methodology was expanded by Heckl et al. for optimal design of multi-period operations. The results show that the production units in multi-periodic systems need to have higher capacity than the average load because of the requirements derived from the time periods. This increase of capacity causes a significant increment in cost as well. In the present work to reduce the gap between the required capacities and the average loads, storage units are involved in the design of the cost optimal process network. Modeling season dependent storage costs and storage losses are incorporated to the multi-periodic P-graph approach.

AUTHOR INDEX

AUTHOR INDEX